Joseph Fourier 250th Birthday

Joseph Fourier 250th Birthday: Modern Fourier Analysis and Fourier Heat Equation in Information Sciences for the XXIst Century

Special Issue Editors

Frédéric Barbaresco
Jean-Pierre Gazeau

MDPI • Basel • Beijing • Wuhan • Barcelona • Belgrade

Special Issue Editors
Frédéric Barbaresco
Department of Advanced Radar Concepts, Thales
Air Systems
France

Jean-Pierre Gazeau
Université Paris-Diderot
France

Editorial Office
MDPI
St. Alban-Anlage 66
4052 Basel, Switzerland

This is a reprint of articles from the Special Issue published online in the open access journal *Entropy* (ISSN 1099-4300) from 2018 to 2019 (available at: https://www.mdpi.com/journal/entropy/special_issues/fourier)

For citation purposes, cite each article independently as indicated on the article page online and as indicated below:

LastName, A.A.; LastName, B.B.; LastName, C.C. Article Title. *Journal Name* **Year**, *Article Number*, Page Range.

ISBN 978-3-03897-746-9 (Pbk)
ISBN 978-3-03897-747-6 (PDF)

Contents

About the Special Issue Editors

Frédéric Barbaresco, Representative of Key Technology Domain PCC (Processing, Control & Cognition), Thales Land & Air Systems.

Jean-Pierre Gazeau, Professor of Université Paris Diderot and Centro Brasileiro de Pesquisas Físicas (CBPF).

entropy

Editorial

Joseph Fourier 250thBirthday: Modern Fourier Analysis and Fourier Heat Equation in Information Sciences for the XXIst Century

Frédéric Barbaresco [1,*] and Jean-Pierre Gazeau [2]

1 Key Technology Domain PCC (Processing, Control & Cognition) Representative, Thales Land & Air Systems, Voie Pierre-Gilles de Gennes, F91470 Limours, France

2 APC (UMR 7164), Department of Physics, Université Paris-Diderot, F75205 Paris, France; gazeau@apc.in2p3.fr

* Correspondence: frederic.barbaresco@thalesgroup.com; Tel.: +1-501-450-5839

Received: 19 February 2019; Accepted: 27 February 2019; Published: 6 March 2019

Abstract: For the 250th birthday of Joseph Fourier, born in 1768 at Auxerre in France, this MDPI special issue will explore modern topics related to Fourier analysis and Fourier Heat Equation. Fourier analysis, named after Joseph Fourier, addresses classically commutative harmonic analysis. The modern development of Fourier analysis during XXth century has explored the generalization of Fourier and Fourier-Plancherel formula for non-commutative harmonic analysis, applied to locally compact non-Abelian groups. In parallel, the theory of coherent states and wavelets has been generalized over Lie groups (by associating coherent states to group representations that are square integrable over a homogeneous space). The name of Joseph Fourier is also inseparable from the study of mathematics of heat. Modern research on Heat equation explores geometric extension of classical diffusion equation on Riemannian, sub-Riemannian manifolds, and Lie groups. The heat equation for a general volume form that not necessarily coincides with the Riemannian one is useful in sub-Riemannian geometry, where a canonical volume only exists in certain cases. A new geometric theory of heat is emerging by applying geometric mechanics tools extended for statistical mechanics, for example, the Lie groups thermodynamics.

Keywords: harmonic analysis on abstract space; heat equation on manifolds and Lie Groups

---

> "The differential equations of the propagation of heat express the most general conditions and the physical questions as a result of the analysis of pure problems, which is properly the object of the theory .... The different forms of body are varied to infinity, to the distribution of heat and penetrations; but all the inequalities fade away quickly and disappear as time goes by. The march of the phenomenon become more regular and simpler, is finally subject to a specific law that is the same for all cases, and that it bears no more any sensible imprint of the initial disposition ... The new theories, explained in our work, are united forever with the mathematical sciences, and rest, like them, on invariable foundations; they will retain all the elements they possess today, and they will acquire, continually, more extension". [Les équations différentielles de la propagation de la chaleur expriment les conditions les plus générales, et ramènent les questions physiques à des problèmes d'analyse pure, ce qui est proprement l'objet de la théorie .... Les formes des corps sont variées à l'infini, la distribution de la chaleur qui les pénètre peut être arbitraire et confuse; mais toutes les inégalités s'effacent rapidement et disparaissent à mesure que le temps s'écoule. La marche du phénomène devenue plus régulière et plus simple, demeure enfin assujettie à une loi déterminée qui est la même pour tous les cas, et qui ne porte plus aucune empreinte sensible de la disposition initiale ... . Les théories nouvelles, expliquées dans notre ouvrage

sont réunies pour toujours aux sciences mathématiques et reposent comme elles sur des fondements invariables; elles conserveront tous les éléments qu'elles possèdent aujourd'hui, et elles acquerront, continuellement plus d'étendue.]—Joseph Fourier (1768–1830), Discours préliminaire à la théorie analytique de la chaleur [1].

For the 250th birthday of Joseph Fourier (Figure 1) [1–6], born in 1768 at Auxerre in France, this MDPI special issue will explore modern topics related to Fourier analysis and Fourier Heat Equation.

Figure 1. Jean-Baptiste-Joseph Fourier (1768–1830) [1].

Fourier analysis, named after Joseph Fourier, who showed that representing a function as a sum of trigonometric functions greatly simplifies the study of heat transfer and addresses classically commutative harmonic analysis. Classical commutative harmonic analysis is restricted to functions defined on a topological locally compact and Abelian group G (Fourier series when $G = R^n/Z^n$, Fourier transform when $G = R^n$, discrete Fourier transform when G is a finite Abelian group). The modern development of Fourier analysis during XXth century has explored the generalization of Fourier and Fourier-Plancherel formula for non-commutative harmonic analysis, applied to locally compact non-Abelian groups. This has been solved by geometric approaches based on "orbits methods" (Fourier-Plancherel formula for G is given by coadjoint representation of G in dual vector space of its Lie algebra) with many contributors (Dixmier, Kirillov, Bernat, Arnold, Berezin, Kostant, Souriau, Duflo, Guichardet, Torasso, Vergne, Paradan, etc.) [7]. It was observed first by Souriau that the coadjoint orbits carry a natural symplectic structure and there is a closed non-degenerate G-invariant 2-form on each orbit, called the Kirillov-Kostant-Souriau symplectic form that plays a central role in geometric quantization and classification of the homogeneous symplectic manifolds. In parallel, theory of coherent states (Klauder, Perelomov, Gilmore, etc.) and wavelets (Grossmann, Daubechies, Meyer, etc.) has been generalized over Lie groups (by associating coherent states to group representations that are square integrable over a homogeneous space) [8]. One should add the developments, over the last 30 years, of the applications of harmonic analysis to the description of the fascinating world of aperiodic structures in condensed matter physics, e.g., quasicrystals and their diffraction spectra [9]. The notions of model set introduced by Y. Meyer, and of almost periodic functions, have revealed themselves as extremely fruitful in this domain of natural sciences.

The name of Joseph Fourier is also inseparable from the study of mathematics of heat, but it took almost a century for the most brilliant scientists of the nineteenth century—Fourier, Biot, Poisson, Lamé, and Boussinesq [10]—to unravel complexity appearances of the propagation of heat in solids, to develop efficient physical concepts and related instruments of mathematics, and from confusion that constitutes the reality of the calorific phenomena, to clarify a new knowledge of diffusion equation in elastic and crystal domains. It is from the study of thermal energy that the very notion of diffusion related to the parabolic-type equation is born with Fourier and Biot. Fourier's first memoir at the Academy of Sciences on this subject dates back to 1807, and completed in 1811 by extensive work that was examined by Malus, Haüy, Laplace, Lagrange, and Legendre. Fourier never adhered to reviewers comments and he reprinted his memoir without taking any account of the critics of these censors. It was in 1822 that the "Analytical Theory of Heat" appeared. The Fourier manuscript must be considered rightly as the foundation of mathematical physics. Modern research on heat equation explores the extension of classical diffusion equation on Riemannian, sub-Riemannian manifolds, and Lie groups (i.e., Hall). The heat equation for a general volume form that not necessarily coincides with the Riemannian one is useful in sub-Riemannian geometry, where a canonical volume only exists in certain cases. Jean-Michel Bismut [11] has introduced the concept of hypoelliptic Laplacian (If X is a Riemannian manifold, the hypoelliptic Laplacian is a family of hypoelliptic operators that interpolates between the ordinary Laplacian and the geodesic flow), with the probabilistic counterpart that is an interpolation between Brownian motion and geodesics. Elliptic heat kernel has infinite propagation speed compared to geodesic flow that has a finite propagation speed. On R^3, Langevin had introduced the Langevin equation to reconcile Brownian motion and classical mechanics. The hypoelliptic diffusion on the total space of the tangent bundle of a Riemannian manifold is a geometric Langevin process that interpolates between the geometric Brownian motion and the geodesic flow. In parallel with Geometric Mechanics, Jean-Marie Souriau [12] has interpreted the temperature vector of Planck as a space-time vector, obtaining, in this way, a phenomenological model of continuous media that presents some interesting properties: The temperature vector and entropy flux are in duality; the positive entropy production is a consequence of Einstein's equations; the Onsager reciprocity relations are generalized; and in the case of a fluid in the non-relativistic approximation, the model unifies heat conduction and viscosity (equations of Fourier and Navier). This work has been extended by Claude Vallée [13], by constructing a relativistic model of a dissipative continuum that complies with the laws of both mechanics and thermodynamics.

A last comment concerns the fundamental contribution of Fourier analysis to quantum physics: Quantum mechanics with the notion of representation based on spectral properties of basic observables, like position, momentum, energy, and spin; the quantum field theory saw the first steps that emerged from solutions of Maxwell equations viewed as assemblies of harmonic vibrations ("modes").

The content of this special issue highlights papers exploring non-commutative Fourier harmonic analysis, hypoelliptic heat equation, and relativistic heat equation in the context of Information Theory and Geometric Science of Information.

> "By scrutinizing the history of these two great thoughts, would we find that the foundation of mathematical thermology by Fourier was less prepared than that of celestial mechanics by Newton?" [En scrutant de près l'histoire de ces deux grandes pensées, trouverait-on que la fondation de la thermologie mathématique par Fourier était moins préparée que celle de la mécanique céleste par Newton].—Auguste Comte, Cours de philosophie positive, t. II, p. 308, published by Bachelier, 1835.

> " We lack this thermodynamics of shapes, needed according to Thom for a true theory of information " [Il nous manque cette thermodynamique des formes nécessaire selon Thom à une véritable théorie de l'information]. Edgard Morin, La méthode, la nature de la nature; points, ed. du seuil, 1977.

We will introduce to each paper the following, structuring the special issue in two main sessions:

- Four papers on modern Fourier Heat Theory;
- Five papers on extension of Fourier Harmonic Analysis.

1. Modern Fourier Heat Theory

The first paper [14], written by F. Barbaresco, deals with Geometric Theory of Heat based on Jean-Marie Souriau Lie Groups Thermodynamics and its extension to define Maximum Entropy (Gibbs) density with higher order moments constraints. In this Souriau model, Planck temperature is described as an element of Lie algebra for the Lie Group acting on the homogeneous Manifold. Souriau has introduced, through the concept of Souriau, non-equivariant coadjoint action of Lie Group on moment map and Souriau cocycle, an invariant metric that is an extension of classical Fisher metric coming from Information Geometry, called, in the paper, Souriau-Fisher metric and vector-valued extension through poly-symplectic model.

The second paper [15], written by Arjan Van der Schaft and Bernhard Maschke, develops a Thermodynamic model initially proposed by Balian and Valentin for symplectization of contact manifolds, and introduces the global geometric definition of a degenerate Riemannian metric on the homogeneous Lagrangian sub-manifold describing the state properties. In the second part of this paper, authors give a geometric formulation of non-equilibrium thermodynamic processes, and the definition of port-thermodynamic systems and interconnection ports.

The third paper of François Gay-Balmaz and Hiroaki Yoshimura [16] presents new results on the variational formulation of nonequilibrium thermodynamics for discrete or continuum systems, and its extension for irreversible processes. These new models are illustrated in the finite dimensional cases, and on the continuum side.

The fourth paper [17], by Tamás Fülöp, Róbert Kovács, Ádám Lovas, Ágnes Rieth, Tamás Fodor, Mátyás Szücs, Péter Ván, and Gyula Gróf, analyzes the non-Fourier heat conduction phenomenon on room temperature and proposes to use the Guyer-Krumhansl equation to replace classical Fourier's law for room-temperature phenomena in the modeling of heterogeneous materials. Then, generalized heat conduction equations are introduced where Fourier heat conduction is coupled to elasticity via thermal expansion, resulting in a particular generalized heat equation for the temperature field. The last model is deduced from pseudo-temperature concept underlying heat conduction mechanics behind non-Fourier phenomena.

2. Extension of Fourier Harmonic Analysis

In the first paper of the second part [18], Hervé Bergeron and Jean Pierre Gazeau implement the so-called covariant integral quantization for Weyl-Heisenberg and affine group symmetries. Any quantization maps linearly function on a phase space to symmetric operators in a Hilbert space, and covariant integral quantization combines operator-valued measure with the symmetry group of the phase space. Covariant means that the quantization map intertwines classical (geometric operation) and quantum (unitary transformations) symmetries. Integral means that all resources of integral calculus are employed when the procedure is applied to singular functions, or distributions, for which the integral calculus is an essential ingredient. This quantization scheme is first reviewed before its specification to the Weyl-Heisenberg and affine groups, and the fundamental role played by Fourier transform in both cases is emphasized. Generalizations of the Wigner-Weyl transform are considered, and many properties of the Weyl integral quantization, commonly viewed as optimal, are shown to actually be shared by a large family of integral quantizations.

The content of the second paper [19], authored by Maurice de Gosson, lies in the continuation of previous works where it was shown that the equivalence of the Heisenberg and Schrödinger pictures of quantum mechanics requires the use of the Born and Jordan quantization rules. It gives further evidence that the Born–Jordan rule is the correct quantization scheme for quantum mechanics. For this purpose, correct short-time approximations to the action functional, initially due to Makri and Miller, are used, and it is shown that they lead to the desired quantization of the classical Hamiltonian.

In the third paper [20], Remco Duits, Erik J. Bekkers, and Alexey Mashtakov consider the Fokker–Planck PDEs (including diffusions) for stable Lévy processes (including Wiener processes) on the joint space of positions and orientations, which play a major role in mechanics, robotics, image analysis, directional statistics, and the probability theory. The exact analytic designs and solutions are known in the 2D case, where they have been obtained using Fourier transform on SE(2). The authors extend these approaches to 3D using Fourier transform on the Lie group SE(3) of rigid body motions. More precisely, they define the homogeneous space of 3D positions and orientations $R^3 \rtimes S^2 := SE(3)/(\{0\} \times SO(2))$ as the quotient in SE(3). In their construction, two group elements are equivalent if they are equal up to a rotation around the reference axis. On this quotient, the authors design a specific Fourier transform and apply it to derive new exact solutions to Fokker–Planck PDEs of α-stable Lévy processes on $R^3 \rtimes S^2$. This reduces classical analysis computations and provides an explicit algebraic spectral decomposition of the solutions. The exact probability kernel for $\alpha = 1$ (the diffusion kernel) is compared to the kernel for $\alpha = 12$ (the Poisson kernel). Stochastic differential equations (SDEs), for the Lévy processes on the quotient, are set up and the corresponding Monte-Carlo methods are derived. The exact probability kernels are shown to arise as the limit of the Monte-Carlo approximations.

In the fourth paper [21], authored by Adam Brus, Jiří Hrivnák, and Lenka Motlochová, sixteen types of the discrete multivariate transforms, induced by the multivariate antisymmetric and symmetric sine functions, are explicitly developed. Provided by the discrete transforms, inherent interpolation methods are formulated. The four generated classes of the corresponding orthogonal polynomials generalize the formation of the Chebyshev polynomials of the second and fourth kinds. Continuous orthogonality relations of the polynomials, together with the inherent weight functions, are deduced. Sixteen cubature rules, including the four Gaussian, are produced by the related discrete transforms. For the three-dimensional case, interpolation tests, unitary transform matrices, and recursive algorithms for calculation of the polynomials are presented.

In the fifth paper [22], Enrico Celeghini, Manuel Gadella, and Mariano A. Del Olmo present recent results in harmonic analysis in the real line R and in the half-line R+, which show a closed relation between Hermite and Laguerre functions, respectively, their symmetry groups and Fourier analysis. This can be done in terms of a unified framework based on the use of rigged Hilbert spaces. A relation is established between the universal enveloping algebra of the symmetry groups with the fractional Fourier transform. The results obtained are relevant to quantum mechanics as well as to signal processing as Fourier analysis has a close relation with signal filters. In addition, some new results concerning a discretized Fourier transform on the circle are presented. The authors introduce new functions on the circle constructed with the use of Hermite functions with interesting properties under Fourier transformations.

Conflicts of Interest: The authors declare no conflict of interest.

References

1. Fourier, J. *Théorie Analytique de la Chaleur*; Firmin Didot: Paris, France, 1822.
2. Fourier, J. *Œuvres publiées par G. Darboux, t. I*; Gauthier-Villars: Paris, France, 1888; t. II, 1890.
3. Dhombres, J.; Robert, J.-B. *Fourier: Créateur de la Physique-Mathématique*; Belin: Paris, France, 1998.
4. Herivel, J. *Joseph Fourier. The Man and the Physicist*; Clarendon Press: Oxford, UK, 1975.
5. Herivel, J. *Joseph Fourier: Face aux Objections Contre sa Théorie de la Chaleur, Lettres Inédites 1808–1816*; Mémoires de la Section des Sciences 8; Bibliothèque Nationale: Paris, France, 1980.
6. Grattan-Guiness, I.; Ravetz, J.R. *Joseph Fourier, 1768–1830*; MIT Press: Cambridge, MA, USA, 1972.
7. Guichardet, A. La méthode des orbites: Historiques, principes, résultats. In *Leçons de Mathématiques D'aujourd'hui, Présentées par Frédéric Bayart et Éric Charpentier*; coll. Le Sel et le Fer; Éditions Cassini: Paris, France, 2010; Volume 4, pp. 33–59.
8. Ali, S.T.; Antoine, J.-P.; Gazeau, J.-P. *Coherent States, Wavelets, and Their Generalizations (Theoretical and Mathematical Physics)*, 2nd ed.; Springer: New York, NY, USA, 2014.
9. Baake, M.; Grimm, U. *Aperiodic Order. Vol. 1. A Mathematical Invitation*; With a foreword by Roger Penrose, Encyclopedia of Mathematics and its Applications; Cambridge University Press: Cambridge, UK, 2013; p. 149.
10. Bachelard, G. *Etude sur L'évolution d'un Problème de Physique: La Propagation Thermique dans les Solides*; Vrin: Paris, France, 1973.
11. Bismut, J.-M. Hypoelliptic Laplacian and probability. *J. Math. Soc. Jpn.* **2015**, *67*, 1317–1357.

12. Souriau, J.-M. Thermodynamique relativiste des fluides. *Rend. Sem. Mat. Univ. Politec. Torino* **1978**, *35*, 21–34.
13. Vallée, C. Relativistic Thermodynamics of Continua. *Int. J. Eng. Sci.* **1981**, *19*, 589–601. [CrossRef]
14. Barbaresco, F. Higher Order Geometric Theory of Information and Heat Based on Poly-Symplectic Geometry of Souriau Lie Groups Thermodynamics and Their Contextures: The Bedrock for Lie Group Machine Learning. *Entropy* **2018**, *20*, 840. [CrossRef]
15. Van der Schaft, A.; Maschke, B. Geometry of Thermodynamic Processes. *Entropy* **2018**, *20*, 925. [CrossRef]
16. Gay-Balmaz, F.; Yoshimura, H. From Lagrangian Mechanics to Nonequilibrium Thermodynamics: A Variational Perspective. *Entropy* **2019**, *21*, 8. [CrossRef]
17. Fülöp, T.; Kovács, R.; Lovas, Á.; Rieth, Á.; Fodor, T.; Szücs, M.; Ván, P.; Gróf, G. Emergence of Non-Fourier Hierarchies. *Entropy* **2018**, *20*, 832. [CrossRef]
18. Bergeron, H.; Gazeau, J. Variations à la Fourier-Weyl-Wigner on Quantizations of the Plane and the Half-Plane. *Entropy* **2018**, *20*, 787. [CrossRef]
19. De Gosson, M. Short-Time Propagators and the Born–Jordan Quantization Rule. *Entropy* **2018**, *20*, 869. [CrossRef]
20. Duits, R.; Bekkers, E.; Mashtakov, A. Fourier Transform on the Homogeneous Space of 3D Positions and Orientations for Exact Solutions to Linear PDEs. *Entropy* **2019**, *21*, 38. [CrossRef]
21. Brus, A.; Hrivnák, J.; Motlochová, L. Discrete Transforms and Orthogonal Polynomials of (Anti)symmetric Multivariate Sine Functions. *Entropy* **2018**, *20*, 938. [CrossRef]
22. Celeghini, E.; Gadella, M.; Del Olmo, M. Hermite Functions, Lie Groups and Fourier Analysis. *Entropy* **2018**, *20*, 816. [CrossRef]

entropy

MDPI

Article

Higher Order Geometric Theory of Information and Heat Based on Poly-Symplectic Geometry of Souriau Lie Groups Thermodynamics and Their Contextures: The Bedrock for Lie Group Machine Learning

Frédéric Barbaresco

Department of Advanced Radar Concepts, Thales Land Air Systems, Voie Pierre-Gilles de Gennes, 91470 Limours, France; frederic.barbaresco@thalesgroup.com

Received: 9 August 2018; Accepted: 9 October 2018; Published: 2 November 2018

check for updates

Abstract: We introduce poly-symplectic extension of Souriau Lie groups thermodynamics based on higher-order model of statistical physics introduced by Ingarden. This extended model could be used for small data analytics and machine learning on Lie groups. Souriau geometric theory of heat is well adapted to describe density of probability (maximum entropy Gibbs density) of data living on groups or on homogeneous manifolds. For small data analytics (rarified gases, sparse statistical surveys, …), the density of maximum entropy should consider higher order moments constraints (Gibbs density is not only defined by first moment but fluctuations request 2nd order and higher moments) as introduced by Ingarden. We use a poly-sympletic model introduced by Christian Günther, replacing the symplectic form by a vector-valued form. The poly-symplectic approach generalizes the Noether theorem, the existence of moment mappings, the Lie algebra structure of the space of currents, the (non-)equivariant cohomology and the classification of G-homogeneous systems. The formalism is covariant, i.e., no special coordinates or coordinate systems on the parameter space are used to construct the Hamiltonian equations. We underline the contextures of these models, and the process to build these generic structures. We also introduce a more synthetic Koszul definition of Fisher Metric, based on the Souriau model, that we name Souriau-Fisher metric. This Lie groups thermodynamics is the bedrock for Lie group machine learning providing a full covariant maximum entropy Gibbs density based on representation theory (symplectic structure of coadjoint orbits for Souriau non-equivariant model associated to a class of co-homology).

Keywords: higher order thermodynamics; Lie groups thermodynamics; homogeneous manifold; poly-symplectic manifold; dynamical systems; non-equivariant cohomology; Lie group machine learning; Souriau-Fisher metric

---

"Inviter les savants géomètres à traiter nos problèmes avec le soucis de la commodité et de l'agrément: qu'ils écartent tout ce qui n'a rien à voir avec la pénétration de l'esprit, seule qualité dont nous faisons grand cas et que nous nous sommes proposé d'éprouver et de couronner" —Blaise Pascal—Deuxième Lettre sur la roulette, Paris, 19 Juillet 1658 [1]

"Nous avons fait de la Dynamique un cas particulier de la Thermodynamique, une Science qui embrasse dans des principes communs tous les changements d'état des corps, aussi bien les changements de lieu que les changements de qualités physiques" —Pierre Duhem, Sur les équations générales de la Thermodynamique, 1891 [2]

"Nous prenons le mot mouvement pour désigner non seulement un changement de position dans l'espace, mais encore un changement d'état quelconque, lors même qu'il ne serait accompagné d'aucun

déplacement ... De la sorte, le mot mouvement s'oppose non pas au mot repos, mais au mot équilibre."
—Pierre Duhem, Commentaire aux principes de la Thermodynamique, 1894 [3]

1. Introduction

These two Pierre Duhem's citations (see [4] for English translations) make reference to Aristotle definition of "motion" (which can be found in *The Physics*) to designate not only a change of position in space, but also any change of state, even if not accompanied by any displacement. In this case, dynamics appears as a special case of General Thermodynamics [2,3,5], to describe in common principles all changes in the state of the body, both changes of place and changes in physical qualities. Making reference to Duhem's *"Energetics"*, Stefano Bordini write in [6]: *"This theoretical design led Duhem to rediscover and reinterpret the tradition of Aristotle's natural philosophy and Pascal's epistemology ... This outcome was surprising and clearly echoed the Aristotelian language and concept of motion as change and transformation: within the framework of Aristotelian natural philosophy, motion in the modern physical sense was actually a special case of the general concept of motion. The mathematisation of thermodynamics coincided with a generalisation of mechanics, and this generalisation led to an unexpected connection between modern mathematical physics and ancient natural philosophy"* (see [7,8] for more developments on the affiliation between Aristotle, Pascal and Duhem philosophies). This conceptual and epistemology point of view was enlightened 75 years later by Jean-Marie Souriau through the symplectic model of geometric mechanics applied to statistical mechanics and used to build a "Lie groups thermodynamics" of dynamical systems, where the Gibbs density is covariant with respect to the action of the Lie group on the system (dynamical groups as Galileo group). This Souriau theory is based on tools related to non-equivariant model associated to a class of co-homology and affine representation of Lie groups and Lie algebra (last approach was independently studied in mathematical domain by Koszul to characterize homogeneous convex cones geometry [9–11]). Duhem [12] and Souriau [13,14] also both studied how to extend Thermodynamics for a continuous media.

In this paper, we will explore and compare the joint geometric contextures shared in information theory (based on Koszul's information geometry) and heat theory (based on Souriau's Lie groups thermodynamics) to highlight their joint elementary structures. Classically, we address analogies between mathematical or physical models by comparing their "structures" defined as the arrangement of and relations between the parts or elements, or as the way in which the parts are arranged or organized. My personal concept of "contexture" is more general and phenomenological and could be defined as the act, process, or manner of weaving parts into a whole. We have then replaced the relations between objects by the act to build these relations. Based on Souriau's general definition of entropy as the Legendre transform of the logarithm of generalized Laplace transform and symplectic structure associated to Lie group coadjoint orbits, we will see how geometric structures of information and heat theories are generated by these Souriau's "generative processes". We will extend theses contextures in the vector-valued case based on poly-symplectic model of higher order Souriau's Lie groups thermodynamics.

In this paper, we identify the Riemanian metric introduced by Souriau based on co-homology, in the framework of "Lie groups thermodynamics" as an extension of classical Fisher metric introduced in information geometry. We have observed that Souriau metric preserves Fisher metric structure as the Hessian of the minus logarithm of a partition function, where the partition function is defined as a generalized Laplace transform on a convex cone. Souriau's definition of Fisher metric extends the classical one in case of Lie groups or homogeneous manifolds. Souriau has developed "Lie groups thermodynamics" in the framework of homogeneous symplectic manifolds in geometric statistical mechanics for dynamical systems, but as observed by Souriau, these model equations are no longer linked to the symplectic manifold but only depend on the Lie group and the associated co-cycle.

This analogy with Fisher metric opens potential applications in machine learning, where the Fisher metric is used by information geometry, to define the "natural gradient" tool to improve ordinary stochastic gradient descent sensitivity to rescaling or changes of variable in parameter space [15–22].

In machine learning revised by natural gradient of information geometry, the ordinary gradient is designed to integrate the Fisher matrix. Amari has theoretically proved the asymptotic optimality of the natural gradient compared to classical gradient. With the Souriau approach, the Fisher metric could be extended, by Souriau-Fisher metric, to design natural gradients for data on homogeneous manifolds.

Information geometry has been derived from invariant geometrical structure involved in statistical inference. The Fisher metric defines a Riemannian metric as the Hessian of two dual potential functions, linked to dually coupled affine connections in a manifold of probability distributions. With the Souriau model, this structure is extended preserving the Legendre transform between two dual potential function parametrized in Lie algebra of the group acting transentively on the homogeneous manifold.

Classically, to optimize the parameter θ of a probabilistic model, based on a sequence of observations y_t, is an online gradient descent:

$$\theta_t \leftarrow \theta_{t-1} - \eta_t \frac{\partial l_t(y_t)^T}{\partial \theta} \tag{1}$$

with learning rate η_t, and the loss function $l_t = -\log p(y_t/\hat{y}_t)$. This simple gradient descent has a first drawback of using the same non-adaptive learning rate for all parameter components, and a second drawback of non invariance with respect to parameter re-encoding inducing different learning rates. Amari has introduced the natural gradient to preserve this invariance to be insensitive to the characteristic scale of each parameter direction. The gradient descent could be corrected by $I(\theta)^{-1}$ where I is the Fisher information matrix with respect to parameter θ, given by:

$$\begin{aligned} & I(\theta) = [g_{ij}] \\ & \text{with } g_{ij} = \left[-E_{y \sim p(y/\theta)} \left[\frac{\partial^2 \log p(y/\theta)}{\partial \theta_i \partial \theta_j} \right] \right]_{ij} = \left[E_{y \sim p(y/\theta)} \left[\frac{\partial \log p(y/\theta)}{\partial \theta_i} \frac{\partial \log p(y/\theta)}{\partial \theta_j} \right] \right]_{ij} \end{aligned} \tag{2}$$

with natural gradient:

$$\theta_t \leftarrow \theta_{t-1} - \eta_t I(\theta)^{-1} \frac{\partial l_t(y_t)^T}{\partial \theta} \tag{3}$$

Amari has proved that the Riemannian metric in an exponential family is the Fisher information matrix defined by:

$$g_{ij} = -\left[\frac{\partial^2 \Phi}{\partial \theta_i \partial \theta_j} \right]_{ij} \text{ with } \Phi(\theta) = -\log \int_{\mathbb{R}} e^{-\langle \theta, y \rangle} dy \tag{4}$$

and the dual potential, the Shannon entropy, is given by the Legendre transform:

$$S(\eta) = \langle \theta, \eta \rangle - \Phi(\theta) \text{ with } \eta_i = \frac{\partial \Phi(\theta)}{\partial \theta_i} \text{ and } \theta_i = \frac{\partial S(\eta)}{\partial \eta_i} \tag{5}$$

In geometric statistical mechanics, Souriau has developed a "Lie groups thermodynamics" of dynamical systems where the (maximum entropy) Gibbs density is covariant with respect to the action of the Lie group. In the Souriau model, previous structures of information geometry are preserved:

$$I(\beta) = -\frac{\partial^2 \Phi}{\partial \beta^2} \text{ with } \Phi(\beta) = -\int_M e^{-\langle \beta, U(\xi) \rangle} d\lambda \tag{6}$$

$$S(Q) = \langle \beta, Q \rangle - \Phi(\beta) \text{ with } Q = \frac{\partial \Phi(\beta)}{\partial \beta} \in \mathfrak{g}^* \text{ and } \beta = \frac{\partial S(Q)}{\partial Q} \in \mathfrak{g} \tag{7}$$

In the Souriau Lie groups thermodynamics model, β is a "geometric" (Planck) temperature, element of Lie algebra $\mathfrak{g}$ of the group, and Q is a "geometric" heat, element of dual Lie algebra $\mathfrak{g}^*$ of the group. Souriau has proposed a Riemannian metric that we have identified as a generalization of the Fisher metric:

$$I(\beta) = [g_\beta] \text{ with } g_\beta([\beta, Z_1], [\beta, Z_2]) = \widetilde{\Theta}_\beta(Z_1, [\beta, Z_2]) \tag{8}$$

$$\text{with } \widetilde{\Theta}_\beta(Z_1, Z_2) = \widetilde{\Theta}(Z_1, Z_2) + \langle Q, ad_{Z_1}(Z_2) \rangle \text{ where } ad_{Z_1}(Z_2) = [Z_1, Z_2] \tag{9}$$

The tensor $\widetilde{\Theta}$ used to define this extended Fisher metric is defined by the moment map $J(x)$, M (homogeneous symplectic manifold) to the dual Lie algebra $\mathfrak{g}^*$, given by:

$$\widetilde{\Theta}(X, Y) = J_{[X,Y]} - \{J_X, J_Y\} \text{ with } J(x) : M \to \mathfrak{g}^* \text{ such that } J_X(x) = \langle J(x), X \rangle,\ X \in \mathfrak{g} \tag{10}$$

This tensor $\widetilde{\Theta}$ is also defined in tangent space of the cocycle $\theta(g) \in \mathfrak{g}^*$ (this cocycle appears due to the non-equivariance of the coadjoint operator Ad_g^*, action of the group on the dual lie algebra):

$$Q(Ad_g(\beta)) = Ad_g^*(Q) + \theta(g) \tag{11}$$

$$\begin{array}{ll} \widetilde{\Theta}(X, Y) : & \mathfrak{g} \times \mathfrak{g} \to \Re \qquad\qquad \text{with } \Theta(X) = T_e\theta(X(e)) \\ & X, Y \mapsto \langle \Theta(X), Y \rangle \end{array} \tag{12}$$

In Souriau's Lie groups thermodynamics, the invariance by re-parameterization in information geometry has been replaced by invariance with respect to the action of the group. When an element of the group g acts on the element $\beta \in \mathfrak{g}$ of the Lie algebra, given by adjoint operator Ad_g. Under the action of the group $Ad_g(\beta)$, the entropy $S(Q)$ and the Fisher metric $I(\beta)$ are invariant:

$$\beta \in \mathfrak{g} \to Ad_g(\beta) \Rightarrow \begin{cases} S[Q(Ad_g(\beta))] = S(Q) \\ I[Ad_g(\beta)] = I(\beta) \end{cases} \tag{13}$$

In the case of small data analytics, we propose to parameterized the (maximum entropy) Gibbs density with higher order "geometric" temperature β_k and higher order heat Q_k, that parameterized higher order entropy $S(Q_1, \ldots, Q_n)$ and dual potential function $\Phi(\beta_1, \ldots, \beta_n)$:

$$\begin{array}{l} S(Q_1, \ldots, Q_n) = \sum\limits_{k=1}^{n} \langle \beta_k, Q_k \rangle - \Phi(\beta_1, \ldots, \beta_n) \\ \text{with } \beta_k = \dfrac{\partial S(Q_1, \ldots, Q_n)}{\partial Q_k} \text{ and } Q_k = \dfrac{\partial \Phi(\beta_1, \ldots, \beta_n)}{\partial \beta_k} \\ \text{where } \Phi(\beta_1, \ldots, \beta_n) = -\log \int\limits_M e^{-\sum\limits_{k=1}^{n} \langle \beta_k, U^k(\xi) \rangle} d\omega \end{array} \tag{14}$$

We will develop in the paper that the geometric approach of statistical thermodynamics, introduced by Souriau, offers an advantage over traditional formulations. Classical thermodynamics has been developed for static systems taking into accound only the time evolution, but in case of dynamical systems (e.g., a centrifuge system), this statistical physics is no longer valid because the Gibbs density (the density of the maximum entropy) is not covariant. In case of only time translation, what is preserved is only the energy, but for dynamical systems where a group is acting, invariants are given by components of the "moment map" (which is a geometrization of the Noether theorem providing invariants if there are symmetries). The "moment map" has been introduced in parallel by Kostant in mathematics and by Souriau in physics. Souriau has developed the non-equivariant case, and has applied it to statistical mechanics. The main advantages of "Lie groups thermodynamics" of dynamical systems, is that this statistical physics is a coordinate-free model preserving invariances with respect to the action of the (dynamical) Lie group acting on the system. We give in appendix the development of the centrifuge thermodynamics with classical approach given by Roger Balian, and prove that with the Souriau approach, the problem is solved by only applying Lie groups thermodunamics equations through moment map computation, where in classical case, we should consider additional terms related to all moments (energy, angular momentum, ...) through additional Lagrange hyper-parameters, that corresponds to components of Souriau's "geometric" (Planck) temperature.

Before developing all these models, and because this topic needs transverse knowledges of many concepts developed in different disciplines as statistical physics & thermodynamics, information geometry, symplectic mechanics and multi-symplectic geometry, we propose to the readers, in the preamble, to study the following books and papers:

- Introduction to Statistical Physics and Thermodynamics: [2,3,23–26]
- Introduction to Higher Order Thermodynamics: [27–40]
- Introduction to Information Geometry: [9–11,15–22,41,42]
- Introduction to Symplectic Mechanics: [43–47]
- Introduction to Multi-Symplectic Geometry: [48–55]

The geometric definition and extension of Fisher metric has been recently studied in the framework of quantum information geometry, but this community seems unaware of Souriau's work on Lie groups thermodynamics for the study of statistical physics of dynamical systems based on symplectic geometry and c-homology tools in the 70s, and in particular the non-equivariant case developed by Souriau and Koszul. We can make reference to the following recent works on the symplectic formulation of the Fisher information theory [56–59].

The structure of the paper is the following:

- In Section 1, we introduce seminal idea on Symplectic geometry used in mechanics and in statistical mechanics, as introduced by Jean-Marie-Souriau during the 60s. From previous work of François Gallissot extending Cartan's results on integral invariant (theorem on types of differential forms generating equations of movement of a material point invariant in the transformations of the Galilean group), we present the Lagrange 2-form and moment map elaborated by Souriau to build a geometric mechanics theory, where a dynamical system is then represented by a foliation of the evolution, determined by an antisymmetric covariant second order tensor. Souriau has applied this tool for mechanical statistics to build a thermodynamics of dynamical systems, where the classical notion of Gibbs canonical ensemble is extended for a homogeneous symplectic manifold on which a Lie group (dynamical group) has a symplectic action. In case of Galileo group, the symmetry is broken, and new "co-homological" Souriau relations should be verified in Lie algebra of the group.
- In Section 2, we synthetize results on higher order thermodynamics based on higher order temperatures and heats, as introduced by Ingarden and Jaworski for mesoscopic systems. This model is based on higher order maximum entropy Gibbs density definition constraining solution with respects to higher order moments.
- In Section 3, we develop "Lie groups thermodynamics" model, developed to describe Gibbs state for dynamical systems, where Souriau introduced the concept of co-adjoint action of a group on its momentum space that allows designing physical observables like energy, heat and momentum or moment as pure geometrical objects. The Souriau model then generalizes the Gibbs equilibrium state to all symplectic manifolds that have a dynamical group, with a "geometric" (Planck) temperature as an element of the Lie algebra and "geometric heat" as an element of the dual Lie algebra. We have observed that Souriau has introduced a symmetric tensor that is an extension of classical Fisher metric in information geometry. This new Fisher-Souriau metric is invariant with respect to the action of the group. These equations are universal, because they are not dependent on the symplectic manifold but only on the dynamical group and its associated two-cocycle. Souriau called it "Lie groups thermodynamics".
- In Section 4, we give an extended Koszul study of Souriau's non-equivariant model associated to a class of co-homology. Koszul has deepened the Souriau model, considering purely algebraic

and geometric developments of geometric mechanics. Koszul has defined a skew symmetric bilinear form by a closed expression depending only on the cocycle and related to the Souriau antisymmetric bilinear map introduced previously in Section 3. This Koszul study of the moment map non-equivariance, and the existence of an affine action of G on g* is at the cornerstone of Souriau theory of Lie groups thermodynamics.

- In Section 5, at the step of the Souriau Lie groups thermodynamics presentation, we will introduce a generalized Souriau definition of entropy, as the Legendre transform of the logarithm of the Laplace transform, making the connection with information geometry. This definition is a general contexture that can be extended to highly abstract spaces preserving Legendre structure, if we are able to generalize the Laplace transform.
- In Section 6, we illustrate Souriau's Lie groups thermodynamics for a centrifuge system. The main Souriau idea was to define the Gibbs states for one-parameter subgroups of the Galilean group, because he proved that the action of the full Galilean group on the space of motions of an isolated mechanical system is not related to any equilibrium Gibbs state (the open subset of the Lie algebra, associated to this Gibbs state, is empty).
- In Section 7, we have defined an higher-order model of Lie groups thermodynamics based on a poly-symplectic vector valued approach. This multi-symplectic extension, is based on a multi-valued one that preserve the notion of (poly-)moment map built by Günther based on an n-symplectic model. We replace the symplectic form of the Souriau model by a vector valued form that is called poly-symplectic. We consider the non-equivariance of poly-moment map by introducing poly-cocycle. We finally conclude with poly-symplectic definition extension of the Fisher-Souriau metric.
- In Section 8, we conclude with potential extension to Lie group machine learning.

To facilitate understanding of previous results, we add some additional complements:

- In Appendix A, we recall a synthesis of Günther's poly-symplectic model with initial notation
- In Appendix B, we develop computation of the Fisher metric for multivariate Gaussian density, to establish links with Souriau's Lie groups Gibbs density model.
- In Appendix C, we give more details on the Legendre transform, the basic tool of information geometry and Souriau Lie groups thermodynamics. More especially, we give a definition of the Legendre transform with projective geometry definition by Chasles as reciprocal polar with respect to a paraboloid.
- In Appendix D, we give solution of a centrifuge system thermodynamics, given by Roger Balian based on a classical approach, to make the link with the Souriau approach.
- In Appendix E, we recall the main proofs of Souriau's Lie groups thermodynamics and its poly-symplectic extension.
- In Appendix F, we present another Souriau statistical physics model, developed for relativistic thermodynamics of continua, which preserves the Legendre transform, where temperature is given by a killing vector.

2. Seminal Idea of Symplectic Geometry in Mechanics and in Statistical Mechanics by Gallissot and Souriau

The symplectic structure has been introduced in mathematics much earlier than the word symplectic, in works of the French physicist Joseph Louis Lagrange (see paper on the slow changes of the orbital elements of planets in the solar system), who showed that this geometry is a fundamental tool in the mathematical model of any problem in mechanics. Jean-Marie Souriau has shown that Lagrange's parentheses (nowdays called Lagranges bracket) are the components of the canonical symplectic 2-form on the manifold of motions of the mechanical system, in the chart of that manifold [60,61].

Jean-Marie Souriau, graduated from ENS ULM 1942, was the nephew of the philosopher Etienne Souriau (graduated from ENS Ulm 1912, ranked 1st at aggregation, a collaborator of Gaston Bachelard in Paris Sorbonne University, PhD supervisor of Film Maker Eric Rohmer), author of *"Les Structures de l'oeuvre d'art"* and grandson of the philosopher Paul Souriau (graduated from ENS Ulm 1873), author of *"Esthétique du mouvement"* and a Latin thesis « De motus perceptione », who both have worked on *"aesthetics"*, and little nephew of literature historian Maurice Souriau, the editor of a critical version Blaise Pascal's *"Pensées"* (awarded by 4 prices of Académie Française). The Souriau family, with Paul, Etienne and Jean-Marie were motivated to explore esthetical issues of "motion structures" (we could summarize by the triptych: the Esthetism of Motion of Paul Souriau, the Structure of Esthetism of Etienne Souriau and the Structure of Motion of Jean-Marie Souriau). Jean-Marie Souriau's book *"Structure des Systems Dynamiques"* (SSD) was elaborated in Carthage and Marseille, where Souriau was installed with his wife Christiane Souriau-Hoebrecht. In 1952 Souriau found a position at Institut des Hautes Études de Tunis (8 rue de Rome, Tunis) (see Figure 1) and was back in Marseille in a position in 1958 at the Faculté des Sciences. The manuscript was given to the editor Dunod in 1969, but only edited in 1970 (2019 is the 50th birthday of this book and tributes will be given in 2 events FGSI'19 [62] and SOURIAU 2019 [63]).

About the source of his book title, we are at the apogee or "acme" of the STRUCTURALISM in anthropology/sociology/linguistic/philosophy/ epistemology in France (Levi-Strauss, Barthes, Foucault, Althusser, Lacan, ...). The word "structure" was in the air of the time, fashionable at the moment, circulating on all the lips as described by François Dosse in *"Histoire du structuralisme I & II"*. After his ONERA PhD Defence in 1953 (I have a copy of his PhD), his PhD supervisor André Lichnerowicz made one comment *"you have many anti-symmetrical forms in your calculations, you should be interested in symplectic structures"*.

Figure 1. Institut des Hautes Etudes de Tunis, 8 rue de Rome where Souriau has developed his theory of Geometric Mechanics and Lie Groups Thermodynamics (http://www.ina.fr/video/AFE01000164).

As early as 1966, influenced by François Gallissot's work, Souriau applied his theory of geometric mechanics to statistical mechanics, developed in Chapter IV of his book *"Structure of Dynamical Systems"* [43,64], what he called "Lie groups thermodynamics". We have discovered that Souriau and Gallissot both attended the 1954 International Congress of Mathematicians (ICM'54) in Moscow. We could assume that they have discussed 1952 Gallissot's paper introducing three types of differential forms generating equations of movement of a material point invariant in the transformations of the Galilean group and their links with Poincaré-Cartan integral invariant. This seminal work of Gallissot helped Souriau to formulate his new geometric mechanics and its extenxion to geometric statistical physics. Using Lagrange's viewpoint, in Souriau statistical mechanics, a statistical state is a probability measure on the manifold of motions. As we can read in his book, Souriau was influenced by François Gallissot to introduce the Lagrange(-Souriau) 2-form.

In place of classical mechanical equations of a material point subjected to a force F, defined by its mass m and its position r at time t, the second order differential equations $m\frac{d^2r}{dt^2} = F$ is rewritten by a system of first order differential equations in phase space $\begin{pmatrix} r \\ v \end{pmatrix}$:

$$m\frac{dv}{dt} = F \text{ and } v = \frac{dr}{dt} \tag{15}$$

If the force F is derived from a potential w, we have classical equations:

$$\begin{cases} L = \frac{1}{2}mv^2 - w \text{ (Lagrangian)} \\ H = \frac{1}{2}mv^2 + w \text{ (Hamiltonian)} \end{cases} \text{ and Hamilton-Jacobi equations } \begin{cases} \frac{dq_i}{dt} = \frac{\partial H}{\partial p_i} \\ \frac{dp_i}{dt} = -\frac{\partial H}{\partial q_i} \end{cases} \text{ with } \begin{cases} r = \begin{bmatrix} q_1 \\ q_2 \\ q_3 \end{bmatrix} \\ mv = \begin{bmatrix} p_1 \\ p_2 \\ p_3 \end{bmatrix} \end{cases} \tag{16}$$
$$\text{with } A = \int_{t_0}^{t_1} L dt$$

This idea of Lagrange, rediscovered by Souriau was to consider time t like the others variables. One should use then the 7-dimensional space V (evolution space) (see Figure 2):

$$y = \begin{pmatrix} t \\ r \\ v \end{pmatrix} \tag{17}$$

Classical system of first order differential equations in phase space can then be rewritten in evolution space V by the homogeneous form:

$$\begin{cases} m\delta v - F\delta t = 0 \\ \delta r - v\delta t = 0 \end{cases} \tag{18}$$

At each point y of V, these equations define the tangent direction to the curve x described by the point y during the evolution of the system. These curves are the leaves (lines of force) of the field of directions defined by the equations of the homogeneous form, as defined for foliated manifolds. See [43], for more details on definition of the different derivatives used.

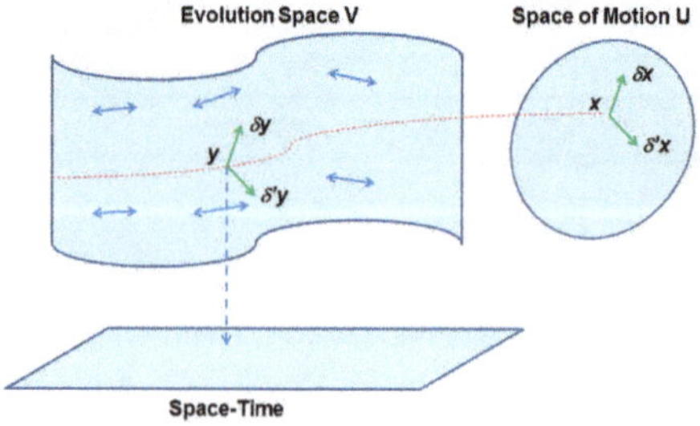

Figure 2. Evolution space V, Space of motions U and classical space time.

A dynamical system is then represented by a foliation of the evolution, where the foliation is determined by an antisymmetric covariant second order tensor, denoted by σ and called

Lagrange-Souriau 2-form. The components of this tensor are expressions known as Lagrange brackets. σ is considered as a bilinear operator on tangent vectors of V. If we choose two such vectors:

$$\delta y = \begin{pmatrix} \delta t \\ \delta r \\ \delta v \end{pmatrix} \text{ and } \delta' y = \begin{pmatrix} \delta' t \\ \delta' r \\ \delta' v \end{pmatrix} \tag{19}$$

σ associates to them an antisymmetric scalar product:

$$\sigma(\delta y)(\delta' y) = \langle m\delta v - F\delta t, \delta' r - v\delta' t \rangle - \langle m\delta' v - F\delta' t, \delta r - v\delta t \rangle \tag{20}$$

In the Souriau-Lagrange model, σ is a 2-form on the evolution space V, and the differential equation of motion $\delta y \in \varepsilon$ implies:

$$\sigma(\delta y)(\delta' y) = 0\,,\ \forall \delta' y \tag{21}$$

which can be written as:

$$\sigma(\delta y) = 0 \ \text{ or } \delta \mathrm{y} \in \ker(\sigma) \tag{22}$$

For study of this Souriau-Lagrange 2-form, readers should see the papers of Obădeanu [65–67].

Souriau has observed that this 2-form was introduced by Lagrange in a different language in his study of celestial mechanics in 1808. Souriau was also influenced by François Gallissot that used this 2-form in [68,69]. We will see in the following the Souriau's "moment map μ" in dual Lie algebra of the group G, and the study of coadjoint orbits of G. For the definition of moment map, we make reference to [45]. Souriau has extended this model for thermodynamics. For this new phenomenological approach of mechanics, thermodynamics and information theory, we can give reference to Souriau introduction of his paper "*Quantique? Alors c'est géométrique*" [70] and a video of his talk [71]:

> *"Plaçons-nous d'abord dans le cadre de la mécanique classique. Étudions un système mécanique isolé, non dissipatif—nous dirons brièvement une «chose». L'ensemble des mouvements de cette «chose» est une variété symplectique. Pourquoi? Il suffit de se reporter à la Mécanique Analytique de Lagrange (1811); l'espace des mouvements y est traité comme variété différentiable; les coordonnées covariantes et contravariantes de la forme symplectique y sont écrites (Ce sont les "parenthèses" et "crochets" de Lagrange). Évoquons maintenant la géométrie du 20 éme siècle. Soit G un groupe difféologique (par exemple un groupe de Lie); μ un moment de G (un moment, c'est une 1-forme invariante à gauche sur G); alors l'action du groupe sur μ engendre canoniquement un espace symplectique (ces groupes pourront avoir une dimension infinie). Présomption épistémologique: derrière chaque «chose» est caché un groupe G (sa "source"), et les mouvements de la «chose» sont simplement des moments de G (doublet latin mnémotechnique : momentum-movimentum). L'isolement de la «chose» indique alors que le groupe de Poincaré (respectivement de Galilée-Bargman) est inséré dans G; voilà l'origine des grandeurs conservées relativistes (respectivement classiques) associées à un mouvement x: elles constituent simplement le moment induit sur le groupe spacio-temporel par le moment-mouvement x."* (In English: *Let's put ourselves first in the framework of classical mechanics. Let's study an isolated, non-dissipative mechanical system—we will briefly say a "thing". The set of movements of this "thing" is a symplectic manifold. Why? It is enough to refer to the Analytical Mechanics of Lagrange (1811); the space of movements is treated as a differentiable manifold; the covariant and contravariant coordinates of the symplectic form are written there (these are the "parentheses" and "brackets" of Lagrange). Let's now talk about the geometry of the 20th century. Let G be a diffeological group (for example a Lie group); μ a moment of G (a moment is a left invariant 1-form on G); then the action of the group on μ canonically generates a symplectic space (these groups can have an infinite dimension). Epistemological presumption: behind each "thing" is hidden a group G (its "source"), and the movements of the "thing" are simply moments of G (mnemonic Latin doublet: momentum-movimentum). The isolation of the "thing" then indicates that the group of Poincaré*

(respectively Galileo-Bargman) is inserted in G; here is the origin of the relativistic (respectively classical) conserved magnitudes associated with a movement x: they simply constitute the moment induced on the spacio-temporal group by the moment-motion x.)

"Il y a un théorème qui remonte au XXème siècle. Si on prend une orbite coadjointe d'un groupe de Lie, elle est pourvue d'une structure symplectique. Voici un algorithme pour produire des variétés symplectiques: prendre des orbites coadjointes d'un groupe. Donc cela laisse penser que derrière cette structure symplectique de Lagrange, il y avait un groupe caché. Prenons le mouvement classique d'un moment du groupe, alors ce groupe est très «gros» pour avoir tout le système solaire. Mais dans ce groupe est inclus le groupe de Galilée, et tout moment d'un groupe engendre des moments d'un sous-groupe. On va retrouver comme cela les moments du groupe de Galilée, et si on veut de la mécanique relativiste, cela va être du groupe de Poincaré. En fait avec le groupe de Galilée, il y a un petit problème, ce ne sont pas les moments du groupe de Galilée qu'on utilise, ce sont les moments d'une extension centrale du groupe de Galilée, qui s'appelle le groupe de Bargman, et qui est de dimension 11. C'est à cause de cette extension, qu'il y a cette fameuse constante arbitraire figurant dans l'énergie. Par contre quand on fait de la relativité restreinte, on prend le groupe de Poincaré et il n'y a plus de problèmes car parmi les moments il y a la masse et l'énergie c'est mc^2. Donc le groupe de dimension 11 est un artéfact qui disparait, quand on fait de la relativité restreinte." (In Engish: *There is a theorem dating back to the twentieth century. If we take a coadjoint orbit of a Lie group, it is provided with a symplectic structure. Here is an algorithm to produce symplectic manifolds: take coadjoint orbits from a group. So it suggests that behind this symplectic structure of Lagrange, there was a hidden group. Take the classic movement of a moment of the group, so this group is very "big" to have the whole solar system. But in this group is included the Galileo group, and any moment of a group generates moments of a subgroup. We will find like that the moments of the group of Galileo, and if we want relativistic mechanics, it will be Poincaré group. In fact with Galileo group, there is a small problem, it is not the moments of the Galileo group that are used, it is the moments of a central extension of the Galileo group, which is called the Bargman group, and that is of dimension 11. It is because of this extension, that there is this famous arbitrary constant appearing in the energy. On the other hand, when we do special relativity, we take Poincaré group and there are no more problems because among the moments there is the mass and the energy is mc^2. So the 11-dimensional group is an artifact that disappears, when we do special relativity.)*

François Gallissot has observed that in his famous lessons on integral invariants, Elie Cartan has shown that all the properties of the differential equations of the dynamics of holonomic systems result from the existence of the integral invariant:

$$\int \omega \text{ with } \omega = \sum_{i} p_i dq_i - Hdt \tag{23}$$

Thus every holonomic system whose forces derive from a force function is associated to a form ω, the equations of motion being the characteristics of the exterior form $d\omega$. Around 1950, the theory of exterior forms on differentiable manifolds has been established on new foundations under the influence of topologists. The question was then to wonder:

- if classical mechanics cannot benefit from these models by placing an exterior form of degree two at its base
- if thanks to the notion of manifold, the notion of connection cannot be introduced in a more natural way
- if the paradoxal indeterminations/impossibilities in the Lagrangian framework could be explained more clearly
- if the problem of integration of equations of motion could be enlightened, generated by a form Ω of degree two.

To reach these various objectives, Gallissot has resumed first the study of the logical bases on which the Galilean mechanics is built. He thus shown that when it is proposed to find generating forms of the equations of motion of a material invariant point in the transformations of the Galilean group, the most interesting form is an exterior form of degree two defined on a variety $E^3 \times E \times T$ (E^3 Euclidean space, T temporal). Gallissot had shown that any holonomic parametric system with n degrees of freedom is associated with a form Ω of degree $2n$ defined on a differentiable manifold whose characteristics are the equations of the movement. This form is expressed by means of $2n$ Pfaff forms and by dt, the Hamiltonian form being a simple special case. He gave a summary of how we can get rid of the servitude of coordinates in the study of dynamical systems and the important role played by the operator $i()$ antiderivative introduced by Cartan, the characteristic field E of the form Ω being defined by the relation $i(E)\Omega = 0$. Gallissot has then introduced the following theorem:

Theorem 1. *There are three types of differential forms generating equations of movement of a material point invariant in the transformations of the Galilean group:*

$$\begin{array}{l} A : \left\{ \begin{array}{l} s = \frac{1}{2m} \sum\limits_{i=1}^{3} (mdv_i - F_i dt)^2 \\ e = \frac{m}{2} \sum\limits_{j=1}^{3} (dx_j - v_j dt)^2 \end{array} \right. \\ B : f = \sum\limits_{1}^{3} \delta_{ij} (dx_i - v_i dt)(mdv_j - F_j dt) \text{ with } \delta_{ij} \text{ krönecker symbol} \\ C : \omega = \sum\limits_{1}^{3} \delta_{ij} (mdv_i - F_i dt) \wedge (dx_j - v_j dt) \end{array} \tag{24}$$

If we consider the last form "C":

$$\omega = \sum_{1}^{3} \delta_{ij} (mdv_i - F_i dt) \wedge (dx_j - v_j dt) = m\delta_{ij} dv_i \wedge dx_j - m\delta_{ij} v_i dv_j \wedge dt + \delta_{ij} F_i dx_j \wedge dt \tag{25}$$

$d\omega = 0$ constraints Pfaff form $\delta_{ij} F_i dx_j$ to be closed, and to reduce the differential of function U:

$$\omega = m\delta_{ij} dv_i \wedge dx_j - dH \wedge dt \tag{26}$$

$$\text{with } H = T - U \text{ and } T = \frac{1}{2} \sum_{i=1}^{3} m(v_i)^2 \tag{27}$$

It proves that the exterior derivative of ω is:

$$d\omega = \sum_{i=1}^{3} mv_i dx_j - Hdt \tag{28}$$

The form $\omega^* = d\omega$ generates Elie Cartan integral invariant.

In Chapter IV of his book, Souriau applied this model based on symplectic geometry for statistical mechanics. Souriau observed that Gibbs equilibrium is not covariant with respect to dynamic groups of physics. To solve this breaking of symmetry, Souriau introduced a new "geometric theory of heat" where the equilibrium states are indexed by a parameter β with values in the Lie algebra of the group, generalizing the Gibbs equilibrium states, where β plays the role of a geometric (Planck) temperature. Souriau observed that the group of time translations of the classical thermodynamics is not a normal subgroup of the Galileo group, proving that if a dynamical system is conservative in an inertial reference frame, it need not be conservative in another. Based on this fact, Souriau generalized the formulation of the Gibbs principle to become compatible with Galileo's relativity in classical mechanics and with Poincaré relativity in relativistic mechanics. The maximum entropy principle is preserved, and the Gibbs density is given by the density of maximum entropy (among the equilibrium states for which the average value of the energy takes a prescribed value, the Gibbs measures are those which have the largest entropy), but with a new principle *"If a dynamical system is invariant under a Lie subgroup G′ of the Galileo group, then the natural equilibria of the system forms the Gibbs ensemble of the dynamical group G′"*. The classical notion of Gibbs canonical ensemble is extended for a homogneous symplectic manifold on which a Lie group (dynamic group) has a symplectic action. In case of a Galileo group, the symmetry is broken, and new "cohomological" relations should be verified in Lie algebra of the group. A natural equilibrium state will thus be characterized by an element of the Lie algebra of the Lie group, determining the equilibrium temperature β. The entropy $s(Q)$, parametrized by Q the geometric heat (mean of energy U, element of the dual Lie algebra) is defined by the Legendre transform of the Massieu potential given by $\Phi(\beta)$, parametrized by β ($\Phi(\beta)$ is the minus logarithm of the partition function $\psi_\Omega(\beta)$):

$$s(Q) = \langle \beta, Q \rangle - \Phi(\beta) \text{ with } \Phi(\beta) = -\log \int_M e^{-\langle \beta, U(\xi) \rangle} d\omega \,,\; Q = \frac{\partial \Phi}{\partial \beta} \in \mathfrak{g}^* \text{ and } \beta = \frac{\partial s}{\partial Q} \in \mathfrak{g} \tag{29}$$

Souriau has proposed to study the statistical mechanics from the new point of view of symplectic geometry, completing the work of Poincaré and Cartan on integral invariant, reinventing the Lagrangian symplectic form in place of classical variational formulation and geometrizing the Noether theorem with a moment map as new conserved quantities. Firstly, Souriau Lie groups thermodynamics gives geometrical status to the (Planck) temperature and the entropy with a new general definition of the Fisher Metric. Secondly, Souriau's relativistic thermodynamics of continua provides a geometrization of the smecond principle by the permanence of the entropy current, whose flux has positive divergence [13,14,72–74]. This 2nd model of Souriau's thermodynamics is described in the Appendix. Other authors have studied this relativistic thermodynamics of continua [75–82].

If some works have been done from the 80s by Ingarden [83,84] and Mrugala [85–89] and Arnold [90] to give a geometric structures to thermodynamics, Souriau's Lie groups thermodynamics was ignored for more than 50 years until recently recovered in [23,91].

3. Higher Order Thermodynamics Based on Higher Order Temperatures

We will generalize Souriau's theory [43,64], reconsidered in [23] and with links to information geometry in [91], in the framework of higher order thermodynamics as introduced by Ingarden [29–31] and Jaworski [32–35] for mesoscopic systems. We can make also reference to other publications of Ingarden [36–40], Jaworsky [92–94] and Nakagomi [95] on higher order thermodynamics. The Gibbs canonical state results from the maximum entropy principle when the statistical mean value of the energy is supposed to be known. A Polish school has studied the maximum entropy inference with higher-order moments of energy (when not only mean values but also statistical moments of higher order of some physical quantities are taken into account). Ingarden in 1963 and Jaworski in 1981 have introduced the concept of second and higher-order temperatures, by assuming a distribution function which includes information not only on the average of the energy but also on higher-order moments, in particular 2nd moment related to fluctuations. This case should be considered in situations where fluctuations are not negligible, such as near phase transitions or critical points, in metastable states in systems with a small number of degrees of freedom. Ingarden's idea is that if we can measure more details, such as the first n cumulants of the energy, we can then introduce n high-order temperature, as the Lagrange multipliers when we maximize the entropy with respect to these values:

$$P_{(\beta_1,\beta_2)} = \frac{1}{Z(\beta_1,\beta_2)} e^{-\beta_1.H-\beta_2(H-U)^2} = e^{\beta_0-\beta_1.H-\beta_2(H-U)^2} \tag{30}$$

Ingarden proposed that if we can measure the second cumulant of the energy (the fluctuation of the energy), the equilibrium state is not the canonical state, but would need two temperatures.

Ingarden argues that for a macroscopic system there is very little difference between the two states, and that we would need a mesoscopic or microscopic system to be able to detect the higher temperature. Jaworski [27,28] has shown that the contribution to the total entropy, arising from the extra information corresponding to the higher-order moments, is $o(N)$, when N tends to infinity and N/V ratio is constant, with N the number of particles and V the volume. The main result of Jaworski is that from a purely thermodynamic point of view, the information corresponding to the higher-order moments of extensive physical quantities is not essential and can be neglected in the maximum entropy procedure. Jaworski showed that the maximum entropy inference has a certain stability property with respect to information corresponding to higher order moments of extensive quantities. It can serve as an argument in favor of the maximum entropy method in statistical physics and to understand better why these methods are successful. Streater [96] has prefered to say that the states with generalized temperatures are not in equilibrium, assuming that the final state, at large times, will be the canonical or grand canonical state depending on mixing properties. Streater [96] intends that this occur even for a mesoscopic system, such as a few atoms, adding that his approach is equivalent to Ingarden model if the relaxation time from the state with generalized temperatures to the final equilibrium is very long.

Some examples of higher order maximum Entropy are given by Ingarden:

- *1st Example of Higher Oder Maximum Entropy Density*:

  Density of maximum Entropy

$$S(P) = -\int_{-\infty}^{+\infty} P(x)\log P(x)dx \tag{31}$$

under the constraints:

$$P(x) \geq 0, \int_{-\infty}^{+\infty} P(x)dx = 1 \text{ and } E\left(x^{2n}\right) = \int_{-\infty}^{+\infty} x^{2n}P(x)dx = \sigma^{2n} \tag{32}$$

is given by:

$$P(x) = \frac{1}{2(2n)^{\frac{1}{2n}}\sigma.\Gamma(1+1/2n)}\exp\left(-\frac{x^{2n}}{2n\sigma^{2n}}\right) = f_n(x) \tag{33}$$

with the following parameters:

$$\beta_n = \frac{1}{2n\sigma^{2n}},\ Z(\beta_n) = \frac{2\Gamma(1+1/2n)}{\beta_n^{1/2n}},\ S(P) = \log Z(\beta_n) + \frac{1}{2n} \tag{34}$$

where:

$$E\left(x^{2k-1}\right) = 0 \text{ and } -\frac{\partial \log Z(\beta_k)}{\partial \beta_k} = \sigma^{2k} = E\left(x^{2k}\right) = \frac{(2n)^{k/n}\sigma^{2k}\Gamma(1+(2k+1)/2n)}{(2k+1)\Gamma(1+1/2n)} \tag{35}$$

We illustrate this higher order maximum entropy density in Figure 3.

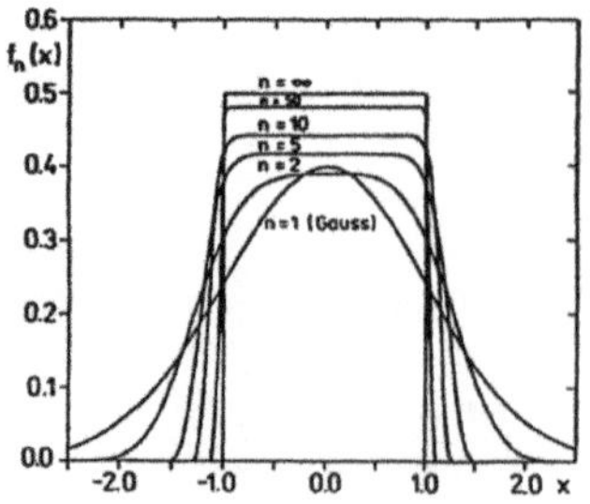

Figure 3. Higher order maximum entropy density for constraints (32) from Ingarden's paper.

- *2nd Example of Higher Oder Maximum Entropy Density:*

Density of maximum Entropy $S(P) = -\int_0^{+\infty} P(x)\log P(x)dx$ under the constraints:

$$P(x) \geq 0, \int_0^{+\infty} P(x)dx = 1 \text{ and } E(x^n) = \int_0^{+\infty} x^n P(x)dx = \sigma^n \tag{36}$$

is given by:

$$P(x) = \frac{1}{n^{\frac{1}{n}}\sigma.\Gamma(1+1/n)}\exp\left(-\frac{x^n}{n\sigma^n}\right) = f_n(x) \tag{37}$$

with the following parameters:

$$\beta_n = \frac{1}{n\sigma^n}, \ Z(\beta_n) = \frac{\Gamma(1+1/n)}{\beta_n^{1/n}}, \ S(P) = \log Z(\beta_n) + \frac{1}{n} \tag{38}$$

where:

$$-\frac{\partial \log Z(\beta_k)}{\partial \beta_k} = \sigma^k = E\left(x^k\right) = \frac{n^{k/n}\sigma^k\Gamma(1+(k+1)/n)}{(k+1)\Gamma(1+1/n)} \tag{39}$$

We illustrate this higher order maximum entropy density in Figure 4.

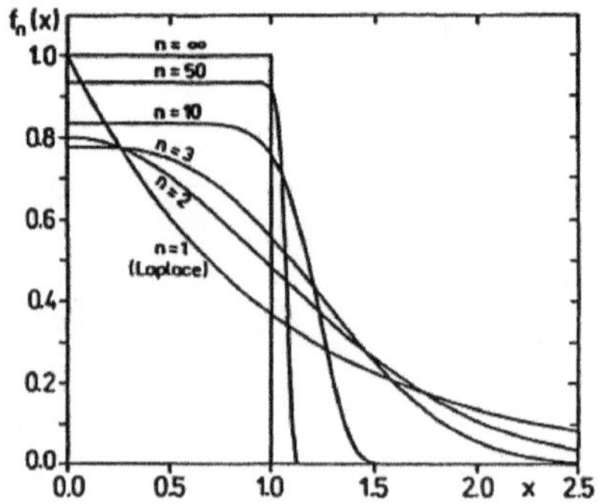

Figure 4. Higher order maximum entropy density for constraints (36) from Ingarden paper.

As soon as 1963, Ingarden has introduced this concept of higher order temperatures for statistical systems such as thermodynamics. In physics, the concept of temperature is connected with the mean value of kinetic energy of molecules in an ideal gas. For a general physical system with interactions among particles (the case of non-ideal gas: liquid or solid), an equilibrium probability distribution depends on temperature T as the only statistical parameter of the Gibbs state: $P_\beta(x) = \frac{1}{Z(\beta)}e^{-\beta.H(x)}$

with $\beta = \frac{1}{k_\beta T}$ and $H(x) = H(p,q)$ where p is position, q the mechanical momentum and k_β the Boltzmann constant (a factor to insure that $\beta.H$ is dimensionless). If there are no stochastic interactions between particles (ideal gas), the partition function Z has the property to be integrable and we can obtain Gauss distribution in the momentum space deduced from the result of the limit theorem for large N. The ideal gas model of Boltzmann can fail if the number of particles is not large enough in the case of mesoscopic systems, and also if the interactions between particles are not weak enough. Gibbs hypothesis can also fail in other cases when stochastic interactions with the environment are not sufficiently weak. As remarked by Ingarden, nobody has ever observed thermal Gibbs equilibrium in large and complex systems (cosmic systems, Earth's atmosphere, biological organisms), but only in cases of turbulence, flows or pumping, by replacing classical approach by local temperature and concept of thermodynamic flows (non-equilibrium thermodynamics and thermo-hydrodynamics), that is non-coherent with the classical concept of temperature which is, by definition, global/intensive and does not depend on position. R.S. Ingarden proposed to consider the stationary case using of the concept of higher order temperatures given by the Gibbs density:

$$P_{(\beta_1,\ldots,\beta_n)}(x) = \frac{1}{Z(\beta_1,\ldots,\beta_n)} e^{-\beta_1.H(x)-\beta_2(H(x)-U)^2-\ldots-\beta_n(H(x)-U)^n} \tag{40}$$

where $U = E(H)$ is the mean energy. This mean energy has been introduced to preserve the the total energy invariance with respect to an arbitrary additive constant, and $\beta_0 = -\log Z(\beta_1,\ldots,\beta_n)$ the constant of normalization. The new constants β_k are said to be β-temperatures of order k. $H(x)$ is usually defined as a quadratic function of x. The probability distribution is uniquely defined from statistical moments which should be measured experimentally. But if values number is too high to make this method practical, we are only able to measure the lowest moments up to some order (if we can neglect the higher orders that do not change the result to a given accuracy), and to fix β-temperatures defined as Lagrange multipliers by maximization of entropy of distribution $S = -\int P_{(\beta_1,\ldots,\beta_n)}(x)\log P_{(\beta_1,\ldots,\beta_n)}(x)dx$, with the given moments as constraints. R.S. Ingarden observed that the entropy maximization randomizes higher moments in a symmetric way, and it cancel any possible bias with respect to their special values, and it gives the best estimate to a given accuracy. The values of β can be found by:

$$E\left(x^k\right) = \frac{\partial\beta_0}{\partial\beta_k} = \frac{\partial\log Z}{\partial\beta_k} \text{ with } E\left(x^k\right) = Z^{-1}\int x^k e^{-\sum\limits_{k=1}^{n}\beta_k x^k}dx = \int x^k P_{(\beta_1,\ldots,\beta_n)}(x)dx \tag{41}$$

$$Z = \int e^{-\sum\limits_{k=1}^{n}\beta_k x^k}dx \text{ and the relation : } S = \sum_{k=1}^{n}\beta_k E\left(x^k\right) + \log Z = \sum_{k=1}^{n}\beta_k\frac{\partial\beta_0}{\partial\beta_k} - \beta_0 \tag{42}$$

Ingarden has applied this model for linguistic statistics, assuming the appearance of higher order temperatures since there occur rather strong statistical correlations between phonemes and words as elements of these statistics. He argued his choice observing that in the case of word statistics, the existence of strong correlations is given by grammatical or semantical studies [9]. Ingarden made the conjecture that his high order thermodynamics is the model of statistically interacting, biological living systems, and small systems although the calculation/observation are more difficult.

Ingarden higher order temperatures could be defined in the case when no variation is considered, but when a probability distribution depending on more than one parameter. It has been observed by Ingarden, that Gibbs assumption can fail if the number of components of the sum goes to infinity and the components of the sum are stochastically independent, and if stochastic interactions with the environment are not sufficiently weak. In all these cases, we never observe absolute thermal equilibrium of Gibbs type but only flows or turbulence. Non-equilibrium thermodynamics could be indirectly addressed by means of high order temperatures.

4. Model of Souriau Lie Groups Thermodynamics

For introduction to symplectic geometry, we make reference to Marle's book [45] and Koszul's book [44]. In 1969, Souriau [43,64] introduced the concept of co-adjoint action of a group on its momentum space, based on the orbit method works, that allows to define physical observables like energy, heat and momentum or moment as pure geometrical objects. The moment map is a constant of the motion and is associated to symplectic cohomology. In a first step to establish new foundations of thermodynamics, Souriau has defined a Gibbs canonical ensemble on a symplectic manifold M for a Lie group action on M. In classical statistical mechanics, a state is given by the solution of Liouville equation on the phase space, the partition function. As symplectic manifolds have a completely continuous measure, invariant by diffeomorphisms, the Liouville measure λ, all statistical states will be the product of the Liouville measure by the scalar function given by the generalized partition function $e^{\Phi(\beta)-\langle \beta, U(\xi)\rangle}$ defined by the energy U (defined in the dual of the Lie algebra of this dynamical group) and the geometric temperature β, where Φ is a normalizing constant such the mass of probability is equal to 1, $\Phi(\beta) = -\log \int\limits_M e^{-\langle \beta, U(\xi)\rangle} d\lambda$. Souriau then generalizes the Gibbs equilibrium state to all symplectic manifolds that have a dynamical group. Souriau has observed that if we apply this theory for a Galileo group, the symmetry has been broken. For each temperature β, element of the Lie algebra $\mathfrak{g}$, Souriau has introduced a tensor $\widetilde{\Theta}_\beta$, equal to the sum of the cocycle $\widetilde{\Theta}$ and the heat coboundary (with [.,.] Lie bracket):

$$\widetilde{\Theta}_\beta(Z_1, Z_2) = \widetilde{\Theta}(Z_1, Z_2) + \langle Q, ad_{Z_1}(Z_2) \rangle \tag{43}$$

This tensor $\widetilde{\Theta}_\beta$ has the following properties: $\widetilde{\Theta}(X,Y) = \langle \Theta(X), Y \rangle$ where the map Θ is the symplectic one-cocycle of the Lie algebra $\mathfrak{g}$ with values in $\mathfrak{g}^*$, with $\Theta(X) = T_e\theta(X(e))$ where θ the one-cocycle of the Lie group G. $\widetilde{\Theta}(X,Y)$ is constant on M and the map $\widetilde{\Theta}(X,Y) : \mathfrak{g} \times \mathfrak{g} \to \Re$ is a skew-symmetric bilinear form, and is called the *symplectic two-cocycle of Lie algebra* $\mathfrak{g}$ associated to the *moment map* J, with the following properties:

$$\widetilde{\Theta}(X,Y) = J_{[X,Y]} - \{J_X, J_Y\} \text{ with } J \text{ the Moment Map} \tag{44}$$

$$\widetilde{\Theta}([X,Y],Z) + \widetilde{\Theta}([Y,Z],X) + \widetilde{\Theta}([Z,X],Y) = 0 \tag{45}$$

where J_X linear application from $\mathfrak{g}$ to differential function on M:$\mathfrak{g} \to C^\infty(M,R), X \to J_X$ and the associated differentiable application J, called moment(um) map:

$$J : M \to \mathfrak{g}^*, x \mapsto J(x) \text{ such that } J_X(x) = \langle J(x), X \rangle,\ X \in \mathfrak{g} \tag{46}$$

The geometric temperature, element of the algebra $\mathfrak{g}$, is in the the kernel of the tensor $\widetilde{\Theta}_\beta$:

$$\beta \in Ker\, \widetilde{\Theta}_\beta \text{ such that } \widetilde{\Theta}_\beta(\beta, \beta) = 0\ ,\ \forall \beta \in \mathfrak{g} \tag{47}$$

The following symmetric tensor $g_\beta([\beta, Z_1], [\beta, Z_2]) = \widetilde{\Theta}_\beta(Z_1, [\beta, Z_2])$, defined on all values of $ad_\beta(.) = [\beta, .]$ is positive definite, and defines extension of the classical Fisher metric in information geometry (as the Hessian of the logarithm of partition function):

$$g_\beta([\beta, Z_1], Z_2) = \widetilde{\Theta}_\beta(Z_1, Z_2)\ ,\ \forall Z_1 \in g, \forall Z_2 \in \mathrm{Im}(ad_\beta(.)) \tag{48}$$

with:

$$g_\beta(Z_1, Z_2) \geq 0\ ,\ \forall Z_1, Z_2 \in \mathrm{Im}(ad_\beta(.)) \tag{49}$$

These equations are universal, because they are not dependent on the symplectic manifold but only on the dynamical group G, the symplectic two-cocycle Θ, the temperature β and the heat Q. Souriau called it "*Lie groups thermodynamics*" (see Figures 5 and 6).

Theorem 2. [Souriau Theorem of Lie Groups Thermodynamics] *Let Ω be the largest open proper subset of $\mathfrak{g}$, Lie algebra of G, such that $\int_M e^{-\langle \beta, U(\xi) \rangle} d\lambda$ and $\int_M \xi . e^{-\langle \beta, U(\xi) \rangle} d\lambda$ are convergent integrals, this set Ω is convex and is invariant under every transformation $Ad_g(.)$.* Then, the fundamental equations of Lie groups thermodynamics are given by the action of the group:

- *Action of Lie group on Lie algebra:*
$$\beta \to Ad_g(\beta) \tag{50}$$
- *Characteristic function after Lie group action:*
$$\Phi \to \Phi - \left\langle \theta\left(g^{-1}\right), \beta \right\rangle \tag{51}$$
- *Invariance of entropy with respect to action of Lie group:*
$$s \to s \tag{52}$$
- *Action of Lie group on geometric heat:*
$$Q \to a(g, Q) = Ad_g^*(Q) + \theta(g) \tag{53}$$

Souriau's equations of Lie groups thermodynamics are summarized in the following figures.

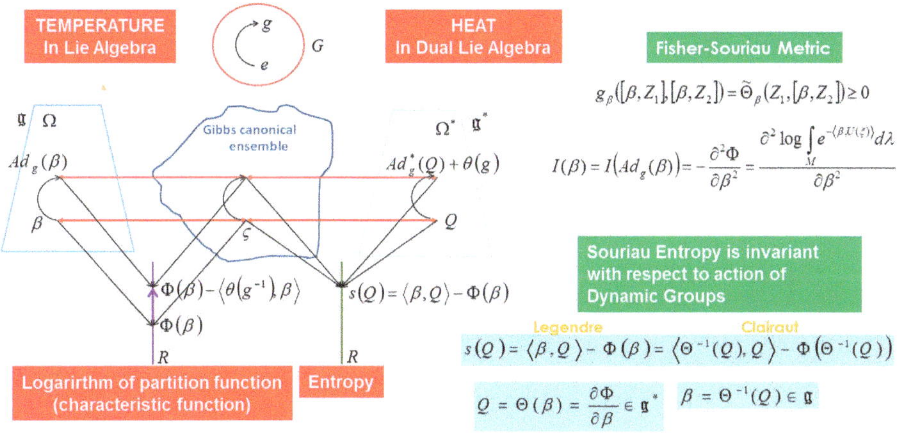

Figure 5. Global Souriau scheme of Lie groups thermodynamics.

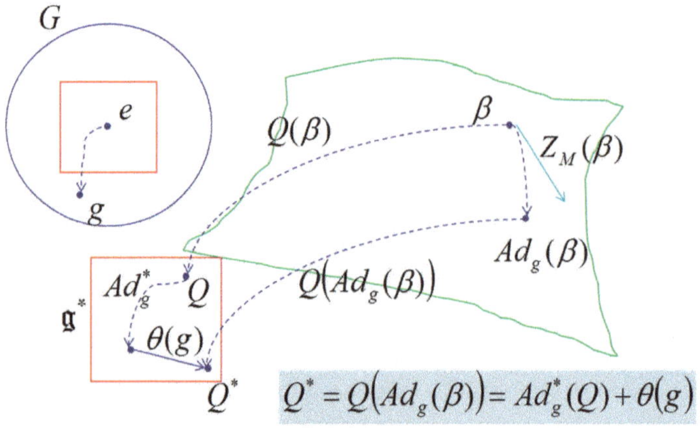

Figure 6. Broken symmetry on geometric heat Q due to adjoint action of the group on temperature β as an element of the Lie algebra.

In the framework of Lie group action on a symplectic manifold, equivariance of moment could be studied to prove that there is a unique action $a(.,.)$ of the Lie group G on the dual $\mathfrak{g}^*$ of its Lie algebra for which the moment map J is equivariant, that means for each $x \in M$:

$$J(\Phi_g(x)) = a(g, J(x)) = Ad_g^*(J(x)) + \theta(g) \tag{54}$$

When the group is not abelian (non-commutative group), the symmetry is broken, and new "cohomological" relations should be verified in Lie algebra of the group. A natural equilibrium state will thus be characterized by an element of the Lie algebra of the Lie group, determining the equilibrium temperature β. The entropy $s(Q)$, parametrized by Q the geometric heat (mean of energy U, element of the dual Lie algebra) is defined by the Legendre transform [97–103] of the Massieu potential $\Phi(\beta)$ parametrized by β ($\Phi(\beta)$ is the minus logarithm of the partition function $\psi_\Omega(\beta)$):

$$s(Q) = \langle \beta, Q \rangle - \Phi(\beta) \text{ with } \begin{cases} Q = \dfrac{\partial \Phi}{\partial \beta} \in \mathfrak{g}^* \\ \beta = \dfrac{\partial s}{\partial Q} \in \mathfrak{g} \end{cases} \tag{55}$$

$$p_{Gibbs}(\xi) = e^{\Phi(\beta) - \langle \beta, U(\xi) \rangle} = \frac{e^{-\langle \beta, U(\xi) \rangle}}{\int\limits_M e^{-\langle \beta, U(\xi) \rangle} d\omega}, \; Q = \frac{\partial \Phi(\beta)}{\partial \beta} = \frac{\int\limits_M U(\xi) e^{-\langle \beta, U(\xi) \rangle} d\omega}{\int\limits_M e^{-\langle \beta, U(\xi) \rangle} d\omega} = \int\limits_M U(\xi) p(\xi) d\omega$$
$$\text{with } \Phi(\beta) = -\log \int\limits_M e^{-\langle \beta, U(\xi) \rangle} d\omega \tag{56}$$

Souriau completed his "geometric heat theory" by introducing a 2-form in the Lie algebra, that is a Riemannian metric tensor in the values of adjoint orbit of β, $[\beta, Z]$ with Z an element of the Lie algebra. This metric is given for (β, Q):

$$g_\beta([\beta, Z_1], [\beta, Z_2]) = \langle \Theta(Z_1), [\beta, Z_2] \rangle + \langle Q, [Z_1, [\beta, Z_2]] \rangle \tag{57}$$

where Θ is a cocycle of the Lie algebra, defined by $\Theta = T_e\theta$ with θ a cocycle of the Lie group defined by $\theta(M) = Q(Ad_M(\beta)) - Ad_M^* Q$.

We observe that Souriau Riemannian metric, introduced with symplectic cocycle, is a generalization of the Fisher metric, that we call the Souriau-Fisher metric, that preserves the property to be defined as a Hessian of the partition function logarithm $g_\beta = -\frac{\partial^2 \Phi}{\partial \beta^2} = \frac{\partial^2 \log \psi_\Omega}{\partial \beta^2}$ as in classical information geometry. We will establish the equality of two terms, between Souriau definition based on Lie group cocycle Θ and parameterized by "geometric heat" Q (element of dual Lie algebra) and "geometric temperature" β (element of Lie algebra) and hessian of characteristic function $\Phi(\beta) = -\log \psi_\Omega(\beta)$ with respect to the variable β:

$$g_\beta([\beta, Z_1], [\beta, Z_2]) = \langle \Theta(Z_1), [\beta, Z_2] \rangle + \langle Q, [Z_1, [\beta, Z_2]] \rangle = \frac{\partial^2 \log \psi_\Omega}{\partial \beta^2} \tag{58}$$

If we differentiate this relation of Souriau theorem $Q\big(Ad_g(\beta)\big) = Ad_g^*(Q) + \theta(g)$, this relation occurs:

$$\frac{\partial Q}{\partial \beta}(-[Z_1, \beta], .) = \widetilde{\Theta}(Z_1, [\beta, .]) + \langle Q, Ad_{.Z_1}([\beta, .]) \rangle = \widetilde{\Theta}_\beta(Z_1, [\beta, .]) \tag{59}$$

$$-\frac{\partial Q}{\partial \beta}([Z_1, \beta], Z_2.) = \widetilde{\Theta}(Z_1, [\beta, Z_2]) + \langle Q, Ad_{.Z_1}([\beta, Z_2]) \rangle = \widetilde{\Theta}_\beta(Z_1, [\beta, Z_2]) \tag{60}$$

$$\Rightarrow -\frac{\partial Q}{\partial \beta} = g_\beta([\beta, Z_1], [\beta, Z_2]) \tag{61}$$

As the entropy is defined by the Legendre transform of the characteristic function, this Souriau-Fisher metric is also equal to the inverse of the hessian of "geometric entropy" $s(Q)$ with respect to the variable Q: $\frac{\partial^2 s(Q)}{\partial Q^2}$.

For the maximum entropy density (Gibbs density), the following three terms coincide: $\frac{\partial^2 \log \psi_\Omega}{\partial \beta^2}$ that describes the convexity of the log-likelihood function, $I(\beta) = -E\left[\frac{\partial^2 \log p_\beta(\xi)}{\partial \beta^2}\right]$ the Fisher metric that describes the covariance of the log-likelihood gradient, whereas $I(\beta) = E\left[(\xi - Q)(\xi - Q)^T\right] = Var(\xi)$ that describes the covariance of the observables.

We can also observe that the Fisher metric $I(\beta) = -\frac{\partial Q}{\partial \beta}$ is exactly the Souriau metric defined through symplectic cocycle:

$$I(\beta) = \widetilde{\Theta}_\beta(Z_1, [\beta, Z_2]) = g_\beta([\beta, Z_1], [\beta, Z_2]) \tag{62}$$

The Fisher metric $I(\beta) = -\frac{\partial^2 \Phi(\beta)}{\partial \beta^2} = -\frac{\partial Q}{\partial \beta}$ has been considered by Souriau as a *generalization of "heat capacity"*. Souriau called it K the *"geometric capacity"*.

We could observe that Souriau Lie groups thermodynamics is compatible with Balian and Valentin's theory of thermodynamics [24], that is obtained by symplectization in dimension $2n + 2$ of contact manifold in dimension $2n + 1$. All elements of the Souriau geometric temperature vector are multiplied by the same gauge parameter. The Balian and Valentin model was first explored in [104] and has been recently developed by der Schaft and Maschke in [26,105].

5. Extended Koszul Study of Souriau Non-Equivariant Model Associated to a Class of Cohomology

Koszul has deepened Souriau's model in his book *"Introduction to symplectic geometry"* [44] as explained in [10]. In the historical foreword of this book, Koszul write *"The development of analytical mechanics provided the basic concepts of symplectic structures. The term symplectic structure is due largely to analytical mechanics. But in this book, the applications of symplectic structure theory to mechanics is not discussed in any detail"*. Koszul considers in this book purely algebraic and geometric developments of geometric/analytic mechanics developed during the 60s, more especially in Jean-Marie Souriau's works detailed in chapters 4 and 5. The originality of this book lies in the fact that Koszul develops new points of view, and demonstrations not considered initially by Souriau and after by the geometrical mechanics community.

To highlight the importance of this Koszul book, we will illustrate the links of the detailed tools, including demonstrations or original Koszul extensions, with Souriau's Lie groups thermodynamics. Koszul originally developed Souriau's model, in the case of non-equivariance, of the action of the group G on the moment map. As explained in [106] by Thomas Delzant at the 2010 CIRM conference "*Action Hamiltoniennes: invariants et classification*", organized with Michel Brion: "*The definition of the moment map is due to Jean-Marie Souriau .... In the book of Souriau, we find a proof of the proposition: the map J is equivariant for an affine action of G on g^* whose linear part is Ad^* .... In Souriau's book, we can also find a study of the non-equivariant case and its applications to classical and quantum mechanics. In the case of the Galileo group operating in the phase space of space-time, obstruction to equivariance (a class of cohomology) is interpreted as the inert mass of the object under study*". We can uniquely define the moment map up to an additive constant of integration, that can always be chosen to make the moment map equivariant (a moment map is G-equivariant, when G acts on g^* via the coadjoint action) if the group is compact or semi-simple. In 1969, Souriau has considered the non-equivariant case where the coadjoint action must be modified to make the map equivariant by a 1-cocycle on the group with values in dual Lie algebra g^*.

The concept and seminal idea of moment map was in the Sophus *Lie's book* 2nd volume published in 1890, developed for homogeneous canonical transformations. Professor Marsden has summarized the development of this concept by Jean-Marie Souriau and Bertram Kostant based on their both testimonials: "*In Kostant's 1965 Phillips lectures at Haverford, and in the 1965 U.S.–Japan Seminar, Kostant introduced the momentum map to generalize a theorem of Wang and thereby classified all homogeneous symplectic manifolds; this is called today 'Kostant's coadjoint orbit covering theorem' .... Souriau introduced the momentum map in his 1965 Marseille lecture notes and put it in print in 1966. The momentum map finally got its formal definition and its name, based on its physical interpretation, by Souriau in 1967. Souriau also studied its properties of equivariance, and formulated the coadjoint orbit theorem. The momentum map appeared as a key tool in Kostant's quantization lectures in 1970 [46], and Souriau discussed in 1970 it at length in his book [43]. Kostant and Souriau realized its importance for linear representations, a fact apparently not foreseen by Lie*". Souriau's book reference date is 1970, but it was published by Dunod in 1969. For information, Jean-Louis Koszul knew very well the Souriau and Kostant works, and as soon as 1958, Koszul made a survey of first Kostant's works at a Bourbaky seminar [47].

In this book in Chapter 4, Koszul calls symplectic G-space a symplectic manifold $(M;\omega)$ on which a Lie group G acts by a symplectic action (an action which leaves unchanged the symplectic form ω). Koszul then introduces and develop properties of the moment map μ (Souriau's invention) of a Hamiltonian action of the Lie algebra g. Koszul also defines the Souriau 2-cocycle, considering that the difference of two moments of the same Hamiltonian action is a locally constant application on M ,showing that when μ is a moment map, for every pair $(a;b)$ of elements of g, the function $c_\mu(a,b) = \{\langle\mu,a\rangle,\langle\mu,b\rangle\} - \langle\mu,\{a,b\}\rangle$ is locally constant on M, defining an antisymmetric bilinear application of gxg in $H^0(M;R)$ which verifies Jacobi's identity. This is the 2-cocycle introduced by Jean-Marie Souriau in Geometric Mechanics, that will play a fundamental role in Souriau Lie Groups Thermodynamics to define an extension of the Fisher Metric from Information Geometry: "Fisher-Souriau metric".

The antisymmetric bilinear map (31) and (32), with definition (27) and (28), introduced by Souriau is exactly equal to the mathematical object extensively studied in Chapter 4 of Koszul's book:

$$c_\mu(a,b) = \{\langle\mu,a\rangle,\langle\mu,b\rangle\} - \langle\mu,\{a,b\}\rangle \tag{63}$$

In this book, Koszul has studied this antisymmetric bilinear map considering the following developments. For any moment map μ, Koszul defines the skew symmetric bilinear form $c_\mu(a,b)$ on Lie algebra by:

$$c_\mu(a,b) = \left\langle \mathrm{d}\theta_\mu(a), b \right\rangle \, , \; a, b \in g \tag{64}$$

Koszul observes that if he uses:

$$\theta_\mu(st) = \mu(stx) - Ad^*_{st}\mu(x) = \theta_\mu(s) + Ad^*_s\mu(tx) - Ad^*_s Ad^*_t \mu(x) = \theta_\mu(s) + Ad^*_s \theta_\mu(t) \tag{65}$$

by developing $d\mu(ax) = {}^t ad_a \mu(x) + d\theta_\mu(a) \, , \; x \in M, a \in \mathfrak{g}$, he obtains:

$$\langle d\mu(ax), b \rangle = \langle \mu(x), [a,b] \rangle + \left\langle \mathrm{d}\theta_\mu(a), b \right\rangle \; = \{\langle \mu, \mathrm{a} \rangle, \langle \mu, b \rangle\}(x) \, , \; x \in M, a, b \in \mathfrak{g} \tag{66}$$

He has then:

$$c_\mu(a,b) = \{\langle \mu, \mathrm{a} \rangle, \langle \mu, b \rangle\} - \langle \mu, [a,b] \rangle = \left\langle \mathrm{d}\theta_\mu(a), b \right\rangle \, , \; a, b \in g \tag{67}$$

and the property:

$$c_\mu([a,b],c) + c_\mu([b,c],a) + c_\mu([c,a],b) = 0 \, , \; a, b, c \in g \tag{68}$$

Koszul concludes by observing that if the moment map is transform as $\mu\prime = \mu + \phi$ then we have:

$$c_{\mu\prime}(a,b) = c_\mu(a,b) - \langle \phi, [a,b] \rangle \tag{69}$$

Finally using $c_\mu(a,b) = \{\langle \mu, \mathrm{a} \rangle, \langle \mu, b \rangle\} - \langle \mu, [a,b] \rangle = \left\langle \mathrm{d}\theta_\mu(a), b \right\rangle \, , \; a, b \in g$, koszul highlights the property that:

$$\{\mu^*(a), \mu^*(b)\} = \{\langle \mu, a \rangle, \langle \mu, b \rangle\} = \mu^*([a,b] + c_\mu(a,b)) = \mu^*\{a,b\}_{c_\mu} \tag{70}$$

In Chapter 4, Koszul introduces the equivariance of the moment map μ. Based on the definitions of the adjoint and coadjoint representations of a Lie group or a Lie algebra, Koszul proves that when $(M; \omega)$ is a connected Hamiltonian G-space and $\mu : M \to \mathfrak{g}^*$ a moment of the action of G, there exists an affine action of G on g^*, whose linear part is the coadjoint action, for which the moment μ is equivariant. This affine action is obtained by modifying the coadjoint action by means of a cocycle. This notion is also developed in Chapter 5 for studying Poisson manifolds.

Defining classical operation $Ad_s a = sas^{-1} \, , \; s \in G, a \in \mathfrak{g}$, $ad_a b = [a,b] \, , \; a \in g, b \in \mathfrak{g}$ and $Ad^*_s = {}^t Ad_{s^{-1}}$, $s \in G$ with classical properties:

$$Ad_{\exp a} = \exp(-ad_a) \, , \; a \in \mathfrak{g} \text{ or } Ad^*_{\exp a} = \exp{}^t(ad_a) \, , \; a \in \mathfrak{g} \tag{71}$$

Koszul considers:

$$x \mapsto sx \, , \; x \in M, \; \mu : M \to \mathfrak{g}^* \tag{72}$$

From which, he obtains:

$$\langle d\mu(v), a \rangle = \omega(ax, v) \tag{73}$$

Koszul then study $\mu \circ s_M - Ad^*_s \circ \mu : M \to \mathfrak{g}^*$, and develops:

$$d\langle Ad^*_s \circ \mu, a \rangle = \langle Ad^*_s d\mu, a \rangle = \langle d\mu, Ad_{s^{-1}} a \rangle \tag{74}$$

$$\langle d\mu(v), Ad_{s^{-1}} a \rangle = \omega\left(s^{-1} asx, v\right) = \omega(asx, sv) = \langle d\mu(sv), a \rangle = (d\langle \mu \circ s_M, a \rangle)(v) \tag{75}$$

$$d\langle Ad^*_s \circ \mu, a \rangle = d\langle \mu \circ s_M, a \rangle \text{ and then proves that } d\langle \mu \circ s_M - Ad^*_s \circ \mu, a \rangle = 0 \tag{76}$$

Koszul considers the cocycle given by $\theta_\mu(s) = \mu(sx) - Ad_s^* \mu(x)$, $s \in G$, and observes that:

$$\theta_\mu(st) = \theta_\mu(s) - Ad_s^* \theta_\mu(\mathrm{t}) \ , \ s, t \in G \tag{77}$$

From this action of the group on dual Lie algebra:

$$G \times \mathfrak{g}^* \to \mathfrak{g}^*, (s, \xi) \mapsto s\xi = Ad_s^* \xi + \theta_\mu(s) \tag{78}$$

Koszul introduces the following properties:

$$\mu(sx) = s\mu(x) = Ad_s^* \mu(x) + \theta_\mu(s) \ , \ \forall s \in G, x \in M \tag{79}$$

$$G \times \mathfrak{g}^* \to \mathfrak{g}^*, (e, \xi) \mapsto e\xi = Ad_e^* \xi + \theta_\mu(e) = \xi + \mu(x) - \mu(x) = \xi \tag{80}$$

$$\begin{aligned} (s_1 s_2)\xi &= Ad_{s_1 s_2}^* \xi + \theta_\mu(s_1 s_2) = Ad_{s_1}^* Ad_{s_2}^* \xi + \theta_\mu(s_1) + Ad_{s_1}^* \theta_\mu(s_2) \\ (s_1 s_2)\xi &= Ad_{s_1}^* \left(Ad_{s_2}^* \xi + \theta_\mu(s_2) \right) + \theta_\mu(s_1) = s_1(s_2 \xi) \ , \ \forall s_1, s_2 \in G, \xi \in \mathfrak{g}^* \end{aligned} \tag{81}$$

This Koszul study of the moment map μ equivariance, and the existence of an affine action of G on g*, whose linear part is the coadjoint action, for which the moment μ is equivariant, is at the cornerstone of Souriau theory of geometric mechanics and Lie groups thermodynamics.

We compare Souriau and Koszul notations in Figure 7.

Souriau model

$$\tilde{\Theta}(X,Y) = J_{[X,Y]} - \{J_X, J_Y\}$$

$J : M \to \mathfrak{g}^*$ with $x \mapsto J(x)$

such that $J_X(x) = \langle J(x), X \rangle, X \in \mathfrak{g}$

$$\tilde{\Theta}([X,Y],Z) + \tilde{\Theta}([Y,Z],X) + \tilde{\Theta}([Z,X],Y) = 0$$

$$\tilde{\Theta}_\beta(Z_1,Z_2) = \tilde{\Theta}(Z_1,Z_2) + \langle Q, ad_{Z_1}(Z_2) \rangle$$

with $ad_{Z_1}(Z_2) = [Z_1, Z_2]$

$$\tilde{\Theta}(\beta,Z) + \langle Q, [\beta, Z] \rangle = 0$$

Koszul model

$$c_\mu(a,b) = \{\langle \mu, a \rangle, \langle \mu, b \rangle\} - \langle \mu, [a,b] \rangle$$

$$c_\mu(a,b) = \langle d\theta_\mu(a), b \rangle \ , \ a, b \in \mathfrak{g}$$

$$c_\mu([a,b],c) + c_\mu([b,c],a) + c_\mu([c,a],b) = 0$$

$$\mu' = \mu + \varphi \Rightarrow c_{\mu'}(a,b) = c_\mu(a,b) - \langle \varphi, [a,b] \rangle$$

Figure 7. Comparison of the Souriau equations (column on the left) and Koszul equations (column on the right).

We have also to make reference to Muriel Casalis' papers [41,42] on this topic.

6. Souriau Model of Generalized Entropy Based on Legendre and Laplace Transforms

At the step of the development of Souriau Lie groups thermodynamics, we will introduce generalized Souriau definition of entropy. Souriau first start to define "Laplace transform":

Let E a vector space of finite size, μ a measure of its dual E^*, then the function given by:

$$\alpha \mapsto \int_{E^*} e^{M\alpha} \mu(M) dM \tag{82}$$

for all $\alpha \in E$ such that the integral is convergent. This function is called (generalized) Laplace transform. This transform F of the measure μ is differentiable inside is definition set $def(F)$. Its p-th derivative is given by the following convergent integral for all point inside $def(F)$:

$$F^{(p)}(\alpha) = \int_{E^*} M \otimes M \ldots \otimes M \mu(M) dM \tag{83}$$

Theorem 3. [Souriau Theorem] *Let E a vector space of finite size, μ a non-zero positive measure of dual space E^*, F its Laplace transform, then:*

- *F is semi-definite convex function,*
$$F(\alpha) > 0, \forall \alpha \in def(F) \tag{84}$$
- *$f = \log F$ is convex and semi-continuous*
- *Let α an interior point of $def(F)$ then:*
$$D^2(f)(\alpha) \geq 0 \tag{85}$$
$$D^2(f)(\alpha) = \int_{E^*} e^{M\alpha}[M - D(f)(\alpha)]^{\otimes^2} \mu(M) dM \tag{86}$$
$$D^2(f)(\alpha) \text{ inversible } \Leftrightarrow \text{ Affine envelop}(\mu)) = E^* \tag{87}$$

See [107], for links between dual convex functions and optimization.

Before introducing Entropy, Souriau introduced the following lemma:

Lemma 1. *Let X be a locally compact space, Let λ a positive measure of X, having X as support, then the following function Φ is convex:*

$$\Phi(h) = \log \int_X e^{h(X)} \lambda(x) dx \,, \; \forall h \in C(X) \tag{88}$$

such that the integral is converging.

The integral is strictly positive when it converges, and then insures existence of its logarithm. The epigraph of Φ is the set of $\begin{pmatrix} h \\ y \end{pmatrix}$ such that $\int_X e^{h(x)-y} \lambda(x) dx \leq 1$. Convexity of exponential shows that this epigraph is convex. Finally, Souriau introduced the "negentropy" as Legendre transform of the function Φ:

Definition 1. [Souriau Entropy Definition] *We call "Boltzmann Law" (relative to λ) all measure μ of X such that the set of real values:*

$$\mu(h) - \Phi(h), h \in def(\Phi) \text{ and h is } \mu\text{-integrable} \tag{89}$$

This definition of entropy by Souriau is a general scheme that can be extended to highly abstract spaces preserving Legendre structure [108], if we can define generalized Laplace transform. These operations of Laplace and Legendre transforms are the core contextures of theory of Information and Heat, generating the well-defined structures, from which we can preserve the definition of "average value". Jean-Marie Souriau explained this contexture property in the following sentence:

> *"Il est évident que l'on ne peut définir de valeurs moyennes que sur des objets appartenant à un espace vectoriel (ou affine); donc—si bourbakiste que puisse sembler cette affirmation—que l'on n'observera et ne mesurera de valeurs moyennes que sur des grandeurs appartenant à un ensemble possédant physiquement une structure affine. Il est clair que cette structure est nécessairement unique—sinon les valeurs moyennes ne seraient pas bien définies."* (In English: *It is obvious that one can only define average values on objects belonging to a vector (or affine) space; Therefore—so this assertion may seem Bourbakist—that we will observe and measure average values only as quantity belonging to a set having physically an affine structure. It is clear that this structure is necessarily unique—if not the average values would not be well defined.*)

See also papers of Kostant [109] and Leray [100] for generalized Laplace transforms.

7. Illustration of Souriau Thermodynamics of a Centrifuge System

Duhem [110–113] and Poincaré [114] have studied statistical mechanics model of centrifuges. We will illustrate Souriau's Lie groups thermodynamics for Souriau Gibbs states for Hamiltonian actions of subgroups of the Galilean group, as illustrated in Souriau's book [43] and more recentltly by Charles-Michel Marle [23].

Consider a Galilean Lie group:

$$\begin{pmatrix} A & \vec{b} & \vec{d} \\ 0 & 1 & e \\ 0 & 0 & 1 \end{pmatrix} \text{ with } \begin{cases} A \in SO(3): \text{ rotation} \\ \vec{b} \in R^3: \text{ boost} \\ \vec{d} \in R^3: \text{space translation} \\ e: \text{ time translation} \end{cases} \tag{90}$$

Galilean Lie algebra:

$$\begin{pmatrix} j(\vec{\omega}) & \vec{\alpha} & \vec{\delta} \\ 0 & 1 & \varepsilon \\ 0 & 0 & 0 \end{pmatrix} \text{ with } \begin{cases} \vec{\omega} = \begin{pmatrix} \omega_x \\ \omega_y \\ \omega_z \end{pmatrix}, \vec{\alpha} \text{ and } \vec{\delta} \in R^3,\ \varepsilon \in R \\ j(\vec{\omega}) = \begin{pmatrix} 0 & -\omega_z & \omega_y \\ \omega_z & 0 & -\omega_x \\ -\omega_y & \omega_x & 0 \end{pmatrix} \in \mathrm{so}(3), j(\vec{\omega})\vec{r} = \vec{\omega} \times \vec{r} \end{cases} \tag{91}$$

Action of Lie group:

$$\begin{pmatrix} A & \vec{b} & \vec{d} \\ 0 & 1 & e \\ 0 & 0 & 1 \end{pmatrix} \begin{pmatrix} \vec{r} \\ t \\ 1 \end{pmatrix} = \begin{pmatrix} A\vec{r} + t\vec{b} + \vec{d} \\ t+e \\ 1 \end{pmatrix} \text{ with } \vec{r} = \begin{pmatrix} x \\ y \\ z \end{pmatrix} \tag{92}$$

Galilean transformation on position and speed is given by:

$$\begin{pmatrix} \vec{r}\prime & \vec{v}\prime \\ t\prime & 1 \\ 1 & 0 \end{pmatrix} = \begin{pmatrix} A & \vec{b} & \vec{d} \\ 0 & 1 & e \\ 0 & 0 & 1 \end{pmatrix} \begin{pmatrix} \vec{r} & \vec{v} \\ t & 1 \\ 1 & 0 \end{pmatrix} = \begin{pmatrix} A\vec{r} + t\vec{b} + \vec{d} & A\vec{v} + \vec{b} \\ t+e & 1 \\ 1 & 0 \end{pmatrix} \tag{93}$$

Souriau has proved that this action is Hamiltonian, with the map J, defined on the evolution space of the particle, with value in the dual g^* of the Lie algebra G, as momentum map:

$$J\left(\vec{r}, t, \vec{v}, m\right) = m \begin{pmatrix} \vec{r} \times \vec{v} & 0 & 0 \\ \vec{r} - t\vec{v} & 0 & 0 \\ \vec{v} & \frac{1}{2}\|\vec{v}\|^2 & 0 \end{pmatrix} = m\left\{\vec{r} \times \vec{v}, \vec{r} - t\vec{v}, \vec{v}, \frac{1}{2}\|\vec{v}\|^2\right\} \in \mathfrak{g}^* \tag{94}$$

where the coupling formula is given by:

$$\begin{aligned} \left\langle J\left(\vec{r}, t, \vec{v}, m\right), \beta\right\rangle &= \left\langle m\left\{\vec{r} \times \vec{v}, \vec{r} - t\vec{v}, \vec{v}, \tfrac{1}{2}\|\vec{v}\|^2\right\}, \left\{\vec{\omega}, \vec{\alpha}, \vec{\delta}, \varepsilon\right\}\right\rangle \\ \left\langle J\left(\vec{r}, t, \vec{v}, m\right), \beta\right\rangle &= m\left(\vec{\omega}.\vec{r} \times \vec{v} - (\vec{r} \times \vec{v}).\vec{\alpha} + \vec{v}.\vec{\delta} - \tfrac{1}{2}\|\vec{v}\|^2 \varepsilon\right) \end{aligned} \tag{95}$$

with:

$$Z = \begin{pmatrix} j(\vec{\omega}) & \vec{\alpha} & \vec{\delta} \\ 0 & 1 & \varepsilon \\ 0 & 0 & 0 \end{pmatrix} = \left\{ \vec{\omega}, \vec{\alpha}, \vec{\delta}, \varepsilon \right\} \in \mathfrak{g} \tag{96}$$

Souriau gave the demonstration for the Galilean moment map for a free particle, considering the definition of moment map:

$$\sigma(dp)(\delta p) = -d\langle J, Z\rangle \, , \forall dp \tag{97}$$

and the definition of tangent vector field:

$$Z_V(p) = \delta[a_V(p)] \tag{98}$$

$$Z = \begin{pmatrix} j(\vec{\omega}) & \vec{\alpha} & \vec{\delta} \\ 0 & 1 & \varepsilon \\ 0 & 0 & 0 \end{pmatrix} \in \mathfrak{g} \underset{Z_V(p)=\delta[a_V(p)]}{\Rightarrow} \begin{cases} \delta t = \varepsilon \\ \delta r_j = \vec{\omega} \times r_j + \vec{\alpha} t + \vec{\delta} \\ \delta v_j = \vec{\omega} \times v_j + \vec{\alpha} \end{cases} \tag{99}$$

Then, as General Lagrange 2 form for a force F is:

$$dp = \begin{pmatrix} dt \\ dr \\ dv \end{pmatrix} \text{ and } \delta p = \begin{pmatrix} \delta t \\ \delta r \\ \delta v \end{pmatrix} \Rightarrow \sigma(dp)(\delta p) = \langle m dv - F dt, \delta r - v\delta t\rangle - \langle m\delta v - F\delta t, dr - v dt\rangle \tag{100}$$

If F is equal to zero, we obtain:

$$\begin{aligned} &\sigma(dp)(\delta p) = \sum_j \left\langle m dv, \vec{\omega} \times r_j + \vec{\alpha} t + \vec{\delta} - v\varepsilon \right\rangle - \left\langle m\left(\vec{\omega} \times v_j + \vec{\alpha}\right), dr - v dt \right\rangle \\ &\sigma(dp)(\delta p) == -d\langle J, Z\rangle = -dJ_Z = -dH \end{aligned} \tag{101}$$

and the co-cycle is given by:

$$\theta(g) = J(Ad_g Z) - Ad_g^*(J(Z)) = \left\{ \vec{d} \times \vec{b}, \vec{d} - \vec{b} e, \vec{b}, \frac{1}{2} \|\vec{b}\|^2 \right\} \tag{102}$$

The main Souriau idea was to define the Gibbs states for one-parameter subgroups of the Galilean group. Souriau has proved that action of the full Galilean group on the space of motions of an isolated mechanical system is not related to any equilibrium Gibbs state (the open subset of the Lie algebra, associated to this Gibbs state, is empty). Then, if we consider the 1-parameter subgroup of the Galilean group generated by b element of Lie algebra, is the set of matrices:

$$\exp(\tau\beta) = \begin{pmatrix} A(\tau) & \vec{b}(\tau) & \vec{d}(\tau) \\ 0 & 1 & \tau\varepsilon \\ 0 & 0 & 1 \end{pmatrix} \text{ with } \begin{cases} A(\tau) = \exp\left(\tau j(\vec{\omega})\right) \text{ and } \vec{b}(\tau) = \left(\sum_{i=1}^{\infty} \frac{\tau^i}{i!} \left(j(\vec{\omega})\right)^{i-1}\right)\vec{\alpha} \\ \vec{d}(\tau) = \left(\sum_{i=1}^{\infty} \frac{\tau^i}{i!} \left(j(\vec{\omega})\right)^{i-1}\right)\vec{\delta} + \varepsilon\left(\sum_{i=2}^{\infty} \frac{\tau^i}{i!} \left(j(\vec{\omega})\right)^{i-2}\right)\vec{\alpha} \end{cases} \tag{103}$$

and:

$$\beta = \begin{pmatrix} j(\vec{\omega}) & \vec{\alpha} & \vec{\delta} \\ 0 & 1 & \varepsilon \\ 0 & 0 & 0 \end{pmatrix} \in \mathfrak{g} \tag{104}$$

Then, Gibbs state defined for a gas enclosed in a moving box could be computed by Souriau formula. If we fix the affine Euclidean reference frame $\left(0, \vec{e}_x, \vec{e}_y, \vec{e}_z\right)$ at $t = 0$, if we set the value $\tau = t/\varepsilon$, moving frame $\left(0, \vec{e}_x(t), \vec{e}_y(t), \vec{e}_z(t)\right)$ velocity and acceleration are given by the vector field related to β element of the Lie algebra. For each point, we can associate a rotation speed $\|\vec{\omega}\|/\varepsilon$, a speed $\vec{\delta}/\varepsilon$ and an acceleration $\vec{\alpha}/\varepsilon$. If we consider a gas made of N point particles, indexed by $i \in \{1,2, \ldots, N\}$, enclosed in a box with rigid and undeformable walls, whose motion is described by the action of the 1-parameter subgroup of the Galilean group, $A(t/\varepsilon)$ where $t \in R$. If we consider $m_i, r_i(t), v_i(t)$, respectively the mass, position vector and velocity vector of the ith particle at time t. If we assume free particle and we neglect contributions given by the collisions of the particles between themselves collisions with the walls, then we can write:

$$\langle J, \beta \rangle = \sum_{i=1}^{N} \langle J_i, \beta \rangle \text{ with } \left\langle J_i\left(\vec{r}_i, t, \vec{v}_i, m_i\right), \beta \right\rangle = m_i\left(\vec{\omega}.\left(\vec{r}_i \times \vec{v}_i\right) - (\vec{r}_i - t\vec{v}_i).\vec{\alpha} + \vec{v}_i.\vec{\delta} - \tfrac{1}{2}\|\vec{v}_i\|^2 \varepsilon\right) \tag{105}$$

The important idea is to observe that $\langle J_i, \beta \rangle$ is invariant by the action of 1-parameter subgroup. The proof of $\langle J_i, \beta \rangle$ invariance is based on Souriau equation for default of equivariance with cocyle. If the action of the 1-parameter subgroup is $\exp\left(\frac{t}{\varepsilon}\beta\right)$, according to Souriau equation:

$$a(g, J) = Ad_g^*(J) + \theta(g) \tag{106}$$

We obtain for:

$$\langle J_i(p), \beta \rangle = \left\langle Ad_g^*(J_i(p_0), \beta \right\rangle + \langle \theta(g), \beta \rangle = \left\langle J_i(p_0), Ad_{g^{-1}}\beta \right\rangle + \langle \theta(g), \beta \rangle$$

that can be reduded by using the properties:

$$\begin{cases} Ad_{g^{-1}}\beta = \beta \\ \langle \theta(g), \beta \rangle = 0 \end{cases} \Rightarrow \langle J_i(p), \beta \rangle = \langle J_i(p_0), \beta \rangle \tag{107}$$

and:

$$\begin{aligned} \text{at t} = 0 \text{ then } \left\langle J_i\left(\vec{r}_i, t, \vec{v}_i, m_i\right), \beta \right\rangle &= m_i\left(\vec{\omega}.\left(\vec{r}_{i0} \times \vec{v}_{i0}\right) - \vec{r}_{i0}.\vec{\alpha} + \vec{v}_{i0}.\vec{\delta} - \tfrac{1}{2}\|\vec{v}_i\|^2 \varepsilon\right) \\ &= m_i\left(\vec{v}_{i0}.\left(\vec{\omega} \times \vec{v}_{i0} + \vec{\delta}\right) - \vec{r}_{i0}.\vec{\alpha} - \tfrac{1}{2}\|\vec{v}_i\|^2 \varepsilon\right) \end{aligned} \tag{108}$$

To obtain Souriau's Gibbs maximum entropy density, we have to use the following change of variables:

$$\vec{U}^* = \frac{1}{\varepsilon}\left(\vec{\omega} \times \vec{v}_{i0} + \vec{\delta}\right) \tag{109}$$

$$\left\langle J_i\left(\vec{r}_i, t, \vec{v}_i, m_i\right), \beta \right\rangle = m_i\varepsilon\left(-\frac{1}{2}\|\vec{v}_{i0} - \vec{U}^*\|^2 - \vec{r}_{i0}.\frac{\vec{\alpha}}{\varepsilon} + \frac{1}{2}\|\vec{U}^*\|^2\right) \tag{110}$$

We can then write:

$$\begin{aligned} &\left\langle J_i\left(\vec{r}_{i0}, \vec{p}_{i0}\right), \beta \right\rangle = -\varepsilon\left(-\tfrac{1}{2m_i}\|\vec{p}_{i0}\|^2 + m_i f_i\left(\vec{r}_{i0}\right)\right) \text{ with } \varepsilon = -\tfrac{1}{\kappa T} \\ &\text{with } \begin{cases} \vec{p}_{i0} = m_i \vec{w}_{i0} = m_i\left(\vec{v}_{i0} - \vec{U}^*\right) \\ f_i\left(\vec{r}_{i0}\right) = \vec{r}_{i0}.\frac{\vec{\alpha}}{\varepsilon} - \frac{1}{2\varepsilon^2}\|\vec{\omega} \times \vec{r}_{i0}\|^2 - \frac{\vec{\delta}}{\varepsilon}.\left(\frac{\vec{\omega}}{\varepsilon} \times \vec{r}_{i0}\right) - \frac{1}{2\varepsilon^2}\|\vec{\delta}\|^2 \end{cases} \end{aligned} \tag{111}$$

and finally, the Souriau Gibbs density is given by:

$$\rho(\beta) = \prod_{i=1}^{N} \rho_i(\beta) \text{ with } \rho_i(\beta) = \frac{1}{P_i(\beta)} \exp(-\langle J_i, \beta \rangle) \tag{112}$$

$$P_i(\beta)=\int_{M_i}\exp(-\langle J_i,\beta\rangle)d\lambda_{\omega_i}\ ,\ Q_i(\beta)=\int_{M_i}J_i\exp(-\langle J_i,\beta\rangle)d\lambda_{\omega_i}\ \text{et}\ P(\beta)=\prod_{i=1}^{N}P_i(\beta) \tag{113}$$

If we consider the case of the centrifuge (as for a butter churn, device used to convert cream into butter), the parameter of Galilean group Lie algebra are reduced to:

$$\begin{array}{l}\vec{\omega}=\omega\vec{e}_z\ ,\ \vec{\alpha}=0 \text{ and } \vec{\delta}=0\\ \textit{Rotation speed}:\frac{\omega}{\varepsilon}\end{array}\quad \text{with } \beta=\begin{pmatrix} j\left(\vec{\omega}\right) & \vec{\alpha} & \vec{\delta}\\ 0 & 1 & \varepsilon\\ 0 & 0 & 0\end{pmatrix}\in\mathfrak{g} \tag{114}$$

with variables:

$$f_i\left(\vec{r}_{i0}\right)=-\frac{\omega^2}{2\varepsilon^2}\left\|\vec{e}_z\times\vec{r}_{i0}\right\|^2 \text{ with } \Delta=\left\|\vec{e}_z\times\vec{r}_{i0}\right\| \text{ distance to axis z} \tag{115}$$

We obtain the closed form for maximum entropy Souriau-Gibbs density:

$$\rho_i(\beta)=\frac{1}{P_i(\beta)}\exp(-\langle J_i,\beta\rangle)=cst.\exp\left(-\frac{1}{2m_i\kappa T}\left\|\vec{p}_{i0}\right\|^2+\frac{m_i}{2\kappa T}\left(\frac{\omega}{\varepsilon}\right)^2\Delta^2\right) \tag{116}$$

This equation describes the behaviour of a gas made of point particles of various masses in a centrifuge rotating at a constant angular velocity and explains the observation that the heavier particles concentrate farther from the rotation axis than the lighter ones. Souriau made reference to thermodynamics of butter churn (see Figure 8).

(a) (b)

Figure 8. Most simple use-case of Souriau's Lie groups thermodynamics: the thermodynamics of the centrifuge of butter churn (device used to convert cream into butter). (**a**) butter churn centrifuge with horizontal axis; (**b**) butter churn centrifuge with vertical axis.

Souriau Lie groups thermodynamics provides right results if we apply it to subgroups of Galileo group, as previous example of a cylindrical box with fluid with an invariance sub-group of size 2 (rotation along the axis, time translation) providing a 2-dimensional Souriau (Planck) temperature-vector. Souriau has observed that the process, by which a refrigerated centrifuge transmits its own temperature-vector to its content, has two names: thermal conduction and viscosity, depending on the temperature-vector component that is considered. Conduction and viscosity should therefore be unified in a fundamental theory of irreversible processes (theory that remains to be constructed).

In the Appendix, we develop a solution given by Roger Balian [25] for the previous case of centrifuge thermodynamics based on classical methods. Balian recover the same Gibbs density but by introducing an additional Lagrange hyper-parameter associated to total angular momentum.

Balian has computed the Boltzmann-Gibbs distribution without knowing the Souriau equations (exercice 7b of). Balian started by considering the constants of motion that are the energy and the component J_z of the total angular momentum $J = \sum_i (r_i \times p_i)$. Balian observed that he must add to the Lagrangian parameter, given by (Planck) temperature β for energy, an additional one associated with J_z. He identifies this additional multiplier with $-\beta\omega$ by evaluating the mean velocity at each point. He then introduced the same results also by changing the frame of reference, the Lagrangian and the Hamiltonian in the rotating frame and by writing down the canonical equilibrium in that frame. He uses the resulting distribution to find, through integration, over the momenta, an expression for the particles density as the function of the distance from the cylinder axis. The main Souriau model advantage is that we can define covariant Gibbs density for dynamical systems, only by applying formulas without any considerations [64].

8. Higher-Order Model of Lie Groups Thermodynamics Based on Poly-Symplectic Vector Valued Model

As observed by Souriau in Chapter IV of [43], the Gausian density is a maximum entropy density of 1st order. Considering multivariate Gaussian density, this remark is clear if we replace classical parameterization z and (m, R) by the new parameterization, linked to information geometry coordinates, ξ and β:

$$\begin{array}{l}
p_{(m,R)}(z) = \frac{1}{(2\pi)^{n/2}\det(R)^{1/2}} e^{-\frac{1}{2}(z-m)^T R^{-1}(z-m)} = \frac{1}{(2\pi)^{n/2}\det(R)^{1/2} e^{\frac{1}{2}m^T R^{-1} m}} e^{-[-m^T R^{-1} z + \frac{1}{2} z^T R^{-1} z]} \\
p_{(m,R)}(z) = p_{\hat{\xi}}(\xi) = \frac{1}{Z} e^{-\langle \beta, \xi \rangle} \text{ with } \xi = \begin{bmatrix} z \\ zz^T \end{bmatrix}, \ \hat{\xi} = \begin{bmatrix} E[z] \\ E[zz^T] \end{bmatrix} = \begin{bmatrix} m \\ R + mm^T \end{bmatrix} \\
\text{and } \beta = \begin{bmatrix} -R^{-1} m \frac{1}{2} R^{-1} \end{bmatrix} = \begin{bmatrix} a \\ H \end{bmatrix} \text{ where } \langle \beta, \xi \rangle = a^T z + z^T H z = Tr\left[z a^T + z z^T H^T \right] \\
\text{with } \log(Z) = \frac{n}{2}\log(2\pi) + \frac{1}{2}\log\det(R) + \frac{1}{2} m^T R^{-1} m \text{ and } S(\hat{\xi}) = \langle \hat{\xi}, \beta \rangle - \Phi(\beta) \\
\hat{\xi} = \Theta(\beta) = \frac{\partial \Phi(\beta)}{\partial \beta} \text{ and } \beta = \Theta^{-1}(\hat{\xi}) \text{ with } \Phi(\beta) = -\log \psi_\Omega(\beta) = -\log \int_{\Omega^*} e^{-\langle \beta, \xi \rangle} d\xi \\
\textit{Fisher: } I(\beta) = \frac{\partial^2 \log \psi_\Omega(\beta)}{\partial \beta^2} = E\left[\frac{\partial \log p_\beta(\xi)}{\partial \beta} \frac{\partial \log p_\beta(\xi)}{\partial \beta}^T \right] = E\left[(\xi - \hat{\xi})(\xi - \hat{\xi})^T \right]
\end{array} \quad (117)$$

We can observe in previous equations that classical multivariate Gaussian density, classically expressed by $p_{(m,R)}(z) = \frac{1}{(2\pi)^{n/2}\det(R)^{1/2}} e^{-\frac{1}{2}(z-m)^T R^{-1}(z-m)}$ could be rewritten in a new parameterization in a Gibbs density form $p_{\hat{\xi}}(\xi) = \frac{1}{Z} e^{-\langle \beta, \xi \rangle}$ with tensor variable $\xi = \begin{bmatrix} z \\ zz^T \end{bmatrix}$, where $\hat{\xi} = E[\xi] = \begin{bmatrix} m \\ R + mm^T \end{bmatrix}$ and tensor parameterization $\beta = \begin{bmatrix} -R^{-1} m \\ \frac{1}{2} R^{-1} \end{bmatrix} = \begin{bmatrix} a \\ H \end{bmatrix}$ with the following definition of duality braket given by $\langle \beta, \xi \rangle = a^T z + z^T H z = Tr\left[z a^T + z z^T H^T \right]$ also written in the initial parameterization $\langle \beta, \xi \rangle = -m^T R^{-1} z + \frac{1}{2} z^T R^{-1} z = Tr\left[-z m^T R^{-1} + \frac{1}{2} z z^T R^{-1} \right]$. To understand the meaning of these tensors, we can consider them as homeomorph to the following respective matrices $\xi = \begin{bmatrix} zz^T & z \\ 0_{1\times n} & 0 \end{bmatrix}$, $\hat{\xi} = \begin{bmatrix} R + mm^T & m \\ 0_{1\times n} & 0 \end{bmatrix}$ and $\beta = \begin{bmatrix} \frac{1}{2} R^{-1} & -R^{-1} m \\ 0_{1\times n} & 0 \end{bmatrix}$ with $\langle \beta, \xi \rangle = Tr\left[\beta \xi^T \right]$ (see [91] for more details).

Z is the classical normalization constant that is equal to $log(Z) = \frac{n}{2}\log(2\pi) + \frac{1}{2}\log\det(R) + \frac{1}{2} m^T R^{-1} m$. In this new parameterization, we can express the entropy by Legendre transform $S(\hat{\xi}) = \langle \hat{\xi}, \beta \rangle - \Phi(\beta)$ of Massieu characteristic function $\Phi(\beta) = -\log \psi_\Omega(\beta) = -\log \int_{\Omega^*} e^{-\langle \beta, \xi \rangle} d\xi$ (minus logarithm of partition function $\psi_\Omega(\beta) = \int_{\Omega^*} e^{-\langle \beta, \xi \rangle} d\xi$), with the Souriay (Planck) geometric temperature given by $\beta = \Theta^{-1}(\hat{\xi})$ where the function $\Theta(.)$ is the inverse of the function given by

$\hat{\xi} = \Theta(\beta) = \dfrac{\partial \Phi(\beta)}{\partial \beta}$ (the temperature is also given by $\beta = \dfrac{\partial S(\hat{\xi})}{\partial \hat{\xi}}$ given by Lagendre transform; where we recover classical definition of entropy by Clausius $dS = \dfrac{dQ}{T}$ when $\beta = \dfrac{1}{T}$ and $\hat{\xi} = Q$ heat). We can also defined Fisher metric of information geometry by $I(\beta) = \dfrac{\partial^2 \log \psi_\Omega(\beta)}{\partial \beta^2}$ or $I(\beta) = -E\left[\dfrac{\partial^2 \log p_\beta(\xi)}{\partial \beta^2}\right] = E\left[\dfrac{\partial \log p_\beta(\xi)}{\partial \beta}\dfrac{\partial \log p_\beta(\xi)}{\partial \beta}^T\right] = E\left[\left(\xi - \hat{\xi}\right)\left(\xi - \hat{\xi}\right)^T\right]$. From this development, we can observe that classical multivariate Gaussian Density $p_{\hat{\xi}}(\xi) = \frac{1}{Z}e^{-\langle \beta, \xi \rangle}$ is a maximum entropy Gibbs density of 1st order with respect to the tensorial variable $\hat{\xi} = E[\xi] = \begin{bmatrix} m \\ R + mm^T \end{bmatrix}$. Classically Gaussian density is considered as a maximum entropy Gibbs density of 2nd order where $p_{(m,R)}(z) = \dfrac{1}{(2\pi)^{n/2}\det(R)^{1/2}}e^{-\frac{1}{2}(z-m)^T R^{-1}(z-m)}$ is solution to $-\int p_{(m,R)}(z)\log p_{(m,R)}(z)dz$ under the constraints that first two moments are known $m = \int z.p_{(m,R)}(z)dz$ and $R = \int (z-m)(z-m)^T.p_{(m,R)}(z)dz$. The question is then, could we define a Gaussian density of higher order?

We have seen that Souriau has replaced classical maximum entropy approach by replacing Lagrange parameters by only one geometric "temperature vector" as element of Lie algebra. In parallel, Ingarden has introduced second and higher order temperature of the Gibbs state that could be extended to Souriau's theory of thermodynamics. The question is then, how to extend the Souriau model to define an higher order Lie groups thermodynamics. For this purpose, we propose to consider multi-symplectic geometry and more particularly poly-symplectic geometry [115]. The variational problems generalization with several variables was developed by Volterra in two papers [116,117] where two different generalizations of the Hamilton system of equations are introduced. In parallel, De Donder [53] has also studied this approach in a geometrical framework based on Elie Cartan's idea of invariant structure with no dependence to local coordinates and based on affine multisymplectic manifold. We can also formalize the multisymplectic geometry with an extension of the Poincaré-Cartan invariant integrals. Frédéric Hélein has observed the fact that different theories could cohabitate was considered jointly by Lepage [54], Dedecker [118,119] and Kijowski [92–94]. The Lepage–Dedecker theory was developed by Hélein [120], and the modern formulation using the multisymplectic $(n + 1)$-form as the fundamental structure of the theory starts with Kijowski's papers. The geometrical multisymplectic approach uses the generalized Legendre correspondence introduced by Lepage and Dedecker and Hamiltonian formalism developed by Hélein [55]. We can also make references to poly-symplectic formulation of physical systems by Carathéodory [121] and Weyl [122].

Among all multi-symplectic models, the more natural multi-valued one that preserve the notion of (poly-)moment map has been initiated by Günther based on n-symplectic model. Günther has shown that the symplectic structure on the phase space remains true, if we replace the symplectic form by a vector valued form, that is called poly-symplectic. The Günther formalism is based on the notion of a poly-symplectic form, which is a vector valued generalization of symplectic forms. Hamiltonian formalism for multiple integral variational problems and field theory is presented in a global geometric setting. Günther has introduced in this poly-symplectic formalism: Hamiltonian equations, canonical transformations, Lagrange systems, symmetries, Field theoretic moment mappings, a classification of G-homogeneous field theoretic systems on a generalization of coadjoint orbits.

Günther has defined six conditions for a multidimensional Hamiltonian formalism:

- *C0*: For each field system, an evolution space can be constructed, which describes the states of the system completely.
- *C1*: The evolution space carries a geometric structure, which assigns to each function (Hamiltonian density) its Hamiltonian equations.

- C2: The geometry of the evolution space gives 'canonical transformations', i.e., the general symmetry group of a system independently of the choice of Hamiltonian density.
- C3: The formalism is covariant, i.e., no special coordinates or coordinate systems on the parameter space are used to construct the Hamiltonian equations.
- C4: There is an equivalence between regular Lagrange systems and certain (regular) Hamiltonian systems.
- C5: For one dimensional parameter space the theory reduces to the ordinary Hamiltonian formalism on symplectic manifolds in classical mechanics.

Günther has observed that Hamiltonian field theory by Marsden is not covariant, because C3 is not verified and causes problems in relativistic theories, and by the multisymplectic approach by Tulczyjew, based on the general theory by Dedecker, does not satisfy C1 and C2.

The key idea of Günther for this generalized Hamiltonian formalism is to replace the symplectic form in classical mechanics by a vector valued, so called poly-symplectic form with the property that:

- the evolution space of a classical field will appear as the dual of a jet bundle, which carries naturally a polysymplectic structure.
- canonical transformations are bundle isomorphisms leaving this poly-symplectic form invariant.

The polysymplectic approach recovers all classical results also generalize the Noether theorem based on canonical transformations and preserve the existence of momentum mappings. Christian Günther's work was inspired by the symplectic formulation of classical mechanics by Souriau and by the work of Edelen [52,123] and Rund [124] on a local Hamiltonian formulation of field theory. Edelen's work is a coordinate version of the local polysymplectic approach of Günther.

Initiated by Gunther [48,49] based on n-symplectic model [50,51], it has been shown that the symplectic structure on the phase space remains true, if we replace the symplectic form by a vector valued form, that is called polysymplectic.

In Günther's poly-symplectic model, we set: P : space of field values , $\phi : U \to P$ and we consider the bundle of linear maps from R^n into the tangent spaces of P:

$$I^n P \cong Hom(R^n, TP) \cong TP \otimes R^{n*} \tag{118}$$

The base of R^n is interpreted as n-tangent vectors of M, there is the isomophy:

$$I^n P \cong \oplus_1^n TP \tag{119}$$

The natural projection is given by:

$$\tau_P^n : I^n P \to P \tag{120}$$

The cojet space $Hom(R^n, TP)$ carries a natural R^n-valued:

- one-form: Θ_0 (canonical one-form):

$$\Theta_0 = \sum_{i=1}^{n} p_i dq \otimes \frac{\partial}{\partial x_i} \tag{121}$$

- two-form: $\Omega_0 = -d\Theta_0$ closed & non-degenerate (canonical polysymplectic form)

$$\Omega_0 = \sum_{i=1}^{n} dq \wedge dp_i \otimes \frac{\partial}{\partial x_i} \tag{122}$$

Definition 2.

- *A closed nondegenerate R^n-valued two-form Ω on a manifold M is called a polysymplectic form. The pair (M, Ω) is a polysymplectic manifold.*
- *A polysymplectic form Ω on a manifold M is called a standard form iff M has an atlas of canonical charts for Ω, i.e., charts in which locally Ω is written as the canonical evaluation form on P x Lin (P,R^n). (M, Ω) is called a standard polysymplectic manifold.*

The classification of symplectic homogeneous spaces by coadjoint orbits by Souriau belong to the major achievements in Hamiltonian mechanics. Günther has extended these results to polysymplectic manifolds. Let $Ad : G \times LG \to LG$ be the adjoint action. We denote by Ad^n induced action on $Lin(R^n, LG)$:

$$\begin{aligned} &Ad_g^n : G \times Lin(R^n, LG) \to Lin(R^n, LG) \\ &Ad_g^n(f)(x) = Ad_g(f(x)) \text{ , } f \in Lin(R^n, LG), x \in R^n, g \in G \end{aligned} \tag{123}$$

The dual of Ad^n is denoted by $Ad_g^{(n)*}$:

$$Ad^{\#} : G \times LG^* \otimes R^n \to LG^* \otimes R^n \tag{124}$$

Corollary 1. [Günther Corollary] *Let the moment map $J^{(n)} : M \to Lin(LG, R^n) = LG^* \otimes R^n$, there is a smooth map $\theta^{(n)}$:*

$$\theta^{(n)} : G \to LG^* \otimes R^n \text{ , } \theta^{(n)}(g) = J^{(n)}(\Phi_g(x)) - Ad_g^{(n)*}\left(J^{(n)}(x)\right) \tag{125}$$

with the following properties: $\theta^{(n)}$ is a 1-cocyle for all $g, h \in G$ then:

$$\theta^{(n)}(gh) = Ad_h^{(n)*}\left(\theta^{(n)}(g)\right) + \theta^{(n)}(h) \tag{126}$$

Theorem 4. [Günther Theorem (Vector-Valued Extension of Souriau Theorem)] *The map:*

$$\begin{aligned} &a : G \times LG^* \otimes R^n \to G \times LG^* \otimes R^n \\ &a(g, \eta) = Ad_g^{(n)*}\eta + \theta^{(n)}(g) \end{aligned} \tag{127}$$

is an affine operation of G on $LG^* \otimes R^n$, and commutes for all $g \in G$.

This extension by Günther defines an action of G over $\mathfrak{g}^* \times \overset{(n)}{\ldots} \times \mathfrak{g}^*$ called n-coadjoint action:

Definition 3.

$$\begin{aligned} Ad_g^{*(n)} : \quad & G \times \left(\mathfrak{g}^* \times \overset{(n)}{\ldots} \times \mathfrak{g}^*\right) \to \mathfrak{g}^* \times \overset{(n)}{\ldots} \times \mathfrak{g}^* \\ & g \times \mu_1 \times \ldots \times \mu_n \mapsto Ad_g^{*(n)}(\mu_1, \ldots, \mu_n) = \left(Ad_g^*\mu_1, \ldots, Ad_g^*\mu_n\right) \end{aligned} \tag{128}$$

Let $\mu = (\mu_1, \ldots, \mu_n)$ a poly-momentum, element of $\mathfrak{g}^* \times \overset{(n)}{\ldots} \times \mathfrak{g}^*$, we can define a n-coadjoint orbit $O_\mu = O_{(\mu_1,\ldots,\mu_n)}$ at the point μ, for which the canonical projection $\mathrm{Pr}_k : \mathfrak{g}^* \times \overset{(n)}{\ldots} \times \mathfrak{g}^* \to \mathfrak{g}^*$, $(\nu_1, \ldots, \nu_n) \mapsto \nu_k$ induces a smooth map between the n-coadjoint orbit O_μ and the coadjoint orbit O_{μ_k}: $\pi_k : O_\mu = O_{(\mu_1,\ldots,\mu_n)} \to O_{\mu_k}$ that is a surjective submersion with $\bigcap_{k=1}^{n} KerT\pi_k = \{0\}$.

Proposition 1. Extending Souriau's approach, equivariance of poly-moment is a unique action a(.,.) of the Lie group G *on* $\mathfrak{g}^* \times \overset{(n)}{\ldots} \times \mathfrak{g}^*$ *for which the polymoment map* $J^{(n)} = (J^1, \ldots, J^n) : M \to \mathfrak{g}^* \times \overset{(n)}{\ldots} \times \mathfrak{g}^*$ *verifies* $x \in M$ *and* $g \in G$:

$$J^{(n)}(\Phi_g(x)) = a(g, J^{(n)}(x)) = Ad_g^{*(n)}\left(J^{(n)}(x)\right) + \theta^{(n)}(g) \tag{129}$$

with:

$$Ad_g^{*(n)}\left(J^{(n)}(x)\right) = \left(Ad_g^* J^1, \ldots, Ad_g^* J^n\right) \tag{130}$$

and:

$$\theta^{(n)}(g) = \left(\theta^1(g), \ldots, \theta^n(g)\right) \tag{131}$$

$\theta^{(n)}(g)$ *is a poly-symplectic one-cocycle.*

Definition 4. *We define a poly-symplectic two-cocycle* $\tilde{\Theta}^{(n)} = \left(\tilde{\Theta}^1, \ldots, \tilde{\Theta}^n\right)$ *with*

$$\tilde{\Theta}^k(X, Y) = \left\langle \Theta^k(X), Y \right\rangle = J^k_{[X,Y]} - \left\{J^k_X, J^k_Y\right\} \tag{132}$$

where:

$$\Theta^k(X) = T_e\theta^k(X(e)) \tag{133}$$

Finally, we propose to define the poly-symplectic Souriau-Fisher metric.

Definition 5.

$$g_\beta([\beta, Z_1], Z_2) = diag\left[\tilde{\Theta}_{\beta_k}(Z_1, Z_2)\right]_k, \ \forall Z_1 \in g, \forall Z_2 \in \mathrm{Im}(ad_\beta(.)), \beta = (\beta_1, \ldots, \beta_n) \tag{134}$$

with

$$\tilde{\Theta}_{\beta_k}(Z_1, Z_2) = -\frac{\partial \Phi(\beta_1, \ldots, \beta_n)}{\partial \beta_k} = \tilde{\Theta}^k(Z_1, Z_2) + \langle Q_k, ad_{Z_1}(Z_2) \rangle \tag{135}$$

is a poly-symplectic extension of Souriau-Fisher Metric.

Compared to the Souriau model, heat is replaced by previous polysymplectic model:

$$Q = (Q_1, \ldots, Q_n) \in \mathfrak{g}^* \times \overset{(n)}{\ldots} \times \mathfrak{g}^* \ \text{ with } Q_k = \frac{\partial \Phi(\beta_1, \ldots, \beta_n)}{\partial \beta_k} = \frac{\int_M U^{\otimes k}(\xi).e^{-\sum_{k=1}^n \langle \beta_k, U^{\otimes k}(\xi)\rangle} d\omega}{\int_M e^{-\sum_{k=1}^n \langle \beta_k, U^{\otimes k}(\xi)\rangle} d\omega} \tag{136}$$

Proposition 2. *The characteristic function:*

$$\Phi(\beta_1, \ldots, \beta_n) = -\log \int_M e^{-\sum_{k=1}^n \langle \beta_k, U^{\otimes k}(\xi)\rangle} d\omega \tag{137}$$

exists.

Proof. We extrapolate Souriau's results, who proved in [1,2] that $\int_M U^{\otimes k}(\xi).e^{-\langle \beta_k, U^{\otimes k}(\xi)\rangle} d\omega$ is locally normally convergent using multi-linear norm $\|U^{\otimes k}\| = \underset{U}{Sup}\langle E, U\rangle^k$ and where $U^{\otimes k} = U \otimes \overset{(k)}{U} \ldots \otimes U$ is defined as a tensorial product [43]. □

Entropy is defined by the Legendre transform of the Souriau-Massieu characteristic function:

Definition 6. *The poly-entropy is given by Legendre transform of the poly-symplectic characteristic function:*

$$S(Q_1,\ldots,Q_n)=\sum_{k=1}^{n}\langle\beta_k,Q_k\rangle-\Phi(\beta_1,\ldots,\beta_n)\text{ where }\beta_k=\frac{\partial S(Q_1,\ldots,Q_n)}{\partial Q_k}\tag{138}$$

The Gibbs density could be then extended with respect to high order temperatures.

Definition 7. *Gibbs density is defined as the maximum entropy density of poly-Entropy:*

$$p_{Gibbs}(\xi)=e^{\Phi(\beta_1,\ldots,\beta_n)-\sum\limits_{k=1}^{n}\langle\beta_k,U^{\otimes k}(\xi)\rangle}=\frac{e^{-\sum\limits_{k=1}^{n}\langle\beta_k,U^{\otimes k}(\xi)\rangle}}{\int\limits_M e^{-\sum\limits_{k=1}^{n}\langle\beta_k,U^{\otimes k}(\xi)\rangle}d\omega}\tag{139}$$

9. Conclusions and Possible Extensions

We have introduced contextures of geometric theory of information and heat based on Souriau's approach, but information geometry is at the interface between different geometries. First, information geometry is at the intersection between "Riemannian geometry", "complex geometry" and "symplectic geometry". Based on seminal work of Cartan on homogeneous domains and other works [125–128], information geometry is jointly founded by (see Figure 9):

- *Geometry of Jean-Marie Souriau*: Study of homogeneous symplectic manifolds geometry with the action of dynamical groups. Introduction of the Lie groups thermodynamics in statistical mechanics [43,44].
- *Geometry of Jean-Louis Koszul*: Study of homogeneous bounded domains geometry, symmetric homogeneous spaces and sharp convex cones. Introduction of an invariant 2-form [9–11,97,98,129].
- *Geometry of Erich Kähler*: Study of differential manifolds geometry equipped with a unitary structure satisfying a condition of integrability. The homogeneous Kähler case studied by André Lichnerowicz [130].

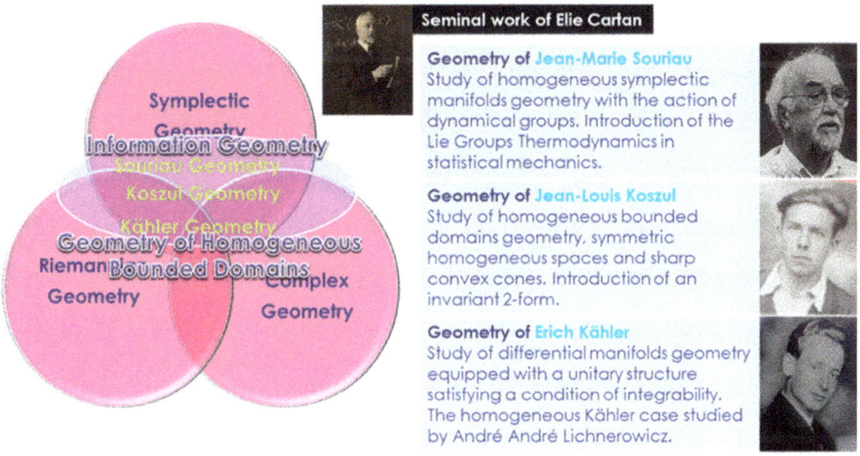

Figure 9. Three Sources of Geometric Structures for Information and Heat.

We have extended Souriau's Lie groups thermodynamics by a vector-valued model based on poly-symplectic geometry, introducing higher order Souriau-Gibbs density with higher order Souriau temperatures, and elements of Lie algebra. This model preserves all contextures of Souriau's thermodynamics with covariance of Gibbs density with respect to dynamical groups in physics. Poly-moment maps are compliant with the Noether theorem generalization in vector-valued cases.

The Jean-Marie Souriau model and equations were extensively studied in the Koszul Lecture given in China in 1986 "*Introduction to Symplectic Geometry*", in Chinese (see Figure 10). This book should be translated in English in 2019. Chuan Yu Ma has written on the Koszul book: "*This beautiful, modern book should not be absent from any institutional library. . . . . During the past eighteen years there has been considerable growth in the research on symplectic geometry. Recent research in this field has been extensive and varied. This work has coincided with developments in the field of analytic mechanics. Many new ideas have also been derived with the help of a great variety of notions from modern algebra, differential geometry, Lie groups, functional analysis, differentiable manifolds and representation theory. [Koszul's book] emphasizes the differential-geometric and topological properties of symplectic manifolds. It gives a modern treatment of the subject that is useful for beginners as well as for experts.*"

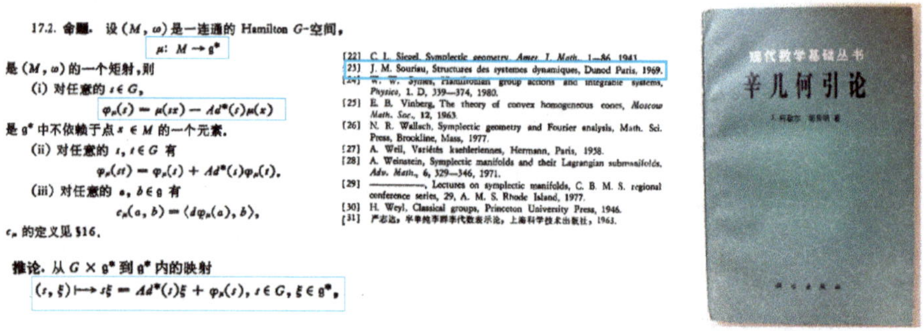

Figure 10. Koszul Lecture on "*Introduction of Symplectic Geometry*" where the Souriau model of non-equivariance is developed.

We have seen that in geometrical mechanics, the Galileo group related to classical mechanics:

$$\begin{bmatrix} \vec{x}\prime \\ t\prime \\ 1 \end{bmatrix} = \begin{bmatrix} R & \vec{u} & \vec{w} \\ 0 & 1 & e \\ 0 & 0 & 1 \end{bmatrix} \begin{bmatrix} \vec{x} \\ t \\ 1 \end{bmatrix}, R \in SO(3),\ \vec{u}, \vec{w} \in R^3, e \in R \tag{140}$$

and its central extension given by the Bargman group:

$$\begin{bmatrix} R & \vec{u} & 0 & \vec{w} \\ 0 & 1 & 0 & e \\ -\vec{u}^t R & -\frac{\|\vec{u}\|^2}{2} & 1 & f \\ 0 & 0 & 0 & 1 \end{bmatrix} \tag{141}$$

and Poincaré group in relativity. We then observe, that affine group or its sub-groups are at cornerstone of different disciplines such as:

- In robotics, the special Euclidean group SE(3) which is the homogeneous Galileo group (robotics also consider the group of similitudes SIM(3)):

$$\begin{bmatrix} Z\prime \\ 1 \end{bmatrix} = \begin{bmatrix} \Omega & t \\ 0 & 1 \end{bmatrix} \begin{bmatrix} Z \\ 1 \end{bmatrix}, \begin{cases} \Omega \in SO(3) \\ t \in R^3 \end{cases} \tag{142}$$

- In information geometry, the general affine group is involved A(n,R) for exponential family:

$$\begin{bmatrix} Z\prime \\ 1 \end{bmatrix} = \begin{bmatrix} A & t \\ 0 & 1 \end{bmatrix} \begin{bmatrix} Z \\ 1 \end{bmatrix}, \begin{cases} A \in GL(n) \\ t \in R^n \end{cases} \quad (143)$$

with particular case of Gaussian density, associated by Cholesky factorisation of covariance matrix, where covariance matrix square root is triangular matrix with positive elements on its diagonal (it is a group):

$$\begin{bmatrix} Y \\ 1 \end{bmatrix} = \begin{bmatrix} R^{1/2} & m \\ 0 & 1 \end{bmatrix} \begin{bmatrix} X \\ 1 \end{bmatrix}, \begin{cases} R^{1/2} \in T_n^+ \\ \left(R^{1/2} : \text{Cholesky de } R\right) \\ m \in R^n \end{cases} \quad (144)$$

- In the study of homogeneous bounded domains, as the simplest one given by Poincaré upper-half plane:

$$\begin{bmatrix} X\prime \\ 1 \end{bmatrix} = \begin{bmatrix} a & b \\ 0 & 1 \end{bmatrix} \begin{bmatrix} X \\ 1 \end{bmatrix}, a \in R_+^* \text{ et } b \in R \quad (145)$$

As illustrated in Figure 11, Jean-Marie Souriau developed these models at Carthage in Tunisia and at Marseilles in France during 50's and 60's. Jean-Marie Souriau was motivated by group invariance, not only in physics but also in neuroscience. Souriau intuition was highly premonitory, because this neuroscience domain has been developed few decades after by Alain Berthoz at College de France (http://public.weconext.eu/academie-sciences/2017-10-03_5a7/video_id_002/index.html) and by Daniel Bennequin (https://www.youtube.com/watch?v=a-ctwxBpJxE) to study the brain sense of movment. We can read in Souriau's text the very interesting remarks on geometry and neuroscience:

> *"Je me suis dit, à force de rencontrer des groupes, il y a quelque chose de caché là-dessous. La catégorie métaphysique des groupes qui plane dans l'empyrée des mathématiques, que nous découvrons et que nous adorons, elle doit se rattacher à quelque chose de plus proche de nous. En écoutant de nombreux exposés faits par des neurophysiologistes, j'ai fini par apprendre le rôle primitif du déplacement des objets. Nous savons manipuler ces déplacements mentalement avec une très grande virtuosité. Ce qui nous permet de nous manipuler nous-même, de marcher, de courir, de sauter, de nous rattraper quand nous tombons,* etc. *Ce n'est pas vrai seulement pour nous, c'est vrai aussi pour les singes ; ils sont beaucoup plus adroits que nous pour anticiper les résultats d'un déplacement. Pour certaines opérations élémentaires de «lecture», ils vont même dix fois plus vite que nous. Beaucoup de neurophysiologistes pensent qu'il y a une structure spéciale génétiquement inscrite dans le cerveau, le câblage d'un groupe ... Lorsque il y un tremblement de terre, nous assistons à la mort de l'Espace. ... Nous vivons avec nos habitudes que nous pensons universelles. ... La neuroscience s'occupe rarement de la géométrie ... Pour les singes qui vivent dans les arbres, certaines propriétés du groupe d'Euclide sont mieux câblées dans leurs cerveaux."* (In Engish: *"I said to myself, because of meeting groups everywhere, there is something hidden there. The metaphysical category of groups that hovers in the empyrean of mathematics, which we discover and adore, must be connected with something closer to us. Listening to many presentations by neurophysiologists, I ended up learning the primitive role of moving objects. We know how to manipulate these movements mentally with great virtuosity. That allows us to manipulate ourselves, to walk, run, jump, catch up when we fall, and so on. This is not true only for us, it is true also for monkeys; they are much more adroit than we are to anticipate the results of a trip. For some basic "reading" operations, they are even ten times faster than us. Many neurophysiologists think that there is a special structure genetically inscribed in the brain, the wiring of a group ... When there is an earthquake, we witness the death of Space. ... We live with our habits that we think universal. ... Neuroscience rarely deals with geometry ... For monkeys living in trees, some of Euclid's group properties are better wired in their brains.*)

Figure 11. Mediterranean sources of Souriau Book on Structure of Dynamical systems at Carthage and Massilia where souriau wrote this text and theory.

Our new research directions will concern extension of *"Le Hasard et la Courbure (Randomness and Curvature)"* (title of Yann Ollivier HDR), that we have synthetized in Souriau-Fisher metric to *"Le Hasard et la Torsion (Randomness and Torsion)"* based on Elie Cartan works founded on Cosserats brothers model of elasticity [125–127,131].

> *"Il est une Cosmologie avec laquelle la Thermodynamique générale présente une analogie non-méconnaissable; cette Cosmologie, c'est la Physique péripatéticienne ... Parmi les attributs de la substance, la Physique péripatéticienne confère une égale importance à la catégorie de la quantité et à la catégorie de la qualité; or, par ses symboles numériques, la Thermodynamique générale représente également les diverses grandeurs des quantités et les diverses intensités des qualités. Le mouvement local n'est, pour Aristote, qu'une des formes du mouvement général, tandis que les Cosmologies cartésienne, atomistique et newtonienne concordent en ceci que le seul mouvement possible est le changement de lieu dans l'espace. Et voici que la Thermodynamique générale traite, en ses formules, d'une foule de modifications telles que les variations de températures, les changements d'état électrique ou d'aimantation, sans chercher le moins du monde à réduire ces variations au mouvement local"* —Pierre Duhem—La théorie Physique: son objet, sa structure [132].

> *"Pour la théorie de la connaissance mais aussi pour les sciences est fondamentale la notion de perspective. Or, les expériences faites dans la géométrie algébriques, dans la théorie des nombres, et dans l'algèbre abstraite m'induisent à tenter une formulation mathématique de cette notion pour surmonter ainsi au moyen de raisonnements d'origine géométrique la géométrie. Il me semble en effet, que la tendance vers l'abstraction observée dans les mathématiques d'aujourd'hui, loin d'être l'ennemi de l'intuition ait le sens profond de quitter l'intuition pour la faire renaitre dans une alliance entre «esprit de géométrie» et «esprit de finesse», alliance rendue possible par les réserves énormes des mathématiques pures dont Pascal et Goethe ne pouvaient pas encore se douter"* —Erich Kähler—Sur la théorie des corps purement algébriques, 1952.

Funding: This research received no external funding.

Conflicts of Interest: The author declares no conflict of interest.

Appendix A. Günther's Polysymplectic Model

We recall in this appendix, a synthesis of Christian Günther Poly-symplectic model with his initial notation [48].

We set:

$$\begin{array}{c} Q: \text{ space of field values} \\ \varphi: U \to Q \end{array} \quad \text{(A1)}$$

The bundle of linear maps from R^n into the tangent spaces of Q.

$$I^nQ \cong Hom(R^n, TQ) \cong TQ \otimes R^{n*} \quad \text{(A2)}$$

If a base of R^n is chosen, can also be interpreted as n-tangent vectors of Q, there is the isomophy:

$$I^nQ \cong \oplus_1^n TQ \quad \text{(A3)}$$

The natural projection is given by:

$$\tau_Q^n : I^nQ \to Q \quad \text{(A4)}$$

In analogy to the canonical forms on the cotangent bundle, the cojet space $Hom(R^n, TQ)$ carries a natural R^n-valued:

- one-form: Θ_0 (canonical one-form)
- two-form: $\Omega_0 = -d\Theta_0$ closed & non-degenerate (canonical polysymplectic form)

In the natural bundle coordinates the canonical forms on $Hom(R^n, TQ)$ have the local representation:

$$\Theta_0 = \sum_{i=1}^{n} p_i dq \otimes \frac{\partial}{\partial x_i} \quad \text{(A5)}$$

$$\Omega_0 = \sum_{i=1}^{n} dq \wedge dp_i \otimes \frac{\partial}{\partial x_i} \quad \text{(A6)}$$

Following diffeomorphism leaves invariant one and two forms:

$$\begin{array}{l} f: Q \to Q \text{ and } I^{n*}f : Hom(TQ, R^n) \to Hom(TQ, R^n) \\ (I^{n*}f)^*\Theta_0 = \Theta_0 \text{ and } (I^{n*}f)^*\Omega_0 = \Omega_0 \end{array} \quad \text{(A7)}$$

Definition A1. *A closed nondegenerate R^n-valued two-form Ω on a manifold M is called a polysymplectic form. The pair (M, Ω) is a polysymplectic manifold.*

The classification of linear polysymplectic forms is not trivial, because two polysymplectic forms are not necessarily locally equivariant.

Definition A2. *A polysymplectic form Ω on a manifold M is called a standard form iff M has an atlas of canonical charts for Ω, i.e., charts in which locally Ω is written as the canonical evaluation form on Q x Lin (Q,R^n). (M, Ω) is called a standard polysymplectic manifold.*

The polysymplectic structure provides the procedure which assigns to a function on M, the Hamiltonian, its associated Hamiltonian equations. Let (M, Ω) a polysymplectic manifold:

$$\left\{ \begin{array}{ll} \Omega^b: & TM \to Hom(TM, R^n) \\ & w_m \mapsto \Omega^b_{(v_m)}(w_m) = \Omega(v_m, w_m) \end{array} \right. \text{ and } \left\{ \begin{array}{l} \Omega^\# : Hom(TM, R^n) \to T^*M \\ X_m \mapsto \Omega^\#(X_m) = tr\left(\Omega^b \circ X_m\right) \\ \text{with } tr\left(\Omega^b \circ X_m\right).v_m = -tr\left(\Omega^b(v_m) \circ X_m\right) \end{array} \right. \quad \text{(A8)}$$

An affine sub bundle of $Hom(R^n, TQ)$ is defined by:

$$\Omega^{\#-1}(dH) = \left\{ X_m \in Hom(R^n, TQ) / \Omega^{\#}(X_m) = dH(m) \right\} \tag{A9}$$

Definition A3. *$\Omega^{\#-1}(dH)$ is called the system of Hamiltonian partial differential equations associated with the Hamiltonian function H. A smooth map $\psi : U \to M$is a solution of $\Omega^{\#-1}(dH)$ iff:*

$$T_u\psi \in \Omega^{\#-1}(dH(\psi(u)))\forall u \in U \tag{A10}$$

Theorem A1. *Let (M, Ω) be a standard polysymplectic manifold, (p,q) canonical coordinates for Ω on M, and H a Hamiltonian function. A smooth map $\psi : U \to M$ is a solution of $\Omega^{\#-1}(dH)$ iff in canonical coordinates:*

$$trdp(u) = -\frac{\partial H}{\partial q}(\psi(u)) \text{ and } Dq(u) = \frac{\partial H}{\partial p}(\psi(u)) \tag{A11}$$

If a base $e_1, \ldots, e_n$ of R^n is chosen and $p(u) = (p_1(u), \ldots, p_n(u))$ with respect to this base, then the equations take the form:

$$\sum_{i=1}^{n} \frac{\partial p_i}{\partial x_i}(u) = -\frac{\partial H}{\partial q}(\psi(u)) \text{ and } \frac{\partial q}{\partial x_i}(u) = \frac{\partial H}{\partial p_i}(\psi(u)) \tag{A12}$$

Proof.

$$\begin{aligned} &X(\psi(u)) = D\psi(u) \in Lin\left(R^n, T_{\psi(u)}M\right) \\ &X(m) = X_q(m) + X_p(m)\ ,\ X_q(m) \in Lin(R^n, Q)\ ,\ X_p(m) \in Lin(R^n, Lin(Q, R^n)) \\ &v(m) = \dot{q}(m) + \dot{p}(m)\ ,\ \dot{q}(m) \in Q\ ,\ \dot{p}(m) \in Lin(Q, R^n) \end{aligned} \tag{A13}$$

$$\begin{aligned} &\Omega^{\#}(X).v = tr\Omega^{b} \circ X(v) = -tr\Omega^{b}(v) \circ X \\ &\Omega^{b}(\dot{q}, \dot{p}).(\dot{q}, \dot{p}) = \dot{p}(\dot{q})\ ,\ (\dot{q}, \dot{p}) \in TM \\ &\Omega^{\#}(X).(\dot{q}, \dot{p}) = -tr(X_p(\dot{q}) - \dot{p} \circ X_q) = dH(\dot{q}, \dot{p}) \\ &dH = \tfrac{\partial H}{\partial q} dq + \sum_{i=1}^{n} \tfrac{\partial H}{\partial p_i} dp_i \Rightarrow -trX_p = \tfrac{\partial H}{\partial q}\ ,\ \tfrac{\partial H}{\partial p} = X_q \end{aligned} \tag{A14}$$

□

Example A1. *Consider a scalar field where $n = 4, Q = R$ and $M = R \times R^4$ with scalar coordinates $(q, p_1, \ldots, p_4)$*

Let $H(q, p_1, \ldots, p_4) = \frac{1}{2} \sum_{i=1}^{4} p_i^2 + mq^2$ an Hamiltonian on M, the canonical polysymplectic form Ω is given by:

$$\Omega = \sum_{i=1}^{4} dq \wedge dp_i \otimes \frac{\partial}{\partial x_i} \tag{A15}$$

The Hamiltonian equations for a scalar field:

$$\psi(x_1, \ldots, x_4) = (q(x_1, \ldots, x_4), p_1(x_1, \ldots, x_4), \ldots, p_4(x_1, \ldots, x_4)) \tag{A16}$$

are:

$$\sum_{i=1}^{4} \frac{\partial p_i}{\partial x_i} = mq \; and \; \frac{\partial q}{\partial x_i} = p_i \tag{A17}$$

Definition A4. *Let (M, Ω) be a polysymplectic manifold, $\Omega^{\#}(X) = dH$, H is called an momentum tensor iff*

$$trd\mathrm{H} = dH \tag{A18}$$

Proposition A1.

$$X\neg\Theta_0 = 0,\ d(tr\mathrm{L}_X\Theta_0) = 0 \text{ and } tr\mathrm{L}_X\Theta_0 = -d(H - tr(X\neg\Theta_0)) \tag{A19}$$

Proof.

$$\begin{aligned} &\Theta_0 = \sum_i p_i dq \otimes \tfrac{\partial}{\partial x_i} \; and \; X = X_q \tfrac{\partial}{\partial q} + \sum_i X_{p_i} \tfrac{\partial}{\partial p_i} \\ &\Rightarrow X\neg\Theta_0 = \sum_i p_i X_q \otimes \tfrac{\partial}{\partial x_i} \end{aligned} \tag{A20}$$

$$\begin{aligned} &tr\mathrm{L}_X\Theta_0 = tr(dX\neg\Theta_0 + X\neg d\Theta_0) \\ &tr(dX\neg\Theta_0 + X\neg d\Theta_0) = -dH + trdX\neg\Theta_0 \end{aligned} \tag{A21}$$

□

The classification of symplectic homogeneous spaces by coadjoint orbits by Souriau belong to the major achievements in Hamiltonian mechanics. C. Günther has extend these results to polysymplectic manifolds. Let $Ad : G \times LG \to LG$ be the adjoint action. We denote by Ad^n induced action on $Lin(R^n, LG)$:

$$\begin{aligned} Ad^n :\ & G \times Lin(R^n, LG) \to Lin(R^n, LG) \\ & Ad^n_g(f)(x) = Ad_g(f(x))\ ,\ f \in Lin(R^n, LG), x \in R^n, g \in G \end{aligned} \tag{A22}$$

The dual of Ad^n is denoted by $Ad^{\#}$:

$$Ad^{\#} : G \times LG^* \otimes R^n \to LG^* \otimes R^n\ ,\ Ad^{\#}_g(\alpha) = \alpha \circ Ad^n_g \tag{A23}$$

$$\lambda(Ad_g u) = \Lambda^*_g(\lambda(u)) \Rightarrow \Lambda^*_g \lambda^n(f) = \lambda^n\left(Ad^n_g f\right) \text{ for all } g \in G, f \in Lin(R^n, LG) \tag{A24}$$

Proposition A2 [Günther Proposition]. *Let $\Lambda : G \times M \to M$ be a strongly polysymplectic group action with momentum map $\mu : M \to Lin(LG, R^n) = LG^* \otimes R^n$. Assume M is connected. Then the map:*

$$\begin{aligned} & M \to LG^* \otimes R^n \\ & m \mapsto \mu(\Lambda_g m) - Ad^{\#}_g(\mu(m)) \end{aligned} \tag{A25}$$

is a constant on M for all $g \in G$.

Corollary A1. *There is a smooth map χ:*

$$\chi : G \to LG^* \otimes R^n\ ,\ \chi(g) = \mu(\Lambda_g m) - Ad^{\#}_g(\mu(m)) \tag{A26}$$

with the following properties:

- *is a 1-cocyle for all $g, h \in G$ then*

$$\chi(gh) = Ad_h^{\#}(\chi(g)) + \chi(h) \tag{A27}$$

- *bilinear map φ* on LG: $\varphi := L_\chi : LG \to LG^* \otimes R^n$, $\varphi : LG \times LG \to R^n$ *is a 2 cocycle*

$$\varphi(u,[v,w]) + \varphi(v,[w,u]) + \varphi(w,[u,v]) = 0\,,\ \forall u,v,w \in LG \tag{A28}$$

Proof.

$$\begin{aligned}
&\chi(hg) = \mu \circ \Lambda_{hg}(m) - Ad_{hg}^{\#}\mu(m) \\
&\chi(hg) = \mu \circ \Lambda_g(\Lambda_h m) - Ad_g^{\#} \circ \mu(\Lambda_h m) + Ad_g^{\#} \circ \mu(\Lambda_h m) - Ad_g^{\#} Ad_h^{\#} \circ \mu(m) \\
&\chi(hg) = \chi(g) + Ad_g^{\#}(\chi(h))
\end{aligned} \tag{A29}$$

□

Theorem A2. [Günther Theorem (Vector-valued extension of Souriau Theorem)] *Let $\Lambda : G \times M \to M$ be a polysymplectic action with momentum map $\mu : M \to LG^* \otimes R^n$. Then the map:*

$$\begin{aligned}
&\Xi : G \times LG^* \otimes R^n \to G \times LG^* \otimes R^n \\
&\Xi(g,\eta) = Ad_g^{\#}\eta + \chi(g)
\end{aligned} \tag{A30}$$

is an affine operation of G on $LG^* \otimes R^n$, and commutes for all $g \in G$ and μ is G-equivariant.

Proof.

$$\begin{aligned}
&\Xi(gh,\eta) = \chi(gh) + Ad_{gh}^{\#}\eta + \chi(h) + \chi(g) \circ Ad_h + Ad_h^{\#} \circ Ad_g^{\#}\eta \\
&\Xi(gh,\eta) = \chi(h) + Ad_h^{\#}\left(\chi(g) + Ad_g^{*}h\right) = \Xi(h, \Xi(g,\eta))
\end{aligned} \tag{A31}$$

Ξ is an action.

$$\begin{aligned}
&\Xi_g \circ \mu(m) = \chi(g) + Ad_g^{\#} \circ \mu(m) \\
&\Xi_g \circ \mu(m) = \mu(\Lambda_g m) - Ad_g^{\#}(\mu(m)) + Ad_g^{\#}\mu(m) = \mu \circ \Lambda_g(m)
\end{aligned} \tag{A32}$$

□

Christian Günther in a never found 1987 paper wrote that "*The mathematical framework developed in this paper is used in a separate publication to provide a rigorous foundation for field theory*". For a more recent study of Günther's poly-symplectic model, we make reference to [133].

Appendix B. Fisher Metric for Multivariate Gaussian Density

We will in the following illustrate information geometry for multivariate Gaussian density:

$$p_{\hat{\xi}}(\xi) = \frac{1}{(2\pi)^{n/2}\det(R)^{1/2}} e^{-\frac{1}{2}(z-m)^T R^{-1}(z-m)} \tag{A33}$$

If we develop:

$$\begin{aligned}
\tfrac{1}{2}(z-m)^T R^{-1}(z-m) &= \tfrac{1}{2}\left[z^T R^{-1} z - m^T R^{-1} z - z^T R^{-1} m + m^T R^{-1} m\right] \\
&= \tfrac{1}{2} z^T R^{-1} z - m^T R^{-1} z + \tfrac{1}{2} m^T R^{-1} m
\end{aligned} \tag{A34}$$

We can write the density as a Gibbs density:

$$p_{\hat{\xi}}(\xi) = \frac{1}{(2\pi)^{n/2}\det(R)^{1/2}e^{\frac{1}{2}m^T R^{-1}m}} e^{-[-m^T R^{-1}z + \frac{1}{2}z^T R^{-1}z]} = \frac{1}{Z}e^{-\langle \xi, \beta \rangle}$$
$$\xi = \begin{bmatrix} z \\ zz^T \end{bmatrix} \text{ and } \beta = \begin{bmatrix} -R^{-1}m \\ \frac{1}{2}R^{-1} \end{bmatrix} = \begin{bmatrix} a \\ H \end{bmatrix} \text{ with } \langle \xi, \beta \rangle = a^T z + z^T H z = Tr\left[za^T + H^T zz^T\right] \tag{A35}$$

We can then rewrite density with canonical variables:

$$p_{\hat{\xi}}(\xi) = \frac{1}{\int\limits_{\Omega^*} e^{-\langle \xi, \beta \rangle}.d\xi} e^{-\langle \xi, \beta \rangle} = \frac{1}{Z}e^{-\langle \xi, \beta \rangle} \text{ with } \log(Z) = n\log(2\pi) + \frac{1}{2}\log\det(R) + \frac{1}{2}m^T R^{-1}m$$
$$\xi = \begin{bmatrix} z \\ zz^T \end{bmatrix}, \hat{\xi} = \begin{bmatrix} E[z] \\ E[zz^T] \end{bmatrix} = \begin{bmatrix} m \\ R + mm^T \end{bmatrix}, \beta = \begin{bmatrix} a \\ H \end{bmatrix} = \begin{bmatrix} -R^{-1}m \\ \frac{1}{2}R^{-1} \end{bmatrix} \text{ with } \langle \xi, \beta \rangle = Tr\left[za^T + H^T zz^T\right] \tag{A36}$$
$$R = E\left[(z-m)(z-m)^T\right] = E\left[zz^T - mz^T - zm^T + mm^T\right] = E\left[zz^T\right] - mm^T$$

The first potential function (free energy/logarithm of characteristic function) is given by:

$$\psi_\Omega(\beta) = \int\limits_{\Omega^*} e^{-\langle \xi, \beta \rangle}.d\xi \text{ and } \Phi(\beta) = -\log \psi_\Omega(\beta) = \tfrac{1}{2}\left[-Tr\left[H^{-1}aa^T\right] + \log\left[(2)^n \det H\right] - n\log(2\pi)\right] \tag{A37}$$

We verify the relation between the first potential function and moment:

$$\frac{\partial \Phi(\beta)}{\partial \beta} = \frac{\partial[-\log \psi_\Omega(\beta)]}{\partial \beta} = \int\limits_{\Omega^*} \xi \frac{e^{-\langle \xi, \beta \rangle}}{\int\limits_{\Omega^*} e^{-\langle \xi, \beta \rangle}.d\xi}.d\xi = \int\limits_{\Omega^*} \xi.p_{\hat{\xi}}(\xi).d\xi = \hat{\xi}$$
$$\frac{\partial \Phi(\beta)}{\partial \beta} = \begin{bmatrix} \frac{\partial \Phi(\beta)}{\partial a} \\ \frac{\partial \Phi(\beta)}{\partial H} \end{bmatrix} = \begin{bmatrix} m \\ R + mm^T \end{bmatrix} = \hat{\xi} \tag{A38}$$

The second potential function (Shannon entropy) is given as a Legendre transform of the first one:

$$S(\hat{\xi}) = \langle \hat{\xi}, \beta \rangle - \Phi(\beta) \text{ with } \frac{\partial \Phi(\beta)}{\partial \beta} = \hat{\xi} \text{ and } \frac{\partial S(\hat{\xi})}{\partial \hat{\xi}} = \beta$$
$$S(\hat{\xi}) = -\int\limits_{\Omega^*} \frac{e^{-\langle \xi, \beta \rangle}}{\int\limits_{\Omega^*} e^{-\langle \xi, \beta \rangle}.d\xi} \log \frac{e^{-\langle \xi, \beta \rangle}}{\int\limits_{\Omega^*} e^{-\langle \xi, \beta \rangle}.d\xi}.d\xi = -\int\limits_{\Omega^*} p_{\hat{\xi}}(\xi) \log p_{\hat{\xi}}(\xi).d\xi \tag{A39}$$

$$S(\hat{\xi}) = -\int\limits_{\Omega^*} p_{\hat{\xi}}(\xi) \log p_{\hat{\xi}}(\xi).d\xi = \tfrac{1}{2}\left[\log (2)^n \det\left[H^{-1}\right] + n\log(2\pi.e)\right] = \tfrac{1}{2}\left[\log\det[R] + n\log(2\pi.e)\right] \tag{A40}$$

This remark was made by Jean-Souriau in his book as soon as 1969. He has observed, as illustrated in following Figure that if we take vector with tensor components $\xi = \begin{pmatrix} z \\ z \otimes z \end{pmatrix}$, components of $\hat{\xi}$ will provide moments of the first and second order of the density of probability $p_{\hat{\xi}}(\xi)$. He used this change of variable $z\prime = H^{1/2}z + H^{-1/2}a$, to compute the logarithm of the characteristic function $\Phi(\beta)$ (see Figure A1 extracted from Souriau Book):

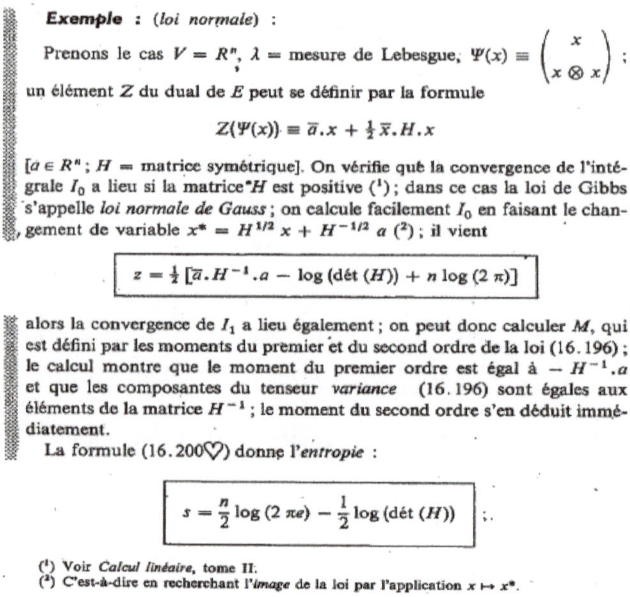

Exemple : (*loi normale*) :

Prenons le cas $V = R^n$, λ = mesure de Lebesgue, $\Psi(x) \equiv \begin{pmatrix} x \\ x \otimes x \end{pmatrix}$;
un élément Z du dual de E peut se définir par la formule

$$Z(\Psi(x)) \equiv \bar{a}.x + \tfrac{1}{2}\bar{x}.H.x$$

[$a \in R^n$; H = matrice symétrique]. On vérifie que la convergence de l'intégrale I_0 a lieu si la matrice H est positive (¹) ; dans ce cas la loi de Gibbs s'appelle *loi normale de Gauss* ; on calcule facilement I_0 en faisant le changement de variable $x^* = H^{1/2}\,x + H^{-1/2}\,a$ (²) ; il vient

$$z = \tfrac{1}{2}\,[\bar{a}.H^{-1}.a - \log(\text{dét}\,(H)) + n \log(2\,\pi)]$$

alors la convergence de I_1 a lieu également ; on peut donc calculer M, qui est défini par les moments du premier et du second ordre de la loi (16.196) ; le calcul montre que le moment du premier ordre est égal à $-H^{-1}.a$ et que les composantes du tenseur *variance* (16.196) sont égales aux éléments de la matrice H^{-1} ; le moment du second ordre s'en déduit immédiatement.

La formule (16.200♡) donne l'*entropie* :

$$s = \frac{n}{2}\log(2\,\pi e) - \frac{1}{2}\log(\text{dét}\,(H))$$;.

(¹) Voir *Calcul linéaire*, tome II.
(²) C'est-à-dire en recherchant l'*image* de la loi par l'application $x \mapsto x^*$.

Figure A1. Introduction of potential function for multivariate Gaussian law in Souriau book.

Appendix C. Geometric Definition of Legendre Transform by Chasles as Reciprocal Polar with Respect to a Paraboloid

The Legendre transform plays a central role related to duality and convexity. Adrien-Marie Legendre [102] has introduced the Legendre transform to solve a minimal surface problem given by Monge (Monge requested him to consolidate its proof), with a link to Poncelet duality [103]. Chasles and Darboux interpreted the Legendre transform as reciprocal polar with respect to a paraboloid (re-used by Hadamard and Fréchet in calculus of variations). Before Legendre, Alexis Clairaut introduced a Clairaut Equation that has been developed by Maurice Fréchet to characterize «distinguished densities» (densities with parameters that have covariance matrix reaching the Fréchet-Cramer-Rao Bound) [9].

Legendre Transform transformes one fonction defined by its value in one point in a fonction defined by its tangent, as illustrated in Figure A2.

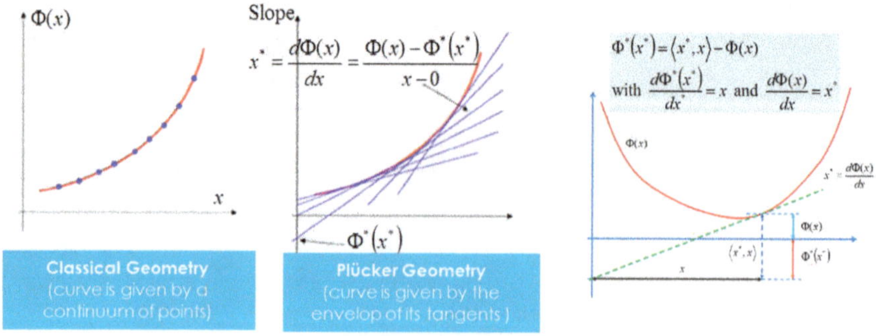

(a) (b) (c)

Figure A2. Legendre Transform and duality. (**a**) Classical Geometry; (**b**) Plücker Geometry; (**c**) Legendre Transform.

Darboux gave in his book one interpretation of Chasles: "*Ce qui revient suivant une remarque de M. Chasles, à substituer à la surface sa polaire réciproque par rapport à un paraboloïde*». In the lecture «*Leçons sur le calcul des variations*", Hadamard, followed by Vessiot, used the reciprocal polar of figurative, and figuratrice. This has also been developed by Belgodère as presented by Cartan on «*Extrémale d'une surface*» [134,135]. Polarity on the plane is a transformation taking points to lines and dually lines to points. A polarity preserves incidence and has degree 2. For a point P (that we name the pole) a conic polarity transforms it to its image which is a line p (that we name the polar) as follows: from P we draw the two tangents to the conic, which touch it in the points Q, R. If we now connect points Q, R with a line p we obtain the polar line of the pole P. A Self-conjugate point Q is incident with its polar q; that is Q lies on q.

Geometric interpretation of the Legendre transform by reciprocal polar with respect to a paraboloid is given by the following simple development. First, let's consider the surface:

$$z = f(x,y) \text{ with } p = \frac{\partial z}{\partial x} \text{ and } q = \frac{\partial z}{\partial y} \tag{A41}$$

We consider the equation of the paraboloid:

$$x^2 + y^2 = 2z \tag{A42}$$

Reciprocal polar with respect to paraboloid has coordinates: X, Y, Z

The polar plan with respect to paraboloid of this reciprocal polar $Xx + Yy - z - Z = 0$ should be equal to tangent plan of the surface at point (x_0, y_0, z_0):

$$z - z_0 = p_0(x - x_0) + q_0(y - y_0) \Rightarrow p_0 x + q_0 y - z - (p_0 x_0 + q_0 y_0 - z_0) = 0 \tag{A43}$$

This equality provides:

$$X = p_0 \,,\; Y = q_0 \,,\; Z = p_0 x_0 + q_0 y_0 - z_0 \tag{A44}$$

This is the Legendre transform. So in classical thermodynamics, the Legendre transform $S(Q) = \langle \beta, Q \rangle - \Phi(\beta)$ is linked with polar reciprocal with respect to the paraboloid:

$$Q^2 = 2S(Q) \tag{A45}$$

We can develop other properties of Legengre transform. Let's $z = f(x,y)$ with $p = \frac{\partial z}{\partial x}$ and $q = \frac{\partial z}{\partial y}$ and $X = p \,,\; Y = q \,,\; Z = px + qy - z$ the Legendre transform.

We compute the first derivative of Z:

$$dZ = PdX + QdY \text{ with } P = \frac{\partial Z}{\partial X} \text{ and } Q = \frac{\partial Z}{\partial Y} \tag{A46}$$

$$Z = px + qy - z \Rightarrow dZ = pdx + qdy - dz + xdp + ydq \underset{\substack{dz = pdx + qdy \\ X = p, Y = q}}{\Rightarrow} dZ = xdX + ydQ \Rightarrow P = x, Q = y \tag{A47}$$

We compute the 2nd derivative of Z:

$$R = \frac{\partial^2 Z}{\partial X^2} = \frac{\partial P}{\partial X} = \frac{\partial x}{\partial X} \,,\; T = \frac{\partial^2 Z}{\partial X \partial Y} = \frac{\partial P}{\partial Y} = \frac{\partial Q}{\partial X} = \frac{\partial x}{\partial Y} = \frac{\partial y}{\partial X} \,,\; S = \frac{\partial^2 Z}{\partial Y^2} = \frac{\partial Q}{\partial Y} = \frac{\partial y}{\partial Y} \tag{A48}$$

$$\begin{cases} dX = rdx + sdy \\ dY = sdx + tdy \end{cases} \Rightarrow \begin{cases} dx = \dfrac{t}{rt-s^2}dX - \dfrac{s}{rt-s^2}dY \\ dy = \dfrac{-s}{rt-s^2}dX + \dfrac{r}{rt-s^2}dY \end{cases}$$
$$r = \frac{\partial^2 z}{\partial x^2}, t = \frac{\partial^2 z}{\partial y^2}, s = \frac{\partial^2 z}{\partial x \partial y}$$
$$\Rightarrow \begin{cases} R = \dfrac{\partial x}{\partial X} = \dfrac{t}{rt-s^2} \\ S = \dfrac{\partial x}{\partial Y} = \dfrac{-s}{rt-s^2} \\ T = \dfrac{\partial y}{\partial Y} = \dfrac{r}{rt-s^2} \end{cases} \Rightarrow \begin{cases} r = \dfrac{T}{RT-S^2} \\ s = \dfrac{-S}{RT-S^2} \\ t = \dfrac{R}{RT-S^2} \end{cases} \tag{A49}$$

The link with with contact transformations is then the following. Considering new variables X,Y,Z and P,Q the derivatives of Z with respect to X and Y, problem of finding in which case this five quantities could be express of x,y,z,p and q est the same problem where we look for five functions X,Y,Z,P and Q of five independant variables x,y,z,p and q satisfying the differential equation:

$$dZ - PdX - QdY = \rho(dz - pdx - qdy) \tag{A50}$$

where ρ is a function of x,y,z,p and q.

Proof.

$$\begin{cases} p = \dfrac{\partial z}{\partial x} \\ q = \dfrac{\partial z}{\partial y} \end{cases} \Rightarrow dz - pdx - qdy = 0 \Rightarrow dZ = PdX + QdY \Rightarrow \begin{cases} P = \dfrac{\partial Z}{\partial X} \\ Q = \dfrac{\partial Z}{\partial Y} \end{cases} \tag{A51}$$

and the reciprocal:

$$\rho = \frac{\partial Z}{\partial z} - P\frac{\partial X}{\partial z} - Q\frac{\partial Y}{\partial z} \tag{A52}$$

Links with Ampere transformation is given then by the following developments. Let's consider Ampere transformation:

$$\begin{aligned} & dz - pdx - qdy = d(z - qy) - pdx + ydq \\ & \text{Set} \begin{cases} Z = z - qy, X = x, Y = q \\ P = p, Q = -y \end{cases} \Rightarrow dZ - pdX - QdY = dz - pdx - qdy \end{aligned} \tag{A53}$$

Then $\rho = 1$, and we have a contact transformation, also valid when Legendre transform is no longer valide (when $rt - s^2 = 0$, p and q are not independant)

The link between Legendre transformation and Ampere transformation is then deduced. Legendre transform is obtained by same equality:

$$\begin{aligned} & dz - pdx - qdy = d(z - qy) - pdx + ydq \\ & \text{Set} \begin{cases} Z = z - qy, X = x, Y = q \\ P = p, Q = -y \end{cases} \Rightarrow dZ - pdX - QdY = dz - pdx - qdy \end{aligned} \tag{A54}$$

We can set:

$$\begin{aligned} & X = p, Y = q, Z = z - px - qy \\ & P = x, Q = y \end{aligned} \tag{A55}$$

□

For complementary studies on the Legendre transform, we can make reference to [99,101].

Appendix D. Centrifuge Thermodynamics by Roger Balian Based on Classical Approach

Balian has studied the case of gas enclosed in a vessel rotating with an angular velocity ω in thermal equilibrium, and proved that the density of the gas is proportional to $e^{\frac{m\omega^2 r^2}{2kT}}$, with classical approach. The density is increased at the periphery due to centrifugal effects.

Balian has computed the Boltzmann-Gibbs distribution without knowing Souriau equations (exercice 7b of [25]). Balian started by considering the constants of motion that are the energy and the component J_z of the total angular momentum $J = \sum_i (r_i \times p_i)$. Balian observed that he must add to the Lagrangian parameter, given by (Planck) temperature β for energy, an additional one associated with J_z. He identifies this additional multiplier with $-\beta\omega$ by evaluating the mean velocity at each point. He then introduced the same results also by changing the frame of reference, the Lagrangian and the Hamiltonian in the rotating frame and by writing down the canonical equilibrium in that frame. He uses the resulting distribution to find, through integration, over the momenta, an expression for the particles density as the function of the distance from the cylinder axis. The fluid carried along by the walls of the rotating vessel acquires a non-vanishing average angular momentum $\langle J_z \rangle$ around the axis of rotation, that is a constant of motion. In order to be able to assign to it a definite value, Balian proposed to associate with it a Lagrangian multiplier λ, in exactly the same way as we classicaly associate the multiplier β with the energy in canonical equilibrium. The average $\langle J_z \rangle$ will be a function of λ. The Gibbs density for rotating gas is given by Balian as:

$$D = \frac{1}{Z} e^{-\beta H - \lambda J_z} = \frac{1}{Z} \exp\left\{ \sum_i \left[\frac{\beta p_i^2}{2m} + \lambda \left(x_i p_{y_i} - y_i p_{x_i} \right) \right] \right\} \tag{A56}$$

With the energy and the average angular momentum given by:

$$U = -\frac{\partial \ln Z}{\partial \beta} = \frac{1}{kT} \text{ and } \langle J_z \rangle = -\frac{\partial \ln Z}{\partial \lambda} \tag{A57}$$

The Lagrangian parameter λ has a mechanical nature. To identify this parameter, Balian compared microscopic and macroscopy descriptions of fluid mechanics. He described the single-particle reduced density by:

$$\begin{aligned} f(r,p) &\propto \exp\left\{ -\frac{\beta p^2}{2m} - \lambda \left(x p_y - y p_x \right) \right\} \\ &= \exp\left\{ -\frac{\beta}{2m} \left(p + \frac{m}{\beta} [\lambda \times r] \right)^2 + \frac{m\lambda^2}{2\beta} \left(x^2 + y^2 \right) \right\} \end{aligned} \text{ and } \langle J_z \rangle = -\frac{\partial \ln Z}{\partial \lambda} \tag{A58}$$

Whence Balian finds the velocity distribution at a point r to be proportional to:

$$\exp\left\{ -\frac{m}{2kT} \left(v + \frac{1}{\beta} [\lambda \times r] \right)^2 \right\} \text{ and } \langle J_z \rangle = -\frac{\partial \ln Z}{\partial \lambda} \tag{A59}$$

The mean velocity of the fluid at the point r is equal to:

$$\langle v \rangle = -\frac{1}{\beta} [\lambda \times r] \text{ and } \langle J_z \rangle = -\frac{\partial \ln Z}{\partial \lambda} \tag{A60}$$

and can be identified with the velocity $[\omega \times r]$ in an uniform rotation with angular velocity ω. By comparison, Balian put $\omega = -\frac{\lambda}{\beta}$. Balian made the remarks that "*The angular momentum is imparted to the gas when the molecules collide with the rotating walls, which changes the Maxwell distribution at every point, shifting its origin. The walls play the role of an angular momentum reservoir. Their motion is characterized by a* *certain angular velocity, and the angular velocities ω of the fluid and of the walls become equal at equilibrium, exactly like the equalization of the temperature through energy exchanges"*.

Considering the invariance principle, Balian observed that the Lagrangian can be taken as remaining under any change of reference frame, because the stationary action principle is independent of the frame. Comparing the Hamiltonian in two frames for a single particle with position $r\prime$ and the velocity $v\prime$ in the rotating frame:

$$L_1 = \frac{1}{2}mv^2 = \frac{1}{2}m(v\prime + [\omega \times r\prime])^2 \tag{A61}$$

Balian then considered the conjugate momentum of $r\prime$:

$$p\prime = \frac{\partial L_1}{\partial v\prime} = m(v\prime + [\omega \times r\prime]) \tag{A62}$$

and the Hamiltonian in the rotating frame:

$$H_1\prime = (p\prime.v\prime) - L_1 = \frac{p\prime^2}{2m} - (\omega.[r\prime \times p\prime]) \tag{A63}$$

The Gibbs density in the rotating frame is then given by:

$$D = \frac{1}{Z}e^{-\beta H\prime} \tag{A64}$$

where H' is the sum over N particles:

$$H\prime = \sum_{i=1}^{N}\left(\frac{p_i\prime^2}{2m_i} - (\omega.[r_i\prime \times p_i\prime])\right) \tag{A65}$$

At this step, Balian observed that to switch back to the original coordinates, $p\prime$ and $[r\prime \times p\prime]$ can be derived from p and $[r \times p]$, respectively, by means of the same change of coordinates that leads from r to $r\prime$. Balian then got:

$$H\prime = H - (\omega.J) \tag{A66}$$

and identified density D with the earlier expression, provided $\lambda = -\beta\omega$.

Balian observed that as in the case of equilibrium of a gas in a gravitational field, the result could have obtained by a macroscopic calculation from Thermodynamics and Fluid Mechanics, using locally the perfect gas law and the balance between the forces, here centrifugal forces and pressure gradients. Balian recalled that we should fix the value of these Lagrangian multipliers by requiring that on the average the angular and linear momenta vanish. For symmetry reasons these quantities vanish at the same time as the corresponding multipliers, and we have:

$$\langle J_z \rangle = -\frac{\partial \ln Z}{\partial \lambda} = Nm\omega R^2 \left[\frac{1}{1-\exp\left(-\frac{m\omega^2 R^2}{2kT}\right)} - \frac{2kT}{m\omega^2 R^2}\right] \underset{\omega \to 0}{\sim} \frac{1}{2}\omega N m R^2 \tag{A67}$$

and the energy:

$$U = -\frac{\partial \ln Z}{\partial \beta} = \frac{3}{2}NkT + \frac{1}{2}\omega\langle J_z \rangle \tag{A68}$$

Balian observed that in the change of frame, the linear momentum $mv\prime$ is no longer equal to the momentum $p\prime$ because the velocity $v = p/m$ in the fixed frame is transformed in $v\prime = p\prime/m - [\omega \times r\prime]$ in the rotating frame. Balian made the analogy with a particle of charge q in a magnetic field characterized by a velocity $(p - qA)/m$.

Balian wrote "*Whereas positions and velocities are physical quantities, momenta have a certain amount of arbitrariness which is connected with the fact that we can change the Lagrangian by adding to it a time derivative without changing the equations of motion.*" Balian gave the example in a Gallilean transformation with velocity u with the procedure where the Lagragian is assumed to be invariant $p_i\prime = p_i$ whereas $v_i\prime = v_i - u$, the Hamiltonian becomes $H\prime = H - \langle u, P\rangle$, where P is the total momentum. Balian observed that another procedure, that better exhibits the Gallielan invariance consists in adding to the Lagrangian the ineffective term:

$$-\sum_i m_i\left((v_i\prime.u) + \frac{1}{2}u^2\right) = \frac{d}{dt}\left(\sum_i m_i\left(\frac{1}{2}u^2 t - (r.u)\right)\right) \tag{A69}$$

When we change coordinates (r_i, v_i) to $(r_i\prime, v_i\prime)$, the momentum which is conjugate to $r_i\prime$ is $p_i\prime\prime = p_i - m_i u = m_i v_i'$ and not $p_i\prime = p_i$ and the Hamiltonian $H\prime\prime = H - (u.P) + \frac{1}{2}Mu^2$ has in terms of the p_i'' exactly the same form as H in terms of the p_i.

Balian presented these argues to be regarded as a microscopic justification of such a calculation and wrote "*As in the case of equilibrium of a gas in a gravitational field, we could have obtained the result by a macroscopic calculation from thermodynamics and fluid mechanics, using locally the perfect gas laws and the balance between the forces, here centrifugal forces and pressure gradients*".

Balian observed that usually no conditions are unquired about the Lagrangian multipliers for dynamical constants of motion sur as the angular or the linear momentum. Balian proposes to fix the values of these multipliers by requiring that on the average the angular and linear momenta vanish. Balian observed that for symmetry reasons, these quantities vanish at the same time as the corresponding multipliers, and we have:

$$\begin{aligned}\langle J_z\rangle &= -\frac{\partial \ln Z}{\partial \lambda} = Nm\omega R^2\left[\frac{1}{1 - e^{-m\omega^2 R^2/2KT}} - \frac{2kT}{m\omega^2 R^2}\right] \\ &\underset{\omega\to 0}{\sim} \tfrac{1}{2}\omega NmR^2\end{aligned} \tag{A70}$$

The angular momentum $\langle J_z\rangle$ is to lowest order in ω the same as for the rotation of a cylinder with uniform density, which has a moment of inertia equal to $\frac{1}{2}NmR^2$. The energy contains a contribution due to the motion, and is given by:

$$\langle J_z\rangle = -\frac{\partial \ln Z}{\partial \beta} = \frac{3}{2}NkT + \frac{1}{2}\omega\langle J_z\rangle \tag{A71}$$

The entropy also depends on the rotational velocity, but only to order ω^4. It decreases with ω, as the rotation produces changes in density which increase the spatial order.

Appendix E. Proof of Convergence for Poly-Symplectic Model Based on Souriau Proof

Jean-Marie Souriau has given the following definition:

Definition A5. [Souriau Generalized Temperature Definition] *Let G a Lie group acting on a symplectic Manifold (M, ω) by an Hamiltonian action $\Gamma : G \times M \to M$, $\mathfrak{g}$ is Lie algebra and $J : M \to \mathfrak{g}^*$ a moment map of the action, a* ***generalized temperature*** *is an element $\beta \in \mathfrak{g}$ such that the integral:*

$$\int_M e^{-\langle \beta, J\rangle} d\lambda_\omega \tag{A72}$$

is normally convergent.

Normal convergence means that there exist an open neighborhood $V$from β to $\mathfrak{g}$, and a function $f : M \to \Re^{+}$integrable on Mrelative to Liouville measure λ_ω, such that:

$$\int\limits_M e^{-\langle \beta, J \rangle} d\lambda_\omega \tag{A73}$$

Lebesgue theorem on dominated convergence gives the proof.

Jean-Marie Souriau then introduced the following proposition:

Proposition A3. [Souriau Differentiability Proposition] *Consider Ω, a non-empty set of generalized temperatures, Ω is a convex open set of Lie algebra $\mathfrak{g}$ that doesn't depend on the choice of the choice of the moment map J associated with the Hamiltonian action. The partition function $I : \Omega \leftrightarrow \Re$ given by $I_0(\beta) = \int\limits_M e^{-\langle \beta, J \rangle} d\lambda_\omega$ is infinitely differentiable on Ω. Its nth differentiation is given by the tensorial integral:*

$$I_n(\beta) = \int\limits_M J^{\otimes n} e^{-\langle \beta, J \rangle} d\lambda_\omega \tag{A74}$$

and is normally convergent.

Let

- $\beta_0, \beta_1 \in \Omega$
- V_0, V_1 neighborhoods respectively of β_0, β_1
- f_0, f_1 positive integrable function on Msuch that:

$$\begin{cases} e^{-\langle \beta_0{\prime}, J \rangle} \leq f_0, \text{ if } \beta_0{\prime} \in V_0 \\ e^{-\langle \beta_1{\prime}, J \rangle} \leq f_1, \text{ if } \beta_1{\prime} \in V_1 \end{cases} \tag{A75}$$

$\forall \lambda \in [0,1], V_\lambda = \{(1-\lambda)\beta_0{\prime} + \lambda\beta_1{\prime} / \beta_0{\prime} \in V_0, \beta_1{\prime} \in V_1\}$ is a neighborhood of β_λ given by $\beta_\lambda = (1-\lambda)\beta_0 + \lambda\beta_1$, and the function $f_\lambda = (1-\lambda)f_0 + \lambda f_1$ is integrable on M and $e^{-\langle \beta_\lambda{\prime}, J \rangle} \leq f_\lambda, \forall \beta_\lambda{\prime} \in V_\lambda$. Then $\beta_\lambda \in \Omega$ proving that Ω is convex.

n-th differential of $e^{-\langle \beta, J \rangle}$ is given:

$$D^n\left(e^{-\langle \beta, J \rangle}\right) = (-1)^n J^{\otimes n} e^{-\langle \beta, J \rangle} \tag{A76}$$

Selecting a norm on Lie algebra $\mathfrak{g}$, and considering Sup Norm on space $L(\mathfrak{g}, \Re)$of n-multilinear forms on $\mathfrak{g}$. We can deduce on $\mathfrak{g}^*$ and on $[\mathfrak{g}^*]^{\otimes n}$ a norm of multi-linear map:

$$\|J^{\otimes n}\| = \underset{\beta}{Sup} |\langle \beta, J \rangle| \tag{A77}$$

Let:

$$\beta \in \Omega, \varepsilon > 0 \text{ and } e^{-\langle \beta, J \rangle} \leq f, \text{ if } \beta{\prime} \in \mathfrak{g} \text{ and } \|\beta{\prime} - \beta\| \leq \varepsilon \tag{A78}$$

Let $\beta{\prime\prime} \in \mathfrak{g}$ and $\|\beta{\prime\prime} - \beta\| \leq \frac{\varepsilon}{2}$, for all $X \in \mathfrak{g}$ and $\|X\| = 1$, then:

$$\|\langle X, J \rangle\| \leq \frac{2n}{\varepsilon} e^{\frac{\varepsilon}{2n}\|\langle X, J \rangle\|} \Rightarrow \|\langle X, J \rangle\|^n e^{-\langle \beta{\prime\prime}, J \rangle} \leq \left(\frac{2n}{\varepsilon}\right)^n e^{-\langle \beta{\prime\prime} \pm \frac{\varepsilon}{2} X, J \rangle} \tag{A79}$$

The last relation is established by considering:

$$\forall \alpha \in \Re, \forall n \in \aleph, \left|\frac{2\alpha}{n}\right|^n \leq \left|2sh\left(\frac{\alpha}{n}\right)\right|^n = \left|e^{\frac{\alpha}{n}} - e^{-\frac{\alpha}{n}}\right|^n = \left|\sum_{p=0}^{n} (-1)^p C_n^p e^{-[-1+2\frac{p}{n}]\alpha}\right| \tag{A80}$$

If we select $X \in \mathfrak{g}$ and $\alpha = \langle X, J \rangle$:

$$\left|\frac{2}{n}\right|^n e^{-\langle \beta, J\rangle} |\langle X, J\rangle|^n \leq \left|\sum_{p=0}^{n} (-1)^p C_n^p e^{-\langle \beta - [2\frac{p}{n} - 1]X\rangle}\right| \tag{A81}$$

$$e^{-\langle \beta, J\rangle} \leq f \Rightarrow e^{-\langle \beta, J\rangle} |\langle X, J\rangle|^n \leq n^n f \text{ , if } \|\beta - \beta_0\| \leq \frac{\varepsilon}{2}, \|V\| \leq \frac{\varepsilon}{2} \tag{A82}$$

For X unitary, and by setting $X = J\frac{\varepsilon}{2}$

$$\|\langle X, J\rangle\|^n e^{-\langle \beta, J\rangle} \leq \left(\frac{2n}{\varepsilon}\right)^n f \tag{A83}$$

In $\|\langle X, J\rangle\|^n e^{-\langle \beta\prime\prime, J\rangle} \leq \left(\frac{2n}{\varepsilon}\right)^n e^{-\langle \beta\prime\prime \pm \frac{\varepsilon}{2} X, J\rangle}$, the sign $\pm$ is selected such that $\langle \pm \varepsilon X, J\rangle \geq 0$.
As $\|\beta - \beta\prime\prime \pm \frac{\varepsilon}{2} X\| \leq \varepsilon$, the final result is deduced:

$$\|D^n\left(e^{-\langle \beta\prime\prime, J\rangle}\right)\| \leq \left[\frac{2n}{\varepsilon}\right]^n f \Rightarrow \|J^{\otimes n} e^{-\langle \beta\prime\prime, J\rangle}\| \leq \left[\frac{2n}{\varepsilon}\right]^n f \tag{A84}$$

It proves that the n-differential of $e^{-\langle \beta, J\rangle}$ is normally integrable on M with respect to Liouville measure, the partition function is infinitely differentiable on Ω.

By considering the taylor expansion of exponential function:

$$e^\alpha - 1 - \alpha = \frac{\alpha^2}{2} e^{\lambda \alpha}, \lambda \in [0, 1] \tag{A85}$$

From which, we deduce that:

$$e^{-\langle \beta - X, J\rangle} J^{\otimes n} - e^{-\langle \beta, J\rangle} J^{\otimes n} - e^{-\langle \beta, J\rangle} J^{\otimes n+1}(X) = \frac{1}{2} e^{\langle \beta - \lambda X, J\rangle} J^{\otimes n+2}(X)(X) \tag{A86}$$

where $T(X)$ means the contraction of a covariant tensor with vector X. Then:

$$\|J^{\otimes n+2} e^{-\langle \beta, J\rangle}\| \leq \left[\frac{2(n+2)}{\varepsilon}\right]^{n+2} f \Rightarrow \frac{1}{2} e^{\langle \beta - \lambda X, J\rangle} J^{\otimes n+2}(X)(X) \leq \frac{1}{2}\left[\frac{2(n+2)}{\varepsilon}\right]^{n+2} f \|X\|^2 \tag{A87}$$

By integration on V and using $\int_V f.Vol = a < +\infty$, we obtain:

$$\|I_n(\beta - X) - I_n(\beta) - I_{n+1}(\beta)\| \leq \frac{a}{2}\left[\frac{2(n+2)}{\varepsilon}\right]^{n+2} \|X\|^2 \text{ if } \beta \in> B\left(\beta_0, \frac{\varepsilon}{4}\right) \text{ and } \|X\| \leq \frac{\varepsilon}{4} \tag{A88}$$

It proves that the function $I_n : \beta \in \mathfrak{g} \rightarrow \Re$ is continuous and derivable in a neighborhood of β_0, and its derivative is given I_{n+1}. Then I_0 is an infinite derivable function with I_n as nth derivable.

These demonstrations can be extended for poly-symplectic model of Souriau Lie groups Thermodynamic by considering the polysymplectic partition function:

$$I_0^{poly} = \int_M e^{-\sum_{k=1}^{n} \langle \beta_k, J^{\otimes k}(\xi)\rangle} d\lambda_\omega \tag{A89}$$

and its n-th derivatices given by:

$$I_{n,i}^{poly} = \frac{\partial^n I_0}{\partial \beta_i^n} = \int_M J^{\otimes nk} e^{-\sum_{k=1}^{n} \langle \beta_k, J^{\otimes k} \rangle} d\lambda_\omega \tag{A90}$$

where $J^{\otimes \mathrm{k}} = J \otimes \overset{(k)}{J} \ldots \otimes J$ is defined as a tensorial product.

Appendix F. Relativistic Souriau Thermodynamics of Continua

We will summarize in this appendix the Souriau relativistic thermodynamics of fluids. This Souriau model about relativistic thermodynamics of continua will give a solution to Duhem's general thermodynanics: *Nous avons fait de la dynamique un cas particulier de la thermodynamique, une Science qui embrasse dans des principes communs tous les changements d'état des corps, aussi bien les changements de lieu que les changements de qualités physiques* ". (In English: *We made dynamics a special case of thermodynamics, and science that embraces common principles in all changes of state bodies, changes of places as well as changes in physical qualities"*.)

The objective is not to make a survey of all literature on this topic. We give this model to compare Souriau's approaches related to invariance and symmetries in thermodynamics. I think that this is the first time that Souriau relativistic model is presented in English. My objective is to underline that Souriau has replaced the geometric temperature of "Lie groups thermodynamics", where the temperature is an element of Lie algebra, by a temperature that is defined as a killing vector. I also underline, that in both models, Souriau was motivated to search solutions where the "Legendre transform" structure is preserved between Massieu thermodynamics potentials.

Kinematics is defined by the vector field Θ and the measurement of number of molecules: using two state functions, Souriau has built a (thermo-)dynamic according to the two principles: conservation of the Noetherian quantities attached to the Poincaré group, positive Entropy production. Such a dissipative fluid has movements in which the entropy production is nil; Θ is then a killing vector; the equations of motion fully integrate; Souriau found in particular the results of kinetic theory at equilibrium. This method can be used to study perfect fluids; Souriau recover the classic Lichnerowicz results; moreover, we can build, even in the non-isentropic case, an space-time 2-Form Ω which is Integral invariant (in the sense of Cartan-Poincaré) of the temperature vector Θ; this provides a generalization of Helmholtz's theorem. In weakly dissipative movements, naturally occur the two viscosity coefficients, as well as the thermal conductivity coefficient; they are accompanied by two other coefficients that may be measurable on actual fluids.

Jean-Marie Sourias has first considered the kinematics of a relativistic simple fluid, considering the following space-time vectors field by temperature vector $X \mapsto \Theta$ with:

$$\Theta = U\varepsilon \begin{cases} U : \text{Unitary quadri-vector} \\ \varepsilon = \frac{1}{T} > 0 \text{ (Boltzman } k = 1) \end{cases} \tag{A91}$$

Θ generates a group with a parameter of diffeomorphisms of space-time E_4; the group's orbits (the current lines of the fluid) form an abstract space V_3 (has a manifold structure of dimension 3, characterized by the fact that the following projection is a restricted submersion:

$$X \in E_4 \mapsto x \in V_3 \tag{A92}$$

Let the metric tensor g Lie derivative (for the vector field $X \in E_4 \mapsto \Theta$):

$$\begin{cases} \gamma = \frac{1}{2}\delta_L g \\ \delta X = \Theta \end{cases} \tag{A93}$$

The Killing formula gives the symmetric tensor:

$$\gamma_{\lambda\mu} = \frac{1}{2}\left[\partial_\lambda \Theta_\mu + \partial_\mu \Theta_\lambda\right] \tag{A94}$$

Let consider positive density n of quotient manifold V_3:

$$x \in V_3 \mapsto n \tag{A95}$$

Integral of n on V_3 gives the number of molecules. Its reciprocal image by projection is defined by:

$$X \in E_4 \mapsto N \tag{A96}$$

Particules conservation is given by:

$$\partial_\lambda N^\lambda = 0 \text{ with } N = Un \tag{A97}$$

Direction of U or Θ defines a foliation of space-time E_4. Leaves are current lines solutions of:

$$\frac{dX}{ds_c} = U \tag{A98}$$

We illustrate in Figure A3, the Souriau's midel of Thermodynamics of continua.

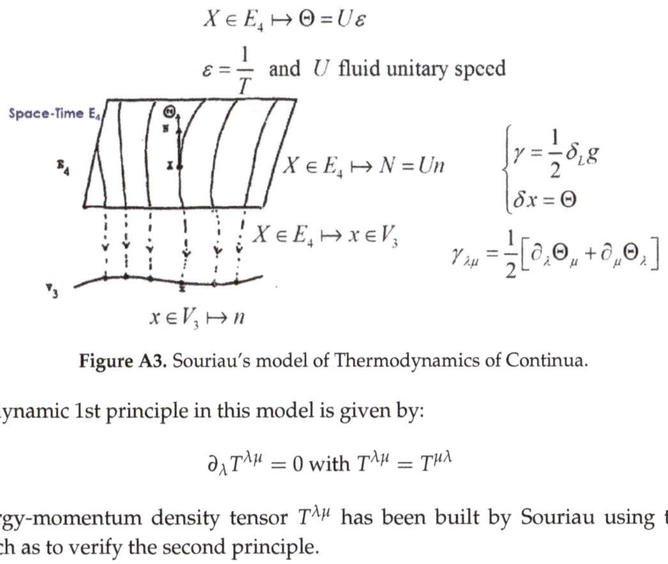

Figure A3. Souriau's model of Thermodynamics of Continua.

Thermodynamic 1st principle in this model is given by:

$$\partial_\lambda T^{\lambda\mu} = 0 \text{ with } T^{\lambda\mu} = T^{\mu\lambda} \tag{A99}$$

The energy-momentum density tensor $T^{\lambda\mu}$ has been built by Souriau using the kinematic quantities, such as to verify the second principle.

Lemma A1 [Souriau Lemma]. *Let* $(n, \varepsilon) \mapsto \zeta$ *a differentiable function, then there is a symmetric tensor* $\overset{\frown}{T}^{\lambda\mu}$ such that:

$$\partial_\lambda \left[N^\lambda \zeta\right] = -\overset{\frown}{T}^{\lambda\mu} \gamma_{\lambda\mu} \text{ with } \Theta = U\varepsilon \text{ et } N = Un \tag{A100}$$

$$\overset{\frown}{T}^{\lambda\mu} = \frac{n^2}{\varepsilon}\frac{\partial \zeta}{\partial n}\left[g^{\lambda\mu} - U^\lambda U^\mu\right] - n\frac{\partial \zeta}{\partial \varepsilon} U^\lambda U^\mu \tag{A101}$$

We assume that there exist $\varphi = \varphi(n, \Theta, \gamma)$ such this function is convex and energy-momentum density are given by:

$$T^{\lambda\mu} = \frac{\partial \varphi}{\partial \gamma_{\lambda\mu}} \tag{A102}$$

If we assume that $\{\gamma_{\lambda\mu} = 0\} \Rightarrow \left\{T^{\lambda\mu} = \overset{\frown}{T}^{\lambda\mu}\right\}$ then the following vector has a positive divergence:

$$S^{\lambda} = N^{\lambda}\zeta + T^{\lambda\mu}\Theta_{\mu} \tag{A103}$$

The Thermodynamic 2nd principle is given by:

$$\partial_{\lambda} S^{\lambda} \geq 0 \tag{A104}$$

Proof is given by:

$$\partial_{\lambda} S^{\lambda} = \left[T^{\lambda\mu} - \overset{\frown}{T}^{\lambda\mu}\right]\gamma_{\lambda\mu}\partial_{\lambda} S^{\lambda} = \left\{\varphi(\gamma) - \varphi(0) - \overset{\frown}{T}^{\lambda\mu}\gamma_{\lambda\mu}\right\} + \{\varphi(0) - \varphi(\gamma) - T^{\lambda\mu}(-\gamma_{\lambda\mu})\} \geq 0 \tag{A105}$$

$\partial_{\lambda} S^{\lambda} \geq 0$ Souriau proposed to define the dynamics of the fluid by means of the two functions ζ and φ which give at each point the energy tensor $T^{\lambda\mu}$ and the entropy flux S^{λ} by following formulas. These functions being determined, we have 5 equations to determine the 5 variables $\left(n, \Theta^{\lambda}\right)$ and, moreover, the S^{λ}; $\partial_{\lambda} S^{\lambda} \geq 0$ will express the 2nd principle.

$$\begin{cases} T^{\lambda\mu} = \dfrac{\partial \varphi(n, \Theta, \gamma)}{\partial \gamma_{\lambda\mu}} \\ S^{\lambda} = N^{\lambda}\zeta(n, \varepsilon) + T^{\lambda\mu}\Theta_{\mu} \text{ with } \Theta = U\varepsilon \text{ and } N = Un \\ \gamma_{\lambda\mu} = \frac{1}{2}\left[\partial_{\lambda}\Theta_{\mu} + \partial_{\mu}\Theta_{\lambda}\right] \\ \partial_{\lambda} T^{\lambda\mu} = 0 \text{ and } \partial_{\lambda} N^{\lambda} = 0 \end{cases} \tag{A106}$$

Souriau has then considered the case of non-dissipative movements. If φ is strictly convex for variable γ then:

$$\partial_{\lambda} S^{\lambda} = 0 \Leftrightarrow \gamma_{\lambda\mu} = 0 \Leftrightarrow \Theta \text{ infinitesimal isometry} \tag{A017}$$

For non-dissipative solution of movement equations, Θ is a Killing vector, associated to an element of Lie algebra of Poincaré group:

$$\Theta = \begin{bmatrix} \Lambda & \Gamma \\ 0 & 0 \end{bmatrix} \tag{A108}$$

with:

$$\left.\begin{matrix} \Theta_{\lambda} = \Lambda_{\lambda\mu}X^{\mu} + \Gamma_{\lambda} \quad (\Lambda_{\lambda\mu} + \Lambda_{\mu\lambda} = 0) \\ \Theta = U\varepsilon \end{matrix}\right\} \Rightarrow U^{\lambda}, \varepsilon \tag{A109}$$

The equations of motion integrate through an arbitrary constant:

$$\zeta + \frac{\partial \zeta}{\partial n} n = Cste \Rightarrow n \tag{A110}$$

Thermodynamics constants are the following:

- specific molecular volume:

$$u = \frac{1}{n} \tag{A111}$$

- specific mass:

$$\rho = -n\frac{\partial \zeta}{\partial \varepsilon} = -\frac{1}{u}\frac{\partial \zeta}{\partial \varepsilon} \tag{A112}$$

- pressure:

$$p = -\frac{n^2}{\varepsilon}\frac{\partial \zeta}{\partial n} = -\frac{1}{\varepsilon}\frac{\partial \zeta}{\partial u} \tag{A113}$$

In case of a nill entropy production:

$$\partial_\lambda S^\lambda = 0 \Rightarrow \left\{ \begin{array}{l} \gamma = 0 \\ \Theta = \Lambda.X + \Gamma \end{array} \right. \Rightarrow \left\{ \begin{array}{l} U^\lambda \partial_\lambda \varepsilon = 0 \Rightarrow \exists \varepsilon, x \in V_3 \mapsto \varepsilon \\ \partial_\lambda U^\lambda = 0 \Rightarrow \left[\partial_\lambda N^\lambda = 0 \Rightarrow U^\lambda \partial_\lambda n = 0 \right] \Rightarrow \exists n, x \in V_3 \mapsto n \\ \varepsilon U^\lambda \partial_\lambda U_\mu + \partial_\mu \varepsilon = 0 \end{array} \right. \tag{A114}$$

$\Rightarrow$ variable n and ε are constant on current lines

We can also deduce the following equations:

$$\left\{ \begin{array}{l} \Theta = \Lambda.X + \Gamma \\ \partial_\lambda N^\lambda = 0 \end{array} \right. \Rightarrow U^\lambda \partial_\lambda \left[\frac{n^2}{\varepsilon}\frac{\partial \zeta}{\partial n} \right] = 0 \text{ and } U^\lambda \partial_\lambda \left[n\frac{\partial \zeta}{\partial n} \right] = 0 \tag{A115}$$

From tensor computation, Souriau has computed the energy-momentum density currents:

$$\begin{array}{l} \partial_\lambda N^\lambda = 0 \Rightarrow \partial_\lambda [N^\lambda \zeta] = N^\lambda \partial_\lambda \zeta = U^\lambda n \left[\frac{\partial \zeta}{\partial n}\partial_\lambda n + \frac{\partial \zeta}{\partial \varepsilon}\partial_\lambda \varepsilon \right] \\ \gamma_{\lambda\mu} = \frac{1}{2}[\partial_\lambda \Theta_\mu + \partial_\mu \Theta_\lambda] = \frac{\varepsilon}{2}[\partial_\lambda U_\mu + \partial_\mu U_\lambda] + \frac{1}{2}[U_\lambda \partial_\mu \varepsilon + U_\mu \partial_\lambda \varepsilon] \\ \Rightarrow g^{\lambda\mu}\gamma_{\lambda\mu} = \varepsilon \partial_\lambda U^\lambda + U^\lambda \partial_\lambda \varepsilon \end{array} \tag{A116}$$

with the following developments:

$$\begin{array}{l} U \text{ unitary} \Rightarrow U_\lambda U^\lambda = g_{\lambda\mu} U^\lambda U^\mu = 1 \Rightarrow U^\lambda \partial_\mu U_\lambda = 0 \Rightarrow U^\lambda U^\mu \gamma_{\lambda\mu} = U^\lambda \partial_\lambda \varepsilon \\ \partial_\lambda N^\lambda = 0 \Rightarrow U^\lambda \partial_\lambda n + \partial_\lambda U^\lambda n = 0 \\ \Rightarrow T^{\lambda\mu} = \frac{n^2}{\varepsilon}\frac{\partial \zeta}{\partial n}[g^{\lambda\mu} - U^\lambda U^\mu] - n\frac{\partial \zeta}{\partial \varepsilon} U^\lambda U^\mu \end{array} \tag{A117}$$

For this non-dissipative movement, we can prove:

$$\left\{ \begin{array}{l} U^\lambda \partial_\lambda \left[\frac{n^2}{\varepsilon}\frac{\partial \zeta}{\partial n} \right] = 0 \\ U^\lambda \partial_\lambda \left[n\frac{\partial \zeta}{\partial n} \right] = 0 \end{array} \right. \text{ and } \left\{ \begin{array}{l} U^\lambda \partial_\lambda \varepsilon = 0 \\ \partial_\lambda U^\lambda = 0 \\ \varepsilon U^\lambda \partial_\lambda U_\mu + \partial_\mu \varepsilon = 0 \end{array} \right. \tag{A118}$$

$$\begin{array}{l} T^{\lambda\mu} = \frac{n^2}{\varepsilon}\frac{\partial \zeta}{\partial n}[g^{\lambda\mu} - U^\lambda U^\mu] - n\frac{\partial \zeta}{\partial \varepsilon} U^\lambda U^\mu \\ \Rightarrow \partial_\lambda T^{\lambda\mu} = g^{\lambda\mu} \left\{ \partial_\lambda \left[\frac{n^2}{\varepsilon}\frac{\partial \zeta}{\partial n} \right] + \frac{\partial_\lambda \varepsilon}{\varepsilon} \left[\frac{n^2}{\varepsilon}\frac{\partial \zeta}{\partial n} + n\frac{\partial \zeta}{\partial \varepsilon} \right] \right\} = \frac{n}{\varepsilon} g^{\lambda\mu} \partial_\lambda \left\{ n\frac{\partial \zeta}{\partial n} + \zeta \right\} \end{array} \tag{A119}$$

$$\Rightarrow \left\{ \begin{array}{l} \partial_\lambda T^{\lambda\mu} = 0 \\ \partial_\lambda N^\lambda = 0 \end{array} \right. \text{integrable on} \left\{ \begin{array}{l} n \text{ constant on current lines} \\ n\frac{\partial \zeta}{\partial n} + \zeta \text{ constant in space-time} \end{array} \right. \tag{A120}$$

Souriau has proved that the entropy vector preserves the Legendre transform:

$$\begin{cases} S^{\lambda} = N^{\lambda}\zeta + T^{\lambda\mu}\Theta_{\mu} \\ T^{\lambda\mu} = \frac{n^2}{\varepsilon}\frac{\partial\zeta}{\partial n}\left[g^{\lambda\mu} - U^{\lambda}U^{\mu}\right] - n\frac{\partial\zeta}{\partial\varepsilon}U^{\lambda}U^{\mu} \\ \Theta = U\varepsilon \text{ and } N = Un \end{cases} \Rightarrow S^{\lambda} = N^{\lambda}\left[\zeta - \varepsilon\frac{\partial\zeta}{\partial\varepsilon}\right] \tag{A121}$$

$$S^{\lambda} = N^{\lambda}s \Rightarrow s = \zeta - \varepsilon\frac{\partial\zeta}{\partial\varepsilon}$$

with the entropy per molecule:

$$s = \zeta + \rho u\varepsilon \tag{A122}$$

ζ is the Massieu potential (Massieu charcateristic function):

$$\begin{array}{l} \zeta = -\frac{F}{T} = -\frac{u\rho - Ts}{T} \text{ with } F: \text{ Helmoltz Free Energy} \\ \zeta + \frac{\partial\zeta}{\partial n}n = -\frac{G}{T} = -\frac{F + pu}{T} \text{ with } G: \text{ Free Gibbs-Duhem Energy} \end{array} \tag{A123}$$

The link with Souriau 2-form and Poincaré-Cartan integral invariant is given by the following developments. Consider the 1-form given by enthalpy:

$$H_{\lambda} = hU_{\lambda} \text{ with } h = \frac{p + \rho}{n} = u[p + \rho] \tag{A124}$$

Its 2-form given by exterior differentiation

$$\Omega_{\lambda\mu} = \partial_{\lambda}H_{\mu} - \partial_{\mu}H_{\lambda} \tag{A125}$$

Movement's equation are replaced by:

$$\begin{cases} \partial_{\lambda}N^{\lambda} = 0 \\ \partial_{\lambda}T^{\lambda\mu} = 0 \end{cases} \Rightarrow \begin{cases} \partial_{\lambda}N^{\lambda} = 0 \\ \Omega_{\lambda\mu}\Theta^{\mu} + \partial_{\lambda}s = 0 \end{cases} \tag{A126}$$

Ω is a Poincaré-Cartan integral invariant of the field:

$$\begin{array}{l} \Omega_{\lambda\mu}\Theta^{\mu} + \partial_{\lambda}s = 0 \Rightarrow \begin{cases} \delta s = 0 \\ \delta_L\Omega = 0 \end{cases} \text{ for } \delta X = \Theta \\ \text{if } \partial_{\lambda}s = 0 \text{ (isentropic movment) } \Rightarrow \Theta \in \ker(\Omega) \end{array} \tag{A127}$$

Souriau has then considered weakly dissipative movements. If we cannot know $\varphi = \varphi(n, \Theta, \gamma)$, it can be approximated by 2nd order development in γ variable:

$$\varphi = \varphi_0 + \widehat{T}^{\lambda\mu}\gamma_{\lambda\mu} + \frac{1}{2}C^{\lambda\mu,vq}\gamma_{\lambda\mu}\gamma_{vq} \Rightarrow T^{\lambda\mu} = T^{\lambda\mu} = \frac{\partial\varphi}{\partial\gamma_{\lambda\mu}} = \widehat{T}^{\lambda\mu} + C^{\lambda\mu,vq}\gamma_{vq} \tag{A128}$$

Entropy production is given by:

$$\begin{array}{c} \partial_{\lambda}S^{\lambda} = \left[T^{\lambda\mu} - \widehat{T}^{\lambda\mu}\right]\gamma_{\lambda\mu} = C^{\lambda\mu,vq}\gamma_{\lambda\mu}\gamma_{vq} \\ \text{Onsager Reciprocity} \Rightarrow C^{\lambda\mu,vq} = C^{vq,\lambda\mu} \end{array} \tag{A129}$$

55 coefficients of Transport coefficients $C^{\lambda\mu,vq}$ are reduced to 5 coefficients (by fluid symetries and Onsager reciprocity): *A*, *B*, *C*, *E & F*.

Souriau then obtained relativistic (Fourier) equation of heat. Let us consider the constraints tensor:

$$\tau_{jk} = -T_{jk} = \delta_{jk}\left[-p + \lambda_{vis}\partial_l v^l - B\frac{\partial \varepsilon}{\partial t}\right] + \mu_{vis}\left[\partial_j v_k + \partial_k v_j\right] \\ (j,k = 1,2,3 \text{ and } v_j \text{ speed, zero at the point considered}) \tag{A130}$$

With the equations given by:

- Heat Flux:

$$T^{j0} = \left\{ F\left[\overrightarrow{grad\varepsilon} - \varepsilon\frac{\partial \vec{v}}{\partial t}\right]\right\}^j \tag{A131}$$

- Specific Mass-Energy:

$$T^{00} = \rho + C\frac{\partial \varepsilon}{\partial t} - B\varepsilon div\left(\vec{v}\right) \tag{A132}$$

with:

$$\lambda_{vis} = \left[A - \frac{2E}{3}\right]\varepsilon,\ \mu_{vis} = E\varepsilon,\ \varepsilon = \frac{1}{T} \text{ and Thermo} - \text{conductivity} : \frac{F}{T^2}$$

Variables A, B, C, E & F are functions of ε and n, and convexity of φ induces:

$$A > 0, C > 0, E > 0, F > 0, |B| < \sqrt{AC} \tag{A133}$$

References

1. Chatelain, J.M. *Pascal, le Coeur et la Raison*; Bibliothèque nationale de France: Paris, France, 2016.
2. Duhem, P. Sur les équations générales de la thermodynamique. In *Annales Scientifiques de l'École Normale Supérieure*; Ecole Normale Supérieure: Paris, France, 1891; Volume 8, pp. 231–266. (In French)
3. Duhem, P. Commentaire aux principes de la Thermodynamique—Troisième partie. *J. Math. Appl.* **1894**, *10*, 207–286. (In French)
4. Needham, P. *Commentary on the Principles of Thermodynamics by Pierre Duhem*; Boston Studies in the Philosophy of Science; Needham, P., Ed.; Springer: Berlin, Germany, 2011.
5. Duhem, P. Commentaire aux principes de la Thermodynamique—Première partie. *J. Math. Appl.* **1892**, *8*, 269–330. (In French)
6. Bordoni, S. From thermodynamics to philosophical tradition: Pierre Duhem's research between 1891 and 1896. *Lettera Matematica* **2017**, *5*, 261–266. [CrossRef]
7. Stoffel, J.-F. Pierre Duhem: Un savant-philosophe dans le sillage de Blaise Pascal. *Revista Portuguese de Filosofia* **2007**, *63*, 275–307. [CrossRef]
8. Le Ferrand, H.; Mazliak, L. *Pierre Duhem (1861–1916) et ses Contemporains Institut Henri Poincaré, 14 Septembre 2016*; Organisée par Hervé Le Ferrand (Dijon)—Laurent Mazliak: Paris, France, 2016.
9. Barbaresco, F. Jean-Louis Koszul and the elementary structures of Information Geometry. In *Geometric Structures of Information Geometry*; Nielsen, F., Ed.; Springer: Berlin, Germany, 2018.
10. Barbaresco, F. Koszul Lecture Contemporaneity: Elementary Structures of Information Geometry and Geometric Heat Theory. In *Introduction to Symplectic Geometry*; Koszul, J.L., Ed.; Springer: Berlin, Germany, 2018.
11. Barbaresco, F. Jean-Louis Koszul et les Structures Elémentaires de la Géométrie de l'Information. Revue SMAI Matapli; SMAI, Ed.; 2018; Volume 116. Available online: https://www.see.asso.fr/pdf_viewer/22381/m (accessed on 2 November 2018).
12. Duhem, P. Recherches sur l'élasticité. *Ann. Ecole Norm.* **1905**, *22*, 143–217. [CrossRef]
13. Souriau, J.M. *Thermodynamique Relativiste des Fluides*; Rendiconti del Seminario Matematico; Università Politecnico di Torino: Torino, Italy, 1978; Volume 35, pp. 21–34.
14. Souriau, J.M. Milieux continus de dimension 1, 2 ou 3: Statique et dynamique. In Proceedings of the 13eme Congrès Français de Mécanique, Poitiers, France, 1–5 September 1997; pp. 41–53.
15. Amari, S.I. Natural gradient works efficiently in learning. *Neural Comput.* **1998**, *10*, 251–276. [CrossRef]

16. Amari, S.I.; Nagaoka, H. *Methods of Information Geometry*; Harada, D., Ed.; Translations of Mathematical Monographs; American Mathematical Society: Providence, RI, USA, 2000; Volume 191.
17. Pascanu, R.; Bengio, Y. Natural gradient revisited. *arXiv* **2013**, arXiv:1301.3584v1.
18. Martens, J. New insights and perspectives on the natural gradient method. *arXiv* **2014**, arXiv:1412.1193.
19. Ollivier, Y. Riemannian metrics for neural networks I: Feedforward networks. *Inf. Inference* **2015**, *4*, 108–153. [CrossRef]
20. Amari, S.I. *Information Geometry and Its Applications*; Applied Mathematical Sciences; Springer: Berlin, Germany, 2016.
21. Ollivier, Y.; Arnold, L.; Auger, A.; Hansen, N. Information-geometric optimization algorithms: A unifying picture via invariance principles. *J. Mach. Learn. Res.* **2017**, *18*, 1–65.
22. Ollivier, Y.; Marceau-Caron, G. Natural Langevin dynamics for neural networks. In *Geometric Science of Information (GSI 2017)*; Nielsen, F., Barbaresco, F., Eds.; Lecture Notes in Computer Science 10589; Springer: Berlin, Germany, 2017; pp. 451–459.
23. Marle, C.-M. From Tools in Symplectic and Poisson Geometry to J.-M. Souriau's Theories of Statistical Mechanics and Thermodynamics. *Entropy* **2016**, *18*, 370. [CrossRef]
24. Balian, R.; Valentin, P. Hamiltonian structure of thermodynamics with gauge. *Eur. Phys. J. B* **2001**, *21*, 269–282. [CrossRef]
25. Balian, R. *From Microphysics to Macrophysics*, 2nd ed.; Springer: Berlin, Germany, 2007; Volume I.
26. Der Schaft, A.; Maschke, B. *Homogeneous Hamiltonian Control Systems Part I: Geometric Formulation*; Elsevier: Amsterdam, The Netherlands, 2018; Volume 51, pp. 1–6.
27. Jaworski, W. Information thermodynamics with the second order temperatures for the simplest classical systems. *Acta Phys. Pol.* **1981**, *60*, 645–659.
28. Jaworski, W. Higher-order moments and the maximum entropy inference: The thermodynamical limit approach. *J. Phys. A Math. Gen.* **1987**, *20*, 915–926. [CrossRef]
29. Ingarden, H.S.; Meller, J. Temperatures in linguistics as a model of thermodynamics. *Open Syst. Inf. Dyn.* **1994**, *2*, 211–230. [CrossRef]
30. Ingarden, R.S.; Nakagomi, T. The second order extension of the Gibbs state. *Open Syst. Inf. Dyn.* **1992**, *1*, 259–268. [CrossRef]
31. Ingarden, R.S.; Kossakowski, A.; Ohya, M. *Information Dynamics and Open Systems*; Classical and Quantum Approach, Fundamental Theories of Physics; Springer: Berlin, Germany, 1997; Volume 86.
32. Jaworski, W.; Ingarden, R.S. On the partition function in information thermodynamics with higher order temperatures. *Bull. Acad. Pol. Sci. Sér. Phys. Astron.* **1980**, *1*, 28–119.
33. Jaworski, W. On Information Thermodynamics with Temperatures of the Second Order. Master's Thesis, Institute of Physics, Nicolaus Copernicus University, Torun, Poland, 1981. (In Polish)
34. Jaworski, W. On the thermodynamic limit in information thermodynamics with higher-order temperatures. *Acta Phys. Pol.* **1983**, *A63*, 3–19.
35. Jaworski, W. Investigation of the Thermodynamic Limit for the States Maximizing Entropy under Auxiliary Conditions for Higher-Order Statistical Moments. Ph.D. Thesis, Institute of Physics, Nicolaus Copernicus University, Torun, Poland, 1983. (In Polish)
36. Ingarden, R.S.; Kossakowski, A. Statistical thermodynamics with higher order temperatures for ideal gases of bosons and fermions. *Acta Phys. Pol.* **1965**, *28*, 499–511.
37. Ingarden, R.S.; Tamassy, L. On parabolic geometry and irreversible macroscopic time. *Rep. Math. Phys.* **1993**, *32*, 11–33. [CrossRef]
38. Ingarden, R.S. Towards mesoscopic thermodynamics: Small systems in higher-order states. *Open Syst. Inf. Dyn.* **1993**, *1*, 75–102. [CrossRef]
39. Ingarden, R.S.; Janyszek, H.; Kossakowski, A.; Kawaguchi, T. Information geometry of quantum statistical systems. *Tensor Ns* **1982**, *37*, 105–111.
40. Ingarden, R.S.; Kossakowski, A. On the connection of nonequilibrium information thermodynamics with non-hamiltonian quantum mechanics of open systems. *Ann. Phys.* **1975**, *89*, 451–485. [CrossRef]
41. Casalis, M. Familles Exponentielles Naturelles Invariantes par un Groupe. Ph.D. Thesis, l'Université Paul Sabatier, Toulouse, France, 1990.
42. Casalis, M. Familles Exponentielles Naturelles sur Rd Invariantes par un Groupe. *Int. Stat. Rev.* **1991**, *59*, 241–262. [CrossRef]

43. Souriau, J.-M. *Structures des Systèmes Dynamiques*; Dunod: Paris, France, 1970.
44. Koszul, J.L. *Introduction to Symplectic Geometry*; Science Press: Beijing, China, 1986. (In Chinese), translated by SPRINGER in English, 2018.
45. Marle, C.M. *Géométrie Symplectique et Géométrie de Poisson*; Mathématiques en Devenir, Calvage & Mounet: Paris, France, 2018.
46. Kostant, B. *Quantization and Unitary Representations*; Lecture Notes in Math. 170; Springer: Berlin, Germany, 1970.
47. Koszul, J.L.; Travaux, D.B. *Kostant sur les Groupes de Lie Semi-Simples*; Séminaire Bourbaki: Paris, France, 1958–1960; pp. 329–337.
48. Gunther, C. The polysymplectic Hamiltonian formalism in field theory and calculus of variations I: The local case. *J. Differ. Geom.* **1987**, *25*, 23–53. [CrossRef]
49. Munteanu, F.; Rey, A.M.; Salgado, M. The Günther's formalism in classical field theory: Momentum map and reduction. *J. Math. Phys.* **2004**, *5*, 1730–1751. [CrossRef]
50. Awane, A. k-symplectic structures. *J. Math. Phys.* **1992**, *33*, 4046–4052. [CrossRef]
51. Awane, A.M. *Goze, Pfaffian Systems, k-Symplectic Systems*; Springer: Berlin, Germany, 2000.
52. Edelen, D.G.B. The invariance group for Hamiltonian systems of partial differential equations. *Arch. Rational Mech. Anal.* **1961**, *5*, 95–176. [CrossRef]
53. De Donder, T. *Théorie Invariante du Calcul des Variations*, Nuov. ed.; Gauthiers–Villars: Paris, France, 1935.
54. Lepage, T. Sur les champs géodésiques du calcul des variations. *Bull. Acad. R. Belg. Classes Sci.* **1936**, *22*.
55. Hélein, F. Multisymplectic formalism and the covariant phase space. In *Variational Problems in Differential Geometry*; Bielawski, R., Houston, K., Speight, M., Eds.; London Mathematical Society Lecture Note Series 394; Cambridge University Press: Cambridge, UK, 2012.
56. Facchi, P.; Kulkarni, R.; Manko, V.I.G.; Marmo, S.E.C.G.; Ventriglia, F. Classical and quantum Fisher information in the geometrical formulation of quantum mechanics. *Phys. Lett. A* **2010**, *374*, 4801. [CrossRef]
57. Contreras, E.; Schiavina, M. On the geometry of mixed states and the quantum information tensor. *J. Math. Phys.* **2016**, *57*, 062209. [CrossRef]
58. Luati, A. Maximum Fisher information in mixed state quantum systems. *Ann. Stat.* **1770**, *32*, 2004. [CrossRef]
59. Contreras, E.; Schiavina, M. Kähler fibrations in quantum information theory. *arXiv* **2018**, arXiv:1801.09793.
60. Souriau, J.-M. La structure symplectique de la mécanique décrite par Lagrange en 1811. *Math. Sci. Hum.* **1986**, *94*, 45–54.
61. Marle, C.M. The inception of Symplectic Geometry: The works of Lagrange and Poisson during the years 1808–1810. *Lett. Math. Phys.* **2009**, *90*, 3. [CrossRef]
62. Barbaresco, F.; Boyom, M. Foundations of Geometric Structure of Information. In Proceedings of the FGSI'19, IMAG lab (Institut Montpelliérain Alexander Grothendieck), Montpellier, France, 4–6 February 2019; Available online: https://fgsi2019.sciencesconf.org/ (accessed on 11 November 2018).
63. Szczeciniarz, J.-J.; Iglesias-Zemmour, P. SOURIAU 2019 Conference, SPHERE, Université Paris-Diderot, Paris, France, 27–31 May 2019. Available online: http://souriau2019.fr/ (accessed on 1 November 2018).
64. Souriau, J.-M. Mécanique statistique, groupes de Lie et cosmologie, Colloques int. du CNRS numéro 237. In Proceedings of the Géométrie Symplectique et Physique Mathématique, Aix-en-Provence, France, 24–28 June 1974; pp. 59–113.
65. Obădeanu, V. Structures géométriques associées a certains systèmes dynamiques. *Balkan J. Geom. Appl.* **2000**, *5*, 81–89.
66. Obădeanu, V. *Systèmes Dynamiques et Structures Géométriques Associées*; Universitatea din Timişoara, Facultatea de Matematică: Timişoara, Romania, 1999.
67. Obădeanu, V. *Systèmes Biodynamiques et Lois de Conservation Applications au Systèmes de Neurones*; Universitatea din Timişoara, Facultatea de Matematică: Timişoara, Romania, 1994.
68. Gallissot, F. Les formes extérieures en mécanique. *Annales de l'Institut Fourier* **1952**, *4*, 145–297. [CrossRef]
69. Gallissot, F. les formes extérieures et la mécaniques des milieux continus. *Annales de l'Institut Fourier* **1958**, *8*, 291–335. [CrossRef]
70. Souriau, J.M. C'est quantique? Donc c'est Géométrique. Feuilletages—Quantification Géométrique: Textes des Journées D'étude des 16 et 17 Octobre 2003. Available online: http://semioweb.msh-paris.fr/f2ds/docs/feuilletages/Jean-Marie_Souriau3.pdf (accessed on 1 November 2018).

71. Souriau, J.M. C'est Quantique? Donc c'est Géométrique. Feuilletages—Quantification Géométrique Video. 2003. Available online: https://www.youtube.com/watch?time_continue=417&v=vZeidrBPljM (accessed on 1 November 2018).
72. Souriau, J.M. *Géométrie et Relativité. Collection Enseignement des Sciences*; Hermann: Paris, France, 1964.
73. Souriau, J.M. Thermodynamique et géométrie. *Lecture Notes Math.* **1976**, *676*, 369–397.
74. Souriau, J.M.; Iglesias, P. *Le Chaud, le Froid et la Géométrie, Groupe de Contact de Géométrie Différentielle et de Topologie Algébrique du FNRS*; Université de Liège: Liège, Belgium, 1980.
75. Stueckelberg, E.C.G.; Wanders, G. Thermodynamique en Relativité Générale. *Helv. Phys. Acta* **1953**, *26*, 307–316.
76. Lichnerowicz, A. *Théories Relativistes de la Gravitation et de L'électromagnétisme*; Relativité Générale et Théories Unitaires; Masson et Cie: Paris, France, 1955.
77. Vallée, C. Lois de Comportement des Milieux Continus Dissipatifs Compatibles avec la Physique Relativiste. Ph.D. Thesis, University of Poitiers, Poitiers, France, 1978.
78. Vallée, C. Relativistic thermodynamics of continua. *Int. J. Eng. Sci.* **1981**, *19*, 589–601. [CrossRef]
79. Garrel, J. Tensorial Local-Equilibrium Axion and Operator of Evolution. *Il Nuovo Cimento* **1986**, *94*, 119–139. [CrossRef]
80. Anile, A.; Choquet-Bruhat, Y. *Relativistic Fluid Dynamics*; Lecture Notes in Mathematics; Springer: Berlin, Germany, 1989.
81. De Saxcé, G.; Vallée, C. *Galilean Mechanics and Thermodynamics of Continua*; Wiley-ISTE: Hoboken, NJ, USA, 2016.
82. de Saxcé, G. 5-Dimensional Thermodynamics of Dissipative Continua. In *Models, Simulation, and Experimental Issues in Structural Mechanics*; Springer: Berlin, Germany, 2017.
83. Ingarden, R.S. Information geometry in functional spaces of classical ad quantum finite statistical systems. *Int. J. Eng. Sci.* **1981**, *19*, 1609–1616. [CrossRef]
84. Ingarden, R.S. Information Geometry of Thermodynamics. *Trans. Tenth Prague Conf.* **1988**, *10*, 421–428.
85. Mrugala, R. On equivalence of two metrics in classical thermodynamics. *Physica* **1984**, *125A*, 631–639. [CrossRef]
86. Mrugala, R. Riemannian and Finslerian geometry in thermodynamics. *Open Syst. Inf. Dyn.* **1992**, *1*, 379–396. [CrossRef]
87. Mrugala, R. On a special family of thermodynamic processes and their invariants. *Rep. Math. Phys.* **2000**, *46*, 461–468. [CrossRef]
88. Mrugała, R. On contact and metric structures on thermodynamic spaces. *RIMS Kokyuroku* **2000**, *1142*, 167–181.
89. Mrugała, R. Structure group U(n) x 1 in thermodynamics. *J. Phys. A Math. Gen.* **2005**, *38*, 10905. [CrossRef]
90. Arnold, V.I. Contact geometry: The geometrical method of Gibbs's thermodynamics. In Proceedings of the Gibbs Symposium, New Haven, CT, USA, 15–17 May 1989; Caldi, D.G., Mostow, G.D., Eds.; Yale University: New Haven, CT, USA, 1989; pp. 163–179.
91. Barbaresco, F. Geometric Theory of Heat from Souriau Lie Groups Thermodynamics and Koszul Hessian Geometry. *Entropy* **2016**, *18*, 386. [CrossRef]
92. Kijowski, W. A finite dimensional canonical formalism in the classical field theory. *Commun. Math. Phys.* **1973**, *30*, 99–128. [CrossRef]
93. Kijowski, W. Multiphase spaces and gauge in the calculus of variations. *Bulletin de L Academie Polonaise des Sciences-Serie des Sciences Mathematiques Astronomiques et Physiques* **1974**, *22*, 1219–1225.
94. Kijowski, W.; Szczyrba, W. A canonical structure for classical field theories. *Commun. Math Phys.* **1976**, *46*, 183–206. [CrossRef]
95. Nakagomi, T. Mesoscopic version of thermodynamic equilibrium condition. Another approach to higher order temperatures. *Open Syst. Inf. Dyn.* **1992**, *1*, 233–241. [CrossRef]
96. Nencka, H.; Streater, R.F. Information Geometry for some Lie algebras. *Infin. Dimens. Anal. Quantum Probab. Relat. Top.* **1999**, *2*, 441–460. [CrossRef]
97. Sampieri, U. Lie group structures and reproducing kernels on homogeneous siegel domains. *Annali di Matematica Pura ed Applicata* **1988**, *152*, 1–19. [CrossRef]

98. Alexeevsky, D. Vinberg's Theory of Homogeneous Convex Cones: Developments and Applications; Transformation groups 2017. Conference dedicated to Prof. Ernest B. Vinberg on the occasion of his 80th birthday, Moscow, December 2017 [Video]. Available online: http://www.mathnet.ru/present19121 (accessed on 1 November 2018).
99. Trépreau, J.-M. Transformation de Legendre et pseudoconvexité avec décalage. *J. Fourier Anal. Appl.* **1995**, *1*, 569–588.
100. Leray, J. Le calcul differentiel et intégral sur une variété analytique complexe. *Bull. Soc. Math. France* **1952**, *87*, 81–180.
101. Brenier, Y. Un algorithme rapide pour le calcul de transformées de Legendre-Fenchel discrètes. *C. R. Acad. Sci. Paris* **1989**, *308*, 587–589.
102. Legendre, A.M. *Mémoire Sur L'intégration de Quelques Equations aux Différences Partielles*; Mémoires de l'Académie des Sciences: Paris, France, 1787; pp. 309–351.
103. Konstantatou, M.; McRobie, A. Reciprocal constructions using conic sections and Poncelet duality. In Proceedings of the IASS 2016 Tokyo Symposium: Spatial Structures in the 21st Century—Graphic Statics, Tokyo, Japan, 26–30 September 2016.
104. Benayoun, L. Méthodes Géométriques pour L'étude des Systèmes Thermodynamiques et la Génération D'équations D'état. Ph.D. Thesis, Institut National Polytechnique de Grenoble, Grenoble, France, 1999.
105. der Schaft, A.; Maschke, B. Homogeneous Hamiltonian Control Systems Part II: Application to thermodynamic systems. *IFAC-PapersOnLine* **2018**, *51*, 7–12.
106. Delzant, T.; Wacheux, C. *Action Hamiltoniennes: Invariants et Classification*; Organisé par Michel Brion et Thomas Delzant, CIRM: Luminy, France, 2010; Volume 1, pp. 23–31.
107. Moreau, J.J. Fonctions convexes duales et points proximaux dans un espace hilbertien. *C. R. Acad. Sci. Paris* **1962**, *255*, 2897–2899.
108. Libermann, P. Legendre foliations on contact manifolds. *Differ. Geom. Appl.* **1991**, *1*, 57–76. [CrossRef]
109. Kostant, B.; Sahi, S. The Capelli identity, tube domains, and the generalized Laplace transform. *Adv. Math.* **1991**, *87*, 71–92. [CrossRef]
110. Duhem, P. Sur la stabilité d'un système animé d'un mouvement de rotation, Comptes rendus, t. CXXXII, séance du 29 Avril 1901. 1021.
111. Duhem, P. Sur la stabilité de l'équilibre d'une masse fluide animée d'un mouvement de rotation. *J. Math.* **1901**, *VII*, 311–330.
112. Duhem, P. Stabilité pour des perturbations quelconques, d'un système animé d'un mouvement de rotation uniforme. *C. R.* **1902**, *CXXXIV*, 23.
113. Duhem, P. Sur la stabilité pour des perturbations quelconques, d'un système animé d'un mouvement de rotation uniforme. *Journal de Mathématiques pures et Appliquées* **1902**, *VIII*, 5.
114. Poincaré, H. Sur l'équilibre d'une masse fluide animée d'un mouvement de rotation, chap. 14, Stabilité des ellipsoïdes. *Acta Mathematica* **1885**, *VII*, 366–367.
115. Barbaresco, F. Poly-symplectic Model of Higher Order Souriau Lie Groups Thermodynamics for Small Data Analytics. In *Geometric Science of Information*; Springer: Berlin, Germany, 2017; Volume 10589, pp. 432–441.
116. Volterra, V. *Sulle Equazioni Differenziali che Provengono da Questiono di Calcolo delle Variazioni*; Serise IV; Tip. della R. Accademia dei Lincei: Roma, Italy, 1890; Volume VI, pp. 42–54.
117. Volterra, V. *Sopra una Estensione della Teoria Jacobi-Hamilton del Calcolo delle Variazioni*; Serise IV; Tip. della R. Accademia dei Lincei: Roma, Italy, 1890; Volume VI, pp. 127–138.
118. Dedecker, P. Calcul des variations, formes différentielles et champs géodésiques. *Géométrie Différentielle* **1953**, *52*, 17.
119. Dedecker, P. On the generalization of symplectic geometry to multiple integrals in the calculus of variations. In *Differential Geometrical Methods in Mathematical Physics*; Bleuler, K., Reetz, A., Eds.; Lect. Notes Maths; Springer: Berlin, Germany, 1977; Volume 570, pp. 395–456.
120. Hélein, F.; Kouneiher, J. Covariant Hamiltonian formalism for the calculus of variations with several variables: Lepage–Dedecker versus De Donder–Weyl. *Adv. Theor. Math. Phys.* **2004**, *8*, 565–601. [CrossRef]
121. Carathéodory, C. Uber die Extremalen und geod ätischen Felder in der Variationsrechnung der mehrfachen Integrale. *Acta Sci. Math. (Szeged)* **1929**, *4*, 193–216.
122. Weyl, H. Geodesic fields in the calculus of variations. *Ann. Math.* **1935**, *36*, 607–629. [CrossRef]
123. Edelen, D.G.B. *Nonlocal Variations and Local Invariance of Fields*; American Elsevier: New York, NY, USA, 1969.

124. Rund, H. *The Hamilton-Jacobi Theory in the Calculus of Variations*; Van Nostrand: Princeton, NJ, USA, 1966.
125. Cartan, E. Sur les espaces à connexion affine et la théorie de la relativité généralisée, partie I. *Ann. Ec. Norm* **1923**, *40*, 325–412.
126. Cartan, E. Sur les espaces à connexion affine et la théorie de la relativité généralisée (suite). *Ann. Ec. Norm.* **1924**, *41*, 1–25.
127. Cartan, E. Sur les espaces connexion affine et la théorie de la relativité généralisée partie II. *Ann. Ec. Norm.* **1925**, *42*, 17–88.
128. Cartan, E. *La Méthode du Repère Mobile, la Théorie des Groupes Continus et les Espaces Généralisés*; Exposés de Géométrie, No. 5; Hermann: Paris, France, 1935.
129. Alekseevsky, D. Vinberg's Theory of Homogeneous Convex Cones: Developments and Applications, Transformation Groups 2017. Conference Dedicated to Prof. Ernest B. Vinberg on the Occasion of His 80th Birthday, Moscow, December 2017. Available online: https://www.mccme.ru/tg2017/slides/alexeevsky.pdf (accessed on 1 November 2018).
130. Lichnerowicz, A.; Medina, A. On Lie groups with left-invariant symplectic or Kählerian structures. *Lett. Math. Phys.* **1988**, *16*, 225–235. [CrossRef]
131. Scholz, E.E. Cartan's attempt at bridge-building between Einstein and the Cosserats—or how translational curvature became to be known as torsion. *arXiv* **2018**, arXiv:1810.03872v1 [math.HO].
132. Duhem, P. La théorie physique: Son objet, sa structure, Vrin Edition ed. 2007. Available online: https://books.openedition.org/enseditions/6077 (accessed on 2 November 2018).
133. Conteras, I.; Alba, N.M. Poly-Poisson sigma models and their relational poly-symplectic groupoids. *J. Math. Phys.* **2018**, *59*, 072901. [CrossRef]
134. Belgodère, P. Courbure moyenne généralisée. *C. R. Acad. Sci. Paris* **1944**, *218*, 739–740.
135. Belgodère, P. Extremales d'une intégrale de surface Sg(p, q)dxdy. *C. R. Acad. Sci. Paris* **1944**, *219*, 272–273.

entropy

MDPI

Article

Geometry of Thermodynamic Processes

Arjan van der Schaft [1,*] and Bernhard Maschke [2]

[1] Bernoulli Institute for Mathematics, Computer Science and Artificial Intelligence, Jan C. Willems Center for Systems and Control, University of Groningen, P.O. Box 407, 9700 AK Groningen, The Netherlands
[2] Laboratoire d'automatique et de génie des procédés (LAGEP) (UMR CNRS 5007), Université Claude Bernard Lyon 1, CNRS, 69622 Villeurbanne, France; bernhard.maschke@univ-lyon1.fr
* Correspondence: a.j.van.der.schaft@rug.nl; Tel.: +31-50-363-3731

Received: 9 November 2018; Accepted: 30 November 2018; Published: 4 December 2018

Abstract: Since the 1970s, contact geometry has been recognized as an appropriate framework for the geometric formulation of thermodynamic systems, and in particular their state properties. More recently it has been shown how the symplectization of contact manifolds provides a new vantage point; enabling, among other things, to switch easily between the energy and entropy representations of a thermodynamic system. In the present paper, this is continued towards the global geometric definition of a degenerate Riemannian metric on the homogeneous Lagrangian submanifold describing the state properties, which is overarching the locally-defined metrics of Weinhold and Ruppeiner. Next, a geometric formulation is given of non-equilibrium thermodynamic processes, in terms of Hamiltonian dynamics defined by Hamiltonian functions that are homogeneous of degree one in the co-extensive variables and zero on the homogeneous Lagrangian submanifold. The correspondence between objects in contact geometry and their homogeneous counterparts in symplectic geometry, is extended to the definition of port-thermodynamic systems and the formulation of interconnection ports. The resulting geometric framework is illustrated on a number of simple examples, already indicating its potential for analysis and control.

Keywords: thermodynamics; symplectization; metrics; non-equilibrium processes; interconnection

1. Introduction

This paper is concerned with the geometric formulation of thermodynamic systems. While the geometric formulation of mechanical systems has given rise to an extensive theory, commonly called geometric mechanics, the geometric formulation of thermodynamics has remained more elusive and restricted.

Starting from Gibbs' fundamental relation, contact geometry has been recognized since the 1970s as an appropriate framework for the geometric formulation of thermodynamics; see in particular [1–8]. More recently, the interest in contact-geometric descriptions has been growing, from different points of view and with different motivations; see, e.g., [9–20].

Despite this increasing interest, the current geometric theory of thermodynamics still poses major challenges. First, most of the work is on the geometric formulation of the equations of state, through the use of Legendre submanifolds [1–3,5,8], while less attention has been paid to the geometric definition and analysis of non-equilibrium dynamics. Secondly, thermodynamic system models commonly appear both in energy and in entropy representation, while in principle, this corresponds to contactomorphic, but different contact manifolds. This is already demonstrated by rewriting Gibbs' equation in energy representation $dE = TdS - PdV$, with intensive variables $T, -P$, into the entropy representation $dS = \frac{1}{T}dE + \frac{P}{T}dV$, with intensive variables $\frac{1}{T}, \frac{P}{T}$. Thirdly, for reasons of analysis and control of composite thermodynamic systems, a geometric description of the interconnection of thermodynamic systems is desirable, but currently largely lacking.

A new viewpoint on the geometric formulation of thermodynamic systems was provided in [21], by exploiting the well-known result in geometry that odd-dimensional contact manifolds can be naturally symplectized to even-dimensional symplectic manifolds with an additional structure of homogeneity; see [22,23] for textbook expositions. While the classical applications of symplectization are largely confined to time-dependent Hamiltonian mechanics [23] and partial differential equations [22], the paper [21] argued convincingly that symplectization provides an insightful angle to the geometric modeling of thermodynamic systems as well. In particular, it yields a clear way to bring together energy and entropy representations, by viewing the choice of different intensive variables as the selection of different homogeneous coordinates.

In the present paper, we aim at expanding this symplectization point of view towards thermodynamics, amplifying our initial work [24,25]. In particular, we show how the symplectization point of view not only unifies the energy and entropy representation, but is also very helpful in describing the dynamics of thermodynamic processes, inspired by the notion of the contact control system developed in [11–13,17–19]; see also [16]. Furthermore, it yields a direct and global definition of a metric on the submanifold describing the state properties, encompassing the locally-defined metrics of Weinhold [26] and Ruppeiner [27], and providing a new angle to the equivalence results obtained in [3,5,7,10]. Finally, it is shown how symplectization naturally leads to a definition of interconnection ports; thus extending the compositional geometric port-Hamiltonian theory of interconnected multi-physics systems (see, e.g., [28–30]) to the thermodynamic realm. All this will be illustrated by a number of simple, but instructive, examples, primarily serving to elucidate the developed framework and its potential.

2. Thermodynamic Phase Space and Geometric Formulation of the Equations of State

The starting point for the geometric formulation of thermodynamic systems throughout this paper is an $(n+1)$-dimensional manifold Q^e, with $n \geq 1$, whose coordinates comprise the extensive variables, such as volume and mole numbers of chemical species, as well as entropy and energy [31]. Emphasis in this paper will be on simple thermodynamic systems, with a single entropy and energy variable. Furthermore, for notational simplicity, and without much loss of generality, we will assume:

$$Q^e = Q \times \mathbb{R} \times \mathbb{R}, \tag{1}$$

with $S \in \mathbb{R}$ the entropy variable, $E \in \mathbb{R}$ the energy variable, and Q the $(n-1)$-dimensional manifold of remaining extensive variables (such as volume and mole numbers).

In composite (i.e., compartmental) systems, we may need to consider multiple entropies or energies; namely for each of the components. In this case, $\mathbb{R} \times \mathbb{R}$ is replaced by $\mathbb{R}^{m_S} \times \mathbb{R}^{m_E}$, with m_S denoting the number of entropies and m_E the number of energies; see Example 3 for such a situation. This also naturally arises in the interconnection of thermodynamic systems, as will be discussed in Section 5.

Coordinates for Q^e throughout will be denoted by $q^e = (q, S, E)$, with q coordinates for Q (the manifold of remaining extensive variables). Furthermore, we denote by $\mathcal{T}^*Q^e$ the $(2n+2)$-dimensional cotangent bundle T^*Q^e without its zero-section. Given local coordinates (q, S, E) for Q^e, the corresponding natural cotangent bundle coordinates for T^*Q^e and $\mathcal{T}^*Q^e$ are denoted by:

$$(q^e, p^e) = (q, S, E, p, p_S, p_E), \tag{2}$$

where the co-tangent vector $p^e := (p, p_S, p_E)$ will be called the vector of co-extensive variables.

Following [21], the thermodynamic phase space $\mathbb{P}(T^*Q^e)$ is defined as the projectivization of $\mathcal{T}^*Q^e$, i.e., as the fiber bundle over Q^e with fiber at any point $q^e \in Q^e$ given by the projective space $\mathbb{P}(T^*_{q^e}Q^e)$. (Recall that elements of $\mathbb{P}(T^*_{q^e}Q^e)$ are identified with rays in $T^*_{q^e}Q^e$, i.e., non-zero multiples of a non-zero cotangent vector.) The corresponding projection will be denoted by $\pi : \mathcal{T}^*Q^e \to \mathbb{P}(T^*Q^e)$.

It is well known [22,23] that $\mathbb{P}(T^*Q^e)$ is a contact manifold of dimension $2n+1$. Indeed, recall [22,23] that a contact manifold is an $(2n+1)$-dimensional manifold N equipped with a maximally non-integrable field of hyperplanes ξ. This means that $\xi = \ker\theta \subset TN$ for a, possibly only locally-defined, one-form θ on N satisfying $\theta \wedge (d\theta)^n \neq 0$. By Darboux's theorem [22,23], there exist local coordinates (called Darboux coordinates) $q_0, q_1, \cdots, q_n, \gamma_1, \cdots, \gamma_n$ for N such that, locally:

$$\theta = dq_0 - \sum_{i=1}^{n} \gamma_i dq_i \tag{3}$$

Then, in order to show that $\mathbb{P}(T^*M)$ for any $(n+1)$-dimensional manifold M is a contact manifold, consider the Liouville one-form α on the cotangent bundle T^*M, expressed in natural cotangent bundle coordinates for T^*M as $\alpha = \sum_{i=0}^{n} p_i dq_i$. Consider a neighborhood where $p_0 \neq 0$, and define the homogeneous coordinates:

$$\gamma_i = -\frac{p_i}{p_0}, \quad i = 1, \cdots, n, \tag{4}$$

which, together with $q_0, q_1, \cdots, q_n$, serve as local coordinates for $\mathbb{P}(T^*M)$. This results in the locally-defined contact form θ as in (3) (with $\alpha = p_0\theta$). The same holds on any neighborhood where one of the other coordinates $p_1, \cdots, p_n$ is different from zero, in which case division by the non-zero p_i results in other homogeneous coordinates. This shows that $\mathbb{P}(T^*M)$ is indeed a contact manifold. Furthermore [22,23], $\mathbb{P}(T^*M)$ is the canonical contact manifold in the sense that every contact manifold N is locally contactomorphic to $\mathbb{P}(T^*M)$ for some manifold M.

Taking $M = Q^e$, it follows that coordinates for the thermodynamical phase space $\mathbb{P}(T^*Q^e)$ are obtained by replacing the coordinates $p^e = (p, p_S, p_E)$ for the fibers $T^*_{q^e}Q^e$ by homogeneous coordinates for the projective space $\mathbb{P}(T^*_{q^e}Q^e)$. In particular, assuming $p_E \neq 0$, we obtain the homogeneous coordinates:

$$\gamma =: \frac{p}{-p_E}, \; \gamma_S := \frac{p_S}{-p_E}, \tag{5}$$

defining the intensive variables of the energy representation. Alternatively, assuming $p_S \neq 0$, we obtain the homogeneous coordinates (see [21] for a discussion of p_S, or p_E, as a gauge variable):

$$\widetilde{\gamma} =: \frac{p}{-p_S}, \; \widetilde{\gamma}_E := \frac{p_E}{-p_S}, \tag{6}$$

defining the intensive variables of the entropy representation.

Example 1. *Consider a mono-phase, single constituent, gas in a closed compartment, with volume $q = V$, entropy S, and internal energy E, satisfying Gibbs' relation $dE = TdS - PdV$. In the energy representation, the intensive variable γ is given by the pressure $-P$, and γ_S is the temperature T. In the entropy representation, the intensive variable $\widetilde{\gamma}$ is equal to $\frac{P}{T}$, while $\widetilde{\gamma}_E$ equals the reciprocal temperature $\frac{1}{T}$.*

In order to provide the geometric formulation of the equations of state on the thermodynamic phase space $\mathbb{P}(T^*Q^e)$, we need the following definitions. First, recall that a submanifold $\mathcal{L}$ of T^*Q^e is called a Lagrangian submanifold [22,23] if the symplectic form $\omega := d\alpha$ is zero restricted to $\mathcal{L}$ and the dimension of $\mathcal{L}$ is equal to the dimension of Q^e (the maximal dimension of a submanifold restricted to which ω can be zero).

Definition 1. *A homogeneous Lagrangian submanifold $\mathcal{L} \subset T^*Q^e$ is a Lagrangian submanifold with the additional property that:*

$$(q^e, p^e) \in \mathcal{L} \Rightarrow (q^e, \lambda p^e) \in \mathcal{L}, \qquad \text{for every } 0 \neq \lambda \in \mathbb{R} \tag{7}$$

In the Appendix A, cf. Proposition A2, homogeneous Lagrangian submanifolds are geometrically characterized as submanifolds $\mathcal{L} \subset \mathcal{T}^*Q^e$ of dimension equal to $\dim Q^e$, on which not only the symplectic form $\omega = d\alpha$, but also the Liouville one-form α is zero.

Importantly, homogeneous Lagrangian submanifolds of $\mathcal{T}^*Q^e$ are in one-to-one correspondence with Legendre submanifolds of $\mathbb{P}(T^*Q^e)$. Recall that a submanifold L of a $(2n+1)$-dimensional contact manifold N is a Legendre submanifold [22,23] if the locally-defined contact form θ is zero restricted to L and the dimension of L is equal to n (the maximal dimension of a submanifold restricted to which θ can be zero).

Proposition 1 ([23], Proposition 10.16). *Consider the projection $\pi : \mathcal{T}^*Q^e \to \mathbb{P}(T^*Q^e)$. Then, $L \subset \mathbb{P}(T^*Q^e)$ is a Legendre submanifold if and only if $\mathcal{L} := \pi^{-1}(L) \subset \mathcal{T}^*Q^e$ is a homogeneous Lagrangian submanifold. Conversely, any homogeneous Lagrangian submanifold $\mathcal{L}$ is of the form $\pi^{-1}(L)$ for some Legendre submanifold L.*

In the contact geometry formulation of thermodynamic systems [1–3,5], the equations of state are formalized as Legendre submanifolds. In view of the correspondence with homogeneous Lagrangian submanifolds, we arrive at the following.

Definition 2. *Consider Q^e and the thermodynamical phase space $\mathbb{P}(T^*Q^e)$. The state properties of the thermodynamic system are defined by a homogeneous Lagrangian submanifold $\mathcal{L} \subset \mathcal{T}^*Q^e$ and its corresponding Legendre submanifold $L \subset \mathbb{P}(T^*Q^e)$.*

The correspondence between Legendre and homogeneous Lagrangian submanifolds also implies the following characterization of generating functions for any homogeneous Lagrangian submanifold $\mathcal{L} \subset \mathcal{T}^*Q^e$. This is based on the fact [22,23] that any Legendre submanifold $L \subset N$ in Darboux coordinates $q_0, q_1, \cdots, q_n, \gamma_1, \cdots, \gamma_n$ for N can be locally represented as:

$$L = \{(q_0, q_1, \cdots, q_n, \gamma_1, \cdots, \gamma_n) \mid q_0 = F - \gamma_J \frac{\partial F}{\partial \gamma_J}, q_J = -\frac{\partial F}{\partial \gamma_J}, \gamma_I = \frac{\partial F}{\partial q_I}\} \tag{8}$$

for some partitioning $I \cup J = \{1, \cdots, n\}$ and some function $F(q_I, \gamma_J)$ (called a generating function for L), while conversely, any submanifold L as given in (8), for any partitioning $I \cup J = \{1, \cdots, n\}$ and function $F(q_I, \gamma_J)$, is a Legendre submanifold.

Given such a generating function $F(q_I, \gamma_J)$ for the Legendre submanifold L, we now define, assuming $p_0 \neq 0$ and substituting $\gamma_J = -\frac{p_J}{p_0}$,

$$G(q_0, \cdots, q_n, p_0, \cdots, p_n) := -p_0 F(q_I, -\frac{p_J}{p_0}) \tag{9}$$

Then a direct computation shows that:

$$-\frac{\partial G}{\partial p_0} = F(q_I, -\frac{p_J}{p_0}) + p_0 \frac{\partial F}{\partial \gamma_J}(q_I, -\frac{p_J}{p_0})\frac{p_J}{p_0^2} = F(q_I, \gamma_J) - \frac{\partial F}{\partial \gamma_J}\gamma_J, \tag{10}$$

implying, in view of (8), that:

$$\pi^{-1}(L) = \{((q_0, \cdots, q_n, p_0, \cdots, p_n) \mid q_0 = -\frac{\partial G}{\partial p_0}, q_J = -\frac{\partial G}{\partial p_J}, p_I = \frac{\partial G}{\partial q_I}\} \tag{11}$$

In its turn, this implies that G as defined in (9) is a generating function for the homogeneous Lagrangian submanifold $\mathcal{L} = \pi^{-1}(L)$. If instead of p_0, another coordinate p_i is different from zero, then by dividing by this $p_i \neq 0$, we obtain a similar generating function. This is summarized in the following proposition.

Proposition 2. *Any Legendre submanifold L can be locally represented as in (8), possibly after renumbering the index set $\{0, 1, \cdots, n\}$, for some partitioning $I \cup J = \{1, \cdots, n\}$ and generating function $F(q_I, \gamma_J)$, and conversely, for any such $F(q_I, \gamma_J)$, the submanifold L defined by (8) is a Legendre submanifold.*

Any homogeneous Lagrangian submanifold $\mathcal{L}$ can be locally represented as in (11) with generating function G of the form (9), and conversely, for any such G, the submanifold (11) is a homogeneous Lagrangian submanifold.

Note that the generating functions G as in (9) are homogeneous of degree one in the variables $(p_0, \cdots, p_n)$; see the Appendix A for further information regarding homogeneity.

The simplest instance of a generating function for a Legendre submanifold L and its homogeneous Lagrangian counterpart $\mathcal{L}$ occurs when the generating F as in (8) only depends on $q_1, \cdots, q_n$. In this case, the generating function G is given by:

$$G(q_0, \cdots, q_n, p_0, \cdots, p_n) = -p_0 F(q_1, \cdots, q_n), \tag{12}$$

with the corresponding homogeneous Lagrangian submanifold $\mathcal{L} = \pi^{-1}(L)$ locally given as:

$$\mathcal{L} = \{(q_0, \cdots, q_n, p_0, \cdots, p_n) \mid q_0 = F(q_1, \cdots, q_n),\ p_1 = -p_0 \frac{\partial F}{\partial q_1}, \cdots, p_n = -p_0 \frac{\partial F}{\partial q_n}\} \tag{13}$$

A particular feature of this case is the fact that exactly one of the extensive variables, in the above q_0, is expressed as a function of all the others, i.e., $q_1, \cdots, q_n$. At the same time, p_0 is unconstrained, while the other co-extensive variables $p_1, \cdots, p_n$ are determined by $p_0, q_1, \cdots, q_n$. For a general generating function G as in (9), this is not necessarily the case. For example, if $J = \{1, \cdots, n\}$, corresponding to a generating function $-p_0 F(\gamma)$, then $q_0, \cdots, q_n$ are all expressed as a function of the unconstrained variables $p_0, \cdots, p_n$.

Remark 1. *In the present paper, crucial use is made of homogeneity in the co-extensive variables (p, p_S, p_E), which is different from homogeneity with respect to the extensive variables (q, q_S, q_E), as occurring, e.g., in the Gibbs–Duhem relations [31].*

The two most important representations of a homogeneous Lagrangian submanifold $\mathcal{L} \subset \mathcal{T}^* Q^e$, and its Legendre counterpart $L \subset \mathbb{P}(T^*Q)$, are the energy representation and the entropy representation. In the first case, $\mathcal{L}$ is represented, as in (12), by a generating function of the form:

$$-p_E E(q, S) \tag{14}$$

yielding the representation:

$$\mathcal{L} = \{(q, S, E, p, p_S, p_E) \mid E = E(q, S), p = -p_E \frac{\partial E}{\partial q}(q, S), p_S = -p_E \frac{\partial E}{\partial S}(q, S)\} \tag{15}$$

In the second case (the entropy representation), $\mathcal{L}$ is represented by a generating function of the form:

$$-p_S S(q, E) \tag{16}$$

yielding the representation:

$$\mathcal{L} = \{(q, S, E, p, p_S, p_E) \mid S = S(q, E), p = -p_S \frac{\partial S}{\partial q}(q, E), p_E = -p_S \frac{\partial S}{\partial E}(q, E)\} \tag{17}$$

Note that in the energy representation, the independent extensive variables are taken to be q and the entropy S, while the energy variable E is expressed as a function of them. On the other hand, in the entropy representation, the independent extensive variables are q and the energy E, with S expressed as a function of them. Furthermore, in the energy representation, the co-extensive variable p_E is "free", while instead in the entropy representation, the co-extensive variable p_S is free. In principle, also other representations could be chosen, although we will not pursue this. For instance, in Example 1, one could consider a generating function $-p_V V(S, E)$ where the extensive variable V is expressed as function of the other two extensive variables S, E.

As already discussed in [1,2], an important advantage of describing the state properties by a Legendre submanifold L, instead of by writing out the equations of state, is in providing a global and coordinate-free point of view, allowing for an easy transition between different thermodynamic potentials. Furthermore, if singularities occur in the equations of state, L is typically still a smooth submanifold. As seen before [21], the description by a homogeneous Lagrangian submanifold $\mathcal{L}$ has the additional advantage of yielding a simple way for switching between the energy and the entropy representation.

Remark 2. *Although the terminology "thermodynamic phase space" for $\mathbb{P}(T^*Q^e)$ may suggest that all points in $\mathbb{P}(T^*Q^e)$ are feasible for the thermodynamic system, this is actually not the case. The state properties of the thermodynamic system are specified by the Legendre submanifold $L \subset \mathbb{P}(T^*Q^e)$, and thus, the actual "state space" of the thermodynamic system at hand is this submanifold L; not the whole of $\mathbb{P}(T^*Q^e)$.*

A proper analogy with the Hamiltonian formulation of mechanical systems would be as follows. Consider the phase space T^*Q of a mechanical system with configuration manifold Q. Then, the Hamiltonian $H : T^*Q \to \mathbb{R}$ defines a Lagrangian submanifold $\mathcal{L}_H$ of $T^*(T^*Q)$ given by the graph of the gradient of H. The homogeneous Lagrangian submanifold $\mathcal{L}$ is analogous to $\mathcal{L}_H$, while the symplectized thermodynamic phase space $\mathcal{T}^*Q^e$ is analogous to $T^*(T^*Q)$.

3. The Metric Determined by the Equations of State

In a series of papers starting with [26], Weinhold investigated the Riemannian metric that is locally defined by the Hessian matrix of the energy expressed as a (convex) function of the entropy and the other extensive variables. (The importance of this Hessian matrix, also called the stiffness matrix, was already recognized in [31,32].) Similarly, Ruppeiner [27], starting from the theory of fluctuations, explored the locally-defined Riemannian metric given by minus the Hessian of the entropy expressed as a (concave) function of the energy and the other extensive variables. Subsequently, Mrugała [3] reformulated both metrics as living on the Legendre submanifold L of the thermodynamic phase space and showed that actually, these two metrics are locally equivalent (by a conformal transformation); see also [9]. Furthermore, based on statistical mechanics arguments, [7] globally defined an indefinite metric on the thermodynamical phase space, which, when restricted to the Legendre submanifold, reduces to the Weinhold and Ruppeiner metrics; thus showing global conformal equivalence. This point of view was recently further extended in a number of directions in [10].

In this section, crucially exploiting the symplectization point of view, we provide a novel global geometric definition of a degenerate pseudo-Riemannian metric on the homogeneous Lagrangian submanifold $\mathcal{L}$ defining the equations of state, for any given torsion-free connection on the space Q^e of extensive variables. In a coordinate system in which the connection is trivial (i.e., its Christoffel symbols are all zero), this metric will be shown to reduce to Ruppeiner's locally-defined metric once we use homogeneous coordinates corresponding to the entropy representation, and to Weinhold's locally-defined metric by using homogeneous coordinates corresponding to the energy representation. Hence, parallel to the contact geometry equivalence established in [3,7,10], we show that the metrics of Weinhold and Ruppeiner are just two different local representations of this same globally-defined degenerate pseudo-Riemannian metric on the homogeneous Lagrangian submanifold of the symplectized thermodynamic phase space.

Recall [33] that a (affine) connection ∇ on an $(n+1)$-dimensional manifold M is defined as an assignment:

$$(X, Y) \longmapsto \nabla_X Y \tag{18}$$

for any two vector fields X, Y, which is $\mathbb{R}$-bilinear and satisfies $\nabla_{fX}Y = f\nabla_X Y$ and $\nabla_X(fY) = f\nabla_X Y + X(f)Y$, for any function f on M. This implies that $\nabla_X Y(q)$ only depends on $X(q)$ and the value of Y along a curve, which is tangent to X at q. In local coordinates q for M, the connection is determined by its Christoffel symbols $\Gamma^a_{bc}(q)$, $a, b, c = 0, \cdots, n$, defined by:

$$\nabla_{\frac{\partial}{\partial q_b}} \frac{\partial}{\partial q_c} = \sum_{a=0}^{n} \Gamma^a_{bc}(q) \frac{\partial}{\partial q_a} \tag{19}$$

The connection is called torsion-free if:

$$\nabla_X Y - \nabla_Y X = [X, Y] \tag{20}$$

for any two vector fields X, Y, or equivalently if its Christoffel symbols satisfy the symmetry property $\Gamma^a_{bc}(q) = \Gamma^a_{cb}(q)$, $a, b, c = 0, \cdots, n$. We call a connection trivial in a given set of coordinates $q = (q_0, \cdots, q_n)$ if its Christoffel symbols in these coordinates are all zero.

As detailed in [34], given a torsion-free connection on M, there exists a natural pseudo-Riemannian ("pseudo" since the metric is indefinite) metric on the cotangent-bundle T^*M, in cotangent bundle coordinates (q, p) for T^*M given as:

$$2 \sum_{i=0}^{n} dq_i \otimes dp_i - 2 \sum_{a,b,c=0}^{n} p_c \Gamma^c_{ab}(q) dq_a \otimes dq_b \tag{21}$$

Let us now consider for M the manifold of extensive variables $Q^e = Q \times \mathbb{R}^2$ with coordinates $q^e = (q, S, E)$ as before, where we assume the existence of a torsion-free connection, which is trivial in the coordinates (q, S, E), i.e., the Christoffel symbols are all zero. Then, the pseudo-Riemannian metric $\mathcal{I}$ on $\mathcal{T}^*Q^e$ takes the form:

$$\mathcal{I} := 2(dq \otimes dp + dS \otimes dp_S + dE \otimes dp_E) \tag{22}$$

Denote by $\mathcal{G}$ the pseudo-Riemannian metric $\mathcal{I}$ restricted to the homogeneous Lagrangian submanifold $\mathcal{L}$ describing the state properties. Consider the energy representation (15) of $\mathcal{L}$, with generating function $-p_E E(q, S)$. It follows that $\frac{1}{2}\mathcal{G}$ equals (in shorthand notation):

$$\begin{aligned}
& dq \otimes d\left(-p_E \tfrac{\partial E}{\partial q}\right) + dS \otimes d\left(-p_E \tfrac{\partial E}{\partial S}\right) + dE \otimes dp_E = \\
& -p_E dq \otimes \left(\tfrac{\partial^2 E}{\partial q^2} dq + \tfrac{\partial^2 E}{\partial q \partial S} dS\right) - dq \otimes \tfrac{\partial E}{\partial q} dp_E \\
& -p_E dS \otimes \left(\tfrac{\partial^2 E}{\partial q \partial S} dq + \tfrac{\partial^2 E}{\partial S^2} dS\right) - dS \otimes \tfrac{\partial E}{\partial S} dp_E \\
& + \tfrac{\partial^T E}{\partial q} dq \otimes dp_E + \tfrac{\partial^T E}{\partial S} dS \otimes dp_E \\
& = -p_E \left(dq \otimes \tfrac{\partial^2 E}{\partial q^2} dq + dq \otimes \tfrac{\partial^2 E}{\partial q \partial S} dS + dS \otimes \tfrac{\partial^2 E}{\partial q \partial S} dq + dS \otimes \tfrac{\partial^2 E}{\partial S^2} dS\right) \\
& =: -p_E \mathcal{W}
\end{aligned} \tag{23}$$

where:

$$\mathcal{W} = dq \otimes \frac{\partial^2 E}{\partial q^2} dq + dq \otimes \frac{\partial^2 E}{\partial q \partial S} dS + dS \otimes \frac{\partial^2 E}{\partial S \partial q} dq + dS \otimes \frac{\partial^2 E}{\partial S^2} dS \tag{24}$$

is recognized as Weinhold's metric [26]; the (positive-definite) Hessian of E expressed as a (strongly convex) function of q and S.

On the other hand, in the entropy representation (17) of $\mathcal{L}$, with generating function $-p_S S(q,E)$, an analogous computation shows that $\frac{1}{2}\mathcal{G}$ is given as $p_S \mathcal{R}$, with:

$$\mathcal{R} = -dq \otimes \frac{\partial^2 S}{\partial q^2} dq - dq \otimes \frac{\partial^2 S}{\partial q \partial E} dE - dE \otimes \frac{\partial^2}{\partial E \partial q} dq - dE \otimes \frac{\partial^2 S}{\partial E^2} dE \tag{25}$$

the Ruppeiner metric [27]; minus the Hessian of S expressed as a (strongly concave) function of q and E. Hence, we conclude that:

$$-p_E \mathcal{W} = p_S \mathcal{R}, \tag{26}$$

implying $\mathcal{W} = -\frac{p_S}{p_E}\mathcal{R} = \frac{\partial E}{\partial S}\mathcal{R} = T\mathcal{R}$, with T the temperature. This is basically the conformal equivalence between $\mathcal{W}$ and $\mathcal{R}$ found in [3]; see also [7,10]. Summarizing, we have found the following.

Theorem 1. *Consider a torsion-free connection on Q^e, with coordinates $q^e = (q, S, E)$, in which the Christoffel symbols of the connection are all zero. Then, by restricting the pseudo-Riemannian metric $\mathcal{I}$ to $\mathcal{L}$, we obtain a degenerate pseudo-Riemannian metric $\mathcal{G}$ on $\mathcal{L}$, which in local energy-representation (15) for $\mathcal{L}$ is given by $-2p_E\mathcal{W}$, with $\mathcal{W}$ the Weinhold metric (24), and in a local entropy representation (17) by $2p_S\mathcal{R}$, with $\mathcal{R}$ the Ruppeiner metric (25).*

We emphasize that the degenerate pseudo-Riemannian metric $\mathcal{G}$ is globally defined on $\mathcal{L}$, in contrast to the locally-defined Weinhold and Ruppeiner metrics $\mathcal{W}$ and $\mathcal{R}$; see also the discussion in [3,5,7,9,10]. We refer to $\mathcal{G}$ as degenerate, since its rank is at most n instead of $n+1$. Note furthermore that $\mathcal{G}$ is homogeneous of degree one in p^e and hence does not project to the Legendre submanifold L.

While the assumption of the existence of a trivial connection appears natural in most cases (see also the information geometry point of view as exposed in [35]), all this can be directly extended to any non-trivial torsion-free connection ∇ on Q^e. For example, consider the following situation.

For the ease of notation, denote $q_S := S, q_E := E$, and correspondingly denote $(q_0, q_1, \cdots, q_{n-2}, q_S, q_E) := (q, S, E)$. Take any torsion-free connection on Q^e given by symmetric Christoffel symbols $\Gamma^c_{ab} = \Gamma^c_{ba}$, with indices $a,b,c = 0,\cdots,n-2,S,E$, satisfying $\Gamma^c_{ab} = 0$ whenever one of the indices a,b,c is equal to the index E. Then, the indefinite metric $\mathcal{I}$ on $\mathcal{T}^*Q^e$ is given by (again in shorthand notation):

$$2\sum_{i=0}^{E} dq_i \otimes dp_i - 2\sum_{a,b,c=0}^{S} p_c \Gamma^c_{ab}(q) dq_a \otimes dq_b \tag{27}$$

It follows that the resulting metric $\frac{1}{2}\mathcal{G}$ on $\mathcal{L}$ is given by the matrix:

$$-p_E \left(\frac{\partial^2 E}{\partial q_a \partial q_b} - \sum_{c=0}^{S} \frac{\partial E}{\partial q_c} \Gamma^c_{ab} \right)_{a,b=0,\cdots,S} \tag{28}$$

Here, the $(n \times n)$-matrix at the right-hand side of $-p_E$ is the globally defined geometric Hessian matrix (see e.g., [36]) with respect to the connection on $Q \times \mathbb{R}$ corresponding to the Christoffel symbols Γ^c_{ab}, $a,b,c = 0,\cdots,n-2,S$.

4. Dynamics of Thermodynamic Processes

In this section, we explore the geometric structure of the dynamics of (non-equilibrium) thermodynamic processes; in other words, geometric thermodynamics. By making crucial use of the symplectization of the thermodynamic phase space, this will lead to the definition of port-thermodynamic systems in Definition 3; allowing for open thermodynamic processes. The definition is illustrated in Section 4.2 on a number of simple examples. In Section 4.3, initial observations will be made regarding the controllability of port-thermodynamic systems.

4.1. Port-Thermodynamic Systems

In Section 2, we noted the one-to-one correspondence between Legendre submanifolds L of the thermodynamic phase space $\mathbb{P}(T^*Q^e)$ and homogeneous Lagrangian submanifolds $\mathcal{L}$ of the symplectized space $\mathcal{T}^*Q^e$. In the present section, we start by noting that there is as well a one-to-one correspondence between contact vector fields on $\mathbb{P}(T^*Q^e)$ and Hamiltonian vector fields X_K on $\mathcal{T}^*Q^e$ with Hamiltonians K that are homogeneous of degree one in p^e (see the Appendix A for further details on homogeneity).

Here, Hamiltonian vector fields X_K on $\mathcal{T}^*Q^e$ with Hamiltonian K are in cotangent bundle coordinates $(q^e, p^e) = (q_0, \cdots, q_n, p_0, \cdots, p_n)$ for T^*Q^e given by the standard expressions:

$$\dot{q}_i = \frac{\partial K}{\partial p_i}(q^e, p^e), \quad \dot{p}_i = -\frac{\partial K}{\partial q_i}(q^e, p^e), \quad i = 0, 1, \cdots, n, \tag{29}$$

while contact vector fields $X_{\widehat{K}}$ on the contact manifold $\mathbb{P}(T^*Q^e)$ are given in local Darboux coordinates $(q^e, \gamma) = (q_0, \cdots, q_n, \gamma_1, \cdots, \gamma_n)$ as: [22,23]

$$\begin{aligned} \dot{q}_0 &= \widehat{K}(q^e,\gamma) - \textstyle\sum_{j=1}^n \gamma_j \frac{\partial \widehat{K}}{\partial \gamma_j}(q^e,\gamma) & & \\ \dot{q}_i &= -\frac{\partial \widehat{K}}{\partial \gamma_i}(q^e,\gamma), & i &= 1, \cdots, n \\ \dot{\gamma}_i &= \frac{\partial \widehat{K}}{\partial q_i}(q^e,\gamma) + \gamma_i \frac{\partial \widehat{K}}{\partial q_0}(q^e,\gamma), & i &= 1, \cdots, n, \end{aligned} \tag{30}$$

for some contact Hamiltonian $\widehat{K}(q^e, \gamma)$.

Indeed, consider any Hamiltonian vector field X_K on $\mathcal{T}^*Q^e$, with K homogeneous of degree one in the co-extensive variables p^e. Equivalently (see Appendix A, Proposition A1), $\mathbb{L}_{X_K}\alpha = 0$, with $\mathbb{L}$ denoting the Lie-derivative. It follows, cf. Theorem 12.5 in [23], that X_K projects under $\pi : \mathcal{T}^*Q^e \to \mathbb{P}(T^*Q^e)$ to a vector field $\pi_* X_K$, satisfying:

$$\mathbb{L}_{\pi_* X_K}\theta = \rho\theta \tag{31}$$

for some function ρ, for all (locally-defined) expressions of the contact form θ on $\mathbb{P}(T^*Q^e)$. This exactly means [23] that the vector field $\pi_* X_K$ is a contact vector field with contact Hamiltonian:

$$\widehat{K} := \theta(\pi_* X_K) \tag{32}$$

Conversely [22,23], any contact vector field $X_{\widehat{K}}$ on $\mathbb{P}(T^*Q^e)$, for some contact Hamiltonian $\widehat{K}$, can be lifted to a Hamiltonian vector field X_K on $\mathcal{T}^*Q^e$ with homogeneous K. In fact, for $\widehat{K}$ expressed in Darboux coordinates for $\mathbb{P}(T^*Q^e)$ as $\widehat{K}(q_0, q_1, \cdots, q_n, \gamma_1, \cdot, \gamma_n)$, the corresponding homogeneous function K is given as, cf. [23] (Chapter V, Remark 14.4),

$$K(q_0, \cdots, q_n, p_0, \cdots, p_n) = p_0 \widehat{K}(q_0, \cdots, q_n, -\frac{p_1}{p_0}, \cdots, -\frac{p_n}{p_0}), \tag{33}$$

and analogously on any other homogeneous coordinate neighborhood of $\mathbb{P}(T^*Q^e)$. This is summarized in the following proposition (N.B.: for brevity, we will from now on refer to a function $K(q^e, p^e)$ that is homogeneous of degree one in the co-extensive variables p^e as a homogeneous function and to a Hamiltonian vector field X_K on $\mathcal{T}^*Q^e$ with K homogeneous of degree one in p^e as a homogeneous Hamiltonian vector field).

Proposition 3. *Any homogeneous Hamiltonian vector field X_K on $\mathcal{T}^*Q^e$ projects under π to a contact vector field $X_{\widehat{K}}$ on $\mathbb{P}(T^*Q^e)$ with $\widehat{K}$ locally given by (32), and conversely, any contact vector field $X_{\widehat{K}}$ on $\mathbb{P}(T^*Q^e)$ lifts under π to a homogeneous Hamiltonian vector field X_K on $\mathcal{T}^*Q^e$ with K locally given by (33).*

Recall, and see also Remark 2, that the equations of state describe the constitutive relations between the extensive and intensive variables of the thermodynamic system, or said otherwise, the state properties of the thermodynamic system. Since these properties are fixed for a given thermodynamic system, any dynamics should leave its equations of state invariant. Equivalently, any dynamics on $\mathcal{T}^*Q^e$ or on $\mathbb{P}(T^*Q^e)$ should leave the homogeneous Lagrangian submanifold $\mathcal{L} \subset \mathcal{T}^*Q^e$, respectively, its Legendre submanifold counterpart $L \subset \mathbb{P}(T^*Q^e)$, invariant. (Recall that a submanifold is invariant for a vector field if the vector field is everywhere tangent to it; and thus, solution trajectories remain on it.)

Furthermore, it is natural to require the dynamics of the thermodynamic system to be Hamiltonian; i.e., homogeneous Hamiltonian dynamics on $\mathcal{T}^*Q^e$ and a contact dynamics on $\mathbb{P}(T^*Q^e)$.

In order to combine the Hamiltonian structure of the dynamics with invariance, we make crucial use of the following properties.

Proposition 4.

1. *A homogeneous Lagrangian submanifold $\mathcal{L} \subset \mathcal{T}^*Q^e$ is invariant for the homogeneous Hamiltonian vector field X_K if and only if the homogeneous $K : \mathcal{T}^*Q^e \to \mathbb{R}$ restricted to $\mathcal{L}$ is zero.*
2. *A Legendre submanifold $L \subset \mathbb{P}(T^*Q^e)$ is invariant for the contact vector field $X_{\widehat{K}}$ if and only if $\widehat{K} : \mathbb{P}(T^*Q^e) \to \mathbb{R}$ restricted to L is zero.*
3. *The homogeneous function $K : \mathcal{T}^*Q^e \to \mathbb{R}$ restricted to $\mathcal{L}$ is zero if and only the corresponding function $\widehat{K} : \mathbb{P}(T^*Q^e) \to \mathbb{R}$ restricted to L is zero.*

Item 2 is well known [22,23], and Item 1 can be found in [23,25], while Item 3 directly follows from the correspondence between K and $\widehat{K}$ in (32) and (33).

Based on these considerations, we define the dynamics of a thermodynamic system as being produced by a homogeneous Hamiltonian function, parametrized by $u \in \mathbb{R}^m$,

$$K := K^a + K^c u : \mathcal{T}^*Q^e \to \mathbb{R}, \quad u \in \mathbb{R}^m, \tag{34}$$

with K^a restricted to $\mathcal{L}$ zero, and K^c an m-dimensional row of functions $K^c_j, j = 1, \cdots, m$, all of which are also zero on $\mathcal{L}$. Then, the resulting dynamics is given by the homogeneous Hamiltonian dynamics on $\mathcal{T}^*Q^e$:

$$\dot{x} = X_{K^a}(x) + \sum_{j=1}^{m} X_{K^c_j}(x) u_j, \quad x = (q^e, p^e), \tag{35}$$

restricted to $\mathcal{L}$. (In [24,25], (35) was called a homogeneous Hamiltonian control system.) By Proposition 3, this dynamics projects to contact dynamics corresponding to the contact Hamiltonian $\widehat{K} = \widehat{K}^a + \widehat{K}^c u$ on the corresponding Legendre submanifold $L \subset \mathbb{P}(T^*Q^e)$.

The invariance conditions on the parametrized Hamiltonian K defining the dynamics on $\mathcal{L}$ and L can be seen to take the following explicit form. Since K is homogeneous of degree one, we can write by Euler's homogeneous function theorem (Theorem A1):

$$\begin{aligned} K^a &= p^T f + p_S f_S + p_E f_E, & f = \tfrac{\partial K^a}{\partial p}, f_S = \tfrac{\partial K^a}{\partial p_S}, f_E = \tfrac{\partial K^a}{\partial p_E} \\ K^c &= p^T g + p_S g_S + p_E g_E, & g = \tfrac{\partial K^c}{\partial p}, g_S = \tfrac{\partial K^c}{\partial p_S}, g_E = \tfrac{\partial K^c}{\partial p_E}, \end{aligned} \tag{36}$$

where the functions f, f_S, f_E, as well as the elements of the m-dimensional row vectors of functions g, g_S, g_E are all homogeneous of degree zero. Now, recall the energy representation (15) of the Lagrangian submanifold $\mathcal{L}$ describing the state properties of the system:

$$\mathcal{L} = \{(q, S, E, p, p_S, p_E) \mid E = E(q, S), p = -p_E \frac{\partial E}{\partial q}(q, S), p_S = -p_E \frac{\partial E}{\partial S}(q, S)\} \tag{37}$$

By substitution of (37) in (36), it follows that K restricted to $\mathcal{L}$ is zero for all u if and only if:

$$\begin{aligned} \left(-p_E \tfrac{\partial E}{\partial q} f - p_E \tfrac{\partial E}{\partial S} f_S + p_E f_E\right)|_{\mathcal{L}} &= 0 \\ \left(-p_E \tfrac{\partial E}{\partial q} g - p_E \tfrac{\partial E}{\partial S} g_S + p_E g_E\right)|_{\mathcal{L}} &= 0 \end{aligned} \tag{38}$$

for all p_E, or equivalently:

$$\left(\frac{\partial E}{\partial q} f + \frac{\partial E}{\partial S} f_S\right)|_{\mathcal{L}} = f_E|_{\mathcal{L}}, \quad \left(\frac{\partial E}{\partial q} g + \frac{\partial E}{\partial S} g\right)|_{\mathcal{L}} = g_E|_{\mathcal{L}} \tag{39}$$

This leads to the following additional requirements on the homogeneous function K^a. The first law of thermodynamics ("total energy preservation") requires that the uncontrolled ($u = 0$) dynamics preserves energy, implying that:

$$f_E|_{\mathcal{L}} = 0 \tag{40}$$

Furthermore, the second law of thermodynamics ("increase of entropy") leads to the following requirement. Writing out $K|_{\mathcal{L}} = 0$ in the entropy representation (17) of $\mathcal{L}$ amounts to:

$$\left(\frac{\partial S}{\partial q} f + \frac{\partial S}{\partial E} f_E\right)|_{\mathcal{L}} = f_S|_{\mathcal{L}}, \quad \left(\frac{\partial S}{\partial q} g + \frac{\partial S}{\partial E} g_E\right)|_{\mathcal{L}} = g_S|_{\mathcal{L}} \tag{41}$$

Plugging in the earlier found requirement $f_E|_{\mathcal{L}} = 0$, this reduces to:

$$\frac{\partial S}{\partial q} f|_{\mathcal{L}} = f_S|_{\mathcal{L}}, \quad \left(\frac{\partial S}{\partial q} g + \frac{\partial S}{\partial E} g_E\right)|_{\mathcal{L}} = g_S|_{\mathcal{L}} \tag{42}$$

Finally, since for $u = 0$, the entropy is non-decreasing, this implies the following additional requirement:

$$f_S|_{\mathcal{L}} \geq 0 \tag{43}$$

All this leads to the following geometric formulation of a port-thermodynamic system.

Definition 3 (Port-thermodynamic system). *Consider the space of extensive variables $Q^e = Q \times \mathbb{R} \times \mathbb{R}$ and the thermodynamic phase space $\mathbb{P}(T^*Q^e)$. A port-thermodynamic system on $\mathbb{P}(T^*Q^e)$ is defined as a pair $(\mathcal{L}, K)$, where the homogeneous Lagrangian submanifold $\mathcal{L} \subset T^*Q^e$ specifies the state properties. The dynamics is given by the homogeneous Hamiltonian dynamics with parametrized homogeneous Hamiltonian $K := K^a + K^c u : T^*Q^e \to \mathbb{R}$, $u \in \mathbb{R}^m$, in the form (36), with K^a, K^c zero on $\mathcal{L}$, and the internal Hamiltonian K^a satisfying (corresponding to the first and second law of thermodynamics):*

$$f_E|_{\mathcal{L}} = 0, \quad f_S|_{\mathcal{L}} \geq 0 \tag{44}$$

This means that, in energy representation (15):

$$\left(\frac{\partial E}{\partial q} f + \frac{\partial E}{\partial S} f_S\right)|_{\mathcal{L}} = 0, \quad \left(\frac{\partial E}{\partial q} g + \frac{\partial E}{\partial S} g_S\right)|_{\mathcal{L}} = g_E|_{\mathcal{L}} \tag{45}$$

and, in entropy representation (17):

$$\frac{\partial S}{\partial q} f|_{\mathcal{L}} = f_S|_{\mathcal{L}} \geq 0, \quad \left(\frac{\partial S}{\partial q} g + \frac{\partial S}{\partial E} g_E\right)|_{\mathcal{L}} = g_S|_{\mathcal{L}} \tag{46}$$

Furthermore, the power-conjugate outputs y_p of the port-thermodynamic system $(\mathcal{L}, K)$ are defined as the row-vector:

$$y_p := g_E|_{\mathcal{L}} \tag{47}$$

Since by Euler's theorem (Theorem A1), all expressions f, f_S, f_E, g, g_S, g_E are homogeneous of degree zero, they project to functions on the thermodynamic phase space $\mathbb{P}(T^*Q^e)$. Hence, the dynamics and the output equations are equally well-defined on the Legendre submanifold $L \subset \mathbb{P}(T^*Q^e)$. Note that as a consequence of the above definition of a port-thermodynamic system:

$$\frac{d}{dt}E|_{\mathcal{L}} = y_p u, \tag{48}$$

expressing that the increase of total energy of the thermodynamic system is equal to the energy supplied to the system by the environment.

Remark 3. *In case f, f_S, f_E, g, g_S, g_E do not depend on p^e (and therefore, are trivially homogeneous of degree zero in p^e), they actually define vector fields on the space of extensive variables Q^e (since they transform as vector fields under a coordinate change for Q^e). In this case, the dynamics on T^*Q^e and $\mathcal{L}$ is equal to the Hamiltonian lift of the dynamics on Q^e; see, e.g., [37].*

Remark 4. *Whenever the dynamics on $\mathcal{L}$ is given as the Hamiltonian lift of dynamics on Q^e (see the previous Remark), the properties (44) can be enforced by formulating the dynamics on Q^e as the sum of a Hamiltonian vector field with respect to the energy E and a gradient vector field with respect to the entropy S, in such a way that S is a Casimir of the Poisson bracket and E is a "Casimir" of the symmetric bracket; see, e.g., [38,39]. The extension of this to the general homogeneous setting employed in Definition 3 is of much interest.*

Remark 5. *Definition 3 is generalized to the compartmental situation $Q^e = Q \times \mathbb{R}^{m_S} \times \mathbb{R}^{m_E}$ by modifying (44) to:*

$$\sum_{i=1}^{m_E} f_{E_i}|_{\mathcal{L}} = 0, \quad \sum_{j=1}^{m_S} f_{S_j}|_{\mathcal{L}} \geq 0, \tag{49}$$

corresponding, respectively, to total energy conservation and total entropy increase; see already Example 3.

Remark 6. *An extension to Definition 3 is to consider a non-affine dependence of K on u, i.e., a general function $K : T^*Q^e \times \mathbb{R}^m \to \mathbb{R}$ that is homogeneous in p^e. See already the damper subsystem in Example 7 and the formulation of Hamiltonian input-output systems as initiated in [40] and continued in, e.g., [37,41,42].*

Defining the vector of outputs as being power-conjugate to the input vector u is the most common option for defining an interaction port (in this case, properly called a power-port) of the thermodynamic system. Nevertheless, there are other possibilities, as well. Indeed, a port representing the rate of entropy flow is obtained by defining the alternative output y_{re} as:

$$y_{re} := g_S|_{\mathcal{L}}, \tag{50}$$

which is the entropy-conjugate to the input vector u, This leads instead to the rate of entropy balance:

$$\frac{d}{dt}S|_{\mathcal{L}} = y_{re}u + f_S|_{\mathcal{L}}, \tag{51}$$

where the second, non-negative, term on the right-hand side is the internal rate of entropy production.

Remark 7. *From the point of view of dissipativity theory [43,44], this means that any port-thermodynamic system, with inputs u and outputs y_p, y_{re}, is cyclo-lossless with respect to the supply rate $y_p u$ and cyclo-passive with respect to the supply rate $y_{re}u$.*

Finally, it is of interest to note that, as illustrated by the examples in the next subsection, the Hamiltonian K generating the dynamics on $\mathcal{L}$ is dimensionless; i.e., its values do not have a physical dimension. Physical dimensions do arise by dividing the homogeneous expression by one of the co-extensive variables.

4.2. Examples of Port-Thermodynamic Systems

Example 2 (Heat compartment). *Consider a simple thermodynamic system in a compartment, allowing for heat exchange with its environment. Its thermodynamic properties are described by the extensive variables S (entropy) and E (internal energy), with E expressed as a function $E = E(S)$ of S. Its state properties (in energy representation) are given by the homogeneous Lagrangian submanifold:*

$$\mathcal{L} = \{(S, E, p_S, p_E) \mid E = E(S), p_S = -p_E E'(S)\}, \tag{52}$$

corresponding to the generating function $-p_E E(S)$. Since there is no internal dynamics, K^a is absent. Hence, taking u as the rate of entropy flow corresponds to the homogeneous Hamiltonian $K = K^c u$ with:

$$K^c = p_S + p_E E'(S), \tag{53}$$

which is zero on $\mathcal{L}$. This yields on $\mathcal{L}$ the dynamics (entailing both the entropy and energy balance):

$$\begin{aligned} \dot{S} &= u & \dot{p}_S &= -p_E E''(S)u \\ \dot{E} &= E'(S)u & \dot{p}_E &= 0, \end{aligned} \tag{54}$$

with power-conjugate output y_p equal to the temperature $T = E'(S)$. Defining the homogeneous coordinate $\gamma = -\frac{p_S}{p_E}$ leads to the contact Hamiltonian $\widehat{K}^c = E'(S) - \gamma$ on $\mathbb{P}(T^*\mathbb{R}^2)$, and the Legendre submanifold:

$$L = \{(S, E, \gamma) \in \mathbb{P}(T^*\mathbb{R}^2) \mid E = E(S), \gamma = E'(S)\} \tag{55}$$

The resulting contact dynamics on L is equal to the projected dynamics $\pi_* X_K = X_{\widehat{K}}$ given as:

$$\begin{aligned} \dot{S} &= u \\ \dot{E} &= E'(S)u \\ \dot{\gamma} &= -\frac{\dot{p}_S}{p_E} = E''(S)u \end{aligned} \tag{56}$$

Here, the third equation corresponds to the energy balance in terms of the temperature dynamics. Note that $E''(S) = \frac{T}{C}$, with C the heat capacitance of the fixed volume.

Alternatively, if we take instead the incoming heat flow as input v, then the Hamiltonian is given by:

$$K = (p_S \frac{1}{E'(S)} + p_E)v, \tag{57}$$

leading to the "trivial" power-conjugate output $y_p = 1$ and to the rate of entropy conjugate output y_{re} given by the reciprocal temperature $y_{re} = \frac{1}{T}$.

Example 3 (Heat exchanger). *Consider two heat compartments as in Example 2, exchanging a heat flow through an interface according to Fourier's law. The extensive variables are S_1, S_2 (entropies of the two compartments) and E (total internal energy). The state properties are described by the homogeneous Lagrangian submanifold:*

$$\mathcal{L} = \{(S_1, S_2, E, p_{S_1}, p_{S_2}, p_E) \mid E = E_1(S_1) + E_2(S_2), p_{S_1} = -p_E E_1'(S_1), p_{S_2} = -p_E E_2'(S_2)\}, \tag{58}$$

corresponding to the generating function $-p_E\left(E_1(S_1)+E_2(S_2)\right)$, *with* E_1, E_2 *the internal energies of the two compartments. Denoting the temperatures* $T_1 = E_1'(S_1)$, $T_2 = E_2'(S_2)$, *the internal dynamics of the two-component thermodynamic system corresponding to Fourier's law is given by the Hamiltonian:*

$$K^a = \lambda(\frac{1}{T_1} - \frac{1}{T_2})(p_{S_1}T_2 - p_{S_2}T_1), \tag{59}$$

with λ *Fourier's conduction coefficient. Note that the total entropy on* $\mathcal{L}$ *satisfies:*

$$\dot{S}_1 + \dot{S}_2 = \lambda(\frac{1}{T_1} - \frac{1}{T_2})(T_2 - T_1) \geq 0, \tag{60}$$

in accordance with (49). *We will revisit this example in the context of the interconnection of thermodynamic systems in Examples* 8 *and* 9.

Example 4 (Mass-spring-damper system). *Consider a mass-spring-damper system in one-dimensional motion, composed of a mass* m *with momentum* π, *linear spring with stiffness* k *and extension* z, *and linear damper with viscous friction coefficient* d. *In order to take into account the thermal energy and the entropy production arising from the heat produced by the damper, the variables of the mechanical system are augmented with an entropy variable* S *and internal energy* $U(S)$ *(for instance, if the system is isothermal, i.e., in thermodynamic equilibrium with a thermostat at temperature* T_0, *the internal energy is* $U(S) = T_0S$*). This leads to the total set of extensive variables* z, π, S, $E = \frac{1}{2}kz^2 + \frac{\pi^2}{2m} + U(S)$ *(total energy). The state properties of the system are described by the Lagrangian submanifold* $\mathcal{L}$ *with generating function (in energy representation):*

$$-p_E\left(\frac{1}{2}kz^2 + \frac{\pi^2}{2m} + U(S)\right) \tag{61}$$

This defines the state properties:

$$\mathcal{L} = \{(z, \pi, S, E, p_z, p_\pi, p_S, p_E) | E = \frac{1}{2}kz^2 + \frac{\pi^2}{2m} + U(S), p_z = -p_E kz, p_\pi = -p_E\frac{\pi}{m}, p_S = -p_E U'(S)\} \tag{62}$$

The dynamics is given by the homogeneous Hamiltonian:

$$K = p_z\frac{\pi}{m} + p_\pi\left(-kz - d\frac{\pi}{m}\right) + p_S\frac{d(\frac{\pi}{m})^2}{U'(S)} + \left(p_\pi + p_E\frac{\pi}{m}\right)u, \tag{63}$$

where u *is an external force. The power-conjugate output* $y_p = \frac{\pi}{m}$ *is the velocity of the mass.*

Example 5 (Gas-piston-damper system). *Consider a gas in an adiabatically-isolated cylinder closed by a piston. Assume that the thermodynamic properties of the system are covered by the properties of the gas (for an extended model, see* [13], *Section 4). Then, the system is analogous to the previous example, replacing* z *by volume* V *and the partial energy* $\frac{1}{2}kz^2 + U(S)$ *by an expression* $U(V, S)$ *for the internal energy of the gas. The dynamics of a force-actuated gas-piston-damper system is defined by the Hamiltonian:*

$$K = p_z\frac{\pi}{m} + p_\pi\left(-\frac{\partial U}{\partial V} - d\frac{\pi}{m}\right) + p_S\frac{d(\frac{\pi}{m})^2}{\frac{\partial U}{\partial S}} + \left(p_\pi + p_E\frac{\pi}{m}\right)u, \tag{64}$$

where the power-conjugate output $y_p = \frac{\pi}{m}$ *is the velocity of the piston.*

Example 6 (Port-Hamiltonian systems as port-thermodynamic systems). *Example 4 can be extended to any input-state-output port-Hamiltonian system [28–30]:*

$$\begin{aligned} \dot{x} &= J(x)e - R(e) + G(x)u, \quad e = \frac{\partial H}{\partial x}(x),\ J(x) = -J^T(x) \\ y &= G^T(x)e \end{aligned} \tag{65}$$

on a state space manifold $x \in \mathcal{X}$, with inputs $u \in \mathbb{R}^m$, outputs $y \in \mathbb{R}^m$, Hamiltonian H (equal to the stored energy of the system), and dissipation $R(e)$ satisfying $e^T R(e) \geq 0$ for all e. Including entropy S as an extra variable, along with an internal energy $U(S)$ (for example, in the isothermal case $U(S) = T_0 S$), the state properties of the port-Hamiltonian system are given by the homogeneous Lagrangian submanifold $\mathcal{L} \subset T^*(\mathcal{X} \times \mathbb{R}^2)$ defined as:

$$\mathcal{L} = \{(x, S, E, p, p_S, p_E) \mid E(x,S) = H(x) + U(S),\ p = -p_E \frac{\partial H}{\partial x}(x), p_S = -p_E U'(S)\}, \tag{66}$$

with generating function $-p_E\,(H(x) + U(S))$. The Hamiltonian K is given by (using the shorthand notation $e = \frac{\partial H}{\partial x}(x)$):

$$K(x, S, E, p, p_S, p_E) = p^T\left(J(x)e - S(e) + G(x)u\right) + p_S \frac{e^T R(e)}{U'(S)} + p_E e^T G(x)u \tag{67}$$

reproducing on $\mathcal{L}$ the dynamics (65) with outputs $y_p = y$. Note that in this thermodynamic formulation of the port-Hamiltonian system, the energy-dissipation term $e^T R(e)$ in the power-balance $\frac{d}{dt}H = -e^T R(e) + y^T u$ is compensated by the equal increase of the internal energy $U(S)$, thus leading to conservation of the total energy $E(x,S) = H(x) + U(S)$.

4.3. Controllability of Port-Thermodynamic Systems

In this subsection, we will briefly indicate how the controllability properties of the port-thermodynamic system $(\mathcal{L}, K)$ can be directly studied in terms of the homogeneous Hamiltonians K^a and K^c_j, $j = 1, \cdots, m$, and their Poisson brackets. First, we note that by Proposition A3, the Poisson brackets of these homogeneous Hamiltonians are again homogeneous. Secondly, we recall the well-known correspondence [22,23,33] between Poisson brackets of Hamiltonians h_1, h_2 and Lie brackets of the corresponding Hamiltonian vector fields:

$$[X_{h_1}, X_{h_2}] = X_{\{h_1,h_2\}} \tag{68}$$

In particular, this property implies that if the homogeneous Hamiltonians h_1, h_2 are zero on the homogeneous Lagrangian submanifold $\mathcal{L}$ and, thus, by Proposition 4, the homogeneous Hamiltonian vector fields X_{h_1}, X_{h_2} are tangent to $\mathcal{L}$, then also $[X_{h_1}, X_{h_2}]$ is tangent to $\mathcal{L}$, and therefore, the Poisson bracket $\{h_1, h_2\}$ is also zero on $\mathcal{L}$. Furthermore, with respect to the projection to the corresponding Legendre submanifold L, we note the following property of homogeneous Hamiltonians:

$$\widehat{\{h_1, h_2\}} = \{\widehat{h}_1, \widehat{h}_2\}, \tag{69}$$

where the bracket on the right-hand side is the Jacobi bracket [22,23] of functions on the contact manifold $\mathbb{P}(T^*Q^e)$. This leads to the following analysis of the accessibility algebra [45] of a port-thermodynamic system, characterizing its controllability.

Proposition 5. *Consider a port-thermodynamic system $(\mathcal{L}, K)$ on $\mathbb{P}(T^*Q^e)$ with homogeneous $K := K^a + \sum_{j=1}^m K^c_j u_j : \mathcal{T}^*Q^e \to \mathbb{R}$, zero on $\mathcal{L}$. Consider the algebra $\mathcal{P}$ (with respect to the Poisson bracket) generated by $K^a, K^c_j, j = 1, \cdots, m$, consisting of homogeneous functions that are zero on $\mathcal{L}$ and the corresponding algebra $\widehat{\mathcal{P}}$ generated by $\widehat{K}^a, \widehat{K}^c_j, j = 1, \cdots, m$, on L. The accessibility algebra [45] is spanned by all contact vector fields $X_{\widehat{h}}$*

on L, with $\widehat{h}$ in the algebra $\widehat{\mathcal{P}}$. It follows that the port-thermodynamic system $(\mathcal{L}, K)$ is locally accessible [45] if the dimension of the co-distribution $d\widehat{\mathcal{P}}$ on L defined by the differentials of $\widehat{h}$, with h in the Poisson algebra $\mathcal{P}$, is equal to the dimension of L. Conversely, if the system is locally accessible, then the co-distribution $d\widehat{\mathcal{P}}$ on L has dimension equal to the dimension of L almost everywhere on L.

Similar statements can be made with respect to local strong accessibility of the port-thermodynamic system; see the theory exposed in [45].

5. Interconnections of Port-Thermodynamic Systems

In this section, we study the geometric formulation of interconnection of port-thermodynamic systems through their ports, in the spirit of the compositional theory of port-Hamiltonian systems [28–30,43]. We will concentrate on the case of power-port interconnections of port-thermodynamic systems, corresponding to power flow exchange (with total power conserved). This is the standard situation in (port-based) physical network modeling of interconnected systems. At the end of this section, we will make some remarks about other types of interconnection; in particular, interconnection by exchange of the rate of entropy.

Consider two port-thermodynamic systems with extensive and co-extensive variables:

$$(q_i, p_i, S_i, p_{S_i}, E_i, p_{E_i}) \in T^*Q_i^e = T^*Q_i \times T^*\mathbb{R}_i \times T^*\mathbb{R}_i, \quad i = 1, 2, \tag{70}$$

and Liouville one-forms $\alpha_i = p_i dq_i + p_{S_i} dS_i + p_{E_i} dE_i$, $i = 1, 2$. With the homogeneity assumption in mind, impose the following constraint on the co-extensive variables:

$$p_{E_1} = p_{E_2} =: p_E \tag{71}$$

This leads to the summation of the one-forms α_1 and α_2 given by:

$$\alpha_{\text{sum}} := p_1 dq_1 + p_2 dq_2 + p_{S_1} dS_1 + p_{S_2} dS_2 + p_E d(E_1 + E_2) \tag{72}$$

on the composed space defined as:

$$T^*Q_1^e \circ T^*Q_2^e := \{(q_1, p_1, q_2, p_2, S_1, p_{S_1}, S_2, p_{S_2}, E, p_E) \in T^*Q_1 \times T^*Q_2 \times T^*\mathbb{R} \times T^*\mathbb{R} \times T^*\mathbb{R}\} \tag{73}$$

Leaving out the zero-section $p_1 = 0, p_2 = 0, p_{S_1} = 0, p_{S_2} = 0, p_E = 0$, this space will be denoted by $\mathcal{T}^*Q_1^e \circ \mathcal{T}^*Q_2^e$ and will serve as the space of extensive and co-extensive variables for the interconnected system. Furthermore, it defines the projectivization $\mathbb{P}(T^*Q_1^e \circ T^*Q_2^e)$, which serves as the composition (through $E_i, p_{E_i}, i = 1, 2$) of the two projectivizations $\mathbb{P}(T^*Q_i^e), i = 1, 2$.

Let the state properties of the two systems be defined by homogeneous Lagrangian submanifolds:

$$\mathcal{L}_i \subset T^*Q_i \times T^*\mathbb{R}_i \times T^*\mathbb{R}_i, \quad i = 1, 2, \tag{74}$$

with generating functions $-p_{E_i} E_i(q_i, S_i), i = 1, 2$. Then, the state properties of the composed system are defined by the composition:

$$\mathcal{L}_1 \circ \mathcal{L}_2 := \{(q_1, q_2, p_1, p_2, S_1, p_{S_1}, S_2, p_{S_2}, E, p_E \mid E = E_1 + E_2, (q_i, p_i, S_i, p_{S_i}, E_i, p_{E_i}) \in \mathcal{L}_i, i = 1, 2\}, \tag{75}$$

with generating function $-p_E\left(E_1(q_1, S_1) + E_2(q_2, S_2)\right)$.

Furthermore, consider the dynamics on $\mathcal{L}_i$ defined by the Hamiltonians $K_i = K_i^a + K_i^c u_i, i = 1, 2$. Assume that K_i does not depend on the energy variable $E_i, i = 1, 2$. Then, the sum $K_1 + K_2$ is well-defined on $\mathcal{L}_1 \circ \mathcal{L}_2$ for all u_1, u_2. This defines a composite port-thermodynamic system, with entropy variables S_1, S_2, total energy variable E, inputs u_1, u_2, and state properties defined by $\mathcal{L}_1 \circ \mathcal{L}_2$.

Next, consider the power-conjugate outputs y_{p1}, y_{p2}; in the sequel, simply denoted by y_1, y_2. Imposing on the power-port variables u_1, u_2, y_1, y_2 interconnection constraints that are satisfying the power-preservation property:

$$y_1 u_1 + y_2 u_2 = 0, \tag{76}$$

yields an interconnected dynamics on $\mathcal{L}_1 \circ \mathcal{L}_2$, which is energy conserving (the p_E-term in the expression for $K_1 + K_2$ is zero by (76)). This is summarized in the following proposition.

Proposition 6. *Consider two port-thermodynamic systems $(\mathcal{L}_i, K_i)$ with spaces of extensive variables Q_i^e, $i = 1, 2$. Assume that K_i does not depend on E_i, $i = 1, 2$. Then, $(\mathcal{L}_1 \circ \mathcal{L}_2, K_1 + K_2)$, with $\mathcal{L}_1 \circ \mathcal{L}_2$ given in (75), defines a composite port-thermodynamic system with inputs u_1, u_2 and outputs y_1, y_2. By imposing interconnection constraints on u_1, u_2, y_1, y_2 satisfying (76), an autonomous (no inputs) port-thermodynamic system is obtained.*

Remark 8. *The interconnection procedure can be extended to the case of an additional open power-port with input vector u and output row vector y, by replacing (76) by power-preserving interconnection constraints on u_1, u_2, u, y_1, y_2, y, satisfying:*

$$y_1 u_1 + y_2 u_2 + yu = 0 \tag{77}$$

Proposition 6 is illustrated by the following examples.

Example 7 (Mass-spring-damper system). *We will show how the thermodynamic formulation of the system as detailed in Example 4 also results from the interconnection of the three subsystems: mass, spring, and damper.*

I. Mass subsystem (leaving out irrelevant entropy). The state properties are given by:

$$\mathcal{L}_m = \{(\pi, \kappa, p_\pi, p_\kappa) \mid \kappa = \frac{\pi^2}{2m}, p_\pi = -p_\kappa \frac{\pi}{m}\}, \tag{78}$$

with energy κ (kinetic energy) and dynamics generated by the Hamiltonian:

$$K_m = (p_\kappa \frac{\pi}{m} + p_\pi) u_m, \tag{79}$$

corresponding to $\dot{\pi} = u_m, y_m = \frac{\pi}{m}$.

II. Spring subsystem (again leaving out irrelevant entropy). The state properties are given by:

$$\mathcal{L}_s = \{(z, P, p_z, p_P) \mid P = \frac{1}{2} k z^2, p_z = -p_P k z\}, \tag{80}$$

with energy P (spring potential energy) and dynamics generated by the Hamiltonian:

$$K_s = (p_P k z + p_z) u_s, \tag{81}$$

corresponding to $\dot{z} = u_s, y_s = kz$.

III. Damper subsystem. The state properties are given by:

$$\mathcal{L}_d = \{(S, U) \mid U = U(S), p_S = -p_U U'(S)\}, \tag{82}$$

involving the entropy S and an internal energy $U(S)$. The dynamics of the damper subsystem is generated by the Hamiltonian:

$$K_d = (p_U + p_S \frac{1}{U'(S)}) d u_d^2 \tag{83}$$

with d the damping constant and power-conjugate output:

$$y_d := d u_d \tag{84}$$

equal to the damping force.

Finally, interconnect, in a power-preserving way, the three subsystems to each other via their power-ports $(u_m, y_m), (u_s, y_s), (u_d, y_d)$ *as:*

$$u_m = -y_s - y_d, \; u_s = y_m = u_d \tag{85}$$

This results (after setting $p_K = p_P = p_U =: p$*) in the interconnected port-thermodynamic system with total Hamiltonian* $K_m + K_s + K_d$ *given as:*

$$\begin{aligned}
&(p\tfrac{\pi}{m} + p_\pi)u_m + (pkz + p_z)u_s + (p + p_S\tfrac{1}{U'(S)})du_d^2 = \\
&(p\tfrac{\pi}{m} + p_\pi)(-kz - d\tfrac{\pi}{m}) + (pkz + p_z)\tfrac{\pi}{m} + (p + p_S\tfrac{1}{U'(S)})d(\tfrac{\pi}{m})^2 = \\
&p_z\tfrac{\pi}{m} + p_\pi(-kz - d\tfrac{\pi}{m}) + p_S\tfrac{d(\frac{\pi}{m})^2}{U'(S)},
\end{aligned} \tag{86}$$

which is equal to the Hamiltonian for $u = 0$ *as obtained before in Example 4, Equation (63).*

Example 8 (Heat exchanger)**.** *Consider two heat compartments as in Example 2, with state properties:*

$$\mathcal{L}_i = \{(S_i, E_i, p_{S_i}, p_{E_i}) \mid E_i = E_i(S_i), p_{S_i} = -p_{E_i}E_i'(S)\}, \; i = 1, 2. \tag{87}$$

The dynamics is given by the Hamiltonians:

$$K_i = (p_{E_i} + p_{S_i}\frac{1}{T_i})v_i, \quad T_i = E_i'(S_i), \quad i = 1, 2, \tag{88}$$

with v_1, v_2 *the incoming heat flows and power-conjugate outputs* y_1, y_2*, which both are equal to one. Consider the power-conserving interconnection:*

$$v_1 = -v_2 = \lambda(T_2 - T_1), \tag{89}$$

with λ *the Fourier heat conduction coefficient. Then, the Hamiltonian of the interconnected port-thermodynamical system is given by:*

$$K_1 + K_2 = \lambda(T_2 - T_1)(\frac{p_{S_1}}{T_1} - \frac{p_{S_2}}{T_2}), \tag{90}$$

which equals the Hamiltonian (59) as obtained in Example 3.

Apart from power-port interconnections as above, we may also define other types of interconnection, not corresponding to the exchange of rate of energy (power), but instead to the exchange of rate of other extensive variables. In particular, an interesting option is to consider interconnection via the rate of entropy exchange. This can be done in a similar way, by considering, instead of the variables $E_i, p_{E_i}, i = 1, 2$, as above, the variables $S_i, p_{S_i}, i = 1, 2$. Imposing alternatively the constraint $p_{S_1} = p_{S_2} =: p_S$ yields a similar composed space of extensive and co-extensive variables, as well as a similar composition $\mathcal{L}_1 \circ \mathcal{L}_2$ of the state properties. By assuming in this case that the Hamiltonians K_i do not depend on the entropies $S_i, i = 1, 2$ and by imposing interconnection constraints on u_1, u_2 and the "rate of entropy" conjugate outputs y_{re1}, y_{re2} leads again to an interconnected port-thermodynamic system. Note however that while it is natural to assume conservation of total energy for the interconnection of two systems via their power-ports, in the alternative case of interconnecting through the rate of entropy ports, the total entropy may not be conserved, but actually increasing.

Example 9. *As an alternative to the previous Example 8, where the heat exchanger was modeled as the interconnection of two heat compartments via power-ports, consider the same situation, but now with outputs* y_i

being the "rate of entropy conjugate" to v_i, i.e., equal (cf. the end of Example 2) to the reciprocal temperatures $\frac{1}{T_i}$ with $T_i = E'(S_i)$, $i = 1, 2$. This results in interconnecting the two heat compartments as, equivalently to (89),

$$v_1 = -v_2 = \lambda(\frac{1}{y_2} - \frac{1}{y_1}) \tag{91}$$

This interconnection is not total entropy conserving, but instead satisfies $y_1 v_1 + y_2 v_2 = \lambda(\frac{1}{y_2} - \frac{1}{y_1})(y_1 - y_2) \geq 0$, corresponding to the increase of total entropy.

6. Discussion

While the state properties of thermodynamic systems have been geometrically formulated since the 1970s through the use of contact geometry, in particular by means of Legendre submanifolds, the geometric formulation of non-equilibrium thermodynamic processes has remained more elusive. Taking up the symplectization point of view on thermodynamics as successfully initiated in [21], the present paper develops a geometric framework based on the description of non-equilibrium thermodynamic processes by Hamiltonian dynamics on the symplectized thermodynamic phase space generated by Hamiltonians that are homogeneous of degree one in the co-extensive variables; culminating in the definition of port-thermodynamic systems in Section 4.1. Furthermore, Section 3 shows how the symplectization point of view provides an intrinsic definition of a metric that is overarching the locally-defined metrics of Weinhold and Ruppeiner and provides an alternative to similar results in the contact geometry setting provided in [3,5,7,10]. The correspondence between objects in contact geometry and corresponding homogeneous objects in symplectic geometry turns out to be very effective. An additional benefit of symplectization is the simplicity of the expressions and computations in the standard Hamiltonian context, as compared to those in contact geometry. This feature is also exemplified by the initial controllability study in Section 4.3. As noted in [38], physically non-trivial examples of mesoscopic dynamics are infinite-dimensional. This calls for an infinite-dimensional extension, following the well-developed theory of infinite-dimensional Hamiltonian systems (but now adding homogeneity) of the presented definition of port-thermodynamic systems, encompassing systems obtained by the Hamiltonian lift of infinite-dimensional GENERIC [38] and dissipative port-Hamiltonian [46] formulations; see also Remark 4. From a control point of view, one of the open problems concerns the stabilization of thermodynamic processes using the developed framework.

Author Contributions: Both authors have made a valuable contribution to the investigation and preparation of this manuscript, and have read and approved the final manuscript.

Funding: The research of the second author was funded by the Agence Nationale de la Recherche, ANR-PRCI project INFIDHEM, ID ANR-16-CE92-0028. **Conflicts of Interest:** The authors declare no conflict of interest.

Appendix A. Homogeneity of Functions, of Hamiltonian Vector Fields, and of Lagrangian Submanifolds

In this section, we use throughout, for notational simplicity, the notation M instead of Q^e. Furthermore, we let $\dim M = n+1$ with $n \geq 0$ denote coordinates for M by $q = (q_0, q_1, \cdots, q_n)$ and co-tangent bundle coordinates for T^*M by $(q, p) = (q_0, q_1, \cdots, q_n, p_0, p_1, \cdots, p_n)$.

The notion of homogeneity in the variables p will be fundamental.

Definition A1. *Let $r \in \mathbb{Z}$. A function $K : \mathcal{T}^*M \to \mathbb{R}$ is called homogeneous of degree r (in the variables $p = (p_0, p_1 \cdots, p_n)$) if:*

$$K(q_0, q_1, \cdots, q_n, \lambda p_0, \lambda p_1, \cdots, \lambda p_n) = \lambda^r K(q_0, q_1, \cdots, q_n, p_0, p_1, \cdots, p_n), \quad \forall \lambda \neq 0 \tag{A1}$$

Note that this definition is independent of the choice of cotangent-bundle coordinates (q, p) for $\mathcal{T}^*M$.

Theorem A1 (Euler's homogeneous function theorem). *A differentiable function $K : \mathcal{T}^*M \to \mathbb{R}$ is homogeneous of degree r (in $p = (p_0, p_1, \cdots, p_n)$) if and only if:*

$$\sum_{i=0}^{n} p_i \frac{\partial K}{\partial p_i}(q,p) = rK(q,p), \quad \textit{for all } (q,p) \in \mathcal{T}^*M \tag{A2}$$

Furthermore, if K is homogeneous of degree r, then its derivatives $\frac{\partial K}{\partial p_i}, i = 0, \cdots, n$, are homogeneous of degree $r-1$.

Geometrically, Euler's theorem can be equivalently formulated as follows. Recall that the Hamiltonian vector field X_h on T^*M with symplectic form $\omega = d\alpha$ corresponding to an arbitrary Hamiltonian $h : T^*M \to \mathbb{R}$ is defined by $i_{X_h}\omega = -dh$. It is immediately verified that $h : T^*M \to \mathbb{R}$ is homogeneous of degree r iff:

$$\alpha(X_h) = r\,h \tag{A3}$$

Define the Euler vector field (also called the Liouville vector field) E on T^*M as the vector field satisfying:

$$d\alpha(E, \cdot) = \alpha \tag{A4}$$

In co-tangent bundle coordinates (q, p) for T^*M, the vector field E is given as $\sum_{i=0}^{n} p_i \frac{\partial}{\partial p_i}$. One verifies that $h : T^*M \to \mathbb{R}$ is homogeneous of degree r iff (with $\mathbb{L}$ denoting Lie-derivative):

$$\mathbb{L}_E h = r\,h \tag{A5}$$

In the sequel, we will only use homogeneity and Euler's theorem for $r = 0$ and $r = 1$. First, it is clear that physical variables defined on the contact manifold $\mathbb{P}(T^*Q^e)$ correspond to functions on $\mathcal{T}^*Q^e$, which are homogeneous of degree zero in p. On the other hand, as formulated in Proposition 3, a Hamiltonian vector field on $\mathcal{T}^*Q^e$ with respect to a Hamiltonian that is homogeneous of degree one in p projects to a contact vector field on the contact manifold $\mathbb{P}(T^*Q)$. Such Hamiltonian vector fields are locally characterized as follows.

Proposition A1. *If $h : T^*M \to \mathbb{R}$ is homogeneous of degree one in p, then $X = X_h$ satisfies:*

$$\mathbb{L}_X \alpha = 0 \tag{A6}$$

Conversely, if a vector field X satisfies (A6), then $X = X_h$ for some locally-defined Hamiltonian h that is homogeneous of degree one in p.

Proof. Note that by Cartan's formula, for any vector field X:

$$\mathbb{L}_X \alpha = i_X d\alpha + d i_X \alpha = i_X d\alpha + d\,(\alpha(X)) \tag{A7}$$

If h is homogeneous of degree one in p, then by (A3), we have $\alpha(X_h) = h$, and thus, $i_{X_h}d\alpha + d\alpha(X_h) = -dh + dh = 0$, implying by (A7) that $\mathbb{L}_{X_h}\alpha = 0$. Conversely, if $\mathbb{L}_X\alpha = 0$, then (A7) yields $i_X d\alpha + d\,(\alpha(X)) = 0$, implying that $X = X_h$, with $h = \alpha(X)$, which by (A3) for $r = 1$ is homogeneous of degree one. □

Summarizing, Hamiltonian vector fields with Hamiltonians that are homogeneous of degree one in p are characterized by (A6); in contrast to general Hamiltonian vector fields X on T^*M, which are characterized by the weaker property $\mathbb{L}_X d\alpha = 0$.

Similar statements as above can be made for homogeneous Lagrangian submanifolds (cf. Definition 1). Recall [22,23,33] that a submanifold $\mathcal{L} \subset T^*M$ is called a Lagrangian submanifold if the symplectic form $\omega := d\alpha$ is zero on $\mathcal{L}$, and $\dim \mathcal{L} = \dim M$.

Proposition A2. *Consider the cotangent bundle T^*M with its canonical one-form α and symplectic form $\omega := d\alpha$. A submanifold $\mathcal{L} \subset \mathcal{T}^*M$ is a homogeneous Lagrangian submanifold if and only if α restricted to $\mathcal{L}$ is zero, and $\dim \mathcal{L} = \dim M$.*

Proof. First of all, note the following. Recall the definition of the Euler vector field E in (A4). In co-tangent bundle coordinates (q, p) for T^*M, the Euler vector field takes the form $E = \sum_{i=0}^{n} p_i \frac{\partial}{\partial p_i}$. Hence, the homogeneity of $\mathcal{L}$ is equivalent to the tangency of E to $\mathcal{L}$.

(If) By Palais' formula (see, e.g., [33], Proposition 2.4.15):

$$d\alpha(X_0, X_1) = \mathbb{L}_{X_0}(\alpha(X_1)) - \mathbb{L}_{X_1}(\alpha(X_0)) - \alpha([X_0, X_1]) \tag{A8}$$

for any two vector fields X_0, X_1. Hence, for any X_1, X_2 tangent to $\mathcal{L}$, we obtain $d\alpha(X_0, X_1) = 0$, implying that $d\alpha$ is zero restricted to $\mathcal{L}$, and thus, $\mathcal{L}$ is a Lagrangian submanifold. Furthermore, by (A4):

$$d\alpha(E, X) = \alpha(X) = 0, \tag{A9}$$

for all vector fields X tangent to $\mathcal{L}$. Because $\mathcal{L}$ is a Lagrangian submanifold, this implies that E is tangent to $\mathcal{L}$ (since a Lagrangian submanifold is a maximal submanifold restricted to $\omega = d\alpha$, whichis zero). Hence, $\mathcal{L}$ is homogeneous.

(Only if) If $\mathcal{L}$ is homogeneous, then E is tangent to $\mathcal{L}$, and thus, since $\mathcal{L}$ is Lagrangian, (A9) holds for all vector fields X tangent to $\mathcal{L}$, implying that α is zero restricted to $\mathcal{L}$. □

Regarding the Poisson brackets of Hamiltonian functions that are either homogeneous of degree one or zero (in p), we have the following proposition.

Proposition A3. *Consider the Poisson bracket $\{h_1, h_2\}$ of functions h_1, h_2 on T^*M defined with respect to the symplectic form $\omega = d\alpha$. Then:*

(a) *If h_1, h_2 are both homogeneous of degree one, then also $\{h_1, h_2\}$ is homogeneous of degree one.*
(b) *If h_1 is homogeneous of degree one and h_2 is homogeneous of degree zero, then $\{h_1, h_2\}$ is homogeneous of degree zero.*
(c) *If h_1, h_2 are both homogeneous of degree zero, then $\{h_1, h_2\}$ is zero.*

Proof.

(a) Since h_1, h_2 are both homogeneous of degree one, we have by Proposition A1, $\mathbb{L}_{X_{h_i}}\alpha = 0, i = 1, 2$. Hence:

$$\mathbb{L}_{X_{\{h_1,h_2\}}}\alpha = \mathbb{L}_{[X_{h_1}, X_{h_2}]}\alpha = \mathbb{L}_{X_{h_1}}(\mathbb{L}_{X_{h_2}}\alpha) - \mathbb{L}_{X_{h_2}}(\mathbb{L}_{X_{h_1}}\alpha) = 0, \tag{A10}$$

implying by Proposition A1 that $\{h_1, h_2\}$ is homogeneous of degree one.

(b) $\alpha(X_{h_2}) = 0$, while by Proposition A1 $\mathbb{L}_{X_{h_1}}\alpha = 0$, implying:

$$0 = \mathbb{L}_{X_{h_1}}(\alpha(X_{h_2})) = (\mathbb{L}_{X_{h_1}}\alpha)(X_{h_2}) + \alpha([X_{h_1}, X_{h_2}]) = \alpha(X_{\{h_1,h_2\}}), \tag{A11}$$

which means that $\{h_1, h_2\}$ is homogeneous of degree zero.

(c) First we note that for any X_h with h homogeneous of degree zero, since $\alpha(X_h) = 0$,

$$\mathbb{L}_{X_h}\alpha = i_{X_h}d\alpha + d(i_{X_h}\alpha) = -dh \tag{A12}$$

Utilizing this property for h_1, we obtain, since $\alpha(X_{h_2}) = 0$,

$$\begin{aligned} 0 = \mathbb{L}_{X_{h_1}}(\alpha(X_{h_2})) = (\mathbb{L}_{X_{h_1}}\alpha)(X_{h_2}) + \alpha(X_{\{h_1,h_2\}}) = \\ -dh_1(X_{h_2}) + \alpha(X_{\{h_1,h_2\}}) = -\{h_1, h_2\} + \alpha(X_{\{h_1,h_2\}}), \end{aligned} \tag{A13}$$

proving that $\{h_1, h_2\}$ is homogeneous of degree one. Hence, by Proposition A1, $\mathbb{L}_{X_{\{h_1,h_2\}}}\alpha = 0$, and thus:

$$\begin{aligned} 0 = \mathbb{L}_{X_{\{h_1,h_2\}}}\alpha = \mathbb{L}_{[X_{h_1}, X_{h_2}]}\alpha = \mathbb{L}_{X_{h_1}}\mathbb{L}_{X_{h_2}}\alpha - \mathbb{L}_{X_{h_2}}\mathbb{L}_{X_{h_1}}\alpha = \\ \mathbb{L}_{X_{h_1}}(-dh_2) - \mathbb{L}_{X_{h_2}}(-dh_1) = -2\{h_1, h_2\} \end{aligned} \tag{A14}$$

where in the fourth equality, we use (A12) for h_1 and h_2.

□

References

1. Hermann, R. *Geometry, Physics and Systems*; Marcel Dekker: New York, NY, USA, 1973.
2. Mrugała, R. Geometric formulation of equilibrium phenomenological thermodynamics. *Rep. Math. Phys.* **1978**, *14*, 419. [CrossRef]

3. Mrugała, R. On equivalence of two metrics in classical thermodynamics. *Physica* **1984**, *125A*, 631–639. [CrossRef]
4. Mrugała, R. Submanifolds in the thermodynamic phase space. *Rep. Math. Phys.* **1985**, *21*, 197. [CrossRef]
5. Mrugała, R. On contact and metric structures on thermodynamic spaces. *RIMS Kokyuroku* **2000**, *1142*, 167–181.
6. Mrugała, R. On a special family of thermodynamic processes and their invariants. *Rep. Math. Phys.* **2000**, *3*, 46. [CrossRef]
7. Mrugała, R.; Nulton, J.D.; Schön, J.C.; Salamon, P. Statistical approach to the geometric structure of thermodynamics. *Phys. Rev. A* **1990**, *41*, 3156. [CrossRef] [PubMed]
8. Mrugała, R.; Nulton, J.D.; Schön, J.C.; Salamon, P. Contact structures in thermodynamic theory. *Rep. Math. Phys.* **1991**, *29*, 109–121. [CrossRef]
9. Bravetti, A.; Lopez-Monsalvo, C.S.; Nettel, F. Contact symmetries and Hamiltonian thermodynamics. *Ann. Phys.* **2015**, *361*, 377–400. [CrossRef]
10. Bravetti, A.; Lopez-Monsalvo, C.S.; Nettel, F. Conformal gauge transformations in thermodynamics. *Entropy* **2015**, *17*, 6150–6168. [CrossRef]
11. Eberard, D.; Maschke, B.M.; van der Schaft, A.J. An extension of Hamiltonian systems to the thermodynamic space: Towards a geometry of non-equilibrium thermodynamics. *Rep. Math. Phys.* **2007**, *60*, 175–198. [CrossRef]
12. Favache, A.; Maschke, B.M.; Santos, V.D.; Dochain, D. Some properties of conservative control systems. *IEEE Trans. Autom. Control* **2009**, *54*, 2341–2351. [CrossRef]
13. Favache, A.; Dochain, D.; Maschke, B.M. An entropy-based formulation of irreversible processes based on contact structures. *Chem. Eng. Sci.* **2010**, *65*, 5204–5216. [CrossRef]
14. Gay-Balmaz, F.; Yoshimura, H. A Lagrangian variational formulation for nonequilibrium thermodynamics. Part I: Discrete systems. *J. Geom. Phys.* **2017**, *111*, 169–193. [CrossRef]
15. Gromov, D. Two approaches to the description of the evolution of thermodynamic systems. *IFAC-Papers OnLine* **2016**, *49*, 34–39. [CrossRef]
16. Merker, J.; Krüger, M. On a variational principle in thermodynamics. *Contin. Mech. Thermodyn.* **2013**, *25*, 779–793. [CrossRef]
17. Ramirez, H.; Maschke, B.; Sbarbaro, D. Feedback equivalence of input-output contact systems. *Syst. Control Lett.* **2013**, *62*, 475–481. [CrossRef]
18. Ramirez, H.; Maschke, B.; Sbarbaro, D. Irreversible port-Hamiltonian systems: A general formulation of irreversible processes with application to the CSTR. *Chem. Eng. Sci.* **2013**, *89*, 223–234. [CrossRef]
19. Ramirez, H.; Maschke, B.; Sbarbaro, D. Partial stabilization of input-output contact systems on a Legendre submanifold. *IEEE Trans. Autom. Control* **2017**, *62*, 1431–1437. [CrossRef]
20. Gromov, D.; Castanos, F. The geometric structure of interconnected thermo-mechanical systems. *IFAC-Papers OnLine* **2017**, *50*, 582–587. [CrossRef]
21. Balian, R.; Valentin, P. Hamiltonian structure of thermodynamics with gauge. *Eur. J. Phys. B* **2001**, *21*, 269–282. [CrossRef]
22. Arnold, V.I. *Mathematical Methods of Classical Mechanics*, 2nd ed.; Springer: Berlin, Germany, 1989.
23. Libermann, P.; Marle, C.-M. *Symplectic Geometry and Analytical Mechanics*; D. Reidel Publishing Company: Dordrecht, The Netherlands, 1987.
24. Maschke, B.M.; van der Schaft, A.J. Homogeneous Hamiltonian control systems, Part II: Application to thermodynamic systems. *IFAC-PapersOnLine* **2018**, *51*, 7–12. [CrossRef]
25. van der Schaft, A.J.; Maschke, B.M. Homogeneous Hamiltonian control systems, Part I: Geometric formulation. *IFAC-Papers OnLine* **2018**, *51*, 1–6. [CrossRef]
26. Weinhold, F. Metric geometry of equilibrium thermodynamics. *J. Chem. Phys.* **1975**, *63*, 2479. [CrossRef]
27. Ruppeiner, G. Thermodynamics: A Riemannian geometric model. *Phys. Rev. A* **1979**, *20*, 1608. [CrossRef]
28. Maschke, B.M.; van der Schaft, A.J. Port controlled Hamiltonian systems: Modeling origins and system theoretic properties. *IFAC Proc. Vol.* **1992**, *25*, 359–365. [CrossRef]
29. Van der Schaft, A.J.; Maschke, B.M. The Hamiltonian formulation of energy conserving physical systems with external ports. *Archiv für Elektronik und Übertragungstechnik* **1995**, *49*, 362–371.
30. Van der Schaft, A.J.; Jeltsema, D. Port-Hamiltonian Systems Theory: An Introductory Overview. *Found. Trends Syst. Control* **2014**, *1*, 173–378. [CrossRef]
31. Callen, H. *Thermodynamics*; Wiley: New York, NY, USA, 1960.
32. Tisza, L. The thermodynamics of phase equilibrium. *Ann. Phys.* **1961**, *13*, 1. [CrossRef]

33. Abraham, R.A.; Marsden, J.E. *Foundations of Mechanics*, 2nd ed.; Benjamin/Cummings: Reading, MA, USA, 1978.
34. Yano, K.; Ishihara, S. *Tangent and Cotangent Bundles*; Marcel Dekker: New York, NY, USA, 1973.
35. Amari, S.I. *Information Geometry and its Applications (Applied Mathematical Sciences)*; Springer: Berlin, Germany, 2016.
36. Bullo, F.; Lewis, A.D. *Geometric Control of Mechanical Systems (Texts in Applied Mathematics)*; Springer: Berlin, Germany, 2005.
37. Crouch, P.E.; van der Schaft, A.J. *Variational and Hamiltonian Control Systems*; Lecture Notes in Control and Information Sciences; Springer: Berlin, Germany, 1987; Volume 101.
38. Grmela, M. Contact geometry of mesoscopic thermodynamics and dynamics. *Entropy* **2014**, *16*, 1652. [CrossRef]
39. Morrison, P.J. A paradigm for joined Hamiltonian and dissipative systems. *Physics D* **1986**, *18*, 410–419. [CrossRef]
40. Brockett, R.W. *Geometric Control Theory*; Volume 7 of Lie Groups: History, Frontiers and Applications, Control Theory and Analytical Mechanics; Martin, C., Hermann, R., Eds.; MathSciPress: Brookline, MA, USA, 1977; pp. 1–46.
41. Van der Schaft, A.J. Hamiltonian dynamics with external forces and observations. *Math. Syst. Theory* **1982**, *15*, 145–168. [CrossRef]
42. Van der Schaft, A.J. System Theory and Mechanics. In *Three Decades of Mathematical System Theory*; Lecture Notes in Control and Information Sciences; Springer: Berlin, Germany, 1989; Volume 135, pp. 426–452.
43. Van der Schaft, A.J. *L_2-Gain and Passivity Techniques in Nonlinear Control*, 3rd ed.; Springer: Berlin, Germany, 2017.
44. Willems, J.C. Dissipative dynamical systems. Part I: General theory. *Arch. Rat. Mech. Anal.* **1972**, *45*, 321–351. [CrossRef]
45. Nijmeijer, H.; van der Schaft, A.J. *Nonlinear Dynamical Control Systems*; Springer-Verlag: New York, NY, USA, 1990.
46. Moses-Badlyan, A.; Maschke, B.; Beattie, C.; Mehrmann, V. Open physical systems: From GENERIC to port-Hamiltonian systems. In Proceedings of the International Symposium on Mathematical Theory of Networks and Systems (MTNS), Hong Kong, China, 16–20 July 2018; pp. 204–211.

entropy

MDPI

Review

From Lagrangian Mechanics to Nonequilibrium Thermodynamics: A Variational Perspective

François Gay-Balmaz [1,*,†] and Hiroaki Yoshimura [2,†]

1 Centre National de la Recherche Scientifique (CNRS), Le Laboratoire de Météorologie Dynamique (LMD), Ecole Normale Supérieure, 75005 Paris, France
2 School of Science and Engineering, Waseda University, Tokyo 169-8050, Japan; yoshimura@waseda.jp
* Correspondence: francois.gay-balmaz@lmd.ens.fr; Tel.: +33-144322266
† These authors contributed equally to this work.

Received: 21 November 2018; Accepted: 16 December 2018; Published: 23 December 2018

Abstract: In this paper, we survey our recent results on the variational formulation of nonequilibrium thermodynamics for the finite-dimensional case of discrete systems, as well as for the infinite-dimensional case of continuum systems. Starting with the fundamental variational principle of classical mechanics, namely, Hamilton's principle, we show, with the help of thermodynamic systems with gradually increasing complexity, how to systematically extend it to include irreversible processes. In the finite dimensional cases, we treat systems experiencing the irreversible processes of mechanical friction, heat, and mass transfer in both the adiabatically closed cases and open cases. On the continuum side, we illustrate our theory using the example of multicomponent Navier–Stokes–Fourier systems.

Keywords: nonequilibrium thermodynamics; variational formulation; nonholonomic constraints; irreversible processes; discrete thermodynamic systems; continuum thermodynamic systems

1. Introduction

This paper reviews our recent work on the development of a variational formulation of nonequilibrium thermodynamics, as established in [1–4]. This formulation extends to nonequilibrium thermodynamics of the Lagrangian formulation of classical and continuum mechanics that include irreversible processes, such as friction, heat, and mass transfer, chemical reactions, and viscosity.

1.1. Some History of the Variational Approaches to Thermodynamics

Thermodynamics was first developed to treat exclusively equilibrium states and the transition from one equilibrium state to another in which a change in temperature plays an important role. In this context, thermodynamics appeared mainly as a theory of heat, and it is viewed today as a branch of *equilibrium thermodynamics*. Such a classical theory, *which does not aim to describe the time evolution* of the system, can be developed in a well-established setting [5] governed by the well-known first and second laws, e.g., [6,7]. It is worth noting that classical mechanics, fluid dynamics, and electromagnetism, being essentially dynamical theories, *cannot* be treated in the context of equilibrium thermodynamics. Although much effort has been applied to the theoretical investigation of nonequilibrium thermodynamics in relation to physics, chemistry, biology, and engineering, the theory of nonequilibrium thermodynamics has not reached the level of completeness. This is in part due to the lack of a general variational formulation for nonequilibrium thermodynamics that would reduce to the classical Lagrangian variational formulation of mechanics in absence of irreversible processes. So far, various variational approaches have been proposed in relation to nonequilibrium thermodynamics. For example, the *principle of least dissipation of energy*, introduced in [8] and later extended in [9,10], underlies the reciprocal relations in linear phenomenological laws, and the *principle of minimum entropy production* by [11,12] sets conditions on steady-state processes. Onsager's approach was generalized in [13] for systems with nonlinear phenomenological laws. We refer to [14] for reviews and developments of Onsager's variational principles and for a study of the relation between Onsager's and Prigogine's principles. We also refer to Section 6 of [15,16] for overviews on variational approaches to irreversible processes. Note that, however, the variational principles developed in these previous works are not natural extensions of Hamilton's principle of classical mechanics, because they do not recover Hamilton's principle for the case in which irreversible processes are not included. Another important work was by [17,18], wherein, in conjunction with thermoelasticity, viscoelasticity, and heat transfer, a *principle of virtual dissipation* as a generalized form of the d'Alembert principle was used with various applications to nonlinear irreversible thermodynamics. In particular, Biot [17] mentioned that the relations between the rate of entropy production and state variables may be given as *nonholonomic constraints*. Nevertheless, this variational approach was restricted to *weakly irreversible systems* or *thermodynamically holonomic and quasi-holonomic* systems. More recently, it was noteworthy that [19] showed a variational formulation for viscoelastic fluids, in which the internal conversion of mechanical power into heat power due to frictional forces was written as a nonholonomic constraint. However, it should be noted that none of the approaches mentioned above present systematic and general variational formulations of nonequilibrium thermodynamics and are hence restricted to a certain class of thermodynamic systems.

Following the initial works of [20–22], the geometry of *equilibrium* thermodynamics has been mainly studied via contact geometry by [23], with further developments by [24–26]. In this geometric setting, thermodynamic properties are encoded by Legendre submanifolds of the thermodynamic phase space. A step toward a geometric formulation of irreversible processes was made in [27] by lifting port-Hamiltonian systems to the thermodynamic phase space. The underlying geometric structure in this construction is again a contact form. A description of irreversible processes using modifications of Poisson brackets was introduced in [28–30]. This was further developed, for instance, in [31–35]. A systematic construction of such brackets from the variational formulation given in the present paper was presented in [36] for the thermodynamics of multicomponent fluids.

1.2. Main Features of Our Variational Formulation

The variational formulation for nonequilibrium thermodynamics developed in [1–4] is distinct from the earlier variational approaches mentioned above, both in its physical meaning and in its mathematical structure, as well as in its goal. Roughly speaking, while most of the earlier variational approaches mainly underlie the equation for the rate of entropy production, in order to justify the expression of the phenomenological laws governing the irreversible processes involved, our variational approach aims to underlie the *complete set of time evolution equations of the system* in such a way that it extends the classical Lagrangian formulation in mechanics to nonequilibrium thermodynamic systems including irreversible processes.

This is accomplished by constructing a generalization of the *Lagrange–d'Alembert principle* of nonholonomic mechanics, where the entropy production of the system, written as the sum of the contribution of each of the irreversible processes, is incorporated into a *nonlinear nonholonomic constraint*. As a consequence, all the *phenomenological laws* are encoded in the nonlinear nonholonomic constraints, to which we naturally associate a *variational constraint* on the allowed variations of the action functional. A natural definition of the variational constraint in terms of the phenomenological constraint is possible thanks to the introduction of the concept of *thermodynamic displacement*, which generalizes the concept of thermal displacement given by [37] to all the irreversible processes.

More concretely, if the system involves internal irreversible processes, denoted by α, and irreversible process at the ports, denoted by β, with thermodynamic fluxes J_α, J_β and thermodynamic affinities X^α, X^β together with a thermodynamic affinity $X^\beta_{\rm ext}$ associated with the exterior, then the thermodynamic displacements $\Lambda^\alpha, \Lambda^\beta$ are such that $\dot{\Lambda}^\alpha = X^\alpha$ and $\dot{\Lambda}^\beta = X^\beta$.

This allows us to formulate the variational constraint associated with the phenomenological constraint in a systematic way, namely, by replacing all the velocities by their corresponding virtual displacement and by removing the external thermodynamic affinity $X^{\beta}_{\rm ext}$ at the exterior of the system as follows:

$$J_{\alpha}\dot{\Lambda}^{\alpha} + J_{\beta}\big(\dot{\Lambda}^{\beta} - X^{\beta}_{\rm ext}\big) \;\;\rightsquigarrow\;\; J_{\alpha}\delta\Lambda^{\alpha} + J_{\beta}\delta\Lambda^{\beta}.$$

Our variational formulation thus has a clear and systematic structure that appears to be common for the macroscopic description of the nonequilibrium thermodynamics of physical systems. It can be applied to the finite-dimensional case of discrete systems, such as classical mechanics, electric circuits, chemical reactions, and mass transfer. Further, our variational approach can be naturally extended to the infinite-dimensional case of continuum systems; for instance, it can be applied to some nontrivial example, such as the *multicomponent Navier–Stokes–Fourier equations*. Again, it is emphasized that our variational formulation consistently recovers Hamilton's principle in classical mechanics when irreversible processes are not taken into account.

1.3. Organization of the Paper

In Section 2, we start with a very elementary review of Hamilton's variational principle in classical mechanics and its extension to the case of mechanical systems with external forces. We also briefly review the variational formulation of mechanical systems with linear nonholonomic constraints by using the Lagrange–d'Alembert principle. Furthermore, we review the extension of Hamilton's principle to continuum systems and illustrate it with the example of compressible fluids in the Lagrangian description. The variational principle in the Eulerian description is then deduced in the context of *symmetry reduction*. In Section 3, we recall the two laws of thermodynamics as formulated by [38], and we present the variational formulation of nonequilibrium thermodynamics for the finite-dimensional case of discrete systems. We first consider adiabatically closed simple systems and illustrate the variational formulation using the case of a movable piston containing an ideal gas and the case of a system consisting of a chemical species experiencing diffusion between several compartments. We then consider adiabatically closed non-simple systems, such as the adiabatic piston with two cylinders and a system with a chemical species experiencing both diffusion and heat conduction between two compartments. Further, we consider the variational formulation for open systems and illustrate it with the example of a piston device with ports and heat sources. In Section 4, we extend the variational formulation of nonequilibrium thermodynamics to the infinite-dimensional case of continuum systems and consider a multicomponent compressible fluid subject to irreversible processes due to viscosity, heat conduction, and diffusion. The variational formulation is first given in the Lagrangian description, from which the variational formulation in the Eulerian description is deduced. This is illustrated with the multicomponent Navier–Stokes–Fourier equations. In Section 5, we make some concluding remarks and mention further developments based on the variational formulation of nonequilibrium thermodynamics, such as variational discretizations, Dirac structures in thermodynamics, reduction by symmetries, and thermodynamically consistent modeling.

2. Variational Principles in Lagrangian Mechanics

2.1. Classical Mechanics

One of the most fundamental statements in classical mechanics is the principle of critical action or Hamilton's principle, according to which the motion of a mechanical system between two given positions is given by a curve that makes the integral of the Lagrangian of the system critical (see, for instance, [39]).

Let us consider a mechanical system with configuration manifold Q. For instance, for a system of N particles moving in the Euclidean 3-space, the configuration manifold is $Q = \mathbb{R}^{3N}$, whereas for a rigid body moving freely in space, $Q = \mathbb{R}^3 \times SO(3)$, the product of the Euclidean 3-space and the rotation group. Let us denote by $(q^1,...,q^n)$ the local coordinates of the manifold Q, also known as generalized coordinates of the mechanical system. Let L be a given Lagrangian of the system, which usually depends only on the position q and velocity v of the system and is hence defined on the tangent bundle or *velocity phase space*, TQ, of the manifold Q. Recall that *tangent bundle* of a manifold Q is the manifold TQ given by the collection of all tangent vectors in Q. As a set, it is given by the disjoint union of the *tangent spaces* of Q, that is, $TQ = \sqcup_{q\in Q} T_qQ$, where T_qQ is the tangent space to Q at q. The elements in T_qQ are denoted by (q,v). The Lagrangian L is usually given by the kinetic minus the potential energy of the system: $L(q,v) = K(q,v) - U(q)$.

Hamilton's principle is written as follows. Suppose that the system occupies the positions q_1 and q_2 at the time t_1 and t_2. Then, the motion $q(t)$ of the mechanical system between these two positions is a solution of the critical point condition

$$\left.\frac{d}{d\epsilon}\right|_{\epsilon=0} \int_{t_1}^{t_2} L(q(t,\epsilon),\dot{q}(t,\epsilon))\mathrm{d}t = 0, \tag{1}$$

where $q(t,\epsilon)$, $t\in[t_1,t_2]$, $\epsilon\in[-a,a]$, is an arbitrary variation of the curve $q(t)$ with fixed endpoints, i.e., $q(t,\epsilon)|_{\epsilon=0} = q(t)$ and $q(t_1,\epsilon) = q(t_1)$, $q(t_2,\epsilon) = q(t_2)$, for all ϵ. The infinitesimal variation associated with a given variation $q(t,\epsilon)$ is denoted by

$$\delta q(t) := \left.\frac{d}{d\epsilon}\right|_{\epsilon=0} q(t,\epsilon).$$

From the fixed endpoint conditions, we have $\delta q(t_1) = \delta q(t_2) = 0$.

The Hamilton principle in Equation (1) is usually written in short form as

$$\delta \int_{t_1}^{t_2} L(q,\dot{q})\mathrm{d}t = 0, \tag{2}$$

for arbitrary infinitesimal variations δq, with $\delta q(t_1) = \delta q(t_2) = 0$. Throughout this paper, we always use this short notation for the variational principles and also simply refer to δq for variations.

The direct application of Equation (1) gives, in local coordinates $q = (q^1,...,q^n)$,

$$\begin{aligned}\delta \int_{t_1}^{t_2} L(q,\dot{q})\mathrm{d}t &= \int_{t_1}^{t_2}\left[\frac{\partial L}{\partial q^i}\delta q^i + \frac{\partial L}{\partial \dot{q}^i}\delta \dot{q}^i\right]\mathrm{d}t \\ &= \int_{t_1}^{t_2}\left[\frac{\partial L}{\partial q^i} - \frac{d}{dt}\frac{\partial L}{\partial \dot{q}^i}\right]\delta q^i\,\mathrm{d}t + \left[\frac{\partial L}{\partial \dot{q}^i}\delta q^i\right]_{t_1}^{t_2},\end{aligned} \tag{3}$$

where Einstein's summation convention is employed. Since δq is arbitrary and since the boundary term vanishes because of the fixed endpoint conditions, we get from Equation (3) the *Euler–Lagrange equations*:

$$\frac{d}{dt}\frac{\partial L}{\partial \dot{q}^i} - \frac{\partial L}{\partial q^i} = 0, \quad i = 1,...,n. \tag{4}$$

We recall that L is called *regular* when the Legendre transform $\mathbb{F}L : TQ \to T^*Q$, locally given by $(q^i,v^i) \mapsto (q^i, \frac{\partial L}{\partial v^i})$, is a local diffeomorphism, where T^*Q denotes the cotangent bundle or *momentum phase space* of Q. Recall that cotangent bundle of a manifold Q is the manifold $T^*Q = \cup_{q\in Q} T_q^*Q$, where T_q^*Q is the *cotangent space* at each q given as the dual space to T_qQ. The elements in T_q^*Q are covectors, denoted by (q,p). When L is regular, the Euler–Lagrange Equation (4) yields a second-order differential equation for the curve $q(t)$.

The energy of a mechanical system with the Lagrangian L is defined on TQ by

$$E(q,v) = \left\langle \frac{\partial L}{\partial v}, v\right\rangle - L(q,v), \tag{5}$$

where $\langle , \rangle$ denotes a dual pairing between the elements in T_q^*Q and T_qQ. It is easy to check that E is conserved along the solutions of the Euler–Lagrange Equation (4), namely,

$$\frac{d}{dt}E(q,\dot{q}) = \left(\frac{d}{dt}\frac{\partial L}{\partial \dot{q}^i} - \frac{\partial L}{\partial q^i}\right)\dot{q}^i = 0.$$

Let us assume that the mechanical system is subject to an external force, given by a map $F^{\text{ext}} : TQ \to T^*Q$ assumed to be fiber preserving, i.e., $F^{\text{ext}}(q,v) \in T_q^*Q$ for all $(q,v) \in T_qQ$. The extension of Equation (2) to forced mechanical systems is given by

$$\delta \int_{t_1}^{t_2} L(q,\dot{q})\mathrm{d}t + \int_{t_1}^{t_2} \left\langle F^{\text{ext}}(q,\dot{q}), \delta q \right\rangle \, \mathrm{d}t = 0, \tag{6}$$

for arbitrary variations δq, with $\delta q(t_1) = \delta q(t_2) = 0$. The second term in Equation (6) is the time integral of the virtual work $\langle F^{\text{ext}}(q,\dot{q}), \delta q\rangle$ done by the force field $F^{\text{ext}} : TQ \to T^*Q$ with a virtual displacement δq in TQ. The principle in Equation (6) leads to the *forced Euler–Lagrange equations*

$$\frac{d}{dt}\frac{\partial L}{\partial \dot{q}^i} - \frac{\partial L}{\partial q^i} = F_i^{\text{ext}}. \tag{7}$$

Systems with Nonholonomic Constraints

Hamilton's principle, as recalled above, is only valid for *holonomic* systems, i.e., systems without constraints or whose constraints are given by functions of the coordinates only, not the velocities. In geometric terms, such constraints are obtained by the specification of a submanifold N of the configuration manifold Q. In this case, the equations of motion are still given by Hamilton's principle for the Lagrangian L restricted to the tangent bundle TN of the submanifold $N \subset Q$.

When the constraints cannot be reduced to relations between the coordinates only, they are called *nonholonomic*. Here, we restrict the discussion to nonholonomic constraints that are linear in velocity. Such constraints are locally given in the form

$$\omega_i^\alpha(q)\dot{q}^i = 0, \quad \alpha = 1,...,k < n, \tag{8}$$

where ω_i^α are functions of local coordinates $q = (q^1,...,q^n)$ on Q. Intrinsically, the functions ω_i^α are the components of k independent one-forms ω^α on Q, i.e., $\omega^\alpha = \omega_i^\alpha dq^i$, for $\alpha = 1,...,k$. Typical examples of linear nonholonomic constraints are those imposed on the motion of rolling bodies, namely, the velocities of the points in contact should be identical.

For systems with nonholonomic constraints (Equation (8)), the corresponding equations of motion can be derived from a modification of the Hamilton principle called the *Lagrange–d'Alembert principle*, which is given by

$$\delta \int_{t_1}^{t_2} L(q,\dot{q})\mathrm{d}t = 0, \tag{9}$$

for variations δq subject to the condition

$$\omega_i^\alpha(q)\delta q^i = 0, \quad \alpha = 1,...,k < n, \tag{10}$$

together with the fixed endpoint conditions $\delta q(t_1) = \delta q(t_2) = 0$. Note the occurrence of two constraints with distinct roles. First, there is the constraint in Equation (8) on the solution curve called the *kinematic constraint*. Second, there is the constraint in Equation (10) on the variations used in the principle, referred to as the *variational constraint*. Later, we show that this distinction becomes more noticeable in nonequilibrium thermodynamics.

A direct application of Equations (9) and (10) yields the *Lagrange–d'Alembert equations*

$$\frac{d}{dt}\frac{\partial L}{\partial \dot{q}^i} - \frac{\partial L}{\partial q^i} = \lambda_\alpha \omega_i^\alpha. \tag{11}$$

These equations, together with the constraints in Equation (8), form a complete set of equations for the unknown curves $q^i(t)$ and $\lambda_\alpha(t)$.

For more information on nonholonomic mechanics, the reader can consult [40–42]. Note that the Lagrange–d'Alembert principle (9) is *not* a critical curve condition for the action integral *restricted to the space of a curve satisfying the constraints*. Such a principle, which imposes the constraint via a Lagrange multiplier, gives equations that are, in general, not equivalent to the Lagrange–d'Alembert Equation (11), see, e.g., [42,43]. Such equations are sometimes referred to as the vakonomic equations.

2.2. Continuum Mechanics

Hamilton's principle permits a natural extension to continuum systems, such as fluid and elasticity. For such systems, the configuration manifold Q is typically a manifold of maps. We shall restrict the discussion here to fluid mechanics in a fixed domain $\mathcal{D} \subset \mathbb{R}^3$ that is assumed to be bounded by a smooth boundary $\partial\mathcal{D}$. Hamilton's principle for fluid mechanics in the Lagrangian description has been discussed at least since the works of [44] for an incompressible fluid and [45,46] for compressible flows (see also [47] for further references on these early developments). Hamilton's principle has since then been an important modeling tool in continuum mechanics.

2.2.1. Configuration Manifolds

For fluid mechanics in a fixed domain and before the occurrence of any shocks, the configuration space can be taken as the manifold $Q = \mathrm{Diff}(\mathcal{D})$ of diffeomorphisms of $\mathcal{D}$. In this paper, we do not describe the functional analytic setting needed to rigorously work in the framework of infinite dimensional manifolds. For example, one can assume that the diffeomorphisms are of some given Sobolev class, regular enough (at least of class C^1) so that $\mathrm{Diff}(\mathcal{D})$ is a smooth infinite-dimensional manifold and a topological group with a smooth right translation. The tangent bundle to $\mathrm{Diff}(\mathcal{D})$ is formally given by the set of vector fields on $\mathcal{D}$ covering a diffeomorphism φ and tangent to the boundary, i.e., for each $\varphi \in \mathrm{Diff}(\mathcal{D})$, we have

$$T_\varphi \mathrm{Diff}(\mathcal{D}) = \{\mathbf{V} : \mathcal{D} \to T\mathcal{D} \mid \mathbf{V}(X) \in T_{\varphi(X)}\mathcal{D},\ \forall\, X \in \mathcal{D},\ \mathbf{V}(X) \in T_{\varphi(X)}\partial\mathcal{D},\ \forall\, X \in \partial\mathcal{D}\}.$$

The motion of the fluid is fully described by a curve $\varphi_t \in \mathrm{Diff}(\mathcal{D})$ defining the position $x = \varphi_t(X)$ at time t of a fluid particle with label $X \in \mathcal{D}$. The vector field $\mathbf{V}_t \in T_{\varphi_t}\mathrm{Diff}(\mathcal{D})$ defined by $\mathbf{V}_t(X) = \frac{d}{dt}\varphi_t(X)$ is the material velocity of the fluid. In local coordinates, we write $x^a = \varphi_t^a(X^A)$ and $\mathbf{V}_t^a(X^A) = \frac{d}{dt}\varphi_t^a(X^A)$.

2.2.2. Hamilton's Principle

Given a Lagrangian $L : TQ \to \mathbb{R}$ defined on the tangent bundle of the infinite-dimensional manifold $Q = \mathrm{Diff}(\mathcal{D})$, Hamilton's principle formally takes the same form as Equation (2), namely,

$$\delta \int_{t_1}^{t_2} L(\varphi, \dot{\varphi})\mathrm{d}t = 0, \tag{12}$$

for variations $\delta\varphi$ such that $\delta\varphi_{t_1} = \delta\varphi_{t_2} = 0$.

Let us consider a Lagrangian of the general form

$$L(\varphi, \dot{\varphi}) = \int_{\mathcal{D}} \mathscr{L}\left(\varphi(X), \dot{\varphi}(X), \nabla\varphi(X)\right) \mathrm{d}^3X,$$

with $\mathscr{L}$ being the Lagrangian density and $\nabla\varphi$ being the Jacobian matrix of φ, known as the deformation gradient in continuum mechanics. The variation of the integral yields

$$\delta \int_{t_1}^{t_2} L(\varphi, \dot{\varphi}) \mathrm{d}t = \int_{t_1}^{t_2} \int_{\mathcal{D}} \left[\frac{\partial \mathscr{L}}{\partial \varphi^a} \delta \varphi^a + \frac{\partial \mathscr{L}}{\partial \dot{\varphi}^a} \delta \dot{\varphi}^a + \frac{\partial \mathscr{L}}{\partial \varphi^a_{,A}} \delta \varphi^a_{,A} \right] \mathrm{d}^3 X \mathrm{d}t$$

$$= \int_{t_1}^{t_2} \int_{\mathcal{D}} \left[\frac{\partial \mathscr{L}}{\partial \varphi^a} \delta \varphi^a - \frac{\partial}{\partial t} \frac{\partial \mathscr{L}}{\partial \dot{\varphi}^a} - \frac{\partial}{\partial A} \frac{\partial \mathscr{L}}{\partial \varphi^a_{,A}} \right] \delta \varphi^a \mathrm{d}^3 X \mathrm{d}t$$

$$+ \int_{\mathcal{D}} \left[\frac{\partial \mathscr{L}}{\partial \dot{\varphi}^a} \delta \varphi^a \right]_{t_1}^{t_2} \mathrm{d}^3 X + \int_{t_1}^{t_2} \int_{\partial \mathcal{D}} \frac{\partial \mathscr{L}}{\partial \varphi^a_{,A}} \mathbf{N}_A \delta \varphi^a \mathrm{d}\mathcal{S} \mathrm{d}t,$$

where $\mathbf{N}$ is the outward-pointing unit normal vector field to the boundary $\partial\mathcal{D}$, and $\mathrm{d}\mathcal{S}$ denotes the area element on the surface $\partial\mathcal{D}$. Hamilton's principle thus yields the Euler–Lagrange equations and the boundary condition

$$\frac{\partial}{\partial t} \frac{\partial \mathscr{L}}{\partial \dot{\varphi}} + \mathrm{DIV} \frac{\partial \mathscr{L}}{\partial \nabla \varphi} = \frac{\partial \mathscr{L}}{\partial \varphi} \quad \text{and} \quad \left. \frac{\partial \mathscr{L}}{\partial \nabla \varphi} \cdot \mathbf{N} \right|_{T\partial\mathcal{D}} = 0 \text{ on } \partial\mathcal{D}, \tag{13}$$

where the divergence operator is defined as $\left(\mathrm{DIV} \frac{\partial \mathscr{L}}{\partial \nabla \varphi}\right)_a = \frac{\partial}{\partial A} \frac{\partial \mathscr{L}}{\partial \varphi^a_{,A}}$. The tensor field

$$\mathbf{P} := -\frac{\partial \mathscr{L}}{\partial \nabla \varphi}, \quad \text{i.e.} \quad \mathbf{P}^A_a = -\frac{\partial \mathscr{L}}{\partial \varphi^a_{,A}} \tag{14}$$

is called the *first Piola–Kirchhoff stress tensor* (see, e.g., [48]).

2.2.3. The Lagrangian of the Compressible Fluid

For a compressible fluid, the Lagrangian has the standard form

$$L(\varphi, \dot{\varphi}) = K(\varphi, \dot{\varphi}) - U(\varphi) = \int_{\mathcal{D}} \left[\frac{1}{2} \varrho_{\mathrm{ref}}(X) |\dot{\varphi}(X)|^2 - \mathscr{E}\left(\varrho_{\mathrm{ref}}(X), S_{\mathrm{ref}}(X), \nabla \varphi(X)\right) \right] \mathrm{d}^3 X, \tag{15}$$

with $\varrho_{\mathrm{ref}}(X)$ and $S_{\mathrm{ref}}(X)$ being the mass density and entropy density in the reference configuration. The two terms in Equation (15) are, respectively, the total kinetic energy of the fluid and minus the total internal energy of the fluid. The function $\mathscr{E}$ is a general expression for the internal energy density written in terms of $\varrho_{\mathrm{ref}}(X)$, $S_{\mathrm{ref}}(X)$, and the deformation gradient $\nabla\varphi(X)$. For fluids, $\mathscr{E}$ depends on the deformation gradient only through the Jacobian of φ, denoted by J_φ. This fact is compatible with the material covariance property of $\mathscr{E}$, written as

$$\mathscr{E}\left(\psi^* \varrho_{\mathrm{ref}}, \psi^* S_{\mathrm{ref}}, \nabla(\varphi \circ \psi)\right) = \psi^* \left[\mathscr{E}\left(\varrho_{\mathrm{ref}}, S_{\mathrm{ref}}, \nabla \varphi\right)\right], \quad \text{for all } \psi \in \mathrm{Diff}(\mathcal{D}), \tag{16}$$

where the pull-back notation is defined as

$$\varphi^* f = (f \circ \varphi) J_\varphi \tag{17}$$

for some function f defined on $\mathcal{D}$. From Equation (16), we deduce the existence of a function ϵ such that

$$\mathscr{E}\left(\varrho_{\mathrm{ref}}, S_{\mathrm{ref}}, \nabla \varphi\right) = \varphi^* \left[\epsilon(\rho, s)\right], \quad \text{for} \quad \rho = \varphi_* \varrho_{\mathrm{ref}}, \quad s = \varphi_* S_{\mathrm{ref}}, \tag{18}$$

(see [48,49]). The function $\epsilon = \epsilon(\rho, s)$ is the internal energy density in the spatial description expressed in terms of the mass density ρ and entropy density s.

For the Lagrangian Equation (15) and with the assumption in Equation (16), the first Piola–Kirchhoff stress tensor (Equation (14)) and its divergence are computed as

$$\mathbf{P}^A_a = \frac{\partial \mathscr{E}}{\partial \varphi^a_{,A}} = -p J_\varphi (\varphi^{-1})^A_{,a} \quad \text{and} \quad \mathrm{DIV}\, \mathbf{P} = (\nabla p \circ \varphi) J_\varphi, \tag{19}$$

where $p = \frac{\partial \epsilon}{\partial \rho}\rho + \frac{\partial \epsilon}{\partial s}s - \epsilon$ is the pressure. Note that for all $\delta\varphi^a$ parallel to the boundary, we have $\mathbf{P}_a^A \mathbf{N}_A \delta\varphi^a = -pJ_\varphi(\varphi^{-1})^A_{,a}\mathbf{N}_A\delta\varphi^a = 0$, since $(\varphi^{-1})^A_{,a}\delta\varphi^a$ is parallel to the boundary. Hence, the boundary condition in Equation (13) is always satisfied. From Equation (19), the Euler–Lagrange Equation (13) becomes

$$\varrho_{\rm ref}\ddot{\varphi} = (\nabla p \circ \varphi)J_\varphi. \tag{20}$$

Equation (20) is the equation of motion for a compressible fluid in the *material (or Lagrangian) description*, which directly follows from the Hamilton principle in Equation (12) applied to the Lagrangian Equation (15). It is, however, highly desirable to have a variational formulation that directly produces the equations of motion in the standard spatial (or Eulerian) description. This is recalled below in Section 2.3 by using Lagrangian reduction by symmetry.

2.3. Lagrangian Reduction by Symmetry

When symmetry is available in a mechanical system, it is often possible to exploit it in order to reduce the dimension of the system and thereby facilitate its study. This process, called *reduction by symmetry*, is presently well understood on both the Lagrangian and Hamiltonian sides (see [50] for an introduction and references).

On the Hamiltonian side, this process is based on the reduction of symplectic or Poisson structures while, on the Lagrangian side, it is usually based on the reduction of variational principles (see [51–53]). Consider a mechanical system with a configuration manifold Q and Lagrangian $L : TQ \to \mathbb{R}$, and consider also the action of a Lie group G on Q, denoted here simply as $q \mapsto g \cdot q$ for $g \in G$, $q \in Q$. This action naturally induces an action on the tangent bundle TQ, denoted here simply as $(q, v) \mapsto (g \cdot q, g \cdot v)$, called the *tangent-lifted action*. We say that the action is a symmetry for the mechanical system if the Lagrangian L is invariant under this tangent-lifted action. In this case, L induces a *symmetry-reduced Lagrangian* $\ell : (TQ)/G \to \mathbb{R}$ defined on the quotient space $(TQ)/G$ of the tangent bundle with respect to the action. The goal of the Lagrangian reduction process is to derive the equations of motion directly on the reduced space $(TQ)/G$. Under standard hypotheses on the action, this quotient space is a manifold, and one obtains the *reduced Euler–Lagrange equations* by computing the *reduced variational principle* for the action integral $\int_{t_1}^{t_2} \ell\, {\rm d}t$ induced by Hamilton's principle (Equation (2)) for the action integral $\int_{t_1}^{t_2} L\, {\rm d}t$. The main difference between the reduced variational principle and Hamilton's principle is the occurrence of constraints on the variations to be considered when computing the critical curves for $\int_{t_1}^{t_2} \ell\, {\rm d}t$. These constraints are uniquely associated with the reduced character of the variational principle and are not due to physical constraints as in Equation (10) earlier.

We now quickly recall the application of Lagrangian reduction for the treatment of fluid mechanics in a fixed domain (see Section 2.2) by following the Euler–Poincaré reduction approach in [54]. In this case, the Lagrangian reduction process encodes the shift from the material (or Lagrangian) description to the spatial (or Eulerian) description.

As we recalled above, in the material description, the motion of the fluid is described by a curve of diffeomorphisms φ_t in the configuration manifold $Q = {\rm Diff}(\mathcal{D})$, and the evolution Equation (20) for φ_t follows from the standard Hamilton principle.

In the spatial description, the dynamics are described by the Eulerian velocity $\mathbf{v}(t, x)$, the mass density $\rho(t, x)$ and the entropy density $s(t, x)$, defined in terms of φ_t as

$$\mathbf{v}_t = \dot{\varphi}_t \circ \varphi_t^{-1}, \quad \rho_t = (\varphi_t)_* \varrho_{\rm ref}, \quad s_t = (\varphi_t)_* S_{\rm ref}. \tag{21}$$

Using these relations and Equation (18), the Lagrangian Equation (15) in the material description induces the following expression in the spatial description:

$$\ell(\mathbf{v}, \rho, s) = \int_{\mathcal{D}} \left[\frac{1}{2}\rho|\mathbf{v}|^2 - \varepsilon(\rho, s)\right] {\rm d}^3x.$$

The symmetry group underlying the Lagrangian reduction process is the subgroup

$$\mathrm{Diff}(\mathcal{D})_{\varrho_{\rm red}, S_{\rm ref}} \subset \mathrm{Diff}(\mathcal{D})$$

of diffeomorphisms that preserve both the mass density $\varrho_{\rm ref}$ and entropy density $S_{\rm ref}$ in the reference configuration. So, we have $Q = \mathrm{Diff}(\mathcal{D})$ and $G = \mathrm{Diff}(\mathcal{D})_{\varrho_{\rm red}, S_{\rm ref}}$ in the general Lagrangian reduction setting described above.

From the relations in Equation (21), we deduce that the variations $\delta\varphi$ used in Hamilton's principle in Equation (12) induce the variations

$$\delta \mathbf{v} = \partial_t \boldsymbol{\zeta} + \mathbf{v} \cdot \nabla \boldsymbol{\zeta} - \boldsymbol{\zeta} \cdot \nabla \mathbf{v}, \quad \delta\rho = -\operatorname{div}(\rho\boldsymbol{\zeta}), \quad \delta s = -\operatorname{div}(s\boldsymbol{\zeta}), \tag{22}$$

where $\boldsymbol{\zeta} = \delta\varphi \circ \varphi^{-1}$ is an arbitrary time-dependent vector field parallel to $\partial\mathcal{D}$. From Lagrangian reduction theory, the Hamilton principle in Equation (12) induces, in the Eulerian description, the (reduced) variational principle

$$\delta \int_{t_1}^{t_2} \ell(\mathbf{v}, \rho, s)\, \mathrm{d}t = 0, \tag{23}$$

for variations $\delta\mathbf{v}$, $\delta\rho$, δs constrained by the relations in Equation (22) with $\boldsymbol{\zeta}(t_1) = \boldsymbol{\zeta}(t_2) = 0$. This principle yields the compressible fluid equations $\rho(\partial_t \mathbf{v} + \mathbf{v} \cdot \nabla \mathbf{v}) = -\nabla p$ in the Eulerian description, while the continuity equations $\partial_t \rho + \operatorname{div}(\rho\mathbf{v}) = 0$ and $\partial_t s + \operatorname{div}(s\mathbf{v}) = 0$ follow from the definition of ρ and s in Equation (21) (see [54]). We refer to [49] for an extension of this Lagrangian reduction approach to the case of fluids with a free boundary.

The variational formulations in Equations (22) and (23) are extended in Section 4 to include irreversible processes and are illustrated using the Navier–Stokes–Fourier system as an example.

3. Variational Formulation for Discrete Thermodynamic Systems

In this section, we present a variational formulation for the finite-dimensional case of discrete thermodynamic systems that reduces to Hamilton's variational principle in Equation (2) in absence of irreversible processes. The form of this variational formulation is similar to that of nonholonomic mechanics recalled earlier (see Equations (8)–(10)) in the sense that the critical curve condition is subject to two constraints: a *kinematic constraint* on the solution curve and a *variational constraint* on the variations to be considered when computing the criticality condition. A major difference, however, with the Lagrange–d'Alembert principle recalled above is that the constraints are *nonlinear* in velocity. This formulation is extended to continuum systems in Section 4.

Before presenting the variational formulation, we recall below the two laws of thermodynamics as formulated in [38].

- *The two laws of thermodynamics*

Let us denote by $\boldsymbol{\Sigma}$ a physical system and by $\boldsymbol{\Sigma}^{\rm ext}$ its exterior. The state of the system is defined by a set of mechanical variables and a set of thermal variables. State functions are functions of these variables. Stueckelberg's formulation of the two laws is given as follows.

- *First law:*

For every system $\boldsymbol{\Sigma}$, there exists an extensive scalar state function E, called *energy*, which satisfies

$$\frac{d}{dt} E(t) = P_W^{\rm ext}(t) + P_H^{\rm ext}(t) + P_M^{\rm ext}(t),$$

where $P_W^{\rm ext}$ is the *power associated with the work done on the system* (here, *work* includes not only *mechanical work* by the action of forces but also other physical work, such as that by the action of electric voltages, etc.), $P_H^{\rm ext}$ is the *power associated with the transfer of heat into the system*, and $P_M^{\rm ext}$ is the *power associated with the transfer of matter into the system.* As we recall below, a transfer of matter into the system is associated with a transfer of work and heat. By convention, $P_W^{\rm ext}$ and $P_H^{\rm ext}$ denote uniquely the power associated with a transfer of work and heat into the system that is *not associated with a transfer of matter.* The power associated with a transfer of heat or work due to a transfer of matter is included in $P_M^{\rm ext}$.

Given a *thermodynamic system*, the following terminology is generally adopted:

- A system is said to be *closed* if there is no exchange of matter, i.e., $P_M^{\rm ext}(t) = 0$. When $P_M^{\rm ext}(t) \neq 0$, the system is said to be *open*.
- A system is said to be *adiabatically closed* if it is closed and there are no heat exchanges, i.e., $P_M^{\rm ext}(t) = P_H^{\rm ext}(t) = 0$.
- A system is said to be *isolated* if it is adiabatically closed and there is no mechanical power exchange, i.e., $P_M^{\rm ext}(t) = P_H^{\rm ext}(t) = P_W^{\rm ext}(t) = 0$.

From the first law, it follows that the *energy of an isolated system is constant.*

- *Second law:*

For every system $\boldsymbol{\Sigma}$, there exists an extensive scalar state function S, called *entropy*, which obeys the following two conditions

(a) Evolution part:

If the system is adiabatically closed, the entropy S is a non-decreasing function with respect to time, i.e.,

$$\frac{d}{dt}S(t) = I(t) \geq 0,$$

where $I(t)$ is the *entropy production rate* of the system accounting for the irreversibility of internal processes.

(b) Equilibrium part:

If the system is isolated, as time tends to infinity, the entropy tends toward a finite local maximum of the function S over all thermodynamic states ρ compatible with the system, i.e.,

$$\lim_{t \to +\infty} S(t) = \max_{\rho \text{ compatible}} S[\rho].$$

By definition, the evolution of an isolated system is said to be *reversible* if $I(t) = 0$, namely, the entropy is constant. In general, the evolution of a system $\boldsymbol{\Sigma}$ is said to be *reversible* if the evolution of the total isolated system with which $\boldsymbol{\Sigma}$ interacts is reversible.

Based on this formulation of the two laws, Stueckelberg and Scheurer [38] developed a systematic approach for the derivation of the equations of motion for thermodynamic systems; it is especially well suited for the understanding of nonequilibrium thermodynamics as an extension of classical mechanics. We refer, for instance, to [55–57] for the applications of Stueckelberg's approach to the derivation of equations of motion for thermodynamical systems.

We present our approach by considering systems with gradually increasing level of complexity. First we treat *adiabatically closed* systems that have only one entropy variable or, equivalently, one temperature. Such systems, called *simple systems*, may involve the irreversible processes of mechanical friction and internal matter transfer. Then, we treat a more general class of finite-dimensional adiabatically closed thermodynamic systems with several entropy variables, which may also involve the irreversible process of heat conduction. We then consider *open* finite-dimensional thermodynamic systems, which can exchange heat and matter with the exterior. Finally, we explain how chemical reactions can be included in the variational formulation.

3.1. *Adiabatically Closed Simple Thermodynamic Systems*

We present below the definition of finite-dimensional and simple systems following [38]. A *finite-dimensional thermodynamic system* $\mathbf{\Sigma}$ is a collection $\mathbf{\Sigma} = \cup_{A=1}^{P} \mathbf{\Sigma}_A$ of a finite number of interacting simple thermodynamic systems $\mathbf{\Sigma}_A$. By definition, a *simple thermodynamic system* is a macroscopic system for which one (scalar) thermal variable and a finite set of nonthermal variables are sufficient to entirely describe the state of the system. From the second law of thermodynamics, we can always choose the entropy S as a thermal variable. A typical example of such a simple system is the one-cylinder problem. We refer to [55] for a systematic treatment of this system via Stueckelberg's approach.

3.1.1. Variational Formulation for Mechanical Systems with Friction

We consider here a simple system which can be described only by a single entropy as a thermodynamic variable, besides mechanical variables. As in Section 2.1 above, let Q be the configuration manifold associated with the mechanical variables of the simple system. The Lagrangian of the simple thermodynamic system is thus a function:

$$L : TQ \times \mathbb{R} \to \mathbb{R}, \quad (q, v, S) \mapsto L(q, v, S),$$

where $S \in \mathbb{R}$ is the entropy. We assume that the system involves external and friction forces given by fiber-preserving maps $F^{\mathrm{ext}}, F^{\mathrm{fr}} : TQ \times \mathbb{R} \to T^*Q$, i.e., such that $F^{\mathrm{fr}}(q, v, S) \in T_q^*Q$, similar to F^{ext}. As stated in [1], the variational formulation for this simple system is given as follows:

Find the curves $q(t)$, $S(t)$ which are critical for the *variational condition*

$$\delta \int_{t_1}^{t_2} L(q, \dot{q}, S) dt + \int_{t_1}^{t_2} \left\langle F^{\mathrm{ext}}(q, \dot{q}, S), \delta q \right\rangle \, dt = 0, \tag{24}$$

subject to the *phenomenological constraint*

$$\frac{\partial L}{\partial S}(q, \dot{q}, S)\dot{S} = \left\langle F^{\mathrm{fr}}(q, \dot{q}, S), \dot{q} \right\rangle, \tag{25}$$

and for variations subject to the *variational constraint*

$$\frac{\partial L}{\partial S}(q, \dot{q}, S)\delta S = \left\langle F^{\mathrm{fr}}(q, \dot{q}, S), \delta q \right\rangle, \tag{26}$$

with $\delta q(t_1) = \delta q(t_2) = 0$.

Taking variations of the integral in Equation (24), integrating by parts, and using $\delta q(t_1) = \delta(t_2) = 0$, it follows that

$$\int_{t_1}^{t_2} \left[\left(\frac{\partial L}{\partial q^i} - \frac{d}{dt}\frac{\partial L}{\partial \dot{q}^i} + F_i^{\mathrm{ext}} \right) \delta q^i + \frac{\partial L}{\partial S} \delta S \right] dt.$$

From the variational constraint in Equation (26), the last term in the integrand of the above equation can be replaced by $F_i^{\mathrm{fr}} \delta q^i$. Hence, using Equation (25), we get the following system of evolution equations for the curves $q(t)$ and $S(t)$:

$$\left\{ \begin{array}{l} \dfrac{d}{dt}\dfrac{\partial L}{\partial \dot{q}} - \dfrac{\partial L}{\partial q} = F^{\mathrm{fr}}(q, \dot{q}, S) + F^{\mathrm{ext}}(q, \dot{q}, S), \\[2ex] \dfrac{\partial L}{\partial S}\dot{S} = \left\langle F^{\mathrm{fr}}(q, \dot{q}, S), \dot{q} \right\rangle. \end{array} \right. \tag{27}$$

This variational formulation is a generalization of Hamilton's principle in Lagrangian mechanics in the sense that it can yield irreversible processes in addition to the Lagrange–d'Alembert equations with external and friction forces. In this generalized variational formulation, the temperature is defined as minus the derivative of L with respect to S, i.e., $T = -\frac{\partial L}{\partial S}$, which is assumed to be positive. When the Lagrangian has the standard form

$$L(q,v,S) = K(q,v) - U(q,S),$$

where the kinetic energy K is assumed to be independent of S, and $U(q,S)$ is the internal energy, then $T = -\frac{\partial L}{\partial S} = \frac{\partial U}{\partial S}$ recovers the standard definition of the temperature in thermodynamics.

When the friction force vanishes, the entropy is constant from the second equation in Equation (27), and hence, the system in Equation (27) reduces to the forced Euler–Lagrange equations in classical mechanics for a Lagrangian depending parametrically on a given constant entropy S_0.

The total energy associated with the Lagrangian is still defined by the same expression as in Equation (5) except that it now depends on S, i.e., we define the total energy $E : TQ \times \mathbb{R} \to \mathbb{R}$ by

$$E(q,v,S) = \left\langle \frac{\partial L}{\partial v}, v \right\rangle - L(q,v,S). \tag{28}$$

Along the solution curve of Equation (27), we have

$$\frac{d}{dt}E = \left(\frac{d}{dt}\frac{\partial L}{\partial \dot{q}^i} - \frac{\partial L}{\partial q^i} \right) \dot{q}^i - \frac{\partial L}{\partial S}\dot{S} = F_i^{\mathrm{ext}} \dot{q}^i = P_W^{\mathrm{ext}},$$

where P_W^{ext} is the power associated with the work done on the system. This is nothing but the statement of the first law for the thermodynamic system, as in Equation (27).

The rate of entropy production of the system is

$$\dot{S} = -\frac{1}{T}\left\langle F^{\mathrm{fr}}, \dot{q} \right\rangle.$$

The second law states that the internal entropy production is always positive, from which the friction force is dissipative, i.e., $\left\langle F^{\mathrm{fr}}(q,\dot{q},S), \dot{q} \right\rangle \leq 0$ for all $(q,\dot{q},S)$. This suggests the phenomenological relation $F_i^{\mathrm{fr}} = -\lambda_{ij}\dot{q}^j$, where λ_{ij}, $i,j = 1,...,n$ are functions of the state variables, with the symmetric part of the matrix λ_{ij} positive semi-definite, which are determined by experiments.

Remark 1 (Phenomenological and variational constraints). *The explicit expression of the constraint in Equation (25) involves phenomenological laws for the friction force* F^{fr}*, which is why we refer to it as a phenomenological constraint. The associated constraint in Equation (26) is called a variational constraint since it is a condition on the variations to be used in Equation (24). Note that the constraint in Equation (25) is nonlinear and also that one shifts from the variational constraint to the phenomenological constraint by formally replacing the time derivatives* $\dot{q}$, $\dot{S}$ *by the variations* δq, δS*:*

$$\frac{\partial L}{\partial S}\dot{S} = \left\langle F^{\mathrm{fr}}, \dot{q} \right\rangle \quad \rightsquigarrow \quad \frac{\partial L}{\partial S}\delta S = \left\langle F^{\mathrm{fr}}, \delta q \right\rangle.$$

Such a systematic correspondence between the phenomenological and variational constraints will hold, in general, for our variational formulation of thermodynamics, as we present in detail below.

Remark 2. *In our macroscopic description, it is assumed that the macroscopically "slow" or collective motion of the system can be described by* $q(t)$*, while the time evolution of the entropy* $S(t)$ *is determined from the microscopically "fast" motions of molecules through statistical mechanics under the assumption of local equilibrium. It follows from statistical mechanics that the internal energy* $U(q,S)$*, given as a potential energy at* *the macroscopic level, is essentially coming from the total kinetic energy associated with the microscopic motion of molecules, which is directly related to the temperature of the system.*

Example 1 (piston). *Consider a gas confined by a piston in a cylinder as in Figure 1. This is an example of a simple adiabatically closed system, whose state can be characterized by (q, v, S).*

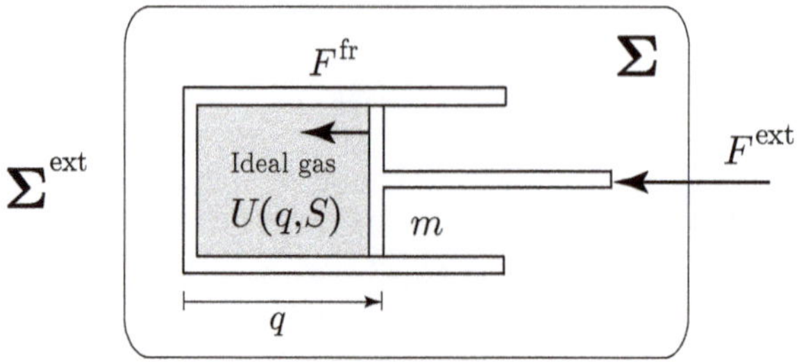

Figure 1. One cylinder.

The Lagrangian is given by $L(q, v, S) = \frac{1}{2}mv^2 - U(q, S)$, where m is the mass of the piston; $U(q, S) := \mathsf{U}(S, V = Aq, N_0)$, where $\mathsf{U}(S, V, N)$ is the internal energy of the gas, N_0 is the constant number of moles, $V = \alpha q$ is the volume, and α is the constant area of the cylinder. Note that we have

$$\frac{\partial U}{\partial S}(q, S) = T(q, S) \quad and \quad \frac{\partial U}{\partial q}(q, S) = -p(q, S)\alpha,$$

where T is temperature and $p = -\frac{\partial \mathsf{U}}{\partial V}$ is the pressure. The friction force reads $F^{\rm fr}(q, \dot{q}, S) = -\lambda(q, S)\dot{q}$, where $\lambda(q, S) \geq 0$ is the phenomenological coefficient, which is determined experimentally.

Following Equations (24)–(26), the variational formulation is given by

$$\delta \int_{t_1}^{t_2} \left[\frac{1}{2}m\dot{q}^2 - U(q, S)\right] {\rm d}t + \int_{t_1}^{t_2} F^{\rm ext}(q, \dot{q}, S)\delta q \, {\rm d}t = 0,$$

subject to the phenomenological constraint

$$\frac{\partial U}{\partial S}(q, S)\dot{S} = \lambda(q, S)\dot{q}^2.$$

and for variations subject to the variational constraint

$$\frac{\partial U}{\partial S}(q, S)\delta S = \lambda(q, S)\dot{q}\delta q.$$

From this principle, we get the equations of motion for the piston-cylinder system as

$$m\ddot{q} = p(q, S)\alpha + F^{\rm ext} - \lambda(q, S)\dot{q}, \qquad T(q, S)\dot{S} = \lambda(q, S)\dot{q}^2,$$

consistent with the equations derived in Section 4 of [55]. We can verify the energy balance, i.e., the first law, as $\frac{d}{dt}E = F^{\rm ext}\dot{q}$, where $E = \frac{1}{2}m\dot{q}^2 + U$ is the total energy.

3.1.2. Variational Formulation for Systems with Internal Mass Transfer

We here extend the previous variational formulation to the finite-dimensional case of discrete systems experiencing internal diffusion processes. Diffusion is particularly important in biology, as many processes depend on the transport of chemical species through bodies. For instance, the setting that we develop is well suited for the description of diffusion across composite membranes, e.g., composed of different elements arranged in a series or parallel array, which occurs frequently in living systems and has remarkable physical properties (see [58–61]).

As illustrated in Figure 2, we consider a thermodynamic system consisting of K compartments that can exchange matter by diffusion across walls (or membranes) of their common boundaries. We assume that the system has a single species, and we denote by N_k the number of moles of the species in the k-th compartment, $k = 1, ..., K$. We assume that the thermodynamic system is simple; i.e., a uniform entropy S, the entropy of the system, is attributed to all the compartments.

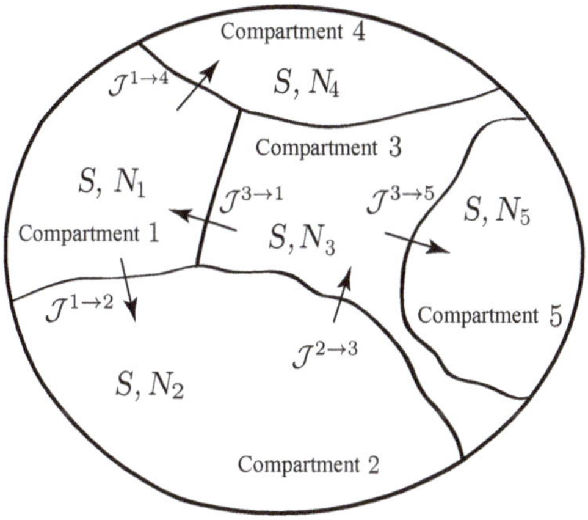

Figure 2. Simple adiabatically closed system with a single chemical species experiencing diffusion among several compartments.

For each compartment $k = 1, ..., K$, the mole balance equation is

$$\frac{d}{dt}N_k = \sum_{\ell=1}^{K} \mathcal{J}^{\ell\to k},$$

where $\mathcal{J}^{\ell\to k} = -\mathcal{J}^{k\to\ell}$ is the molar flow rate from compartment ℓ to compartment k due to diffusion of the species. We assume that the simple system also involves mechanical variables, friction, and exterior forces F^{fr} and F^{ext}, as in (A). The Lagrangian of the system is thus a function:

$$L : TQ \times \mathbb{R} \times \mathbb{R}^K \to \mathbb{R}, \quad (q, v, S, N_1, ..., N_K) \mapsto L(q, v, S, N_1, ..., N_K).$$

Thermodynamic displacements associated with matter exchange. The variational formulation involves the new variables W^k, $k = 1, ..., K$, which are examples of *thermodynamic displacements* and play a central role in our formulation. In general, we define the *thermodynamic displacement associated with an irreversible process* as the primitive in time of the thermodynamic force (or affinity) of the process. This force (or affinity) thus becomes the rate of change of the thermodynamic displacement. In the case of matter transfer, $\dot{W}^k$ corresponds to the chemical potential of N_k.

The variational formulation for a simple system with an internal diffusion process is stated as follows.

Find the curves $q(t)$, $S(t)$, $W^k(t)$, $N_k(t)$ which are critical for the *variational condition*

$$\delta \int_{t_1}^{t_2} \left[L(q, \dot{q}, S, N_1, ..., N_K) + \dot{W}^k N_k \right] \mathrm{d}t + \int_{t_1}^{t_2} \left\langle F^{\mathrm{ext}}, \delta q \right\rangle \mathrm{d}t = 0, \tag{29}$$

subject to the *phenomenological constraint*

$$\frac{\partial L}{\partial S}\dot{S} = \left\langle F^{\rm fr}, \dot{q} \right\rangle + \sum_{k,\ell=1}^{K} \mathcal{J}^{\ell\to k}\dot{W}^k, \tag{30}$$

and for variations subject to the *variational constraint*

$$\frac{\partial L}{\partial S}\delta S = \left\langle F^{\rm fr}, \delta q \right\rangle + \sum_{k,\ell=1}^{K} \mathcal{J}^{\ell\to k}\delta W^k, \tag{31}$$

with $\delta q(t_1) = \delta q(t_2) = 0$ and $\delta W^k(t_1) = \delta W^k(t_2) = 0$, $k = 1, ..., K$.

Taking variations of the integral in Equation (29), integrating by parts, and using $\delta q(t_1) = \delta q(t_2) = 0$ and $\delta W^k(t_1) = \delta W^k(t_2) = 0$, it follows that

$$\int_{t_1}^{t_2} \left[\left(\frac{\partial L}{\partial q^i} - \frac{d}{dt}\frac{\partial L}{\partial \dot{q}^i} + F_i^{\rm ext} \right)\delta q^i + \frac{\partial L}{\partial S}\delta S + \left(\frac{\partial L}{\partial N_k} + \dot{W}^k \right)\delta N_k - \dot{N}_k \delta W^k \right] dt.$$

Then, using the variational constraint in Equation (31), we get the following conditions:

$$\begin{aligned}
\delta q^i:&\quad \frac{d}{dt}\frac{\partial L}{\partial \dot{q}^i} - \frac{\partial L}{\partial q^i} = F_i^{\rm fr} + F_i^{\rm ext}, \quad i = 1, ..., n,\\
\delta N_k:&\quad \frac{d}{dt}W^k = -\frac{\partial L}{\partial N_k}, \quad k = 1, ..., K,\\
\delta W^k:&\quad \frac{d}{dt}N_k = \sum_{\ell=1}^{K} \mathcal{J}^{\ell\to k}, \quad k = 1, ..., K.
\end{aligned} \tag{32}$$

These conditions, combined with the phenomenological constraint in Equation (30), yield the system of evolution equations for the curves $q(t)$, $S(t)$, and $N^k(t)$:

$$\begin{cases}
\dfrac{d}{dt}\dfrac{\partial L}{\partial \dot{q}} - \dfrac{\partial L}{\partial q} = F^{\rm fr} + F^{\rm ext},\\[2mm]
\dfrac{d}{dt}N_k = \displaystyle\sum_{\ell=1}^{K} \mathcal{J}^{\ell\to k}, \quad k = 1, ..., K,\\[2mm]
\dfrac{\partial L}{\partial S}\dot{S} = \left\langle F^{\rm fr}, \dot{q} \right\rangle - \displaystyle\sum_{k<\ell} \mathcal{J}^{\ell\to k}\left(\dfrac{\partial L}{\partial N_k} - \dfrac{\partial L}{\partial N_\ell} \right).
\end{cases} \tag{33}$$

The total energy is defined as in Equations (5) and (28) and depends here on the mechanical variables $(q, v) \in TQ$, the entropy S, and the number of moles N_k, $k = 1, ..., K$, i.e., we define $E : TQ \times \mathbb{R} \times \mathbb{R}^K \to \mathbb{R}$ as

$$E(q, v, S, N_1, ..., N_K) = \left\langle \frac{\partial L}{\partial v}, v \right\rangle - L(q, v, S, N_1, ..., N_K). \tag{34}$$

On the solutions of Equation (33), we have

$$\frac{d}{dt}E = \left(\frac{d}{dt}\frac{\partial L}{\partial \dot{q}^i} - \frac{\partial L}{\partial q^i} \right)\dot{q}^i - \frac{\partial L}{\partial S}\dot{S} - \frac{\partial L}{\partial N_k}\dot{N}_k = F_i^{\rm ext}\dot{q}^i = P_W^{\rm ext},$$

where $P_W^{\rm ext}$ is the power associated with the work done on the system. This is the statement of the first law for the thermodynamic system in Equation (33).

For a given Lagrangian L, the temperature and chemical potentials of each compartment are defined as

$$T := -\frac{\partial L}{\partial S} \quad \text{and} \quad \mu^k := -\frac{\partial L}{\partial N_k}, \ k = 1, ..., K.$$

The last equation in Equation (33) yields the rate of entropy production of the system as

$$\dot{S} = -\frac{1}{T}\left\langle F^{\mathrm{fr}}, \dot{q} \right\rangle + \frac{1}{T}\sum_{k<\ell} \mathcal{J}^{k\to\ell}(\mu^k - \mu^\ell),$$

where the two terms correspond, respectively, to the rate of entropy production due to mechanical friction and to matter transfer. The second law suggests the phenomenological relations

$$F_i^{\mathrm{fr}} = -\lambda_{ij}\dot{q}^j \quad \text{and} \quad \mathcal{J}^{k\to\ell} = G^{kl}(\mu^k - \mu^\ell),$$

where λ_{ij}, $i, j = 1, ..., n$ and $G^{k\ell}$, $k, \ell = 1, ..., K$ are functions of the state variables, with the symmetric part of the matrix λ_{ij} positive semi-definite and with $G^{k\ell} \geq 0$ for all k, ℓ.

Example 2 (mass transfer associated with nonelectrolyte diffusion through a homogeneous membrane). *We consider a system with diffusion due to internal matter transfer through a homogeneous membrane separating two reservoirs. We suppose that the system is simple (so it is described by a single entropy variable) and involves a single chemical species. We assume that the membrane consists of three regions, namely, the central layer denotes the membrane capacitance in which energy is stored without dissipation, while the outer layers indicate transition regions in which dissipation occurs with no energy storage. We denote by N_m the number of mole of this chemical species in the membrane and by N_1 and N_2 the numbers of mole in reservoirs 1 and 2, as shown in Figure 3. Define the Lagrangian by $L(S, N_1, N_2, N_m) = -U(S, N_1, N_2, N_m)$, where $U(S, N_1, N_2, N_m)$ denotes the internal energy of the system, and assume that the volumes are constant and the system is isolated. We denote by $\mu^k = \frac{\partial U}{\partial N_k}$ the chemical potential of the chemical species in the reservoirs ($k = 1, 2$) and in the membrane ($k = m$). The flux from reservoir 1 into the membrane is denoted by $\mathcal{J}^{1\to m}$, and the flux from the membrane into reservoir 2 is denoted by $\mathcal{J}^{m\to 2}$.*

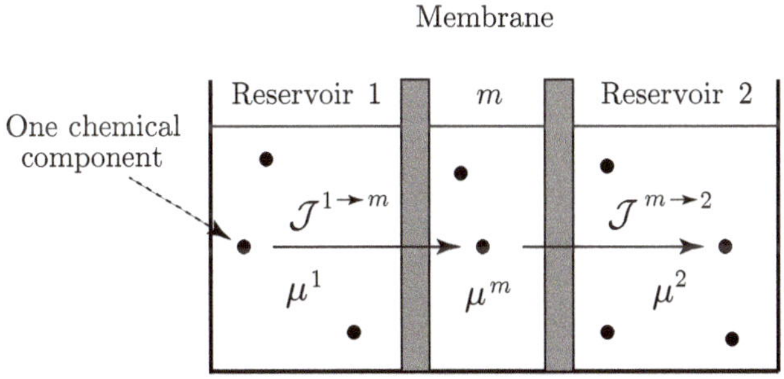

Figure 3. Nonelectrolyte diffusion through a homogeneous membrane.

The variational condition for the diffusion process is provided by

$$\delta \int_{t_1}^{t_2} \left[L(S, N_1, N_2, N_m) + \dot{W}^1 N_1 + \dot{W}^2 N_2 + \dot{W}^m N_m \right] dt = 0, \tag{35}$$

subject to the phenomenological constraint

$$\frac{\partial L}{\partial S}\dot{S} = \mathcal{J}^{m\to 1}(\dot{W}^1 - \dot{W}^m) + \mathcal{J}^{m\to 2}(\dot{W}^2 - \dot{W}^m) \tag{36}$$

and for variations subject to the variational constraint

$$\frac{\partial L}{\partial S}\delta S = \mathcal{J}^{m\to 1}(\delta W^1 - \delta W^m) + \mathcal{J}^{m\to 2}(\delta W^2 - \delta W^m), \tag{37}$$

with $\delta W^k(t_i) = 0$ *for* $k = 1, 2, m$ *and* $i = 1, 2$.

Thus, it follows that

$$\dot{N}_1 = \mathcal{J}^{m\to 1}, \quad \dot{N}_m = \mathcal{J}^{1\to m} + \mathcal{J}^{2\to m}, \quad \dot{N}_2 = \mathcal{J}^{m\to 2} \tag{38}$$

and $\dot{W}^1 = \mu^1$, $\dot{W}^2 = \mu^2$, $\dot{W}^m = \mu^m$. *The constraint in Equation (36) becomes*

$$-T\dot{S} = \mathcal{J}^{m\to 1}(\mu^1 - \mu^m) + \mathcal{J}^{m\to 2}(\mu^2 - \mu^m), \tag{39}$$

where $T = -\frac{\partial L}{\partial S}$. *Equations (38) and (39) are equivalent to those derived in ([61] Section 2.2). From Equations (38) and (39), we have energy conservation* $\frac{d}{dt}U = 0$, *which is consistent with the fact that the system is isolated.*

3.2. Adiabatically Closed Non-Simple Thermodynamic Systems

We now consider a general finite-dimensional system $\boldsymbol{\Sigma} = \cup_{A=1}^{P} \boldsymbol{\Sigma}_A$ composed of interconnected simple thermodynamic systems $\boldsymbol{\Sigma}_A$, as illustrated in Figure 4. This class of non-simple interconnected systems extends the class of *interconnected mechanical systems* (see [62]) to include the irreversible processes. In addition to the irreversible processes of friction and mass transfer described earlier, these systems can also involve the process of heat conduction.

The main difference from the previous cases is the occurrence of several entropy variables, namely, each subsystem $\boldsymbol{\Sigma}_A$ has an entropy denoted by S_A, $A = 1, ..., P$. Besides the variables S_A, each subsystem $\boldsymbol{\Sigma}_A$ may also be described by mechanical variables $q^A \in Q_A$ and number of moles $(N_{A,1}, ..., N_{A,K_A}) \in \mathbb{R}^{K_A}$, where Q_A is a configuration manifold for a mechanical variable associated with $\boldsymbol{\Sigma}_A$ and where K_A is the number of compartments in a simple system $\boldsymbol{\Sigma}_A$. For simplicity, we assume that independent mechanical coordinates $q \in Q$ have been chosen to represent the mechanical configuration of the interconnected system $\boldsymbol{\Sigma}$. The state variables needed to describe this system are

$$(q, v) \in TQ, \qquad S_A,\ A = 1, ..., P, \qquad N_{A,k},\ k = 1, ..., K_A,\ A = 1, ..., P. \tag{40}$$

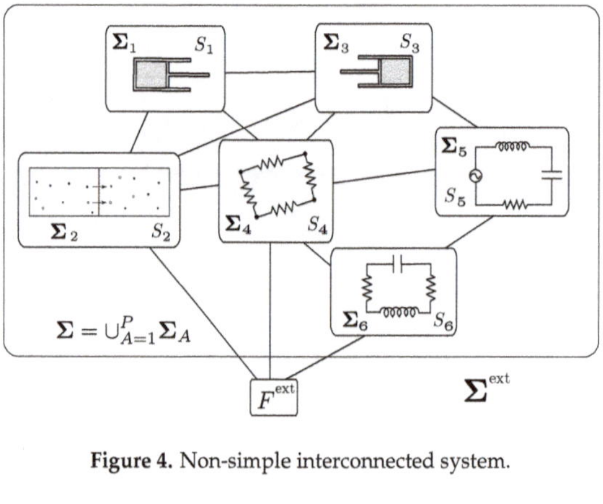

Figure 4. Non-simple interconnected system.

We present the variational formulation for these systems in two steps, exactly as in Section 3.1, by first considering the case without any transfer of mass.

3.2.1. Variational Formulation for Systems with Friction and Heat Conduction

Besides the entropies S_A, $A = 1, ..., P$, these systems only involve mechanical variables. The Lagrangian of the system is thus a function:

$$L : TQ \times \mathbb{R}^P \to \mathbb{R}, \quad (q, v, S_1, ..., S_P) \mapsto L(q, v, S_1, ..., S_P).$$

We denote by $F^{\text{ext}\to A} : T^*Q \times \mathbb{R}^P \to T^*Q$ the external force acting on subsystem $\mathbf{\Sigma}_A$. Consistent with the fact that the mechanical variables $q = (q^1, ..., q^n)$ describe the configuration of the entire interconnected system $\mathbf{\Sigma}$, only the total exterior force $F^{\text{ext}} = \sum_{A=1}^{P} F^{\text{ext}\to A}$ appears explicitly in the variational condition in Equation (42). We denote by $F^{\text{fr}(A)} : T^*Q \times \mathbb{R}^P \to T^*Q$ the friction forces experienced by subsystem $\mathbf{\Sigma}_A$. This friction force is at the origin of an entropy production for subsystem $\mathbf{\Sigma}_A$ and appears explicitly in the phenomenological constraint (Equation (43)) and the variational constraint (Equation (44)) of the variational formulation. We also introduce the fluxes J_{AB}, $A \neq B$ associated with the heat exchange between subsystems $\mathbf{\Sigma}_A$ and $\mathbf{\Sigma}_B$ and such that $J_{AB} = J_{BA}$. The relation between the fluxes J_{AB} and the heat power exchange $P_H^{A\to B}$ are given later. For the construction of variational structures, it is convenient to define the flux J_{AB} for $A = B$ as

$$J_{AA} := -\sum_{B \neq A} J_{AB},$$

so that we have

$$\sum_{A=1}^{P} J_{AB} = 0, \quad \text{for all } B. \tag{41}$$

Thermodynamic displacements associated with heat exchange. To incorporate heat exchange into our variational formulation, the new variables Γ^A, $A = 1, ..., P$ are introduced. These are again examples of *thermodynamic displacements* in the same way as we defined W^k before. For the case of heat exchange, $\dot{\Gamma}^A$ corresponds to the temperature of the subsystem $\mathbf{\Sigma}_A$, where Γ^A is identical to the *thermal displacement* employed in [37], which was originally introduced by [63]. The introduction of Γ^A is accompanied by the introduction of an entropy variable Σ_A whose meaning will be clarified later.

Now, the variational formulation for a system with friction and heat conduction is stated as follows:

Find the curves $q(t)$, $S_A(t)$, $\Gamma^A(t)$, $\Sigma_A(t)$ which are critical for the *variational condition*

$$\delta \int_{t_1}^{t_2} \left[L(q, \dot{q}, S_1, ..., S_K) + \dot{\Gamma}^A (S_A - \Sigma_A) \right] dt + \int_{t_1}^{t_2} \langle F^{\text{ext}}, \delta q \rangle \, dt = 0, \tag{42}$$

subject to the *phenomenological constraint*

$$\frac{\partial L}{\partial S_A} \dot{\Sigma}_A = \left\langle F^{\text{fr}(A)}, \dot{q} \right\rangle + J_{AB} \dot{\Gamma}^B, \quad \text{for } A = 1, ..., P, \tag{43}$$

and for variations subject to the *variational constraint*

$$\frac{\partial L}{\partial S_A} \delta \Sigma_A = \left\langle F^{\text{fr}(A)}, \delta q \right\rangle + J_{AB} \delta \Gamma^B, \quad \text{for } A = 1, ..., P, \tag{44}$$

with $\delta q(t_1) = \delta q(t_2) = 0$ and $\delta \Gamma^A(t_1) = \delta \Gamma^A(t_2) = 0$, $A = 1, ..., P$.

Taking variations of the integral in Equation (42), integrating by parts, and using $\delta q(t_1) = \delta(t_2) = 0$ and $\delta\Gamma_A(t_1) = \delta\Gamma_A(t_2) = 0$, it follows that

$$\int_{t_1}^{t_2} \left[\left(\frac{\partial L}{\partial q^i} - \frac{d}{dt}\frac{\partial L}{\partial \dot{q}^i} + F_i^{\text{ext}} \right) \delta q^i + \frac{\partial L}{\partial S_A}\delta S_A - (\dot{S}_A - \dot{\Sigma}_A)\delta\Gamma^A + \dot{\Gamma}^A(\delta S_A - \delta\Sigma_A) \right] \text{d}t = 0.$$

Then, using the variational constraint (Equation (44)), we get the following conditions:

$$\begin{aligned}
\delta q^i: \quad & \frac{\partial L}{\partial q^i} - \frac{d}{dt}\frac{\partial L}{\partial \dot{q}^i} - \sum_{A=1}^{P} \frac{\dot{\Gamma}^A}{\frac{\partial L}{\partial S_A}} F_i^{\text{fr}(A)} + F_i^{\text{ext}} = 0, \quad i = 1, \ldots, n, \\
\delta S_A: \quad & \frac{\partial L}{\partial S_A} + \dot{\Gamma}^A = 0, \quad A = 1, \ldots, P, \\
\delta\Gamma^A: \quad & -\dot{S}_A + \dot{\Sigma}_A - \sum_{B=1}^{P} \frac{\dot{\Gamma}^A}{\frac{\partial L}{\partial S_A}} J_{BA} = 0, \quad A = 1, \ldots, P.
\end{aligned}$$

The second equation yields

$$\dot{\Gamma}^A = -\frac{\partial L}{\partial S_A} =: T^A, \tag{45}$$

where T^A is the temperature of the subsystem $\boldsymbol{\Sigma}_A$. This implies that Γ_A is a thermal displacement. Because of Equation (41), the last equation yields $\dot{S}_A = \dot{\Sigma}_A$. Hence, using Equation (43), we get the following system of evolution equations for the curves $q(t)$ and $S_A(t)$:

$$\begin{cases}
\dfrac{d}{dt}\dfrac{\partial L}{\partial \dot{q}} - \dfrac{\partial L}{\partial q} = \displaystyle\sum_{A=1}^{P} F^{\text{fr}(A)} + F^{\text{ext}}, \\
\dfrac{\partial L}{\partial S_A}\dot{S}_A = \left\langle F^{\text{fr}(A)}, \dot{q} \right\rangle - \displaystyle\sum_{B=1}^{P} J_{AB}\left(\dfrac{\partial L}{\partial S_B} - \dfrac{\partial L}{\partial S_A} \right), \quad A = 1, \ldots, P.
\end{cases} \tag{46}$$

As before, we have $\frac{d}{dt}E = \left\langle F^{\text{ext}}, \dot{q} \right\rangle = P_W^{\text{ext}}$, where the total energy E is defined in the same way as before. Since the system is *non-simple*, it is instructive to analyze the energy behavior of each subsystem. This can be done if the Lagrangian is given by the sum of the Lagrangians of the subsystems, i.e.,

$$L(q, v, S_1, \ldots, S_P) = \sum_{A=1}^{P} L_A(q, v, S_A).$$

The mechanical equation for $\boldsymbol{\Sigma}_A$ is given as

$$\frac{d}{dt}\frac{\partial L_A}{\partial \dot{q}} - \frac{\partial L_A}{\partial q} = F^{\text{fr}(A)} + F^{\text{ext}\to A} + \sum_{B=1}^{P} F^{B\to A},$$

where $F^{B\to A} = -F^{A\to B}$ is the internal force exerted by $\boldsymbol{\Sigma}_B$ on $\boldsymbol{\Sigma}_A$. Denoting E_A as the total energy of $\boldsymbol{\Sigma}_A$, we have

$$\begin{aligned}
\frac{d}{dt}E_A &= \left\langle F^{\text{ext}\to A}, \dot{q} \right\rangle + \sum_{B=1}^{P} \left\langle F^{B\to A}, \dot{q} \right\rangle + \sum_{B=1}^{P} J_{AB}\left(\frac{\partial L}{\partial S_B} - \frac{\partial L}{\partial S_A} \right) \\
&= P_W^{\text{ext}\to A} + \sum_{B=1}^{P} P_W^{B\to A} + \sum_{B=1}^{P} P_H^{B\to A},
\end{aligned} \tag{47}$$

where $P_W^{\text{ext}\to A}$ and $P_W^{B\to A}$ denote the power associated with the work done on $\boldsymbol{\Sigma}_A$ by the exterior and that by the subsystem $\boldsymbol{\Sigma}_B$, respectively, and where $P_H^{B\to A}$ is the power associated with the heat transfer from $\boldsymbol{\Sigma}_B$ to $\boldsymbol{\Sigma}_A$. The link between the flux J_{AB} and the power exchange is thus

$$P_H^{B\to A} = J_{AB}(T^A - T^B).$$

Since entropy is an extensive variable, the total entropy of the system is $S = \sum_{A=1}^{P} S_A$. From Equation (46), it follows that the rate of total entropy production $\dot{S} = \sum_{A=1}^{P} \dot{S}_A$ of the system is given by

$$\dot{S} = -\sum_{A=1}^{P} \frac{1}{T^A} \left\langle F^{\mathrm{fr}(A)}, \dot{q} \right\rangle + \sum_{A<B}^{K} J_{AB} \left(\frac{1}{T^B} - \frac{1}{T^A} \right) (T^B - T^A). \tag{48}$$

The second law suggests the phenomenological relations

$$F_i^{\mathrm{fr}(A)} = -\lambda_{ij}^A \dot{q}^j \quad \text{and} \quad J_{AB} \frac{T^A - T^B}{T^A T^B} = \mathcal{L}_{AB}(T^B - T^A), \tag{49}$$

where λ_{ij}^A and $\mathcal{L}_{AB}$ are functions of the state variables, with the symmetric part of the matrices λ_{ij}^A positive semi-definite and with $\mathcal{L}_{AB} \geq 0$ for all A, B. From the second relation, we deduce $J_{AB} = -\mathcal{L}_{AB} T^A T^B = -\kappa_{AB}$, with $\kappa_{AB} = \kappa_{AB}(q, S_A, S_B)$ being the heat conduction coefficients between subsystem $\mathbf{\Sigma}_A$ and subsystem $\mathbf{\Sigma}_B$.

Example 3 (The adiabatic piston). *We consider a piston-cylinder system composed of two cylinders connected by a rod, each of which contains a fluid (or an ideal gas) and is separated by a movable piston, as shown in Figure 5. We assume that the system is isolated. Despite its apparent simplicity, this system has attracted a lot of attention in the literature because there has been some controversy about the final equilibrium state of this system when the piston is adiabatic. We refer to [55] for a review of this challenging problem and for the derivation of the time evolution of this system, based on the approach of [38].*

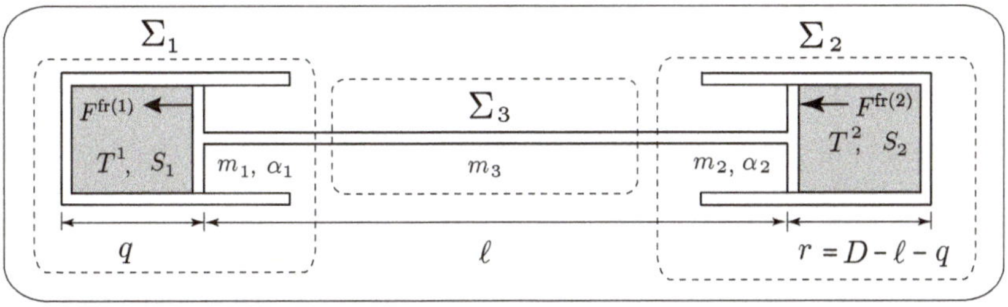

Figure 5. The two-cylinder problem.

The system $\mathbf{\Sigma}$ may be regarded as an interconnected system consisting of three simple systems; namely, the two pistons $\mathbf{\Sigma}_1, \mathbf{\Sigma}_2$ of mass m_1, m_2 and the connecting rod $\mathbf{\Sigma}_3$ of mass m_3. As illustrated in Figure 5, q and $r = D - \ell - q$ denote, respectively, the distance between the bottom of each piston to the top, where D is a constant. In this setting, we choose the variables (q, v, S_1, S_2) (the entropy associated with $\mathbf{\Sigma}_3$ is constant) to describe the dynamics of the interconnected system, and the Lagrangian is given by

$$L(q, v, S_1, S_2) = \frac{1}{2} M v^2 - U_1(q, S_1) - U_2(q, S_2), \tag{50}$$

where $M := m_1 + m_2 + m_3$, and

$$U_1(q, S_1) := \mathsf{U}_1(S_1, V_1 = \alpha_1 q, N_1), \quad U_2(q, S_2) := \mathsf{U}_2(S_2, V_2 = \alpha_2 r, N_2),$$

with $\mathsf{U}_i(S_i, V_i, N_i)$ as the internal energies of the fluids, N_i as the constant number of moles, and α_i as the constant areas of the cylinders, $i = 1, 2$.

As in Equation (49), we have $F^{\mathrm{fr}(A)}(q, \dot{q}, S_A) = -\lambda^A \dot{q}$, with $\lambda^A = \lambda^A(q, S^A) \geq 0$, $A = 1, 2$ and $J_{AB} = -\kappa_{AB} =: -\kappa$, where $\kappa = \kappa(S_1, S_2, q) \geq 0$ is the heat conductivity of the connecting rod.

From the variational formulations (Equations (42)–(44)), we get the following system for $q(t)$, $S_1(t)$, $S_2(t)$, in light of Equation (46), as

$$\begin{cases} M\ddot{q} = p_1(q,S_1)\alpha_1 - p_2(q,S_2)\alpha_2 - (\lambda^1+\lambda^2)\dot{q}, \\ T^1(q,S_1)\dot{S}_1 = \lambda^1\dot{q}^2 + \kappa\left(T^2(q,S_2) - T^1(q,S_1)\right), \\ T^2(q,S_2)\dot{S}_2 = \lambda^2\dot{q}^2 + \kappa\left(T^1(q,S_1) - T^2(q,S_2)\right), \end{cases}$$

where we used $\frac{\partial U_i}{\partial S_i}(q,S_i) = T^i(q,S_i)$, $\frac{\partial U_1}{\partial q} = -p_1(q,S_1)\alpha_1$, and $\frac{\partial U_2}{\partial q} = p_2(q,S_2)\alpha_2$.

These equations recover those derived in [55], (51)–(53). We have $\frac{d}{dt}E = 0$, where $E = \frac{1}{2}M\dot{q}^2 + U_1(q,S_1) + U(q,S_2)$, consistent with the fact that the system is isolated. The rate of total entropy production is

$$\frac{d}{dt}S = \left(\frac{\lambda^1}{T^1} + \frac{\lambda^2}{T^2}\right)\dot{q}^2 + \kappa\frac{(T^2-T^1)^2}{T^1T^2} \geq 0.$$

The equations of motion for the adiabatic piston are obtained by setting $\kappa = 0$.

3.2.2. Variational Formulation for Systems with Friction, Heat Conduction, and Internal Mass Transfer

We extend the previous case to one in which the subsystems $\mathbf{\Sigma}_A$ not only exchange work and heat but also exchange matter. In general, each subsystem may itself have several compartments, in which case the variables are those listed in Equation (40). For simplicity, we assume that each subsystem has only one compartment. The reader can easily extend this approach to the general case. The Lagrangian is thus a function:

$$L : TQ \times \mathbb{R}^P \times \mathbb{R}^P \to \mathbb{R}, \quad (q,v,S_1,...,S_P,N_1,...,N_P) \mapsto L(q,v,S_1,...,S_P,N_1,...,N_P),$$

where S_A and N_A are the entropy and number of moles of subsystem $\mathbf{\Sigma}_A$, $A = 1,...,P$. Since the previous cases are presented in detail above, we here just present the variational formulation and the resulting equations of motion.

Find the curves $q(t)$, $S_A(t)$, $\Gamma^A(t)$, $\Sigma_A(t)$, $W^A(t)$, $N_A(t)$ which are critical for the *variational condition*

$$\delta\int_{t_1}^{t_2}\left[L(q,\dot{q},S_1,...,S_P,N_1,...,N_P) + \dot{W}^A N_A + \dot{\Gamma}^A(S_A - \Sigma_A)\right]dt + \int_{t_1}^{t_2}\langle F^{\text{ext}},\delta q\rangle\, dt = 0, \tag{51}$$

subject to the *phenomenological constraint*

$$\frac{\partial L}{\partial S_A}\dot{\Sigma}_A = \left\langle F^{\text{fr}(A)},\dot{q}\right\rangle + J_{AB}\dot{\Gamma}^B + \mathcal{J}^{B\to A}\dot{W}^A, \quad \text{for } A = 1,...,P, \tag{52}$$

and for variations subject to the *variational constraint*

$$\frac{\partial L}{\partial S_A}\delta\Sigma_A = \left\langle F^{\text{fr}(A)},\delta q\right\rangle + J_{AB}\delta\Gamma^B + \mathcal{J}^{B\to A}\delta W^A, \quad \text{for } A = 1,...,P, \tag{53}$$

with $\delta q(t_i) = \delta W^A(t_i) = \delta\Gamma^A(t_i) = 0$, $i = 1,2$, $A = 1,...,P$.

From Equations (51)–(53), we obtain the following system of evolution equations for the curves $q(t)$, $S_A(t)$, and $N_A(t)$:

$$\begin{cases} \dfrac{d}{dt}\dfrac{\partial L}{\partial \dot{q}} - \dfrac{\partial L}{\partial q} = \sum\limits_{A=1}^{P} F^{\mathrm{fr}(A)} + F^{\mathrm{ext}}, \\ \dfrac{d}{dt} N_A = \sum\limits_{B=1}^{P} \mathcal{J}^{B\to A}, \quad A = 1, ..., P, \\ \dfrac{\partial L}{\partial S_A}\dot{S}_A = \left\langle F^{\mathrm{fr}(A)}, \dot{q} \right\rangle - \sum\limits_{B=1}^{P} J_{AB}\left(\dfrac{\partial L}{\partial S_B} - \dfrac{\partial L}{\partial S_A} \right) - \sum\limits_{B=1}^{P} \mathcal{J}^{B\to A}\dfrac{\partial L}{\partial N_A}, \quad A = 1, ..., P. \end{cases} \tag{54}$$

We also obtain the conditions

$$\dot{\Gamma}^A = -\frac{\partial L}{\partial S_A} =: T^A, \qquad \dot{W}^A = -\frac{\partial L}{\partial N_A} =: \mu^A, \qquad \dot{\Sigma}_A = \dot{S}_A, \quad A = 1, ..., P,$$

where we defined the temperature T^A and the chemical potential μ^A of the subsystem $\mathbf{\Sigma}_A$. The variables Γ^A and W^A are again the thermodynamic displacements associated with the processes of heat and matter transfer.

The total energy satisfies $\frac{d}{dt}E = P_W^{\mathrm{ext}}$ and the detailed energy balances can be carried out as in Equation (47) and yields here

$$P_{H+M}^{B\to A} = J_{AB}(T^A - T^B).$$

The rate of total entropy production of the system is computed as

$$\dot{S} = -\sum_{A=1}^{P} \frac{1}{T^A}\left\langle F^{\mathrm{fr}(A)}, \dot{q} \right\rangle + \sum_{A<B} J_{AB}\left(\frac{1}{T^B} - \frac{1}{T^A} \right)(T^B - T^A) + \sum_{A<B} \mathcal{J}^{B\to A}\left(\frac{\mu^B}{T^B} - \frac{\mu^A}{T^A} \right).$$

From the second law of thermodynamics, the total entropy production must be positive and hence suggests the phenomenological relations

$$F_i^{\mathrm{fr}(A)} = -\lambda_{ij}^A \dot{q}^j, \qquad \begin{bmatrix} \frac{T^A - T^B}{T^A T^B} J_{AB} \\ \mathcal{J}^{B\to A} \end{bmatrix} = \mathcal{L}_{AB} \begin{bmatrix} T^B - T^A \\ \frac{\mu^B}{T^B} - \frac{\mu^A}{T^A} \end{bmatrix}, \tag{55}$$

where the symmetric part of the $n \times n$ matrices λ^A and of the 2×2 matrices $\mathcal{L}_{AB}$ are positive. The entries of these matrices are phenomenological coefficients determined experimentally, which may generally depend on the state variables. From Onsager's reciprocal relations, the 2×2 matrix

$$\mathcal{L}_{AB} = \begin{bmatrix} \mathcal{L}_{AB}^{HH} & \mathcal{L}_{AB}^{HM} \\ \mathcal{L}_{AB}^{MH} & \mathcal{L}_{AB}^{MM} \end{bmatrix}$$

is symmetric for all A, B. The matrix elements $\mathcal{L}_{AB}^{HH}$ and $\mathcal{L}_{AB}^{MM}$ are related to the processes of heat conduction and diffusion between $\mathbf{\Sigma}_A$ and $\mathbf{\Sigma}_B$. The coefficients $\mathcal{L}_{AB}^{MH}$ and $\mathcal{L}_{AB}^{HM}$ describe the cross-effects, and hence are associated with discrete versions of the process of thermal diffusion and the Dufour effect. Thermal diffusion is the process of matter diffusion due to the temperature difference between the compartments. The Dufour effect is the process of heat transfer due to difference of chemical potentials between the compartments.

Example 4 (Heat conduction and diffusion between two compartments). *We consider a closed system consisting of two compartments, as illustrated in Figure 6. The compartments are separated by a permeable wall* *through which heat conduction and diffusion is possible. The system is closed and, therefore, there is no matter transfer with exterior, while we have heat and mass transfer between the compartments.*

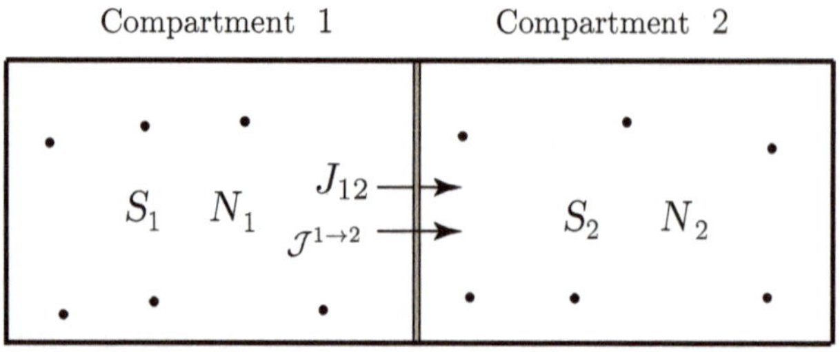

Figure 6. Non-simple closed system with a single chemical species, experiencing diffusion and heat conduction between two compartments.

The Lagrangian of this system is

$$L(S_1, S_2, N_1, N_2) = -U_1(S_1, N_1) - U_2(S_2, N_2),$$

where $U_i(S_i, N_i)$ is the internal energy of the ith chemical species and the volume is assumed to be constant. In this case, the system in Equation (54) specifies

$$\begin{cases} \dot{N}_1 = \mathcal{J}^{2\to 1}, \qquad \dot{N}_2 = \mathcal{J}^{1\to 2}, \\ T^1 \dot{S}_1 = -J^{12}(T^2 - T^1) - \mathcal{J}^{2\to 1}\mu^1, \\ T^2 \dot{S}_2 = -J^{12}(T^1 - T^2) - \mathcal{J}^{1\to 2}\mu^2, \end{cases} \tag{56}$$

where

$$T^A = \frac{\partial U}{\partial S_A}, \qquad \mu^A = \frac{\partial U}{\partial N_A}, \quad A = 1, 2$$

are the temperatures and chemical potentials of the Ath compartments. From Equation (56), it follows that the equation for the total entropy $S = S_1 + S_2$ of the system is

$$\dot{S} = J^{12}(T^1 - T^2)\left(\frac{1}{T^1} - \frac{1}{T^2}\right) + \mathcal{J}^{1\to 2}\left(\frac{\mu^1}{T^1} - \frac{\mu^2}{T^2}\right) \geq 0,$$

from which the phenomenological relations are obtained as in Equation (55). The energy balance in each compartment is

$$\frac{d}{dt}U_1 = -J^{12}(T^2 - T^1), \qquad \frac{d}{dt}U_2 = -J^{12}(T^1 - T^2),$$

which shows the relation between the flux J^{12} and the power $P^{1\to 2} = J^{12}(T^2 - T^1)$ exchanged between the two compartments due to heat conduction, diffusion, and their cross-effects. The total energy $E = U_1 + U_2$ is conserved.

Remark 3 (General structure of the variational formulation for adiabatically closed systems). *In each of the situation considered, the variational constraint can be systematically obtained from the phenomenological constraint by replacing the time derivative by the delta variation for each process. For the most general case treated above, we have the following correspondence:*

$$\frac{\partial L}{\partial S_A}\dot{\Sigma}_A = \left\langle F^{\mathrm{fr}(A)}, \dot{q} \right\rangle + J_{AB}\dot{\Gamma}^B + \mathcal{J}^{B\to A}\dot{W}^A \quad \rightsquigarrow \quad \frac{\partial L}{\partial S_A}\delta\Sigma_A = \left\langle F^{\mathrm{fr}(A)}, \delta q \right\rangle + J_{AB}\delta\Gamma^B + \mathcal{J}^{B\to A}\delta W^A.$$

In the above, the quantities to be determined from the state variables by phenomenological laws are $F^{\mathrm{fr}(A)}$, J_{AB}, *and* $\mathcal{J}^{B\to A}$.

The structure of our variational formulation is better explained by adopting a general point of view. If we denote by $\mathcal{Q}$ *the thermodynamic configuration manifold and by* $x \in \mathcal{Q}$ *the collection of all the variables of the thermodynamic system, for instance,* $x = (q, S_A, N_A, W^A, \Gamma^A, \Sigma_A)$, $A = 1, ..., P$ *in the preceding case, then the variational formulation for an adiabatically closed system falls into the following abstract setting. Given a Lagrangian* $\mathcal{L} : T\mathcal{Q} \to \mathbb{R}$, *an external force* $\mathcal{F}^{\mathrm{ext}} : T\mathcal{Q} \to T^*\mathcal{Q}$, *and fiber-preserving maps* $A^\alpha : T\mathcal{Q} \to T^*\mathcal{Q}$, $A^\alpha(x,v) \in T^*_x\mathcal{Q}$, $\alpha = 1, ..., k$, *the variational formulation reads as follows:*

$$\delta \int_{t_1}^{t_2} \mathcal{L}(x(t), \dot{x}(t))\mathrm{d}t + \int_{t_1}^{t_2} \left\langle \mathcal{F}^{\mathrm{ext}}(x(t), \dot{x}(t)), \delta x(t) \right\rangle \mathrm{d}t = 0, \tag{57}$$

where the curve $x(t)$ *satisfies the phenomenological constraint*

$$A^\alpha(x, \dot{x}) \cdot \dot{x} = 0, \quad \text{for } \alpha = 1, ..., k, \tag{58}$$

and for variations δx *subject to the variational constraint*

$$A^\alpha(x, \dot{x}) \cdot \delta x = 0, \quad \text{for } \alpha = 1, ..., k, \tag{59}$$

with $\delta x(t_1) = \delta x(t_2) = 0$.

This yields the system of equations

$$\begin{cases} \dfrac{d}{dt}\dfrac{\partial \mathcal{L}}{\partial \dot{x}} - \dfrac{\partial \mathcal{L}}{\partial x} - \mathcal{F}^{\mathrm{ext}} = \lambda_\alpha A^\alpha(x, \dot{x}), \\ A^\alpha(x, \dot{x}) \cdot \dot{x} = 0, \quad \alpha = 1, ..., k. \end{cases} \tag{60}$$

It is clear that all the variational formulations for the adiabatically closed system considered above fall into this category by appropriately choosing x, $\mathcal{L}(x, \dot{x})$, $\mathcal{F}^{\mathrm{ext}}(x, \dot{x})$, *and* $A^\alpha(x, \dot{x})$. *The energy defined by* $\mathcal{E}(x, v) = \left\langle \frac{\partial \mathcal{L}}{\partial v}, v \right\rangle - \mathcal{L}(x, v)$ *satisfies* $\frac{d}{dt}\mathcal{E} = \langle \mathcal{F}^{\mathrm{ext}}, \dot{x} \rangle$.

The constraints involved in this variational formulation admit an intrinsic geometric description. The variational constraint (Equation (59)) defines the subset $C_V \subset T\mathcal{Q} \times_{\mathcal{Q}} T\mathcal{Q}$ *given by*

$$C_V = \{(x, v, \delta x) \in T\mathcal{Q} \times_{\mathcal{Q}} T\mathcal{Q} \mid A^\alpha(x, v) \cdot \delta x = 0, \ \text{for } \alpha = 1, ..., k\},$$

so that $C_V(x,v) := C_V \cap (\{(x,v)\} \times T_x\mathcal{Q})$ *is a vector subspace of* $T_x\mathcal{Q}$ *for all* $(x,v) \in T\mathcal{Q}$. *The phenomenological constraint (Equation (58)) defines the subset* $C_K \subset T\mathcal{Q}$ *given by*

$$C_K = \{(x, v) \in T\mathcal{Q} \mid A^\alpha(x, v) \cdot v = 0, \ \text{for } \alpha = 1, ..., k\}.$$

Then, one notes that the constraint C_K *can be intrinsically defined from* C_V *as*

$$C_K = \{(x, v) \in T\mathcal{Q} \mid (x, v) \in C_V(x, v)\}.$$

Constraints C_V *and* C_K *related in this way are called nonlinear nonholonomic constraints of thermodynamic type (see [1,64]).*

3.3. Open Thermodynamic Systems

The thermodynamic systems that we considered so far are restricted to the adiabatically closed cases. For such systems, interaction with the exterior is only through the exchange of mechanical work, and hence the first law for such systems reads

$$\frac{d}{dt}E = \langle F^{\rm ext}, \dot{q} \rangle = P^{\rm ext}_W.$$

We now consider the more general case of open systems exchanging work, heat, and matter with the exterior. In this case, the first law reads

$$\frac{d}{dt}E = P^{\rm ext}_W + P^{\rm ext}_H + P^{\rm ext}_M,$$

where $P^{\rm ext}_H$ is the power associated with the transfer of heat into the system and $P^{\rm ext}_M$ is the power associated with the transfer of matter into the system. As we recall below, the transfer of matter into or out of the system is associated with a transfer of work and heat. By convention, $P^{\rm ext}_W$ and $P^{\rm ext}_H$ denote uniquely the power associated with work and heat that is not associated with a transfer of matter. The power associated with a transfer of heat or work due to a transfer of matter is included in $P^{\rm ext}_M$.

In order to get a concrete expression for $P^{\rm ext}_M$, let us consider an open system with several ports, $a = 1, ..., A$, through which matter can flow into or out of the system. We suppose, for simplicity, that the system involves only one chemical species and denote by N the number of moles of this species. The mole balance equation is

$$\frac{d}{dt}N = \sum_{a=1}^{A} \mathcal{J}^a,$$

where $\mathcal{J}^a$ is the molar flow rate *into* the system through the ath port so that $\mathcal{J}^a > 0$ indicates the flow into the system and $\mathcal{J}^a < 0$ indicates the flow out of the system.

As matter enters or leaves the system, it carries its internal, potential, and kinetic energy. This energy flow rate at the ath port is the product $\mathsf{E}^a \mathcal{J}^a$ of the energy per mole (or molar energy) E^a and the molar flow rate $\mathcal{J}^a$ at the ath port. In addition, as matter enters or leaves the system, it also exerts work on the system that is associated with pushing the species into or out of the system. The associated energy flow rate is given at the a-th port by $\mathcal{J}^a p^a \mathsf{V}^a$, where p^a and V^a are the pressure and the molar volume of the substance flowing through the ath port. From this, we get the expression

$$P^{\rm ext}_M = \sum_{a=1}^{A} \mathcal{J}^a (\mathsf{E}^a + p^a \mathsf{V}^a). \tag{61}$$

We refer, for instance, to [65,66] for the detailed explanations of the first law for open systems.

We present below an extension of the variational formulation to the case of open systems. In order to motivate the form of the constraints that we use, we first consider a particular case of simple open system, namely, the case of a system with a single chemical species N in a single compartment with constant volume V and without mechanical effects. In this particular situation, the energy of the system is given by the internal energy written as $U = U(S, N)$, since $V = V_0$ is constant. The balance of moles and energy are respectively given by

$$\frac{d}{dt}N = \sum_{a=1}^{A} \mathcal{J}^a, \qquad \frac{d}{dt}U = \sum_{a=1}^{A} \mathcal{J}^a (\mathsf{U}^a + p^a \mathsf{V}^a) = \sum_{a=1}^{A} \mathcal{J}^a \mathsf{H}^a$$

(see Equation (61)), where $\mathsf{H}^a = \mathsf{U}^a + p^a \mathsf{V}^a$ is the molar enthalpy at the ath port and where U^a, p^a, and V^a are, respectively, the molar internal energy, the pressure, and the molar volume at the ath port. From these equations and the second law, one obtains the equations for the rate of change of the entropy of the system as

$$\frac{d}{dt}S = I + \sum_{a=1}^{A} \mathcal{J}^a \mathsf{S}^a, \tag{62}$$

where S^a is the molar entropy at the ath port and I is the rate of internal entropy production of the system given by

$$I = \frac{1}{T}\sum_{a=1}^{A} \mathcal{J}^a \left(\mathsf{H}^a - T\mathsf{S}^a - \mu\right), \tag{63}$$

with $T = \frac{\partial U}{\partial S}$ being the temperature and $\mu = \frac{\partial U}{\partial N}$ being the chemical potential. For our variational treatment, it is useful to rewrite the rate of internal entropy production as

$$I = \frac{1}{T}\sum_{a=1}^{A} \left[\mathcal{J}_S^a (T^a - T) + \mathcal{J}^a(\mu^a - \mu)\right],$$

where we define the entropy flow rate $\mathcal{J}_S^a := \mathcal{J}^a \mathsf{S}^a$ and also use the relation $\mathsf{H}^a = \mathsf{U}^a + p^a \mathsf{V}^a = \mu^a + T^a \mathsf{S}^a$. The thermodynamic quantities known at the ath port are usually the pressure p^a and the temperature T^a, from which the other thermodynamic quantities, such as $\mu^a = \mu^a(p^a, T^a)$ or $\mathsf{S}^a = \mathsf{S}^a(p^a, T^a)$, are deduced in light of the state equations of the gas.

Here, we only show the variational formulation for a simplified case of open systems, namely, an open system with only one entropy variable and one compartment with a single species. So, the open system is a simple system. The reader is referred to [3] for the more general cases of open systems, such as the extensions of Equations (29)–(31) and (51)–(53) to open systems, as well as for the case when the mechanical energy of the species is taken into account.

The state variables needed to describe the system are $(q, v, S, N) \in TQ$, and the Lagrangian is a map

$$L : TQ \times \mathbb{R} \times \mathbb{R} \to \mathbb{R}, \quad (q, v, S, N) \mapsto L(q, v, S, N),$$

We assume that the system has A ports, through which species can flow out of or into the system, and B heat sources. As above, μ^a and T^a denote the chemical potential and temperature at the ath port, and T^b denotes the temperature of the bth heat source.

Find the curves $q(t), S(t), \Gamma(t), \Sigma(t), W(t), N(t)$ which are critical for the *variational condition*

$$\delta \int_{t_1}^{t_2} \left[L(q, \dot{q}, S, N) + \dot{W}N + \dot{\Gamma}(S - \Sigma)\right] dt + \int_{t_1}^{t_2} \langle F^{\text{ext}}, \delta q\rangle \, dt = 0, \tag{64}$$

subject to the *phenomenological constraint*

$$\frac{\partial L}{\partial S}\dot{\Sigma} = \left\langle F^{\text{fr}}, \dot{q}\right\rangle + \sum_{a=1}^{A}\left[\mathcal{J}^a(\dot{W} - \mu^a) + \mathcal{J}_S^a(\dot{\Gamma} - T^a)\right] + \sum_{b=1}^{B} \mathcal{J}_S^b(\dot{\Gamma} - T^b), \tag{65}$$

and for variations subject to the *variational constraint*

$$\frac{\partial L}{\partial S}\delta\Sigma = \left\langle F^{\text{fr}}, \delta q\right\rangle + \sum_{a=1}^{A}\left[\mathcal{J}^a \delta W + \mathcal{J}_S^a \delta\Gamma\right] + \sum_{b=1}^{B} \mathcal{J}_S^b \delta\Gamma, \tag{66}$$

with $\delta q(t_1) = \delta q(t_2) = 0$, $\delta W(t_1) = \delta W(t_2) = 0$, and $\delta\Gamma(t_1) = \delta\Gamma(t_2) = 0$.

We note that the variational constraint (Equation (66)) follows from the phenomenological constraint (Equation (65)) by formally replacing the time derivatives $\dot{\Sigma}$, $\dot{q}$, $\dot{W}$, $\dot{\Gamma}$ by the corresponding virtual displacements $\delta\Sigma$, δq, δW, $\delta\Gamma$ and by removing all the terms that depend uniquely on the exterior, i.e., the terms $\mathcal{J}^a\mu^a$, $\mathcal{J}_S^a T^a$, and $\mathcal{J}_S^b T^b$. Such a systematic correspondence between the phenomenological and variational constraints extends to open systems the correspondence for adiabatically closed systems verified in Equations (25) $\rightsquigarrow$ (26), (30) $\rightsquigarrow$ (31), (43) $\rightsquigarrow$ (44), (52) $\rightsquigarrow$ (53); see also Remarks 1 and 3. Note that the action functional in Equation (64) has the same form as that in the case of adiabatically closed systems.

Taking variations of the integral in Equation (64), integrating by parts, and using $\delta q(t_1) = \delta(t_2) = 0$, $\delta W(t_1) = \delta W(t_2) = 0$, and $\delta\Gamma(t_1) = \delta\Gamma(t_2) = 0$ and using the variational constraint (Equation (66)), we obtain the following conditions:

$$\begin{aligned}
\delta q:&\quad \frac{d}{dt}\frac{\partial L}{\partial \dot{q}^i} - \frac{\partial L}{\partial q^i} = F_i^{\mathrm{fr}} + F_i^{\mathrm{ext}}, \quad i = 1, \ldots, n,\\
\delta S:&\quad \dot{\Gamma} = -\frac{\partial L}{\partial S},\\
\delta W:&\quad \dot{N} = \sum_{a=1}^{A} \mathcal{J}^a,\\
\delta N:&\quad \dot{W} = -\frac{\partial L}{\partial N},\\
\delta\Gamma:&\quad \dot{S} = \dot{\Sigma} + \sum_{a=1}^{A} \mathcal{J}_S^a + \sum_{b=1}^{B} \mathcal{J}_S^b.
\end{aligned} \tag{67}$$

By the second and fourth equations, the variables Γ and W are thermodynamic displacements as before. The main difference from the earlier cases is that now $\dot{S}$ and $\dot{\Sigma}$ are no longer equal. The physical interpretation of Σ is given below. From Equation (65), it follows that the system of evolution equations for the curves $q(t)$, $S(t)$, $N(t)$ is defined by

$$\begin{cases}
\dfrac{d}{dt}\dfrac{\partial L}{\partial \dot{q}} - \dfrac{\partial L}{\partial q} = F^{\mathrm{fr}} + F^{\mathrm{ext}}, \qquad \dfrac{d}{dt}N = \displaystyle\sum_{a=1}^{A} \mathcal{J}^a,\\
\dfrac{\partial L}{\partial S}\Big(\dot{S} - \displaystyle\sum_{a=1}^{A} \mathcal{J}_S^a - \sum_{b=1}^{B} \mathcal{J}_S^b\Big)\\
\quad = \left\langle F^{\mathrm{fr}}, \dot{q}\right\rangle - \displaystyle\sum_{a=1}^{A}\Big[\mathcal{J}^a\Big(\frac{\partial L}{\partial N} + \mu^a\Big) + \mathcal{J}_S^a\Big(\frac{\partial L}{\partial S} + T^a\Big)\Big] - \sum_{b=1}^{B} \mathcal{J}_S^b\Big(\frac{\partial L}{\partial S} + T^b\Big).
\end{cases} \tag{68}$$

The energy balance for this system is computed as

$$\frac{d}{dt}E = \underbrace{\left\langle F^{\mathrm{ext}}, \dot{q}\right\rangle}_{=P_W^{\mathrm{ext}}} + \underbrace{\sum_{b=1}^{B} \mathcal{J}_S^b T^b}_{=P_H^{\mathrm{ext}}} + \underbrace{\sum_{a=1}^{A}\left(\mathcal{J}^a\mu^a + \mathcal{J}_S^a T^a\right)}_{=P_M^{\mathrm{ext}}}.$$

From the last equation in Equation (68), the rate of entropy of the system is found by the equation

$$\dot{S} = I + \sum_{a=1}^{A} \mathcal{J}_S^a + \sum_{b=1}^{B} \mathcal{J}_S^b, \tag{69}$$

where I is the rate of internal entropy production given by

$$I = \underbrace{-\frac{1}{T}\left\langle F^{\mathrm{fr}}, \dot{q}\right\rangle}_{\text{mechanical friction}} + \underbrace{\frac{1}{T}\sum_{a=1}^{A}\Big[\mathcal{J}^a\big(\mu^a - \mu\big) + \mathcal{J}_S^a\big(T^a - T\big)\Big]}_{\text{mixing of matter flowing into the system}} + \underbrace{\frac{1}{T}\sum_{b=1}^{B} \mathcal{J}_S^b\big(T^b - T\big)}_{\text{heating}}.$$

From the last Equations (67) and (69), *we notice that* $\dot{\Sigma} = I$ *is the rate of internal entropy production.* The second and third terms in Equation (69) represent the entropy flow rate into the system associated with the ports and the heat sources. The second law requires $I \geq 0$, whereas the sign of the rate of entropy flow into the system is arbitrary.

Example 5 (A piston device with ports and heat sources Figure 7). *We consider a piston with mass* m *moving in a cylinder containing a species with internal energy* $U(S, V, N)$. *We assume that the cylinder has two external heat sources with entropy flow rates* $\mathcal{J}^{b_i}$, $i = 1, 2$, *and two ports through which the species is injected into or flows out of the cylinder with molar flow rates* $\mathcal{J}^{a_i}$, $i = 1, 2$. *The entropy flow rates at the ports are given by* $\mathcal{J}_S^{a_i} = \mathcal{J}^{a_i} \mathsf{S}^{a_i}$.

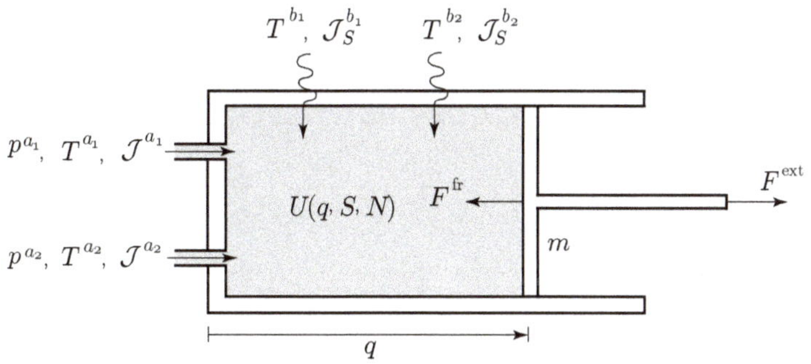

Figure 7. A piston device with ports and heat sources.

The variable q *characterizes the one-dimensional motion of the piston such that the volume occupied by the species is* $V = \alpha q$, *with* α *the sectional area of the cylinder. The Lagrangian of the system is*

$$L(q, \dot{q}, S, N) = \frac{1}{2} m \dot{q}^2 - U(S, Aq, N).$$

The variational formulations (Equations (64)–(66)) yield the evolution equations for $q(t)$, $S(t)$, $N(t)$

$$m\ddot{q} = p(q, S, N)\alpha + F^{\mathrm{fr}} + F^{\mathrm{ext}}, \qquad \dot{N} = \sum_{a=1}^{A} \mathcal{J}^a, \qquad \dot{S} = I + \sum_{i=1}^{2} \mathcal{J}_S^{a_i} + \sum_{j=1}^{2} \mathcal{J}_S^{b_j},$$

where $p(q, S, N) = -\frac{\partial U}{\partial V}$ *is the pressure and* $I = \dot{\Sigma}$ *is the internal entropy production given by*

$$I = -\frac{1}{T} F^{\mathrm{fr}} \dot{q} + \frac{1}{T} \sum_{i=1}^{2} \Big[(\mu^{a_i} - \mu) + \mathsf{S}^{a_i}(T^{a_i} - T) \Big] \mathcal{J}^{a_i} + \frac{1}{T} \sum_{j=1}^{2} \mathcal{J}_S^{b_j} (T^{b_j} - T).$$

The first term represents the entropy production associated with the friction experienced by the moving piston, the second term is the entropy production associated with the mixing of gas flowing into the cylinder at the two ports a_1, a_2, *and the third term denotes the entropy production due to the external heating. The second law requires that each of these terms is positive. The energy balance holds as*

$$\frac{d}{dt} E = \underbrace{F^{\mathrm{ext}} \dot{q}}_{= P_W^{\mathrm{ext}}} + \underbrace{\sum_{j=1}^{2} \mathcal{J}_S^{b_j} T^{b_j}}_{= P_H^{\mathrm{ext}}} + \underbrace{\sum_{i=1}^{2} (\mathcal{J}^{a_i} \mu^{a_i} + \mathcal{J}_S^{a_i} T^{a_i})}_{= P_M^{\mathrm{ext}}}.$$

Remark 4 (Inclusion of chemical reactions). *The variational formulations presented so far can be extended to include several chemical species undergoing chemical reactions. Let us denote by $I = 1, ..., R$ the chemical species and by $a = 1, ..., r$ the chemical reactions. Chemical reactions may be represented by*

$$\sum_I \nu'^a_I I \underset{a_{(2)}}{\overset{a_{(1)}}{\rightleftarrows}} \sum_I \nu''^a_I I, \quad a = 1, ..., r,$$

where $a_{(1)}$ and $a_{(2)}$ are the forward and backward reactions associated with reaction a, and ν''^a_I, ν'^a_I are the forward and backward stoichiometric coefficients for component I in reaction a. Mass conservation during each reaction is given by

$$\sum_I m_I \nu^a_I = 0 \quad \textit{for } a = 1, ..., r \textit{ (Lavoisier law)},$$

where $\nu^a_I := \nu''^a_I - \nu'^a_I$, and m_I is the molecular mass of species I. The affinity of reaction a is the state function defined by $\mathcal{A}^a = -\sum_{I=1}^R \nu^a_I \mu^I$, $a = 1, ..., r$, where μ^I is the chemical potential of the chemical species I. The thermodynamic flux associated with reaction a is the rate of extent denoted J_a.

The thermodynamic displacements are W^I and ν^a such that

$$\dot{W}^I = \mu^I, \quad I = 1, ..., R \quad \textit{and} \quad \dot{\nu}^a = -\mathcal{A}^a, \quad a = 1, ..., r. \tag{70}$$

For chemical reactions in a single compartment assumed to be adiabatically closed and without mechanical components, the variational formulation is given as follows.

Find the curves $S(t)$, $N_I(t)$, $W^I(t)$, $\nu^a(t)$, $I = 1, ..., R$, $a = 1, ..., r$, which are critical for the variational condition

$$\delta \int_{t_1}^{t_2} \left[L(N_1, ..., N_R, S) + \dot{W}^I N_I \right] \mathrm{d}t = 0, \tag{71}$$

subject to the phenomenological and chemical constraints

$$\frac{\partial L}{\partial S} \dot{S} = J_a \dot{\nu}^a \quad \textit{and} \quad \dot{\nu}^a = \nu^a_I \dot{W}^I, \quad a = 1, ..., r, \tag{72}$$

and for variations subject to the variational constraints

$$\frac{\partial L}{\partial S} \delta S = J_a \delta \nu^a \quad \textit{and} \quad \delta \nu^a = \nu^a_I \delta W^I, \quad a = 1, ..., r, \tag{73}$$

with $\delta W^I(t_1) = \delta W^I(t_2) = 0$, $I = 1, ..., R$.

The variational formulations (Equations (71)–(73)) yield the evolution equations for chemical reactions

$$\dot{N}_I = J_a \nu^a_I, \quad I = 1, ..., R \quad \textit{and} \quad T\dot{S} = J_a \mathcal{A}^a,$$

together with the conditions in Equation (70).

Chemical reactions can be included in of all the thermodynamic systems considered previously by combining the variational formulations given by Equations (71)–(73) for chemical reactions with the variational formulations given by Equations (29)–(31), (51)–(53), and (64)–(66).

Remark 5 (General structure of the variational formulation for open systems). *As opposed to the adiabatically closed case, the phenomenological and variational constraints depend explicitly on time $t \in \mathbb{R}$ for the case of open systems. In addition, the phenomenological constraint involves an affine term that depends only on the properties at the ports. From a general point of view, letting Q be the configuration manifold,* *these constraints are defined by the maps* $A^\alpha : \mathbb{R} \times TQ \to T^*Q$, $A(t,x,v) \in T_x^*Q$, *with* $A^\alpha(t,x,v) \in T_x^*Q$, *and* $B^\alpha : \mathbb{R} \times TQ \to \mathbb{R}$, $\alpha = 1,...,k$, *where* $t \in \mathbb{R}$ *and* $(x,v) \in TQ$.

Given a time-dependent Lagrangian $\mathcal{L} : \mathbb{R} \times TQ \to \mathbb{R}$ *and an external force* $\mathcal{F}^{\text{ext}} : \mathbb{R} \times TQ \to T^*Q$, *the variational formulations in Equations (57)–(59) are extended as follows.*

$$\delta \int_{t_1}^{t_2} \mathcal{L}(t, x(t), \dot{x}(t))\mathrm{d}t + \int_{t_1}^{t_2} \left\langle \mathcal{F}^{\text{ext}}(t, x(t), \dot{x}(t)), \delta x(t) \right\rangle \mathrm{d}t = 0, \tag{74}$$

where the curve $x(t)$ *satisfies the phenomenological constraint*

$$A^\alpha(t,x,\dot{x})\cdot\dot{x} + B^\alpha(t,x,\dot{x}) = 0, \;\; \textit{for } \alpha = 1,...,k. \tag{75}$$

and for variations δx *subject to the variational constraint*

$$A^\alpha(t,x,\dot{x})\cdot\delta x = 0, \;\; \textit{for } \alpha = 1,...,k. \tag{76}$$

with $\delta x(t_1) = \delta x(t_2) = 0$.

This yields the system of equations

$$\begin{cases} \dfrac{d}{dt}\dfrac{\partial \mathcal{L}}{\partial \dot{x}} - \dfrac{\partial \mathcal{L}}{\partial x} - \mathcal{F}^{\text{ext}} = \lambda_\alpha A^\alpha(t,x,\dot{x}) \\ A^\alpha(t,x,\dot{x})\cdot\dot{x} + B^\alpha(t,x,\dot{x}) = 0, \;\; \alpha = 1,...,k. \end{cases} \tag{77}$$

The variational formulation for open systems falls into this category by appropriately choosing x *and* $\mathcal{L}$. *For instance, for Equations (64)–(66), one has* $x = (q, S, N, W, \Gamma, \Sigma)$, *and* $\mathcal{L}$ *is the integrand in Equation (64). Note that in Equation (74), we chose the Lagrangian to be time-dependent for the sake of generality. In fact, all the variational formulations for thermodynamics presented above generalize easily to time-dependent Lagrangians. We refer to [3] for a full treatment.*

The energy defined by $\mathcal{E}(t,x,v) = \left\langle \frac{\partial \mathcal{L}}{\partial v}, v \right\rangle - \mathcal{L}(t,x,v)$ *satisfies the energy balance equation*

$$\frac{d}{dt}\mathcal{E} = \left\langle \mathcal{F}^{\text{ext}}, \dot{x} \right\rangle - \lambda_\alpha B^\alpha - \frac{\partial \mathcal{L}}{\partial t}. \tag{78}$$

In the application to open thermodynamic systems, the first term is identified with P_W^{ext}, *the second term is identified with* P_{H+M}^{ext}, *while the third term is due to the explicit dependence of the Lagrangian on the time.*

4. Variational Formulation for Continuum Thermodynamic Systems

In this section, we extend Hamilton's principle of continuum mechanics (12) to nonequilibrium continuum thermodynamics, in the same way as Hamilton's principle of classical mechanics (Equation (2)) was extended to the finite-dimensional case of discrete thermodynamic systems in Section 3.

We consider a multicomponent compressible fluid subject to the irreversible processes of viscosity, heat conduction, and diffusion. In presence of irreversible processes, we impose no-slip boundary conditions, hence, the configuration manifold for the fluid motion is the manifold $Q = \text{Diff}_0(\mathcal{D})$ of diffeomorphisms that keep the boundary $\partial\mathcal{D}$ pointwise fixed.

We assume that the fluid has P components with mass densities $\varrho_A(t,X)$, $A = 1,...,P$ in the material description, and we denote by $S(t,X)$ the entropy density in the material description. The motion of the multicomponent fluid is given as before by a curve of diffeomorphisms $\varphi_t \in \text{Diff}_0(\mathcal{D})$, but now $\dot{\varphi}_t$ is interpreted as the barycentric material velocity of the multicomponent fluid. The Lagrangian of the multicomponent fluid with irreversible processes is

$$L : T\,\text{Diff}_0(\mathcal{D}) \times \mathcal{F}(\mathcal{D}) \times \mathcal{F}(\mathcal{D})^P \to \mathbb{R}, \quad (\varphi, \dot{\varphi}, S, \varrho_1, ..., \varrho_P) \mapsto L(\varphi, \dot{\varphi}, S, \varrho_1, ..., \varrho_P),$$

where $\mathcal{F}(\mathcal{D})$ denotes a space of functions on $\mathcal{D}$ and is given by

$$\begin{aligned} L(\varphi,\dot{\varphi},S,\varrho_1,...,\varrho_P) &= K(\varphi,\dot{\varphi},\varrho_1,...,\varrho_P) - U(\varphi,S,\varrho_1,...,\varrho_P) \\ &= \int_{\mathcal{D}} \left[\frac{1}{2}\varrho(X)|\dot{\varphi}(X)|^2 - \mathscr{E}\big(\varrho_1(X),...,\varrho_P(X),S(X),\nabla\varphi(X)\big)\right] \mathrm{d}^3X. \end{aligned} \tag{79}$$

The first term is the total kinetic energy of the fluid, where $\varrho := \sum_{A=1}^{P} \varrho_A$ is the total mass density. The second term is minus the total internal energy of the fluid, where $\mathscr{E}$ is a general expression for the internal energy density written in terms of $\varrho_A(X)$, $S(X)$, and the deformation gradient $\nabla\varphi(X)$. As in Equation (16), $\mathscr{E}$ satisfies the material covariance assumption and depends on the deformation gradient only through the Jacobian J_φ. As in Equation (18), there is a function ϵ, the internal energy density in the spatial representation, such that

$$\mathscr{E}\big(\varrho_1,...,\varrho_P,\nabla\varphi)\big) = \varphi^*\left[\epsilon(\rho_1,...,\rho_P,s)\right], \quad \text{for} \quad \rho_A = \varphi_*\varrho_A, \quad s = \varphi_*S. \tag{80}$$

In the spatial description, the Lagrangian Equation (79) reads as

$$\ell(\mathbf{v},s,\rho_1,...,\rho_P) = \int_{\mathcal{D}} \left[\frac{1}{2}\rho|\mathbf{v}|^2 - \varepsilon(\rho_1,...,\rho_P,s)\right] \mathrm{d}^3x.$$

Note that in absence of irreversible process, the Lagrangian (79) would just be defined on the tangent bundle $T\,\mathrm{Diff}(\mathcal{D})$ with $\varrho_A = \varrho_{\mathrm{ref}A}$, $A = 1,...,P$ and $S = S_{\mathrm{ref}}$ seen as fixed parameters, exactly as in Equation (15) for the single-component case.

Remark 6 (Material vs spatial variational principle). *As we present below, the variational formulation for continuum thermodynamical systems in the material description is the natural continuum (infinite-dimensional) version of that of discrete (finite-dimensional) thermodynamical systems described in Section 3. This is analogous to the conservative reversible case recalled earlier, namely, the Hamilton principle (Equation (12)); associated with the material description of continuum systems is the natural continuum version of the classical Hamilton principle Equation (2). This is why we first consider below in Section 4.1 the variational formulation of continuum systems in the material description and deduce from it the variational formulation in the spatial description later in Section 4.2. The latter is more involved since it contains additional constraints, as we have seen in the conservative reversible case in Section 2.3.*

4.1. Variational Formulation in the Lagrangian Description

The variational formulation of a multicomponent fluid subject to the irreversible processes of viscosity, heat conduction, and diffusion is the continuum version of the variational formulations (Equations (51)–(53)) for finite-dimensional thermodynamic systems with friction, heat, and mass transfer. Analogous to the thermodynamic fluxes F^{fr}, J_{AB}, $\mathcal{J}^{B\to A}$ are the viscous stress, the entropy flux density, and the diffusive flux density given by $\mathbf{P}^{\mathrm{fr}}$, $\mathbf{J}_S$, $\mathbf{J}_A$ in the material description. Total mass conservation imposes the condition $\sum_{A=1}^{P} \mathbf{J}_A = 0$.

We give below the variational formulation for a general Lagrangian with density $\mathscr{L}$, i.e.,

$$L(\varphi,\dot{\varphi},S,\varrho_1,...,\varrho_P) = \int_{\mathcal{D}} \mathscr{L}(\varphi,\dot{\varphi},\nabla\varphi,S,\varrho_1,...,\varrho_P)\mathrm{d}^3X. \tag{81}$$

The continuum version of the variational formulations (Equations (51)–(53)) that we propose are the following.

Find the curves $\varphi(t), S(t), \Gamma(t), \Sigma(t), W^A(t), \varrho_A(t)$ which are critical for the *variational condition*:

$$\int_0^T \int_{\mathcal{D}} \left[\mathscr{L}(\varphi, \dot{\varphi}, \nabla\varphi, S, \varrho_1, ..., \varrho_P) + \dot{W}^A \varrho_A + \dot{\Gamma}(S - \Sigma) \right] d^3X dt = 0 \tag{82}$$

subject to the *phenomenological constraint*

$$\frac{\partial \mathscr{L}}{\partial S} \dot{\Sigma} = -\mathbf{P}^{\mathrm{fr}} : \nabla\dot{\varphi} + \mathbf{J}_S \cdot \nabla\dot{\Gamma} + \mathbf{J}_A \cdot \nabla\dot{W}^A \tag{83}$$

and for variations subject to the *variational constraint*

$$\frac{\partial \mathscr{L}}{\partial S} \delta\Sigma = -\mathbf{P}^{\mathrm{fr}} : \nabla\delta\varphi + \mathbf{J}_S \cdot \nabla\delta\Gamma + \mathbf{J}_A \cdot \nabla\delta W^A \tag{84}$$

with $\delta\varphi(t_i) = \delta\Gamma(t_i) = \delta W^A(t_i) = 0, i = 1, 2$, and with $\delta\varphi|_{\partial\mathcal{D}} = 0$.

Taking variations of the integral in Equation (82), integrating by parts, and using $\delta\varphi(t_i) = \delta\Gamma(t_i) = \delta W^A(t_i) = 0, i = 1, 2$, and $\delta\varphi|_{\partial\mathcal{D}} = 0$, it follows that

$$\int_{t_1}^{t_2} \int_{\mathcal{D}} \left[\left(\frac{\partial \mathscr{L}}{\partial \varphi^a} \delta\varphi^a - \frac{\partial}{\partial t} \frac{\partial \mathscr{L}}{\partial \dot{\varphi}^a} - \frac{\partial}{\partial A} \frac{\partial \mathscr{L}}{\partial \varphi^a_{,A}} \right) \delta\varphi^a + \frac{\partial \mathscr{L}}{\partial S} \delta S + \left(\frac{\partial \mathscr{L}}{\partial \varrho_A} + \dot{W}^A \right) \delta\varrho_A \right.$$
$$\left. - \dot{\varrho}_A \delta W^A - (\dot{S} - \dot{\Sigma}) \delta\Gamma + \dot{\Gamma}(\delta S - \delta\Sigma) \right] d^3X dt = 0.$$

Using the variational constraint (Equation (84)), integrating by parts, and collecting the terms proportional to $\delta\varphi, \delta\Gamma, \delta S, \delta W^A$, and $\delta\varrho_A$, we get

$$\begin{aligned}
&\delta\varphi : \quad \frac{d}{dt} \frac{\partial \mathscr{L}}{\partial \dot{\varphi}} + \mathrm{DIV}\left(\frac{\partial \mathscr{L}}{\partial \nabla\varphi} + \dot{\Gamma} \frac{\partial \mathscr{L}}{\partial S}^{-1} \mathbf{P}^{\mathrm{fr}} \right) - \frac{\partial \mathscr{L}}{\partial \varphi} = 0 \\
&\delta\Gamma : \quad \dot{S} = \mathrm{DIV}\left(\dot{\Gamma} \frac{\partial \mathscr{L}}{\partial S}^{-1} \mathbf{J}_S \right) + \dot{\Sigma}, \qquad \delta S : \quad \dot{\Gamma} = -\frac{\partial \mathscr{L}}{\partial S}, \\
&\delta W_A : \quad \dot{\varrho}_A = \mathrm{DIV}\left(\dot{\Gamma} \frac{\partial \mathscr{L}}{\partial S}^{-1} \mathbf{J}_A \right), \qquad \delta\varrho_A : \quad \dot{W}^A = -\frac{\partial \mathscr{L}}{\partial \varrho_A},
\end{aligned} \tag{85}$$

together with the boundary conditions

$$\int_{\partial\mathcal{D}} \mathbf{P}^{\mathrm{fr}}{}_a^B \mathbf{N}_B \delta\varphi^a d\mathcal{S} = 0, \qquad \int_{\partial\mathcal{D}} \mathbf{J}_S \cdot \mathbf{N} \delta\Gamma d\mathcal{S} = 0, \qquad \int_{\partial\mathcal{D}} \mathbf{J}_A \cdot \mathbf{N} \delta W^A d\mathcal{S} = 0,$$

where $\mathbf{N}$ is the outward-pointing unit normal vector field to $\partial\mathcal{D}$. The first boundary term vanishes since $\delta\varphi|_{\partial\mathcal{D}} = 0$ from the no-slip boundary condition. The second and third conditions give

$$\mathbf{J}_S \cdot \mathbf{N} = 0 \quad \text{and} \quad \mathbf{J}_A \cdot \mathbf{N} = 0, \; A = 1, ..., P, \quad \text{on} \quad \partial\mathcal{D},$$

i.e., the fluid is adiabatically closed.

From the third and fifth conditions in Equation (85), we have $\dot{\Gamma} = -\frac{\partial \mathscr{L}}{\partial S} = \mathfrak{T}$, the temperature in the material representation, and $\dot{W}^A = -\frac{\partial \mathscr{L}}{\partial \varrho_A} = \Upsilon^A$, a generalization of the chemical potential of component A in the material representation. The second equation in Equation (85) thus reads as

$\dot{S} + \mathrm{DIV}\,\mathbf{J}_S = \dot{\Sigma}$ and attributes to Σ the meaning of *entropy generation rate density*. From the first and fourth equation and the constraint, we get the system

$$\begin{cases} \dfrac{d}{dt}\dfrac{\partial \mathscr{L}}{\partial \dot{\varphi}} + \mathrm{DIV}\left(\dfrac{\partial \mathscr{L}}{\partial \nabla \varphi} - \mathbf{P}^{\mathrm{fr}}\right) - \dfrac{\partial \mathscr{L}}{\partial \varphi} = 0 \\ \dot{\varrho}_A + \mathrm{DIV}\,\mathbf{J}_A = 0, \quad A = 1, ..., P \\ \mathfrak{T}(\dot{S} + \mathrm{DIV}\,\mathbf{J}_S) = \mathbf{P}^{\mathrm{fr}} : \nabla\dot{\varphi} - \mathbf{J}_S \cdot \nabla\mathfrak{T} - \mathbf{J}_A \cdot \nabla \Upsilon^A, \end{cases} \tag{86}$$

for the fields $\varphi(t,X)$, $\varrho_A(t,X)$, and $S(t,X)$. The parameterization of the thermodynamic fluxes $\mathbf{P}^{\mathrm{fr}}$, $\mathbf{J}_S$, $\mathbf{J}_A$ in terms of the thermodynamic forces are discussed in the Eulerian description below.

4.2. Variational Formulation in the Eulerian Description

While the variational formulation is simpler in the material description, the resulting equations of motion are usually written and studied in the spatial description. It is therefore useful to have an Eulerian version of the variational formulations (Equations (82)–(84)). In order to obtain such a variational formulation, all the variables used in Equations (82)–(84) must be converted to their Eulerian analogue. We have already seen the relations $s = \varphi_* S$ and $\rho_A = \varphi_* \varrho_A$ between the Eulerian and Lagrangian mass densities and entropy densities, where the pull-back notation is defined in Equation (17). The Eulerian quantities associated with Σ, Γ, and W^A are defined as follows

$$\sigma = \varphi_* \Sigma, \quad \gamma = \Gamma \circ \varphi^{-1}, \quad w^A = W^A \circ \varphi^{-1}.$$

The Eulerian version of the Piola–Kirchhoff viscous stress tensor $\mathbf{P}^{\mathrm{fr}}$ is the viscous stress tensor σ^{fr} obtained via the Piola transform (see [2,48]).

From the material covariance assumption, the Lagrangian (81) can be rewritten exclusively in terms of spatial variables as

$$\ell(\mathbf{v}, s, \rho_1, ..., \rho_P) = \int_{\mathcal{D}} \mathcal{L}(\mathbf{v}, s, \rho_1, ..., \rho_P)\mathrm{d}^3x,$$

where the Lagrangian density is defined by

$$\mathcal{L}(\mathbf{v}, s, \rho_1, ..., \rho_P) = \varphi_* \left[\mathscr{L}(\mathbf{v} \circ \varphi, \varphi^* \rho_1, ..., \varphi^* \rho_P, \varphi^* s)\right].$$

Using all the preceding relations between Lagrangian and Eulerian variables, we can rewrite the variational formulations Equations (82)–(84) in the following purely Eulerian form.

Find the curves $\mathbf{v}(t)$, $s(t)$, $\gamma(t)$, $\sigma(t)$, $w^A(t)$, $\rho_A(t)$ which are critical for the *variational condition*

$$\int_0^T \int_{\mathcal{D}} \Big[\mathcal{L}(\mathbf{v}, s, \rho_1, ..., \rho_P) + D_t w^A \rho_A + D_t \gamma (s - \sigma)\Big] \mathrm{d}^3x\mathrm{d}t = 0 \tag{87}$$

subject to the *phenomenological constraint*

$$\frac{\partial \mathcal{L}}{\partial s} \bar{D}_t \sigma = -\sigma^{\mathrm{fr}} : \nabla \mathbf{v} + \mathbf{j}_S \cdot \nabla D_t \gamma + \mathbf{j}_A \cdot \nabla D_t w^A \tag{88}$$

and for variations $\delta \mathbf{v} = \partial_t \boldsymbol{\zeta} + \mathbf{v} \cdot \nabla \boldsymbol{\zeta} - \boldsymbol{\zeta} \cdot \nabla \mathbf{v}$, $\delta \rho_A$, δw^A, δs, $\delta \sigma$, and $\delta \gamma$ subject to the *variational constraint*

$$\frac{\partial \mathcal{L}}{\partial s} \bar{D}_\delta \sigma = -\sigma^{\mathrm{fr}} : \nabla \boldsymbol{\zeta} + \mathbf{j}_S \cdot \nabla D_\delta \gamma + \mathbf{j}_A \cdot \nabla D_\delta w^A \tag{89}$$

with $\boldsymbol{\zeta}(t_i) = \delta\gamma(t_i) = \delta w^A(t_i) = 0, i = 1, 2$, and with $\boldsymbol{\zeta}|_{\partial\mathcal{D}} = 0$.

In Equations (87)–(89), we use the notations $D_t f = \partial_t f + \mathbf{v} \cdot \nabla f$, $\bar{D}_t f = \partial_t f + \operatorname{div}(f\mathbf{v})$, $D_\delta f = \delta f + \boldsymbol{\zeta} \cdot \nabla f$, and $\bar{D}_\delta f = \delta f + \operatorname{div}(f\boldsymbol{\zeta})$ for the Lagrangian time derivatives and variations of functions and densities.

The variational formulations (Equations (87)–(89)) yield the system

$$\begin{cases} (\partial_t + \pounds_{\mathbf{v}})\dfrac{\partial \mathcal{L}}{\partial \mathbf{v}} = \rho_A \nabla \dfrac{\partial \mathcal{L}}{\partial \rho_A} + s\nabla \dfrac{\partial \mathcal{L}}{\partial s} + \operatorname{div} \boldsymbol{\sigma}^{\mathrm{fr}} \\ \bar{D}_t \rho_A + \operatorname{div} \mathbf{j}_A = 0, \quad A = 1, ..., P \\ \dfrac{\partial \mathcal{L}}{\partial s}(\bar{D}_t s + \operatorname{div} \mathbf{j}_s) = -\boldsymbol{\sigma}^{\mathrm{fr}} : \nabla \mathbf{v} - \mathbf{j}_s \cdot \nabla \dfrac{\partial \mathcal{L}}{\partial s} - \mathbf{j}_A \cdot \nabla \dfrac{\partial \mathcal{L}}{\partial \rho_A}, \end{cases} \tag{90}$$

together with the conditions

$$\bar{D}_t \sigma = \bar{D}_t s + \operatorname{div} \mathbf{j}_s, \quad D_t \gamma = -\frac{\partial \mathcal{L}}{\partial s}, \quad D_t w_A = -\frac{\partial \mathcal{L}}{\partial \rho_A}.$$

In Equation (90), $\pounds_{\mathbf{v}}$ denotes the Lie derivative defined as $\pounds_{\mathbf{v}} \mathbf{m} = \mathbf{v} \cdot \nabla \mathbf{m} + \nabla \mathbf{v}^{\mathsf{T}} \cdot \mathbf{m} + \mathbf{m} \operatorname{div} \mathbf{v}$. We refer to [2] for a detailed derivation of these equations from the variational formulations (Equations (87)–(89)).

The multicomponent Navier–Stokes–Fourier equations. For the Lagrangian

$$\ell(\mathbf{v}, s, \rho_1, ..., \rho_P) = \int_{\mathcal{D}} \left[\frac{1}{2}\rho|\mathbf{v}|^2 - \epsilon(\rho_1, ..., \rho_P, s)\right] \mathrm{d}^3 x$$

we get

$$\begin{cases} \rho(\partial_t \mathbf{v} + \mathbf{v} \cdot \nabla \mathbf{v}) = -\nabla p + \operatorname{div} \boldsymbol{\sigma}^{\mathrm{fr}} \\ \bar{D}_t \rho_A + \operatorname{div} \mathbf{j}_A = 0, \quad A = 1, ..., P \\ T(\bar{D}_t s + \operatorname{div} \mathbf{j}_s) = \boldsymbol{\sigma}^{\mathrm{fr}} : \nabla \mathbf{v} - \mathbf{j}_s \cdot \nabla T - \mathbf{j}_A \cdot \nabla \mu^A \end{cases} \tag{91}$$

with $\mu^A = \frac{\partial \epsilon}{\partial \rho_A}$, $T = \frac{\partial \epsilon}{\partial s}$, and $p = \mu^A \rho_A + Ts - \epsilon$.

The system of Equation (91) needs to be supplemented with phenomenological expressions for the *thermodynamic fluxes* $\boldsymbol{\sigma}^{\mathrm{fr}}$, $\mathbf{j}_S$, $\mathbf{j}_A$ in terms of the *thermodynamic affinities* $\operatorname{Def} \mathbf{v}$, ∇T, $\nabla \mu^A$ compatible with the second law. It is empirically accepted that for a large class of irreversible processes and under a wide range of experimental conditions, the thermodynamic fluxes J_α are linear functions of the thermodynamic affinities X^α, i.e., $J_\alpha = \mathcal{L}_{\alpha\beta} X^\beta$, where the transport coefficients $\mathcal{L}_{\alpha\beta}(...)$ are state functions that must be determined by experiments or, if possible, by kinetic theory. Besides defining a positive quadratic form, the coefficients $\mathcal{L}_{\alpha\beta}(...)$ must also satisfy *Onsager's reciprocal relations* [8] due to the microscopic time reversibility and the *Curie principle* associated with material invariance (see, for instance, [67–70]). In the case of the multicomponent fluid, writing the traceless part of $\boldsymbol{\sigma}^{\mathrm{fr}}$ and $\operatorname{Def} \mathbf{v}$ as $(\boldsymbol{\sigma}^{\mathrm{fr}})^{(0)} = \boldsymbol{\sigma}^{\mathrm{fr}} - \frac{1}{3}(\operatorname{Tr} \boldsymbol{\sigma}^{\mathrm{fr}})\delta$ and $(\operatorname{Def} \mathbf{v})^{(0)} = \operatorname{Def} \mathbf{v} - \frac{1}{3}(\operatorname{div} \mathbf{v})\delta$, we have the following phenomenological linear relations

$$-\begin{bmatrix} \mathbf{j}_S \\ \mathbf{j}_A \end{bmatrix} = \begin{bmatrix} \mathcal{L}_{SS} & \mathcal{L}_{SB} \\ \mathcal{L}_{AS} & \mathcal{L}_{AB} \end{bmatrix} \begin{bmatrix} \nabla T \\ \nabla \mu^B \end{bmatrix}, \qquad \frac{1}{3} \operatorname{Tr} \boldsymbol{\sigma}^{\mathrm{fr}} = \zeta \operatorname{div} \mathbf{v}, \qquad (\boldsymbol{\sigma}^{\mathrm{fr}})^{(0)} = 2\mu (\operatorname{Def} \mathbf{v})^{(0)},$$

where all the coefficients may depend on $(s, \rho_1, ..., \rho_P)$. The first linear relation describes the vectorial phenomena of heat conduction (Fourier law), diffusion (Fick law), and their cross-effects (Soret and Dufour effects); the second relation describes the scalar processes of bulk viscosity with coefficient

$\zeta \geq 0$, and the third relation is the tensorial process of shear viscosity with coefficient $\mu \geq 0$. The associated friction stress reads

$$\sigma^{\text{fr}} = 2\mu \operatorname{Def} \mathbf{v} + \left(\zeta - \frac{2}{3}\mu\right)(\operatorname{div} \mathbf{v})\delta.$$

All these phenomenological considerations take place with the phenomenological constraint (Equation (88)) and the associated variational constraint (Equation (89)), but they are not involved in the variational condition (87). Note that our variational formulation holds independently on the linear character of the phenomenological laws.

Remark 7. *For simplicity, we chose the fluid domain $\mathcal{D}$ as a subset of $\mathbb{R}^3$ endowed with the Euclidean metric. More generally, the variational formulation can be intrinsically written on Riemannian manifolds (see [2]). Making the dependence of the Riemannian metric explicit, even if it is given by the standard Euclidean metric, is important for the study of the covariance properties [49].*

5. Concluding Remarks

In this paper, we survey our recent developments on the Lagrangian variational formulation for nonequilibrium thermodynamics developed in [1–3], which is a natural extension of Hamilton's principle in mechanics to include irreversible processes.

Before going into details, we briefly review Hamilton's principle as it applies to (finite-dimensional) discrete systems in classical mechanics, as well as to (infinite-dimensional) continuum systems. Then, in order to illustrate our variational formulation for nonequilibrium thermodynamics, we first start with the finite dimensional case of adiabatically closed systems together with representative examples, such as a piston containing an ideal gas, a system with a chemical species experiencing diffusion between several compartments, an adiabatic piston with two cylinders, and a system with a chemical species experiencing diffusion and heat conduction between two compartments. Then, we extend the variational formulation to open finite-dimensional systems that can exchange heat and matter with the exterior. This case is illustrated with the help of a piston device with ports and heat sources. We also demonstrate how chemical reactions can be naturally incorporated into our variational formulation.

Second, we illustrate the variational formulation with the infinite-dimensional case of continuum systems by focusing on a compressible fluid with the irreversible processes due to viscosity, heat conduction, and diffusion. The formulation is first given in the Lagrangian (or material) description because it is in this description that the variational formulation is a natural continuum extension of the one for discrete systems. The variational formulation in the Eulerian (or spatial) description is then deduced by Lagrangian reduction and yields the multicomponent Navier–Stokes–Fourier equations.

One of the key issue of our variational formulation is the introduction and the use of the concept of thermodynamic displacement, whose time derivative corresponds to the affinity of the process. Thermodynamic displacement allows for systematically developing the variational constraints associated with the nonlinear phenomenological constraints. The variational formulations presented in this paper use the entropy as an independent variable, but a variational approach based on the temperature can be also developed by considering free energy Lagrangians (see [4]).

Further Developments

Associated with our variational formulation of nonequilibrium thermodynamics, there are the following interesting and important topics, which we have not described here due to lack of space, but they are quite relevant to the variational formulation of nonequilibrium thermodynamics, reviewed in this paper.

- *Dirac structures and Dirac systems:* It is well known that when the Lagrangian is regular, the equations of classical mechanics can be transformed into the setting of Hamiltonian systems. The underlying geometric object for this formulation is the canonical symplectic form on the phase space T^*Q of the configuration manifold. When irreversible processes are included, this geometric formulation is lost because of the degeneracy of the Lagrangians and the presence of the nonlinear nonholonomic constraints. Hence, one may ask: what is the appropriate geometric object that generalizes the canonical symplectic form in the formulation of thermodynamics? In [64,71], it was shown that the evolution equations for both adiabatically closed and open systems can be geometrically formulated in terms of various classes of Dirac structures induced by the phenomenological constraint and from the canonical symplectic form on T^*Q or on $T^*(Q \times \mathbb{R})$.
- *Reduction by symmetry:* When symmetries are available, reduction processes can be applied to the variational formulation of thermodynamics, thereby extending the process of Lagrangian reduction from classical mechanics to thermodynamics. This is illustrated in Section 4.2 for the Navier–Stokes–Fourier equation, but it can be carried out in general for all the variational formulations presented in this paper. For instance, we refer to [72] for the case of simple thermodynamic systems on Lie groups with symmetries.
- *Variational discretization:* Associated with the variational formulation in this paper, there exist variational integrators for the nonequilibrium thermodynamics of simple adiabatically closed systems (see [72,73]). These integrators are structure-preserving numerical schemes that are obtained by a discretization of the variational formulation. The structure-preserving property of the flow of such systems is an extension of the symplectic property of the flow of variational integrators for Lagrangian mechanics.
- *Modeling of thermodynamically consistent models:* The variational formulation for thermodynamics can be also used to derive new models, which are automatically thermodynamically consistent. We refer to [74] for an application of the variational formulation to atmospheric thermodynamics and its pseudo-incompressible approximation.

Funding: F.G.-B. is partially supported by the ANR project GEOMFLUID, ANR-14-CE23-0002-01; H.Y. is partially supported by JSPS Grant-in-Aid for Scientific Research (16KT0024, 24224004), the MEXT "Top Global University Project" and Waseda University (SR 2018K-195).

Conflicts of Interest: The authors declare no conflict of interest.

References

1. Gay-Balmaz, F.; Yoshimura, H. A Lagrangian variational formulation for nonequilibrium thermodynamics. Part I: Discrete systems. *J. Geom. Phys.* **2017**, *111*, 169–193. [CrossRef]
2. Gay-Balmaz, F.; Yoshimura, H. A Lagrangian variational formulation for nonequilibrium thermodynamics. Part II: Continuum systems. *J. Geom. Phys.* **2017**, *111*, 194–212. [CrossRef]
3. Gay-Balmaz, F.; Yoshimura, H. A variational formulation of nonequilibrium thermodynamics for discrete open systems with mass and heat transfer. *Entropy* **2018**, *20*, 163. [CrossRef]
4. Gay-Balmaz, F.; Yoshimura, H. A free energy Lagrangian variational formulation of the Navier-Stokes-Fourier system. *Int. J. Geom. Methods Mod. Phys.* **2018**, in press. [CrossRef]
5. Gibbs, J.W. *Collected Works*; Scribner: New York, NY, USA, 1902.
6. Callen, H.B. *Thermodynamics and an Introduction to Thermostatistics*, 2nd ed.; John Wiley & Sons Inc.: New York, NY, USA, 1985.
7. Landau, L.D.; Lifshitz, E.M. *Statistical Physics*; Volume 5 of A Course of Theoretical Physics; Pergamon Press: Oxford, UK, 1969.
8. Onsager, L. Reciprocal relations in irreversible processes I. *Phys. Rev.* **1931**, *37*, 405–426. [CrossRef]
9. Onsager, L.; Machlup, S. Fluctuations and irreversible processes. *Phys. Rev.* **1953**, *91*, 1505–1512. [CrossRef]
10. Onsager, L.; Machlup, S. Fluctuations and irreversible processes II. Systems with kinetic energy. *Phys. Rev.* **1953**, *91*, 1512–1515. [CrossRef]
11. Prigogine, I. *Etude Thermodynamique des Phénomènes Irréversibles*; Bruxelles: Brussels, Belgium, 1947.

12. Glansdorff, P.; Prigogine, I. *Thermodynamic Theory of Structure, Stability, and Fluctuations*; Wiley-Interscience: New York, NY, USA, 1971.
13. Ziegler, H. A possible generalization of Onsager's theory. In *Irreversible Aspects of Continuum Mechanics*; Barkus, H., Sedov, L.I., Eds.; Springer: New York, NY, USA, 1968.
14. Gyarmati, I. *Nonequilibrium Thermodynamics: Field Theory and Variational Principles*; Springer: New York, NY, USA, 1970.
15. Lavenda, B.H. *Thermodynamics of Irreversible Processes*; Macmillan: London, UK, 1978.
16. Ichiyanagi, M. Variational principles in irreversible processes. *Phys. Rep.* **1994**, *243*, 125–182. [CrossRef]
17. Biot, M.A. A virtual dissipation principle and Lagrangian equations in non-linear irreversible thermodynamics. *Acad. R. Belg. Bull. Cl. Sci.* **1975**, *5*, 6–30.
18. Biot, M.A. New variational-Lagrangian irreversible thermodynamics with application to viscous flow, reaction-diffusion, and solid mechanics. *Adv. Appl. Mech.* **1984**, *24*, 1–91.
19. Fukagawa, H.; Fujitani, Y. A variational principle for dissipative fluid dynamics. *Prog. Theor. Phys.* **2012**, *127*, 921–935. [CrossRef]
20. Gibbs, J.W. Graphical methods in the thermodynamics of fluids. *Trans. Conn. Acad.* **1873**, *2*, 309–342.
21. Gibbs, J.W. A method of geometrical representation of the thermodynamic properties of substances by means of surfaces. *Trans. Conn. Acad.* **1873**, *2*, 382–404.
22. Carathéodory, C. Untersuchungen über die Grundlagen der Thermodynamik. *Math. Ann.* **1909**, *67*, 355–386. [CrossRef]
23. Hermann, R. *Geometry, Physics and Systems*; Dekker: New York, NY, USA, 1973.
24. Mrugala, R. Geometrical formulation of equilibrium phenomenological thermodynamics. *Rep. Math. Phys.* **1978**, *14*, 419–427. [CrossRef]
25. Mrugala, R. A new representation of Thermodynamic Phase Space. *Bull. Pol. Acad. Sci.* **1980**, *28*, 13–18.
26. Mrugala, R.; Nulton, J.D.; Schon, J.C.; Salamon, P. Contact structure in thermodynamic theory. *Rep. Math. Phys.* **1991**, *29*, 109–121. [CrossRef]
27. Eberard, D.; Maschke, B.M.; van der Schaft, A.J. An extension of Hamiltonian systems to the thermodynamic phase space: Towards a geometry of nonreversible processes. *Rep. Math. Phys.* **2007**, *60*, 175–198. [CrossRef]
28. Grmela, M. Bracket formulation of dissipative fluid mechanics equations. *Phys. Lett. A* **1984**, *102*, 355–358. [CrossRef]
29. Kaufman, A. Dissipative Hamiltonian systems: A unifying principle. *Phys. Lett. A* **1984**, *100*, 419–422. [CrossRef]
30. Morrison, P. Bracket formulation for irreversible classical fields. *Phys. Lett. A* **1984**, *100*, 423–427. [CrossRef]
31. Edwards, B.J.; Beris, A.N. Noncanonical Poisson bracket for nonlinear elasticity with extensions to viscoelasticity. *Phys. A Math. Gen.* **1991**, *24*, 2461–2480. [CrossRef]
32. Edwards, B.J.; Beris, A.N. Unified view of transport phenomena based on the generalized bracket formulation. *Ind. Eng. Chem. Res.* **1991**, *30*, 873–881. [CrossRef]
33. Grmela, M.; Öttinger, H.-C. Dynamics and thermodynamics of complex fluids. I. Development of a general formalism. *Phys. Rev. E* **1997**, *56*, 6620–6632. [CrossRef]
34. Morrison, P. A paradigm for joined Hamiltonian and dissipative systems. *Physica D* **1986**, *18*, 410–419. [CrossRef]
35. Öttinger, H.C.; Grmela, M. Dynamics and thermodynamics of complex fluids. II. Illustrations of a general formalism. *Phys. Rev. E* **1997**, *56*, 6633–6655. [CrossRef]
36. Eldred, C.; Gay-Balmaz, F. Single and double generator bracket formulations of geophysical fluids with irreversible processes. *arXiv* **2018**, arXiv:1811.11609v1.
37. Green, A.E.; Naghdi, P.M. A re-examination of the basic postulates of thermomechanics. *Proc. R. Soc. Lond. Ser. A* **1991**, *432*, 171–194. [CrossRef]
38. Stueckelberg, E.C.G.; Scheurer, P.B. *Thermocinétique Phénoménologique Galiléenne*; Birkhäuser: Basel, Switzerland, 1974.
39. Landau, L.D.; Lifshitz, E.M. *Mechanics*; Volume 1 of A Course of Theoretical Physics; Pergamon Press: Oxford, UK, 1969.
40. Neimark, J.I.; Fufaev, N.I. *Dynamics of Nonholonomic Systems*; Volume 33 of Translations of Mathematical Monographs; American Mathematical Society: Providence, RI, USA, 1972.

41. Arnold, V.I.; Kozlov, V.V.; Neishtadt, A.I. *Dynamical Systems III*; Encyclopedia of Mathematics; Springer: New York, NY, USA, 1988; Volume 3.
42. Bloch, A.M. *Nonholonomic Mechanics and Control*; Volume 24 of Interdisciplinary Applied Mathematics; Baillieul, J., Crouch, P., Marsden, J., Krishnaprasad, P.S., Murray, R.M., Zenkov, D., Eds.; Springer: New York, NY, USA, 2003.
43. Lewis, A.; Murray, R.M. Variational principles in constrained systems: Theory and experiments. *Int. J. Nonlinear Mech.* **1995**, *30*, 793–815. [CrossRef]
44. Herivel, J.W. The derivation of the equations of motion of an ideal fluid by Hamilton's principle. *Proc. Camb. Philos. Soc.* **1955**, *51*, 344–349. [CrossRef]
45. Serrin, J. Mathematical principles of classical fluid mechanics. In *Handbuch der Physik VIII-I*; Springer: Berlin, Germany, 1959; pp. 125–263.
46. Eckart, C. Variation principles of hydrodynamics. *Phys. Fluids* **1960**, *3*, 421–427. [CrossRef]
47. Truesdell, C.; Toupin, R. The classical field theories. In *Handbuch der Physik III-I*; Springer: Berlin, Germany, 1960; pp. 226–790.
48. Marsden, J.E.; Hughes, T.J.R. *Mathematical Foundations of Elasticity*; Prentice Hall: New York, NY, USA, 1983.
49. Gay-Balmaz, F.; Marsden, J.E.; Ratiu, T.S. Reduced variational formulations in free boundary continuum mechanics. *J. Nonlinear Sci.* **2012**, *22*, 553–597. [CrossRef]
50. Marsden, J.E.; Ratiu, T.S. *Introduction to Mechanics and Symmetry*, 2nd ed.; Texts in Applied Mathematics; Springer: New York, NY, USA, 1999; Volume 17.
51. Marsden, J.E.; Scheurle, J. Lagrangian reduction and the double spherical pendulum. *ZAMP* **1993**, *44*, 17–43. [CrossRef]
52. Marsden, J.E.; Scheurle, J. The reduced Euler–Lagrange equations. *Fields Inst. Commun.* **1993**, *1*, 139–164.
53. Cendra, H.; Marsden, J.E.; Ratiu, T.S. *Lagrangian Reduction by Stages*; Memoirs of the AMS: Providence, RI, USA, 2001; Volume 152.
54. Holm, D.D.; Marsden, J.E.; Ratiu, T.S. The Euler-Poincaré equations and semidirect products with applications to continuum theories. *Adv. Math.* **1998**, *137*, 1–8. [CrossRef]
55. Gruber, C. Thermodynamics of systems with internal adiabatic constraints: Time evolution of the adiabatic piston. *Eur. J. Phys.* **1999**, *20*, 259–266. [CrossRef]
56. Ferrari, C.; Gruber, C. Friction force: From mechanics to thermodynamics. *Eur. J. Phys.* **2010**, *31*, 1159–1175. [CrossRef]
57. Gruber, C.; Brechet, S.D. Lagrange equation coupled to a thermal equation: Mechanics as a consequence of thermodynamics. *Entropy* **2011**, *13*, 367–378. [CrossRef]
58. Kedem, O.; Katchalsky, A. Permeability of composite membranes. Part 1. Electric current, volume flow and flow of solute through membranes. *Trans. Faraday Soc.* **1963**, *59*, 1918–1930. [CrossRef]
59. Kedem, O.; Katchalsky, A. Permeability of composite membranes. Part 2. Parallel elements. *Trans. Faraday Soc.* **1963**, *59*, 1931–1940. [CrossRef]
60. Kedem, O.; Katchalsky, A. Permeability of composite membranes. Part 3. Series array of elements. *Trans. Faraday Soc.* **1963**, *59*, 1941–1953. [CrossRef]
61. Oster, G.F.; Perelson, A.S.; Katchalsky, A. Network thermodynamics: Dynamic modelling of biophysical systems. *Q. Rev. Biophys.* **1973**, *6*, 1–134. [CrossRef] [PubMed]
62. Jacobs, H.; Yoshimura, H. Tensor products of Dirac structures and interconnection in Lagrangian mechanics. *J. Geom. Mech.* **2014**, *6*, 67–98. [CrossRef]
63. von Helmholtz, H. *Studien zur Statik Monocyklischer Systeme*; Sitzungsberichte der Königlich Preussischen Akademie der Wissenschaften zu Berlin: Berlin, Germany, 1884; pp. 159–177.
64. Gay-Balmaz, F.; Yoshimura, H. Dirac structures in nonequilibrium thermodynamics. *J. Math. Phys.* **2018**, *59*, 012701. [CrossRef]
65. Sandler, S.I. *Chemical, Biochemical, and Engineering Thermodynamics*; John Wiley & Sons: New York, NY, USA, 2006.
66. Klein, S.; Nellis, G. *Thermodynamics*; Cambridge University Press: Cambridge, UK, 2011.
67. de Groot, S.R.; Mazur, P. *Nonequilibrium Thermodynamics*; North-Holland: Amsterdam, The Netherlands, 1969.
68. Kondepudi, D.; Prigogine, I. *Modern Thermodynamics*; John Wiley & Sons: New York, NY, USA, 1998.
69. Landau, L.D.; Lifshitz, E.M. *Fluid Mechanics*; Volume 6 of A Course of Theoretical Physics; Pergamon Press: Oxford, UK, 1969.
70. Woods, L.C. *The Thermodynamics of Fluid Systems*; Clarendon Press: Oxford, UK, 1975.

71. Gay-Balmaz, F.; Yoshimura, H. Dirac structures in nonequilibrium thermodynamics for open systems. *Preprint* **2018**, unpublished.
72. Couéraud, B.; Gay-Balmaz, F. Variational discretization of simple thermodynamical systems on Lie groups. *Disc. Cont. Dyn. Syst. Ser. S* **2018**, in press.
73. Gay-Balmaz, F.; Yoshimura, H. Variational discretization for the nonequilibrium thermodynamics of simple systems. *Nonlinearity* **2018**, *31*, 1673. [CrossRef]
74. Gay-Balmaz, F. A variational derivation of the nonequilibrium thermodynamics of a moist atmosphere with rain process and its pseudoincompressible approximation. *arXiv* **2018**, arXiv:1701.03921v2.

entropy

MDPI

Article

Emergence of Non-Fourier Hierarchies

Tamás Fülöp [1,2], Róbert Kovács [1,2,3,*], Ádám Lovas [1], Ágnes Rieth [1], Tamás Fodor [1], Mátyás Szücs [1,2], Péter Ván [1,2,3] and Gyula Gróf [1]

1 Department of Energy Engineering, Faculty of Mechanical Engineering, BME, 1521 Budapest, Hungary; fulop@energia.bme.hu (T.Fü.); lovas.adam91@gmail.com (Á.L.); rieth.agnes@gmail.com (Á.R.); tamas.fdr@gmail.com (T.Fo.); szucsmatyas@energia.bme.hu (M.S.); van.peter@wigner.mta.hu (P.V.); grof@energia.bme.hu (G.G.)

2 Montavid Thermodynamic Research Group, 1112 Budapest, Hungary

3 Department of Theoretical Physics, Institute for Particle and Nuclear Physics, Wigner Research Centre for Physics, 1525 Budapest, Hungary

* Correspondence: kovacsrobert@energia.bme.hu

Received: 21 August 2018; Accepted: 26 October 2018; Published: 30 October 2018

Abstract: The non-Fourier heat conduction phenomenon on room temperature is analyzed from various aspects. The first one shows its experimental side, in what form it occurs, and how we treated it. It is demonstrated that the Guyer-Krumhansl equation can be the next appropriate extension of Fourier's law for room-temperature phenomena in modeling of heterogeneous materials. The second approach provides an interpretation of generalized heat conduction equations using a simple thermo-mechanical background. Here, Fourier heat conduction is coupled to elasticity via thermal expansion, resulting in a particular generalized heat equation for the temperature field. Both aforementioned approaches show the size dependency of non-Fourier heat conduction. Finally, a third approach is presented, called pseudo-temperature modeling. It is shown that non-Fourier temperature history can be produced by mixing different solutions of Fourier's law. That kind of explanation indicates the interpretation of underlying heat conduction mechanics behind non-Fourier phenomena.

Keywords: non-Fourier heat conduction; thermal expansion; heat pulse experiments; pseudo-temperature; Guyer-Krumhansl equation

1. Introduction

The Fourier's law [1]

$$\mathbf{q} = -k\vec{\nabla} T \tag{1}$$

is one of the most applicable, well-known elementary physical laws in engineering practice. Here, $\mathbf{q}$ is the heat flux vector, T is absolute temperature, k is thermal conductivity. However, as all the constitutive equations, it also has limits of validation. Phenomena that do not fit into these limits, called non-Fourier heat conduction, appear in many different forms. Some of them occur at low temperature such as the so-called second sound and ballistic (thermal expansion induced) propagation [2–7]. These phenomena have been experimentally measured several times [8–11] and many generalized heat equations exist to simulate them [12–20]. The success in low-temperature experiments resulted in the extension of this research field to find the deviation at room temperature as well. One of the most celebrated result is related to Mitra et al. [21,22] where the measured temperature history was very similar to a wave-like propagation. However, these results have not been reproduced by anyone and undoubtedly demanded for further investigation.

In most of the room-temperature measurements, the existence of Maxwell-Cattaneo-Vernotte (MCV) type behavior attempted to be proved [23,24]. It is this MCV equation that is used to model the aforementioned second sound, the dissipative wave propagation form of heat [3,25,26]. The validity of MCV equation for room-temperature behavior has not yet been justified, despite of the numerous experiments. It is important to note that many other extensions of Fourier equation exist beyond the MCV one, such as the Guyer-Krumhansl (GK) equation [27–32], the dual-phase-lag model [33], and their modifications, too [7,34,35]. Some of these possess stronger physical background, some others not [36–38]. Here we would like to emphasize that we restrict ourselves to the GK equation that shows the simplest hierarchical arrangement of Fourier's law and applicable for room-temperature problems.

The simplest extension of MCV equation is the GK model, which reads:

$$\tau\dot{\mathbf{q}} + \mathbf{q} + k\vec{\nabla}T - \kappa^2\triangle\mathbf{q} = 0, \tag{2}$$

where the coefficient τ is called relaxation time and κ^2 is regarded as a dissipation parameter and the dot denotes the time derivative. This GK-type constitutive equation contains the MCV-type by considering $\kappa^2 = 0$ and the Fourier equation taking $\tau = \kappa^2 = 0$. This feature of GK equation allows to model both wave-like temperature history and over-diffusive one. This is more apparent when one applies the balance equation of internal energy to eliminate $\mathbf{q}$:

$$\rho c\dot{T} + \vec{\nabla}\cdot\mathbf{q} = 0, \tag{3}$$

with mass density ρ, specific heat c and volumetric source neglected, one obtains

$$\tau\ddot{T} + \dot{T} = a\triangle T + \kappa^2\triangle\dot{T}, \tag{4}$$

with thermal diffusivity $a = k/(\rho c)$. One can realize that Equation (4) contains the Fourier heat equation

$$\dot{T} = a\triangle T \tag{5}$$

as well as its time derivative, with different coefficients. It becomes more visible after rearranging Equation (4):

$$\tau\left(\dot{T} - \frac{\kappa^2}{\tau}\triangle T\right)^{\cdot} + \dot{T} - a\triangle T = 0. \tag{6}$$

when the so-called [39,40] Fourier resonance condition $\kappa^2/\tau = a$ holds, the solutions of the Fourier Equation (5) are covered by the solutions of (4). Meanwhile, when $\kappa^2 < a\tau$ the wave-like behavior is recovered, and this domain is known as under-damped region. In the opposite case ($\kappa^2 > a\tau$), there is no visible wave propagation and it is called over-diffusive (or over-damped) region. We measured the corresponding over-diffusive effect several times in various materials such as metal foams, rocks and in a capacitor, too [39,40]. Furthermore, a similar temperature history was observed in a biological material [38]. It is also important to note that originally the GK equation is derived from Boltzmann equation applying phonon hydrodynamics in the background. Here, we would like to emphasize that in non-equilibrium thermodynamics it can also be derived without assuming any phonon interaction in the material [6,7] keeping the GK equation applicable for room-temperature heat conduction.

In this paper, further aspects of over-diffusive propagation are discussed. In the following sections the size dependence of the observed over-damped phenomenon is discussed both experimentally and theoretically. Moreover, the approach of pseudo-temperature is presented to provide one concrete possible interpretation for non-Fourier heat conduction.

2. Size Dependence

Our measurements reported here are performed on basalt rock samples with three different thicknesses, 1.86, 2.75 and 3.84 mm, respectively. We have applied the same apparatus of heat pulse experiment as described in [39,40], schematically depicted in Figure 1 below.

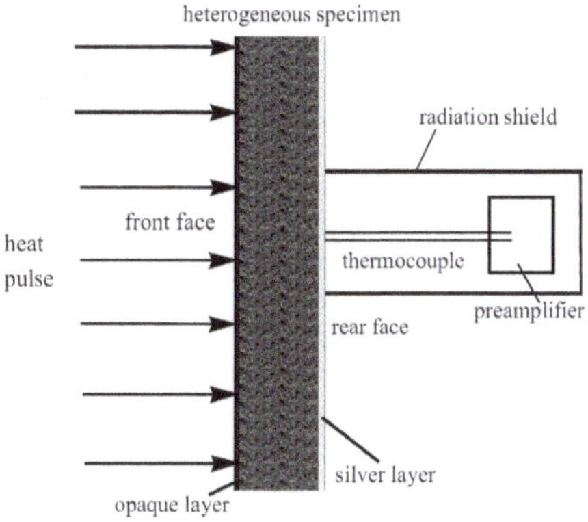

Figure 1. Setup of our heat pulse experiment [40].

In each case, the rear-side temperature history was measured and numerically evaluated solving the GK equation with constant coefficients, i.e., they do not depend on the temperature due to its small change. It is also assumed that the GK equation characterizes the whole sample. We choose the GK equation as the simplest thermodynamically consistent one that can predict signal shapes observed in room-temperature measurements. (The heat pulse setup—a widely used one for transient heat conduction measurements—is not capable of obtaining space dependence of temperature along the sample but even such measurement data would be insufficient to determine an underlying partial differential equation - any experimental data can only refute or support an equation (at some confidence level).) The GK coefficients used below are best fits. The recorded dimensionless temperature signals are plotted in Figures 2–4. In these figures, the dashed line shows the solution of Fourier equation using thermal diffusivity corresponding to the initial part of temperature rising on the rear side. The measured signal deviates from the Fourier-predicted one even when considering non-adiabatic (cooling) boundary condition. That deviation weakens with increasing sample thickness; for the thickest one it is hardly visible, and the prediction of Fourier's law is almost acceptable.

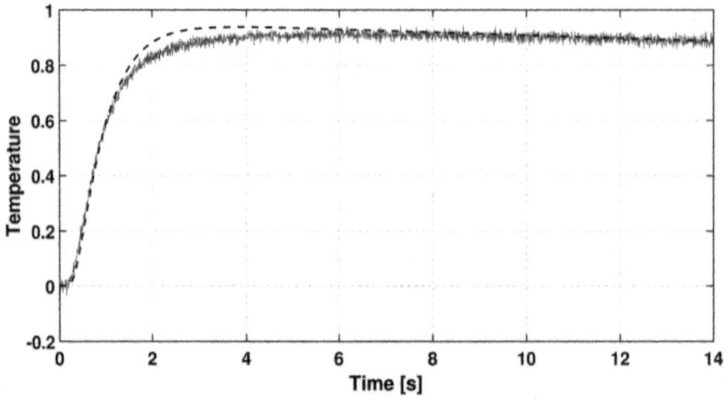

Figure 2. Data recorded for basalt rock sample with thickness of 1.86 mm. The dashed line shows the prediction of Fourier's law.

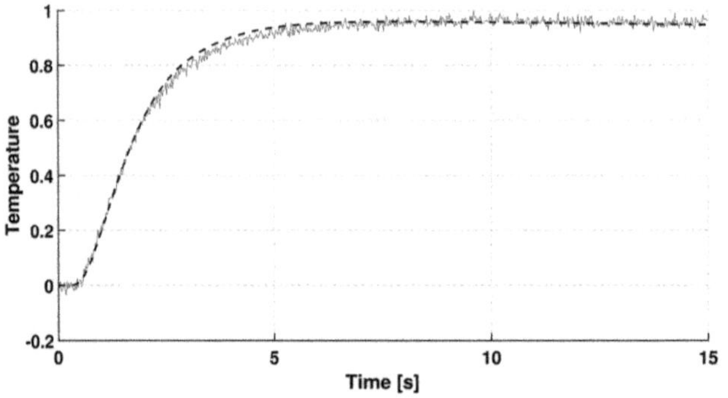

Figure 3. Data recorded for basalt rock sample with thickness of 2.75 mm. The dashed line shows the prediction of Fourier's law.

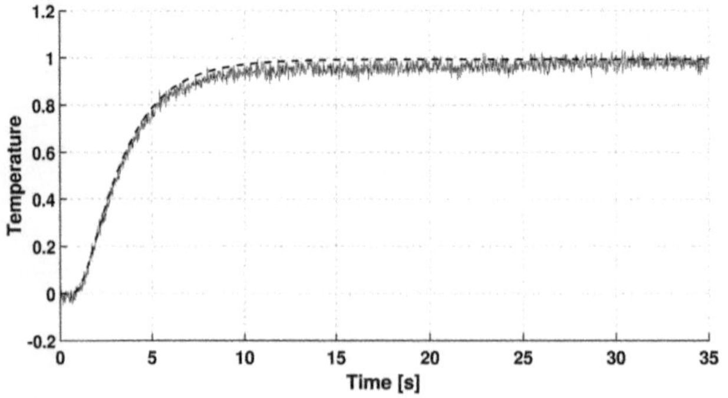

Figure 4. Data recorded for basalt rock sample with thickness of 3.84 mm. The dashed line shows the prediction of Fourier's law.

The evaluation of the thinnest sample using the GK equation is shown in Figure 5. The fitted coefficients are summarized in Table 1. It is important to mention that MCV equation using the presented parameters would show a wave-like propagation that is not observed in the experiments.

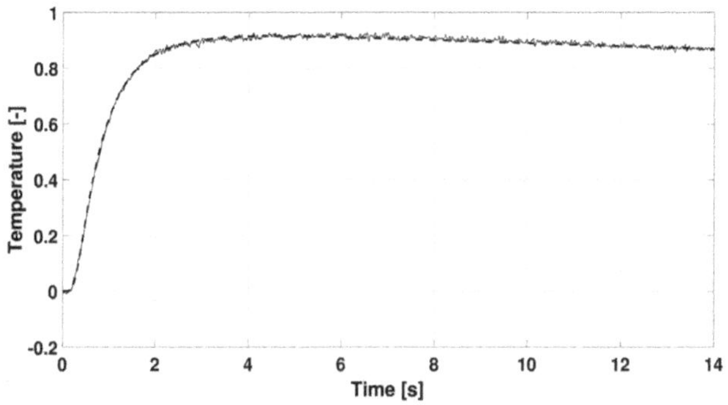

Figure 5. Data recorded using the basalt with thickness of 1.86 mm. The dashed line shows the prediction of GK equation.

Table 1. Summarized results of fitted coefficients in Fourier and GK equations.

Thickness L, [mm]	Fourier Thermal Diffusivity a_F,$\cdot 10^{-6}\left[\frac{m^2}{s}\right]$	Guyer-Krumhansl Thermal Diffusivity a_{GK}, $\cdot 10^{-6}\left[\frac{m^2}{s}\right]$	Relaxation Time τ, [s]	Dissipation Parameter κ^2, $\cdot 10^{-6}[m^2]$
1.86	0.62	0.55	0.738	0.509
2.75	0.67	0.604	0.955	0.67
3.84	0.685	0.68	0.664	0.48

Deviation from the Fourier prediction is weak but is clearly present, and has size dependent attributes. Concerning the ratio of parameters, i.e., investigating how considerably the Fourier resonance condition $a\tau/\kappa^2 = 1$ is violated, the outcome can be seen in Table 2. As analysis of the results, it is remarkable to note the deviation of the GK fitted thermal diffusivity from the Fourier fitted one, and that this deviation is size dependent. For the thickest sample, which can be well described by Fourier's law, the fitted thermal diffusivity values are practically equal, and the ratio of parameters is very close to the Fourier resonance value 1.

Table 2. Ratio of the fitted coefficients.

Thickness L, [mm]	Ratio of Parameters $\frac{a_{GK}\tau}{\kappa^2}$
1.86	0.804
2.75	0.854
3.84	0.943

The next section is devoted to a possible explanation for the emergence of a generalized heat equation with higher time and space derivatives. All coefficients of the higher time and space derivative terms are related to well-known material parameters. The result also features size dependent non-Fourier deviation.

3. Seeming Non-Fourier Heat Conduction Induced by Elasticity Coupled via Thermal Expansion

While, in general, one does not have a direct physical interpretation of the phenomenon that leads to, at the phenomenological level, non-Fourier heat conduction here follows a case where we do know this background phenomenon. Namely, in case of heat conduction in solids, a plausible possibility is provided by an interplay between elasticity and thermal expansion. Namely, without thermal expansion, elasticity—a tensorial behavior—is not coupled to Fourier heat conduction—a vectorial one—in isotropic materials. However, with nonzero thermal expansion, strains and displacements must be in accord both with what elastic mechanics dictates and with what position dependent temperature imposes. The coupled set of equations of Fourier heat conduction, of elastic mechanics and of kinematic relationships, after eliminating the kinematic and mechanical quantities, leads to an equation for temperature only that contains higher derivative corrections to Fourier's equation. It is important to check how remarkable these corrections are. In the following section we present this derivation and investigation.

The Basic Equations

In all respects involved, we choose the simplest assumptions: the small-strain regime, a Hooke-elastic homogeneous and isotropic solid material, with constant thermal expansion coefficient, essentially being at rest with respect to an inertial reference frame. Kinematic, mechanical and thermodynamical quantities and their relationships are considered along the approach detailed in [41–43].

The Hooke-elastic homogeneous and isotropic material model states, at any position $\mathbf{r}$, the constitutive relationship

$$\sigma^{\mathrm{d}} = E^{\mathrm{d}}\mathbf{D}^{\mathrm{d}}, \quad \sigma^{\mathrm{s}} = E^{\mathrm{s}}\mathbf{D}^{\mathrm{s}}, \qquad E^{\mathrm{d}} = 2G, \quad E^{\mathrm{s}} = 3K, \tag{7}$$

$$\sigma = E^{\mathrm{d}}\mathbf{D}^{\mathrm{d}} + E^{\mathrm{s}}\mathbf{D}^{\mathrm{s}} = E^{\mathrm{d}}\mathbf{D} + \left(E^{\mathrm{s}} - E^{\mathrm{d}}\right)\mathbf{D}^{\mathrm{s}} \tag{8}$$

between stress tensor σ and elastic deformedness tensor $\mathbf{D}$ (which, in many cases, coincides with the strain tensor), where ^{d} and ^{s} denote the deviatoric (traceless) and spherical (proportional to the unit tensor $\mathbf{1}$) parts, i.e.,

$$\mathbf{D}^{\mathrm{s}} = \frac{1}{3}\left(\mathrm{tr}\,\mathbf{D}\right)\mathbf{1}, \quad \mathbf{D}^{\mathrm{d}} = \mathbf{D} - \mathbf{D}^{\mathrm{s}}; \qquad \text{hence, e.g.,} \quad \mathbf{1}^{\mathrm{s}} = \mathbf{1}, \quad \mathbf{1}^{\mathrm{d}} = \mathbf{0}. \tag{9}$$

Stress induces a time derivative in the velocity field $\mathbf{v}$ of the solid medium, according to the equation

$$\varrho\dot{\mathbf{v}} = \sigma \cdot \overleftarrow{\nabla} \tag{10}$$

with mass density ϱ being constant in the in the small-strain regime; hereafter $\overleftarrow{\nabla}$ and $\overrightarrow{\nabla}$ denote derivative of the function standing to the left and to the right, respectively, to display the tensorial order (tensorial index order) properly for vector/tensor valued functions. For the velocity gradient $\mathbf{L}$ and its symmetric part, one has

$$\mathbf{L} = \mathbf{v} \otimes \overleftarrow{\nabla}, \quad \mathrm{tr}\,\mathbf{L}^{\mathrm{sym}} = \mathrm{tr}\,\mathbf{L} = \mathbf{v} \cdot \overleftarrow{\nabla}, \quad \left(\mathbf{L}^{\mathrm{sym}}\right)^{\mathrm{s}} = \frac{1}{3}\left(\mathrm{tr}\,\mathbf{L}^{\mathrm{sym}}\right)\mathbf{1} = \frac{1}{3}\left(\mathbf{v} \cdot \overleftarrow{\nabla}\right)\mathbf{1}, \tag{11}$$

$$\left(\mathbf{L}^{\mathrm{sym}} \cdot \overleftarrow{\nabla}\right) \cdot \overleftarrow{\nabla} = \frac{1}{2}\partial_i\partial_j\left(\partial_i v_j + \partial_j v_i\right) = \frac{1}{2}\left[\triangle\left(\overleftarrow{\nabla} \cdot \mathbf{v}\right) + \triangle\left(\overrightarrow{\nabla} \cdot \mathbf{v}\right)\right] = \triangle\left(\mathbf{v} \cdot \overleftarrow{\nabla}\right), \tag{12}$$

$$\left(\mathbf{L} \cdot \overleftarrow{\nabla}\right) \cdot \overleftarrow{\nabla} = \triangle\left(\mathbf{v} \cdot \overleftarrow{\nabla}\right), \tag{13}$$

where the Einstein summation convention for indices has also been applied. Again, using this convention, and the Kronecker delta notation, to any scalar field f,

$$\partial_j\left(f\delta_{ij}\right) = \delta_{ij}\partial_j f = \partial_i f, \qquad (f\mathbf{1}) \cdot \overleftarrow{\nabla} = \overrightarrow{\nabla} f \tag{14}$$

follow, which are also to be used below.

The small-deformedness relationship among the kinematic quantities, with linear thermal expansion coefficient α considered constant, and absolute temperature T, is

$$\mathbf{L}^{\mathrm{sym}} = \dot{\mathbf{D}} + \alpha \dot{T} \mathbf{1} . \tag{15}$$

For specific internal energy e,

$$e = cT + \frac{E^{\mathrm{s}}\alpha}{\varrho} T \operatorname{tr} \mathbf{D}^{\mathrm{s}} + e_{\mathrm{el}} , \quad e_{\mathrm{el}} = \frac{E^{\mathrm{d}}}{2\varrho} \operatorname{tr} \left[\left(\mathbf{D}^{\mathrm{d}} \right)^2 \right] + \frac{E^{\mathrm{s}}}{2\varrho} \operatorname{tr} \left[(\mathbf{D}^{\mathrm{s}})^2 \right] , \tag{16}$$

its balance,

$$\varrho \dot{e} = \operatorname{tr} (\sigma \mathbf{L}) - \mathbf{q} \cdot \overleftarrow{\nabla} , \tag{17}$$

after subtracting the contribution $\varrho \dot{e}_{\mathrm{el}}$ coming from specific elastic energy e_{el} and the corresponding elastic part $\operatorname{tr}(\sigma \dot{\mathbf{D}})$ of the mechanical power $\operatorname{tr}(\sigma \mathbf{L})$, is

$$\varrho \left(e - e_{\mathrm{el}} \right)^{\cdot} = \varrho c \dot{T} + E^{\mathrm{s}} \alpha T_0 \operatorname{tr} \dot{\mathbf{D}}^{\mathrm{s}} = - \mathbf{q} \cdot \overleftarrow{\nabla} , \qquad \text{with} \quad \mathbf{q} = -k \overrightarrow{\nabla} T , \tag{18}$$

where c is specific heat corresponding to constant zero stress (or pressure), temperature has been approximated in one term of (18) by an initial homogeneous absolute temperature value T_0 to stay in accord with the linear (small-strain) approximation, and heat flux $\mathbf{q}$ follows the Fourier heat conduction constitutive relationship with thermal conductivity k also treated as a constant.

The Derivation

The strategy is to eliminate σ in favor of (with the aid of) $\mathbf{D}$, then $\mathbf{D}$ is eliminated in favor of $\mathbf{L}^{\mathrm{sym}}$, after which we can realize that both from the mechanical direction and from the thermal one we obtain relationship between $\mathbf{v} \cdot \overleftarrow{\nabla}$ and T, which, eliminating $\mathbf{v} \cdot \overleftarrow{\nabla}$, yields an equation for T only.

Starting with the thermal side,

$$\begin{aligned} \varrho c \dot{T} + E^{\mathrm{s}} \alpha T_0 \operatorname{tr} \left(\mathbf{L}^{\mathrm{sym}} - \alpha \dot{T} \mathbf{1} \right)^{\mathrm{s}} &= \varrho c \dot{T} + E^{\mathrm{s}} \alpha T_0 \left(\mathbf{v} \cdot \overleftarrow{\nabla} \right) - E^{\mathrm{s}} \alpha^2 T_0 \dot{T} \cdot 3 = \\ &= \underbrace{\left(\varrho c - 3 E^{\mathrm{s}} \alpha^2 T_0 \right)}_{\gamma_1} \dot{T} + E^{\mathrm{s}} \alpha T_0 (\mathbf{v} \cdot \overleftarrow{\nabla}) , \qquad (19) \\ &= - \mathbf{q} \cdot \overleftarrow{\nabla} = - \left(-k \overrightarrow{\nabla} T \right) \cdot \overleftarrow{\nabla} = k \triangle T \quad \Longrightarrow \end{aligned}$$

$$E^{\mathrm{s}} \alpha T_0 \left(\mathbf{v} \cdot \overleftarrow{\nabla} \right) = k \triangle T - \gamma_1 \dot{T} . \tag{20}$$

Meanwhile, from the mechanical direction, aiming at being in tune with (20):

$$
\begin{aligned}
E^{s}\alpha T_0\left(\ddot{\mathbf{v}}\cdot\overleftarrow{\nabla}\right) &= E^{s}\alpha T_0\frac{1}{\varrho}\left(\dot{\boldsymbol{\sigma}}\cdot\overleftarrow{\nabla}\right)\cdot\overleftarrow{\nabla} = \\
&= \frac{E^{s}\alpha T_0}{\varrho}\left\{\left[E^{d}\dot{\mathbf{D}}+\left(E^{s}-E^{d}\right)\dot{\mathbf{D}}^{s}\right]\cdot\overleftarrow{\nabla}\right\}\cdot\overleftarrow{\nabla} = \\
&= \frac{E^{s}\alpha T_0}{\varrho}\left\{\left[E^{d}\left(\mathbf{L}^{\mathrm{sym}}-\alpha\dot{T}\mathbf{1}\right)+\right.\right. \\
&\qquad\left.\left.+\left(E^{s}-E^{d}\right)\left(\mathbf{L}^{\mathrm{sym}}-\alpha\dot{T}\mathbf{1}\right)^{s}\right]\cdot\overleftarrow{\nabla}\right\}\cdot\overleftarrow{\nabla} = \\
&= \frac{E^{s}\alpha T_0}{\varrho}\left\{\left[E^{d}\mathbf{L}^{\mathrm{sym}}-E^{d}\alpha\dot{T}\mathbf{1}+\left(E^{s}-E^{d}\right)\frac{1}{3}\left(\mathbf{v}\cdot\overleftarrow{\nabla}\right)\mathbf{1}-\right.\right. \\
&\qquad\left.\left.-\left(E^{s}-E^{d}\right)\alpha\dot{T}\mathbf{1}\right]\cdot\overleftarrow{\nabla}\right\}\cdot\overleftarrow{\nabla} = \\
&= \frac{E^{s}\alpha T_0}{\varrho}\left[E^{d}\triangle\left(\mathbf{v}\cdot\overleftarrow{\nabla}\right)+\frac{E^{s}-E^{d}}{3}\triangle\left(\mathbf{v}\cdot\overleftarrow{\nabla}\right)-E^{s}\alpha\triangle\dot{T}\right] = \\
&= \frac{E^{s}\alpha T_0}{\varrho}\left[\frac{E^{s}+2E^{d}}{3}\triangle\left(\mathbf{v}\cdot\overleftarrow{\nabla}\right)-E^{s}\alpha\triangle\dot{T}\right] = \\
&= \frac{E^{s}+2E^{d}}{3\varrho}\triangle\left[E^{s}\alpha T_0\left(\mathbf{v}\cdot\overleftarrow{\nabla}\right)\right]-\frac{(E^{s}\alpha)^2T_0}{\varrho}\triangle\dot{T} = \\
&= \underbrace{\frac{E^{s}+2E^{d}}{3\varrho}}_{c_{\parallel}^{2}}\triangle\left(k\triangle T-\gamma_1\dot{T}\right)-\frac{(E^{s}\alpha)^2T_0}{\varrho}\triangle\dot{T}; \quad \text{in parallel,} \\
&= \left(k\triangle T-\gamma_1\dot{T}\right)^{\cdot\cdot} = k\triangle\ddot{T}-\gamma_1\dddot{T} \qquad [\text{cf. } (20)]
\end{aligned}
\tag{21}
$$

(where $c_{\parallel}$ is the longitudinal elastic wave propagation velocity); hence, summarizing the final result in two equivalent forms,

$$
\left(\gamma_1\dot{T}-k\triangle T\right)^{\cdot\cdot} = c_{\parallel}^{2}\triangle\left(\underline{\gamma_1}\dot{T}-k\triangle T\right)+\frac{(E^{s}\alpha)^2T_0}{\varrho}\triangle\dot{T}, \tag{22}
$$

$$
\gamma_1\left(\ddot{T}-\underline{c_{\parallel}^{2}}\triangle T\right)^{\cdot} = k\triangle\gamma_1\left(\ddot{T}-c_{\parallel}^{2}\triangle T\right)+\frac{(E^{s}\alpha)^2T_0}{\varrho}\triangle\dot{T}. \tag{23}
$$

The first form here tells us that we have here the wave equation of a heat conduction equation, the last term on the r.h.s. somewhat detuning the heat conduction equation of the r.h.s. with respect to the one on the l.h.s. (the underlined coefficient is the one becoming modified when its term is melted together with the last term). In the meantime, the second form shows the heat conduction equation of a wave equation, the last term on the r.h.s. detuning the underlined coefficient.

Both forms show that coupling, after elimination, leads to a hierarchy of equations, with an amount of detuning that is induced by the coupling—for similar further examples, see [44].

We close this section by rewriting the final result in a form that enables to estimate the contribution of thermal expansion coupled elasticity to heat conduction:

$$
\frac{1}{c_{\parallel}^{2}}\left(\gamma_1\dot{T}-k\triangle T\right)^{\cdot\cdot} = \triangle\left[\left(\gamma_1+\frac{(E^{s}\alpha)^2T_0}{\varrho c_{\parallel}^{2}}\right)\dot{T}-k\triangle T\right], \tag{24}
$$

i.e.,

$$\frac{1}{c_{\parallel}^2}\left(\gamma_1 \dot{T} - k\triangle T\right)^{\cdot\cdot} = \triangle\left[\Big(\underbrace{\varrho c - \frac{6E^{\mathrm{d}}E^{\mathrm{s}}\alpha^2 T_0}{E^{\mathrm{s}} + 2E^{\mathrm{d}}}}_{\gamma_2}\Big)\dot{T} - k\triangle T\right]. \tag{25}$$

One message here is that, thermal expansion coupled elasticity modifies the thermal diffusivity $a = k/(\varrho c)$ to an effective one $a_2 = k/\gamma_2 = (\varrho c/\gamma_2) \cdot a$ (see the heat conduction on the r.h.s.). For metals, this means a few percent shift (1% for steel and copper, and 6% for aluminum) at room temperature.

The other is that, for a length scale (e.g., characteristic sample size) ℓ and the corresponding Fourier time scale ℓ^2/a, the r.h.s. is, to a (very) rough estimate, $1/\ell^2$ times a heat conduction equation while the l.h.s. is (similarly roughly)

$$\frac{1}{(\ell^2/a)^2} \cdot \frac{1}{c_{\parallel}^2} \tag{26}$$

times the (nearly) same heat conduction equation (a one with $a_1 = k/\gamma_1$). In other words, the l.h.s. provides a contribution to the r.h.s. via a dimensionless factor

$$\frac{\ell^2}{(\ell^2/a)^2} \cdot \frac{1}{c_{\parallel}^2} = \frac{a^2}{\ell^2 c_{\parallel}^2}. \tag{27}$$

This dimensionless factor is about 10^{-10} to 10^{-13} for metals, 10^{-14} for rocks and 10^{-15} for plastics with $\ell = 3\,\mathrm{mm}$, a typical size for flash experiments. Therefore, the effect of the l.h.s. appears to be negligible with respect to the r.h.s.

It is important to point out that the first phenomenon—the emergence of effective thermal diffusivity—would remain unnoticed in the analogous one space dimensional calculation:

$$\sigma = ED, \qquad \varrho\dot{v} = \sigma', \qquad L = v' = \dot{D} + \alpha\dot{T}, \tag{28}$$

$$q = -kT', \qquad e = cT + \frac{E\alpha}{\varrho}TD + \frac{E}{2\varrho}D^2 \qquad \Longrightarrow \tag{29}$$

$$\frac{\varrho}{E}\left[(\varrho c - E\alpha^2 T_0)\dot{T} - kT''\right]^{\cdot\cdot} = \left[\varrho c\dot{T} - kT''\right]'' \tag{30}$$

[no detuning of ϱc on the r.h.s.]. It is revealed only in the full 3D treatment, which reveals possible pitfalls of 1D considerations in general as well.

As conclusion of this section, thermal expansion coupled elasticity may introduce a few percent effect (a material dependent but sample size independent value) in determining thermal diffusivity from flash experiments or other transient processes (while its other consequences may be negligible).

4. Pseudo-Temperature Approach

The experimental results serve to check whether a certain theory used for describing the observed phenomenon is acceptable or not. The heat pulse (flash) experiment results may show various temperature histories. Generally, the flash measurement results are according to the Fourier theory. In some cases, as reported in [39,40] the temperature histories show "irregular" characteristics, especially these histories could be described by the help of various non-Fourier models [7,34,45,46]. Some kind of non-Fourier behavior could be constructed as it is shown in the following. This is only an illustration how two parallel Fourier mechanisms could result a non-Fourier-like temperature history. The idea is strongly motivated by the hierarchy of Fourier equations in the GK model [44] as mentioned previously; however, their interaction is not described in detail.

The sample that we investigate now is only a hypothetic one, we may call it as a "pseudo-matter". We consider in the following that the pseudo-matter formed by parallel material strips is wide enough that the interface effects might be neglected, i.e., they are like insulated parallel channels. We also consider that only the thermal conductivities are different, and the strips have the same mass density and specific heat. During the flash experiment after the front side energy input, a simple temperature equalization process happens in the sample in case of adiabatic boundary conditions. Since the flash method is widely developed, the effects of the real measurement conditions (heat losses, heat gain, finite pulse time, etc.) are well treated in the literature.

Figure 6 shows two temperature histories with thermal diffusivities of different magnitude, both are the solution of Fourier heat equation.

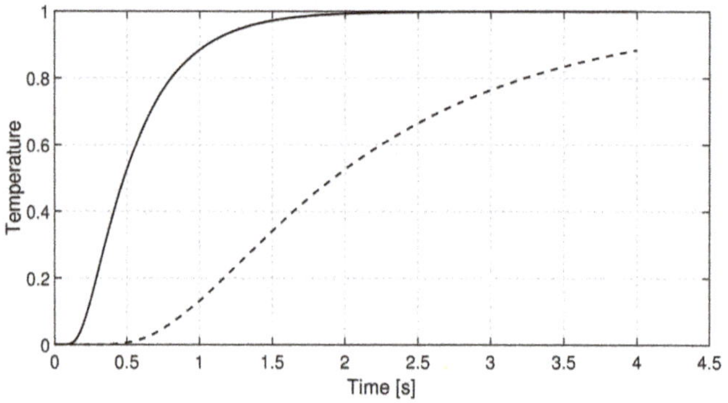

Figure 6. Rear-side temperature history; solid line: $a = 10^{-6}$ m^2/s, dashed line: $a = 2.5 \cdot 10^{-7}$ m^2/s, $L = 2$ mm.

The mathematical formula that expresses the temperature history of the rear side in the adiabatic case is [47]:

$$\nu(\xi = 1, Fo) = 1 + 2\sum_{m=1}^{\infty}(-1)^m e^{-(m^2\pi^2 Fo)}, \tag{31}$$

where ν is the dimensionless temperature, i.e., $\nu = \frac{T-T_0}{T_{\max}-T_0}$, where T_0 is the initial temperature and $T_{\max}$ is the asymptotic temperature corresponding to equilibrium with adiabatic boundary conditions, ξ is the normalized spatial coordinate ($\xi = 1$ corresponds to the rear-side) and $Fo = a \cdot t/(L^2)$ stands for the Fourier number (dimensionless time variable). This is an infinite series with property of slow convergence for short initial time intervals. An alternative formula derived using the Laplace theorem to obtain faster convergence for $Fo < 1$ [48]:

$$p(Fo) = \frac{2}{\sqrt{\pi Fo}}\sum_{n=0}^{\infty} e^{-\frac{(2n+1)^2}{4Fo}}, \tag{32}$$

wherein p is the Laplace transform of ν. In the further analysis we use Equation (32) to calculate the rear-side temperature history.

So far, we described two parallel heat-conducting layers without direct interaction among them; however, let us suppose that they can change energy only at their rear side through a very thin layer with excellent conduction properties. Eventually, that models the role of the silver layer used in our experiments to close the thermocouple circuit and assure that we measure the temperature of that layer instead of any internal one from the material. Actually, the silver layer averages the rear-side temperature histories of the parallel strips. We considered the mixing of temperature histories using the formula:

$$p(Fo) = \Theta p_1(a = 10^{-6}\,\mathrm{m^2/s}, Fo_1) + (1-\Theta)p_2(a = 2.5\cdot 10^{-7}\,\mathrm{m^2/s}, Fo_2), \tag{33}$$

that is, taking the convex combination of different solutions of Fourier heat Equation (5). Figure 7 shows a few possible cases of mixing.

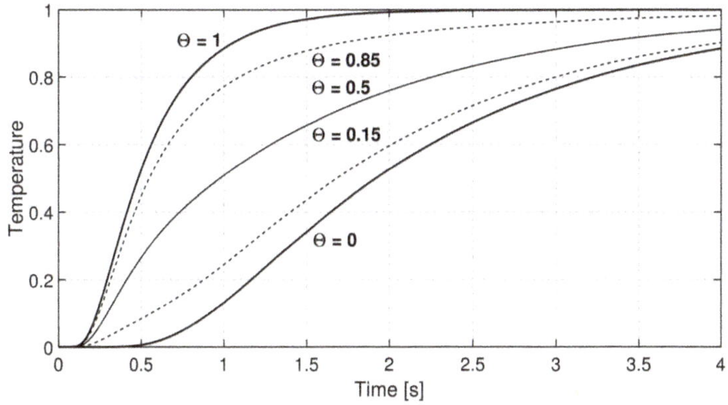

Figure 7. Rear-side temperature histories.

5. Outlook and Summary

This pseudo-material virtual experiment is only to demonstrate that there might be several effects causing non-Fourier behavior of the registered temperature data. Here, the assumed mixing of "Fourier-temperatures" is analogous with the GK equation in sense of the hierarchy of Fourier equation: dual heat-conducting channels are present and interact with each other. However, the GK equation is more general, there is no need to assume some mechanism to derive the constitutive equation.

Comparing Equations (6) to (25), the hierarchy of Fourier equation appears in a different way. While (6) contains the zeroth and first order time derivatives of Fourier equation, the (25) instead contains its second order time and spaces derivatives. Recalling that Equation (25)

$$\frac{1}{c_{\parallel}^2}\left(\gamma_1 \dot{T} - k\triangle T\right)^{\cdot\cdot} = \triangle\left[\Big(\underbrace{\varrho c - \frac{6E^{\mathrm{d}}E^{\mathrm{s}}\alpha^2 T_0}{E^{\mathrm{s}} + 2E^{\mathrm{d}}}}_{\gamma_2}\Big)\dot{T} - k\triangle T\right]. \tag{34}$$

is derived using the assumption that thermal expansion is present beside heat conduction, it becomes obvious to compare it to a ballistic (i.e., thermal expansion induced) heat conduction model. Let us consider such model from [7]:

$$\tau_1\tau_2\dddot{T} + (\tau_1+\tau_2)\ddot{T} + \dot{T} = a\triangle T + (\kappa^2 + a\tau_2)\triangle\dot{T}, \tag{35}$$

where τ_1 and τ_2 are relaxation times. Equation (35) have been tested on experiments, too [16]. Eventually, the GK equation is extended with a third order time derivative and the coefficients are modified by presence of τ_2. On contrary to Equation (34), it does not contain any fourth order derivative. Actually, the existing hierarchy of Fourier equation is extended, instead of τ and κ^2 the terms $(\tau_1+\tau_2)$ and $(\kappa^2 + a\tau_2)$ appear within (35).

Although it is still not clear exactly what leads to over-diffusive heat conduction, the presented possible interpretations and approaches can be helpful to understand the underlying mechanism. It is not the first time to experimentally measure the over-diffusive propagation, but it is to consider its size dependence. The simplest thermo-mechanical coupling predicts size dependence of material coefficients that can be relevant in certain cases. All three approaches lead to a system of partial differential equations, which can be called hierarchical.

Author Contributions: T.Fü. developed the thermo-mechanical model presented in Section 3. Section 4 is suggested by G.G. and with P.V. they designed the experiments. Á.L., Á.R., T.Fo., M.S. performed and analyzed the experimental data. R.K. compiled and composed the paper. All the authors contributed equally to the paper.

Funding: The work was supported by the grants National Research, Development and Innovation Office - NKFIH 116197, NKFIH 116375, NKFIH 124366 (124508) and NKFIH 123815.

Acknowledgments: Authors thank to Ákos Gyenis for doing improvements on the measurement equipment.

Conflicts of Interest: The authors declare no conflict of interest.

References

1. Fourier, J. *Theorie Analytique de la Chaleur*; Firmin Didot: Paris, France, 1822.
2. Tisza, L. Transport phenomena in Helium II. *Nature* **1938**, *141*, 913. [CrossRef]
3. Joseph, D.D.; Preziosi, L. Heat waves. *Rev. Mod. Phys.* **1989**, *61*, 41. [CrossRef]
4. Joseph, D.D.; Preziosi, L. Addendum to the paper on heat waves. *Rev. Mod. Phys.* **1990**, *62*, 375–391. [CrossRef]
5. Chen, G. Ballistic-diffusive heat-conduction equations. *Phys. Rev. Lett.* **2001**, *86*, 2297–2300. [CrossRef] [PubMed]
6. Ván, P.; Fülöp, T. Universality in Heat Conduction Theory—Weakly Nonlocal Thermodynamics. *Annalen der Physik (Berlin)* **2012**, *524*, 470–478. [CrossRef]
7. Kovács, R.; Ván, P. Generalized heat conduction in heat pulse experiments. *Int. J. Heat Mass Transf.* **2015**, *83*, 613–620. [CrossRef]
8. Ackerman, C.C.; Bertman, B.; Fairbank, H.A.; Guyer, R.A. Second sound in solid Helium. *Phys. Rev. Lett.* **1966**, *16*, 789–791. [CrossRef]
9. Jackson, H.E.; Walker, C.T. Thermal conductivity, second sound and phonon-phonon interactions in NaF. *Phys. Rev. B* **1971**, *3*, 1428–1439. [CrossRef]
10. Peshkov, V. Second sound in Helium II. *J. Phys. (Moscow)* **1944**, *8*, 381.
11. McNelly, T.F. Second Sound and Anharmonic Processes in Isotopically Pure Alkali-Halides. Ph.D. Thesis, Cornell University, Ithaca, NY, USA, 1974.
12. Dreyer, W.; Struchtrup, H. Heat pulse experiments revisited. *Contin. Mech. Thermodyn.* **1993**, *5*, 3–50. [CrossRef]
13. Müller, I.; Ruggeri, T. *Rational Extended Thermodynamics*; Springer: New York, NY, USA, 1998.
14. Frischmuth, K.; Cimmelli, V.A. Numerical reconstruction of heat pulse experiments. *Int. J. Eng. Sci.* **1995**, *33*, 209–215. [CrossRef]
15. Kovács, R.; Ván, P. Models of Ballistic Propagation of Heat at Low Temperatures. *Int. J. Thermophys.* **2016**, *37*, 95. [CrossRef]
16. Kovács, R.; Ván, P. Second sound and ballistic heat conduction: NaF experiments revisited. *Int. J. Heat Mass Transf.* **2018**, *117*, 682–690. [CrossRef]
17. Bargmann, S.; Steinmann, P. Finite element approaches to non-classical heat conduction in solids. *Comput. Model. Eng. Sci.* **2005**, *9*, 133–150.
18. Herwig, H.; Beckert, K. Fourier versus non-Fourier heat conduction in materials with a nonhomogeneous inner structure. *Trans.-Am. Soc. Mech. Eng. J. Heat Transf.* **2000**, *122*, 363–364. [CrossRef]
19. Zhang, Y.; Ye, W. Modified ballistic–diffusive equations for transient non-continuum heat conduction. *Int. J. Heat Mass Transf.* **2015**, *83*, 51–63. [CrossRef]
20. Zhukovsky, K.V.; Srivastava, H.M. Analytical solutions for heat diffusion beyond Fourier law. *Appl. Math. Comput.* **2017**, *293*, 423–437. [CrossRef]
21. Mitra, K.; Kumar, S.; Vedevarz, A.; Moallemi, M.K. Experimental evidence of hyperbolic heat conduction in processed meat. *J. Heat Transf.* **1995**, *117*, 568–573. [CrossRef]

22. Roetzel, W.; Putra, N.; Das, S.K. Experiment and analysis for non-Fourier conduction in materials with non-homogeneous inner structure. *Int. J. Therm. Sci.* **2003**, *42*, 541–552. [CrossRef]
23. Cattaneo, C. Sur une forme de lequation de la chaleur eliminant le paradoxe dune propagation instantanee. *C. R. Hebd. Seances Acad. Sci.* **1958**, *247*, 431–433.
24. Vernotte, P. Les paradoxes de la théorie continue de léquation de la chaleur. *C. R. Hebd. Seances Acad. Sci.* **1958**, *246*, 3154–3155.
25. Tisza, L. The theory of liquid Helium. *Phys. Rev.* **1947**, *72*, 838–877. [CrossRef]
26. Landau, L. On the theory of superfluidity of Helium II. *J. Phys.* **1947**, *11*, 91–92.
27. Guyer, R.A.; Krumhansl, J.A. Solution of the Linearized Phonon Boltzmann Equation. *Phys. Rev.* **1966**, *148*, 766–778. [CrossRef]
28. Guyer, R.A.; Krumhansl, J.A. Thermal Conductivity, Second Sound and Phonon Hydrodynamic Phenomena in Nonmetallic Crystals. *Phys. Rev.* **1966**, *148*, 778–788. [CrossRef]
29. Ván, P. Weakly Nonlocal Irreversible Thermodynamics—The Guyer-Krumhansl and the Cahn-Hilliard Equations. *Phys. Lett. A* **2001**, *290*, 88–92. [CrossRef]
30. Zhukovsky, K.V. Exact solution of Guyer–Krumhansl type heat equation by operational method. *Int. J. Heat Mass Transf.* **2016**, *96*, 132–144. [CrossRef]
31. Zhukovsky, K.V. Operational Approach and Solutions of Hyperbolic Heat Conduction Equations. *Axioms* **2016**, *5*, 28. [CrossRef]
32. Kovács, R. Analytic solution of Guyer-Krumhansl equation for laser flash experiments. *Int. J. Heat Mass Transf.* **2018**, *127*, 631–636. [CrossRef]
33. Tzou, D.Y. *Macro- to Micro-Scale Heat Transfer: The Lagging Behavior*; CRC Press: Boca Raton, FL, USA, 1996.
34. Sellitto, A.; Cimmelli, V.A.; Jou, D. Nonequilibrium Thermodynamics and Heat Transport at Nanoscale. In *Mesoscopic Theories of Heat Transport in Nanosystems*; Springer: Cham, Switzerland, 2016; pp. 1–30.
35. Rogolino, P.; Kovács, R.; Ván, P.; Cimmelli, V.A. Generalized heat-transport equations: Parabolic and hyperbolic models. *Contin. Mech. Thermodyn.* **2018**, 1–14. [CrossRef]
36. Fabrizio, M.; Lazzari, B.; Tibullo, V. Stability and Thermodynamic Restrictions for a Dual-Phase-Lag Thermal Model. *J. Non-Equilib. Thermodyn.* **2017**, *42*, 243–252. [CrossRef]
37. Rukolaine, S.A. Unphysical effects of the dual-phase-lag model of heat conduction: Higher-order approximations. *Int. J. Therm. Sci.* **2017**, *113*, 83–88. [CrossRef]
38. Kovács, R.; Ván, P. Thermodynamical consistency of the Dual Phase Lag heat conduction equation. *Contin. Mech. Thermodyn.* **2017**, 1–8, doi:10.1007/s00161-017-0610-x. [CrossRef]
39. Both, S.; Czél, B.; Fülöp, T.; Gróf, Gy.; Gyenis, Á.; Kovács, R.; Ván, P.; Verhás, J. Deviation from the Fourier law in room-temperature heat pulse experiments. *J. Non-Equilib. Thermodyn.* **2016**, *41*, 41–48. [CrossRef]
40. Ván, P.; Berezovski, A.; Fülöp, T.; Gróf, Gy.; Kovács, R.; Lovas, Á.; Verhás, J. Guyer-Krumhansl-type heat conduction at room temperature. *EPL* **2017**, *118*, 50005. [CrossRef]
41. Asszonyi, C.; Csatár, A.; Fülöp, T. Elastic, thermal expansion, plastic and rheological processes—theory and experiment. *Period. Polytech. Civ. Eng.* **2016**, *60*, 591–601. [CrossRef]
42. Fülöp, T.; Ván, P. Kinematic quantities of finite elastic and plastic deformation. *Math. Methods Appl. Sci.* **2012**, *35*, 1825–1841. [CrossRef]
43. Fülöp, T. Objective thermomechanics. *arXiv* **2015**, arXiv:1510.08038.
44. Ván, P.; Kovács, R.; Fülöp, T. Thermodynamics hierarchies of evolution equations. *Proc. Est. Acad. Sci.* **2015**, *64*, 389–395. [CrossRef]
45. Jou, D.; Carlomagno, I.; Cimmelli, V.A. A thermodynamic model for heat transport and thermal wave propagation in graded systems. *Phys. E Low-Dimens. Syst. Nanostruct.* **2015**, *73*, 242–249. [CrossRef]
46. Jou, D.; Cimmelli, V.A. Constitutive equations for heat conduction in nanosystems and non-equilibrium processes: An overview. *Commun. Appl. Ind. Math.* **2016**, *7*, 196–222.

47. Parker, W.J.; Jenkins, R.J.; Butler, C.P.; Abbott, G.L. Flash method of determining thermal diffusivity, heat capacity, and thermal conductivity. *J. Appl. Phys.* **1961**, *32*, 1679–1684. [CrossRef]
48. James, H.M. Some extensions of the flash method of measuring thermal diffusivity. *J. Appl. Phys.* **1980**, *51*, 4666–4672. [CrossRef]

entropy

MDPI

Article

Variations à la Fourier-Weyl-Wigner on Quantizations of the Plane and the Half-Plane

Hervé Bergeron [1,*] and Jean-Pierre Gazeau [2]

1 ISMO, UMR 8214 CNRS, Université Paris-Sud, 91405 Orsay, France
2 APC, UMR 7164 CNRS, Université Paris Diderot, Sorbonne Paris Cité, 75205 Paris, France; gazeau@apc.univ-paris7.fr
* Correspondence: herve.bergeron@u-psud.fr

Received: 1 September 2018; Accepted: 8 October 2018; Published: 13 October 2018

Abstract: Any quantization maps linearly function on a phase space to symmetric operators in a Hilbert space. Covariant integral quantization combines operator-valued measure with the symmetry group of the phase space. *Covariant* means that the quantization map intertwines classical (geometric operation) and quantum (unitary transformations) symmetries. *Integral* means that we use all resources of integral calculus, in order to implement the method when we apply it to singular functions, or distributions, for which the integral calculus is an essential ingredient. We first review this quantization scheme before revisiting the cases where symmetry covariance is described by the Weyl-Heisenberg group and the affine group respectively, and we emphasize the fundamental role played by Fourier transform in both cases. As an original outcome of our generalisations of the Wigner-Weyl transform, we show that many properties of the Weyl integral quantization, commonly viewed as optimal, are actually shared by a large family of integral quantizations.

Keywords: Weyl-Heisenberg group; affine group; Weyl quantization; Wigner function; covariant integral quantization

---

1. Introduction: A Historical Overview

More than one century after the publication by Fourier of his "Théorie analytique de la chaleur" [1,2], the Fourier transform revealed its tremendous importance at the advent of quantum mechanics with the setting of its specific formalism, especially with the seminal contributions of Weyl (1927) [3] on phase space symmetry, and Wigner (1932) [4] on phase space distribution. The phase space they were concerned with is essentially the Euclidean plane $\mathbb{R}^2 = \{(q,p)\,,\, q,p,\in\mathbb{R}\}$, q (mathematicians prefer to use x) for *position* and p for *momentum*. It is the phase space for the motion on the line and its most immediate symmetry is translational invariance: no point is privileged and so every point can be chosen as the origin. Non-commutativity relation $[Q,P] = \mathrm{i}\hbar I_{\mathcal{H}}$ between the self-adjoint quantum position Q and momentum P, the QM key stone, results from this symmetry through the Weyl projective unitary irreducible representation U [5] of the abelian group $\mathbb{R}^2$ in some separable Hilbert space $\mathcal{H}$,

$$\mathbb{R}^2 \ni (q,p) \mapsto \mathsf{U}(q,p) = e^{\frac{\mathrm{i}}{\hbar}(pQ-qP)}\,, \quad \mathsf{U}(q,p)\,\mathsf{U}(q',p') = e^{-\frac{\mathrm{i}}{2\hbar}(qp'-q'p}\,\mathsf{U}(q+q',p+p') \tag{1}$$

or equivalently the true representation of the so-called Weyl-Heisenberg group, central extension with parameter ϑ of the above one,

$$\mathbb{R}\times\mathbb{R}^2 \ni (\vartheta,q,p) \mapsto \mathcal{U}_{WH}(\vartheta,q,p) = e^{\mathrm{i}\vartheta/\hbar}\mathsf{U}(q,p)\,. \tag{2}$$

In 1932, Wigner introduced his function (or quasidistribution) to study quantum corrections to classical statistical mechanics, originally in view of associating the wavefunction $\psi(x)$, i.e., the pure state $\rho_\psi = |\psi\rangle\langle\psi|$, with a probability distribution in phase space. It is a Fourier transform, up to a constant factor, for all spatial autocorrelation functions of $\psi(x)$:

$$\mathcal{W}_{\rho_\psi}(q,p) = 2\int_{-\infty}^{+\infty} \mathrm{d}x\, \overline{\psi(q+x)}\, \psi(q-x)\, e^{\frac{2\mathrm{i}}{\hbar}px} = \mathrm{tr}\left(\mathsf{U}(q,p) 2\mathsf{P}\mathsf{U}^\dagger(q,p)\rho_\psi\right). \tag{3}$$

The alternative expression using in the above the parity operator $(\mathsf{P}\psi)(x) = \psi(-x)$ [6] allows us to extend this transform to any density operator ρ, and in fact to any traceclass operator A in $\mathcal{H}$

$$A \mapsto \mathcal{W}_A(q,p) = \mathrm{tr}\left(\mathsf{U}(q,p)2\mathsf{P}\mathsf{U}^\dagger(q,p)A\right). \tag{4}$$

One of the most attractive aspects of the above Wigner transform is that it is one-to-one. The inverse is precisely the Weyl quantization, more precisely the integral Weyl-Wigner quantization, defined as the map (with $\hbar = 1$)

$$f(q,p) \mapsto A_f = \int_{\mathbb{R}^2} \frac{\mathrm{d}q\,\mathrm{d}p}{2\pi}\, f(q,p)\, \mathsf{U}(q,p)2\mathsf{P}\mathsf{U}^\dagger(q,p) = \int_{\mathbb{R}^2} \frac{\mathrm{d}q\,\mathrm{d}p}{2\pi} \mathsf{U}(q,p)\, \overline{\mathfrak{F}_\mathfrak{s}}[f](q,p). \tag{5}$$

Hence, $\mathcal{W}_{A_f}(q,p) = f(q,p)$, with mild conditions on f. In the second expression of the Weyl-Wigner quantization (5) is introduced the dual of the symplectic Fourier transform. The latter is defined as

$$\mathfrak{F}_\mathfrak{s}[f](q,p) = \int_{\mathbb{R}^2} \frac{\mathrm{d}q\,\mathrm{d}p}{2\pi}\, e^{-\mathrm{i}(qp'-q'p)}\, f(q',p'). \tag{6}$$

It is involutive, $\mathfrak{F}_\mathfrak{s}\left[\mathfrak{F}_\mathfrak{s}[f]\right] = f$ like its dual defined as $\overline{\mathfrak{F}_\mathfrak{s}}[f](q,p) = \mathfrak{F}_\mathfrak{s}[f](-q,-p)$.

Hence, we observe that the Fourier transform lies at the heart of the above interplay of Weyl and Wigner approaches. Please note that both the maps (46) and (5) allow one to set up a *quantum mechanics in phase space*, as was developed at a larger extent in the 1940s by Groenewold [7] and Moyal [8]. This feature became so popular that it led some people to claim that if one seeks a single consistent quantization procedure mapping functions on the classical phase space to operators, the Weyl quantization is the "best" option. Actually, we will see below that this claimed preponderance should be somewhat attenuated, for various reasons.

The organisation of the paper is as follows. In Section 2 we give a general presentation of what we call covariant integral quantization associated with a Lie group, and its *semi-classical* side. The content of this section should be viewed as a shortened reiteration of a necessary material present in previous publications by one of or both the authors, essentially [9–13]. The original content of the paper is found in the next sections, namely the fact that many properties of the Weyl integral quantization, commonly viewed as optimal, are actually shared by a large family of integral quantizations. In Section 3 we revisit the Weyl-Heisenberg symmetry and the related Wigner-Weyl transform and Wigner function by inserting in their integral definition a kernel which allows to preserve one of their fundamental properties, the one-to-one character of the corresponding quantization. In Section 4 we devote a similar study to the case of the half-plane, for which the affine symmetry replaces the translational symmetry, and we compare our results with some previous works. We summarize the main points of the content in Section 5. Detailed proofs of two of our results are given in Appendix A.

2. Covariant Integral Quantization: A Summary

Integral quantization [9–13] is a generic name for approaches to quantization based on operator-valued measures. It includes the so-called Berezin-Klauder-Toeplitz quantization, and more generally coherent state quantization [10,14,15]. The integral quantization framework includes as well quantizations based on Lie groups. In the sequel we will refer to this case as *covariant integral quantization*. We mentioned in the introduction its most famous example, namely the covariant integral quantization based on the Weyl-Heisenberg group (WH), like Weyl-Wigner [3,6,16–18] and (standard) coherent states quantizations [14]. It is well established that the WH group underlies the canonical commutation rule, a paradigm of quantum physics. However, one should be aware that there is a world of quantizations that follow this rule [9,13]. Another basic example of covariant integral quantization concerns the half-plane viewed as the phase space for the motion on the half-line. The involved Lie group is the group of affine transformations $x \mapsto (q,p)\cdot x := x/q + p$, $q > 0$, of the real line [9,11]. The latter has been proven essential in a series of recent works devoted to quantum cosmology [19–23]. Let us notice that the affine group and related coherent states were also used for the quantization of the half-plane in works by J. R. Klauder, although from a different point of view (see [24–26] with references therein).

2.1. General Settings

We first proceed with a necessary repetition of the material needed to understand the method and found in the previously quoted [9–13]. Let X be a set equipped with some structures, e.g., measure, topology, manifold, etc. In this paper X will be viewed as a phase space for a mechanical system. Let $\mathcal{C}(X)$ be a vector space of complex-valued functions $f(x)$ on X, defined through some functional or distributional constraints, and viewed here as classical observables. A quantization of elements of $\mathcal{C}(X)$ is a linear map $\mathfrak{Q} : f \in \mathcal{C}(X) \mapsto \mathfrak{Q}(f) \equiv A_f \in \mathcal{A}(\mathcal{H})$ to a vector space $\mathcal{A}(\mathcal{H})$ of linear operators on some Hilbert space $\mathcal{H}$. Furthermore this map must fulfill the following conditions:

(i) To $f = 1$ there corresponds $A_f = I_{\mathcal{H}}$, where $I_{\mathcal{H}}$ is the identity in $\mathcal{H}$,
(ii) To a real function $f \in \mathcal{C}(X)$ there corresponds a(n) (essentially) self-adjoint operator A_f in $\mathcal{H}$.

From a physical point of view it will be necessary to add to this minimal material an interpretative measurement context.

Let us now assume that $X = G$ is a Lie group with left Haar measure $d\mu(g)$. Let $g \mapsto U_g$ be a unitary irreducible representation (UIR) of G as operators in $\mathcal{H}$. Let M be a bounded self-adjoint operator on $\mathcal{H}$ and let us define U_g-translations of M as

$$M(g) = U_g M U_g^{\dagger}\,. \tag{7}$$

The application of Schur's Lemma under mild conditions allows to infer that there exists a real constant $c_M \in \mathbb{R}$ such that the following resolution of the identity holds (in the weak sense of bilinear forms)

$$\int_G M(g)\,\frac{d\mu(g)}{c_M} = I_{\mathcal{H}}\,. \tag{8}$$

For instance, in the case of a square-integrable unitary irreducible representation $U : g \mapsto U_g$ (see Chapters 7 and 8 in [10] for details and references), let us pick a unit vector $|\psi\rangle$ for which $c_M = \int_G d\mu(g)|\langle\psi|U_g\psi\rangle|^2 < \infty$, i.e., $|\psi\rangle$ is an admissible unit vector for U. With $M = |\psi\rangle\langle\psi|$ the resolution of the identity (8) provided by the family of states $|\psi_g\rangle = U_g|\psi\rangle$ reads

$$\int_G |\psi_g\rangle\langle\psi_g|\,\frac{d\mu(g)}{c_M} = I_{\mathcal{H}}\,. \tag{9}$$

Vectors $|\psi_g\rangle$ are named (generalized) coherent states (or wavelets) for the group G.

With the resolution (8) in hand one can proceed with the integral quantization of complex-valued functions or distributions on the group G as follows

$$f \mapsto A_f = \int_G M(g) f(g)\,\frac{d\mu(g)}{c_M}\,. \tag{10}$$

Of course, some conditions have to be imposed to f in order to ensure the existence of the operator, or *quantum observable*, A_f. With such conditions, the quantization (10) is covariant in the sense that $U_g A_f U_g^\dagger = A_F$ where $F(g') = (\mathcal{U}_g f)(g') = f(g^{-1}g')$.

To be more precise about the existence of the operator-valued integral in (10), the latter should be understood in a weak sense. Precisely, the sesquilinear form

$$\mathcal{H} \ni \psi_1, \psi_2 \mapsto B_f(\psi_1, \psi_2) = \int_G \langle \psi_1 | M_g | \psi_2 \rangle f(g) \frac{\mathrm{d}\mu(g)}{c_M}\,, \tag{11}$$

is assumed to be defined on a dense subspace of $\mathcal{H}$. If f is a complex bounded function, B_f is a bounded sesquilinear form, and from the Riesz lemma we deduce that there exists a unique bounded operator A_f associated with B_f. If f is real and semi-bounded, and if M is a positive operator, Friedrich's extension of B_f ([27], Thm. X23) univocally defines a self-adjoint operator. However, if f is real but not semi-bounded, there is no natural choice for a self-adjoint operator associated with B_f. In this case, one can consider directly the symmetric operator A_f enabling us to obtain a possible self-adjoint extension (an example of this kind of mathematical study is presented in [28]).

2.2. Semi-Classical Framework With Probabilistic Interpretation

Integral quantization allows to develop what is commonly viewed as a semi-classical analysis/interpretation of quantum observables. If $M = \rho$ and $\tilde{\rho}$ are two non-negative ("density operator") unit trace operators, we obtain the classical-like expectation value formula

$$\mathrm{tr}(\tilde{\rho} A_f) = \int_G f(g) w(g) \frac{\mathrm{d}\mu(g)}{c_M}\,. \tag{12}$$

Indeed, resolution of the identity, non-negativeness and unit-trace conditions imply that $w(g) = \mathrm{tr}(\tilde{\rho}\,\rho(g)) \geq 0$ is, up to the coefficient c_M, a classical probability distribution on the group. Moreover, we consider the map

$$f \mapsto \check{f}(g) = \int_G \mathrm{tr}\,(\tilde{\rho}(g)\,\rho(g'))\;f(g') \frac{\mathrm{d}\mu(g)}{c_M}\,. \tag{13}$$

as a generalization of Berezin or heat kernel or Segal-Bargmann transforms [29] on G. Given f, the new function $\check{f}$ is called lower or covariant symbol of the operator A_f. It may be viewed as one of its semi-classical representations.

In the case of coherent states $|\psi_g\rangle$ (i.e., $M = \rho = |\psi\rangle\langle\psi|$), Equation (12) reads

$$\mathrm{tr}(\tilde{\rho} A_f) = \int_G f(g)\, \langle \psi_g | \tilde{\rho} | \psi_g \rangle \frac{\mathrm{d}\mu(g)}{c_M}\,, \tag{14}$$

where $w(g) = \langle \psi_g | \tilde{\rho} | \psi_g \rangle \geq 0$ is viewed here as a classical probability distribution on the group (up to the coefficient c_M). Similarly assuming $\tilde{\rho} = |\tilde{\psi}\rangle\langle\tilde{\psi}|$, the lower symbol $\check{f}(g)$ involved in (13) reads

$$\check{f}(g) = \int_G |\langle \tilde{\psi}_g | \psi_{g'} \rangle|^2 f(g') \frac{\mathrm{d}\mu(g')}{c_M} \tag{15}$$

2.3. Semi-Classical Picture Without Probabilistic Interpretation

A semi-classical framework similar to (13) can be also developed if the operators M and $\tilde{M}$ are not positive:

$$f \mapsto \check{f}(g) = \mathrm{tr}\left(\tilde{M}(g) A_f\right) = \int_G \mathrm{tr}\,(\tilde{M}(g)\,M(g'))\;f(g') \frac{\mathrm{d}\mu(g')}{c_M} \tag{16}$$

Then the probabilistic interpretation is lost in general due to the loss of positiveness of the map $g' \mapsto \mathrm{tr}\left(\tilde{M}(g)M(g')\right)$. However, in some special cases Equation (16) allows one to obtain an inverse of the quantization map (10). Namely for special pairs $(M, \tilde{M})$ we obtain

$$\mathrm{tr}\left(\tilde{M}(g)\, A_f\right) = f(g) \tag{17}$$

In the sequel we analyze different examples of this kind in the case of the quantization of the plane (Weyl-Heisenberg group) and the half-plane (affine group).

3. Quantization of the Plane: Generalizations of the Wigner-Weyl Transform

3.1. The Group Background

Let us first recall some definitions with more details about the Weyl-Heisenberg (WH) group G_{WH}, that we have already mentioned in the introduction. More details can be found for instance in [10,13]. It is a central extension of the group of translations of the two-dimensional euclidean plane. In classical mechanics the latter is viewed as the phase space for the motion of a particle on the real line. The UIR we are concerned with is the unitary representation of G_{WH}, acting in some separable Hilbert space $\mathcal{H}$, which integrates the canonical commutation rule (CCR) of quantum mechanics, $[Q, P] = \mathrm{i}\hbar I_{\mathcal{H}}$. Forgetting about physical dimensions ($\hbar = 1$), an arbitrary element g of G_{WH} is of the form

$$g = (\vartheta, q, p), \quad \vartheta \in \mathbb{R},\ (q, p) \in \mathbb{R}^2, \tag{18}$$

with multiplication law

$$g_1 g_2 = (\vartheta_1 + \vartheta_2 + \zeta[(q_1, p_1), (q_2, p_2)], q_1 + q_2, p_1 + p_2)\,, \tag{19}$$

where ζ is the multiplier function $\zeta[(q_1, p_1), (q_2, p_2)] = \frac{1}{2}(p_1 q_2 - p_2 q_1)$. Any infinite dimensional UIR $\mathcal{U}^{\lambda}_{WH}$ of G_{WH} is characterized by a real number $\lambda \neq 0$ (in addition, there are also degenerate, one-dimensional, UIR's corresponding to $\lambda = 0$, but they are irrelevant here). These UIR's may be realized on the same Hilbert space $\mathcal{H}$, as the one carrying an irreducible representation of the CCR:

$$\mathcal{U}^{\lambda}_{WH}(\vartheta, q, p) = e^{\mathrm{i}\lambda\vartheta}\mathrm{U}^{\lambda}(q, p) = e^{\mathrm{i}\lambda(\theta - qp/2)} e^{\mathrm{i}\lambda pQ} e^{-\mathrm{i}\lambda qP}\,. \tag{20}$$

If $\mathcal{H} = L^2(\mathbb{R}, \mathrm{d}x)$ corresponding to the spectral decomposition $Q = \int_{\mathbb{R}} x\, |x\rangle\langle x|\, \mathrm{d}x$ of the essentially self-adjoint position operator Q, the action of $\mathcal{U}^{\lambda}_{WH}$ reads as

$$\left(\mathcal{U}^{\lambda}_{WH}(\vartheta, q, p)\phi\right)(x) = e^{\mathrm{i}\lambda\vartheta} e^{\mathrm{i}\lambda p(x - q/2)}\phi(x - q), \quad \phi \in L^2(\mathbb{R}, \mathrm{d}x)\,. \tag{21}$$

Thus, the three operators $I_{\mathcal{H}}$, Q, P appear as the generators of this representation and are realized as:

$$(Q\phi)(x) = x\phi(x), \quad (P\phi)(x) = -\frac{\mathrm{i}}{\lambda}\phi'(x), \quad [Q, P] = \frac{\mathrm{i}}{\lambda} I_{\mathcal{H}}\,. \tag{22}$$

For our purpose we take $\lambda = 1/\hbar = 1$ and simply write $\mathcal{U}_{WH}$ for the corresponding representation.

3.2. Hyperbolic W-H Covariant Integral Quantization

3.2.1. General Settings

We investigate special cases of the Weyl-Heisenberg covariant integral quantization that have remarkable properties. They are included in our general framework as a special case. Namely let us choose some function $F \in L^1(\mathbb{R}, \mathrm{d}x)$ and define its Fourier transform $\hat{F}$ as

$$\hat{F}(\omega) = \int_{\mathbb{R}} F(u) e^{-\mathrm{i}\omega u}\, \mathrm{d}u\,. \tag{23}$$

This framework will be extended to distributions when necessary. We define the operator $\mathcal{P}_0^{(F)}$ (corresponding to the operator (denoted by M in Section 2.1) as the Weyl transform of $\hat{F}$:

$$\mathcal{P}_0^{(F)} = \int_{\mathbb{R}^2} \frac{\mathrm{d}q\mathrm{d}p}{2\pi} \hat{F}(qp)\, e^{\mathrm{i}(pQ-qP)}\,. \tag{24}$$

The associate quantization is named *hyperbolic* because of this special dependence through a function of qp. The operator $\mathcal{P}_0^{(F)}$ is bounded if $F \in L^1\left(\mathbb{R}, |u^2 - 1/4|^{-1/2}\, du\right)$ (see Appendix A for the proof). The main interest of this choice at the physical level is that all quantizations of this kind involve solely the Planck constant $\hbar$ as a dimensional parameter. In fact, $\hbar$ can be restored as follows

$$\mathcal{P}_0^{(F)} = \int_{\mathbb{R}^2} \frac{\mathrm{d}q\mathrm{d}p}{2\pi\hbar} \hat{F}(qp/\hbar)\, e^{\mathrm{i}(pQ-qP)/\hbar}\,. \tag{25}$$

The already mentioned canonical Wigner-Weyl transform or the Born-Jordan quantization [30–32] are special cases, but the above generalisation of the latter offers a large freedom in the choice of F with no need for introducing extra dimensional parameters.

In terms of the Dirac *kets* $|x\rangle$ such that $Q\,|x\rangle = x\,|x\rangle$, the kernel $\langle x|\mathcal{P}_0^{(F)}|y\rangle$ reads as:

$$\langle x|\mathcal{P}_0^{(F)}|y\rangle = \frac{1}{|x-y|} \int_{\mathbb{R}} \frac{\mathrm{d}u}{2\pi} \hat{F}(u) \exp\left(\mathrm{i}u \frac{x+y}{2(x-y)}\right) \tag{26}$$

which gives

$$\langle x|\mathcal{P}_0^{(F)}|y\rangle = \frac{1}{|x-y|} F\left(\frac{x+y}{2(x-y)}\right)\,. \tag{27}$$

The bounded operator $\mathcal{P}_0^{(F)}$ is self-adjoint if F verifies the hilbertian symmetry $\overline{F(u)} = F(-u)$. We assume this condition to be fulfilled in the sequel.

The kernel of the operator $\mathcal{P}_{q,p}^{(F)}$ corresponding to the WH transported operators $M(g)$ as in Equation (7) reads

$$\langle x|\mathcal{P}_{q,p}^{(F)}|y\rangle = \frac{1}{|x-y|} F\left(\frac{x+y-2q}{2(x-y)}\right) e^{\mathrm{i}p(x-y)}\,. \tag{28}$$

While the variable p appears in this formula as the Fourier reciprocal variable, the variable q appears as a translation parameter from the *arithmetic* mean of the variables x and y. Such an observation will take its real importance when we will deal with the affine symmetry in the next part of this paper.

3.2.2. Resolution of the Identity

From the Weyl-Heisenberg covariance and Schur's lemma, we obtain the resolution of unity as

$$\int_{\mathbb{R}^2} \frac{\mathrm{d}q\,\mathrm{d}p}{2\pi} \mathcal{P}_{q,p}^{(F)} = c\, I_{\mathcal{H}} \tag{29}$$

where $c = \int_{\mathbb{R}} F(u)\mathrm{d}u$. Therefore we assume in the sequel $\int_{\mathbb{R}} F(u)\mathrm{d}u = 1$.

At this point it is valuable to give a direct proof of (29). Due to the polarization identity, it is sufficient to prove that for any $\psi \in \mathcal{H}$:

$$\int_{\mathbb{R}^2} \frac{\mathrm{d}q\,\mathrm{d}p}{2\pi} \langle\psi|\mathcal{P}_{q,p}^{(F)}|\psi\rangle = c\langle\psi|\psi\rangle\,. \tag{30}$$

First

$$\langle\psi|\mathcal{P}_{q,p}^{(F)}|\psi\rangle = \int_{\mathbb{R}^2} \mathrm{d}x\,\mathrm{d}y\, \overline{\psi(x)}\, \psi(y)\, \frac{1}{|x-y|} F\left(\frac{x+y-2q}{2(x-y)}\right) e^{\mathrm{i}p(x-y)}\,. \tag{31}$$

By performing the change of variables $X = (x + y)/2$, $z = x - y$, we obtain

$$\langle\psi|\mathcal{P}^{(F)}_{q,p}|\psi\rangle = \int_{\mathbb{R}^2} \mathrm{d}X\,\mathrm{d}z\,\overline{\psi(X + z/2)}\,\psi(X - z/2)\frac{1}{|z|}F\left(\frac{X - q}{z}\right)e^{\mathrm{i}pz}\,. \tag{32}$$

Then we keep z and we change X in $u = (X - q)/z$. This leads to

$$\langle\psi|\mathcal{P}^{(F)}_{q,p}|\psi\rangle = \int_{\mathbb{R}^2} \mathrm{d}u\,\mathrm{d}z\,F(u)\,e^{\mathrm{i}pz}\,\overline{\psi(q + (u + 1/2)z)}\,\psi(q + (u - 1/2)z)\,. \tag{33}$$

We remark that this equation is in fact a generalization of the Wigner function. The latter is recovered with $F(u) = \delta(u)$. In this sense, the function F is a Cohen kernel [33,34], but its interpretation in the present quantization context is different of the role it was given by this author and others, like [35]. Now the integral over p gives

$$\int_{\mathbb{R}} \frac{\mathrm{d}p}{2\pi}\langle\psi|\mathcal{P}^{(F)}_{q,p}|\psi\rangle = \int_{\mathbb{R}} \mathrm{d}u\,F(u)|\psi(q)|^2\,. \tag{34}$$

and finally

$$\int_{\mathbb{R}^2} \frac{\mathrm{d}q\,\mathrm{d}p}{2\pi}\langle\psi|\mathcal{P}^{(F)}_{q,p}|\psi\rangle = \langle\psi|\psi\rangle\int_{\mathbb{R}} \mathrm{d}u\,F(u)\,. \tag{35}$$

Assuming $\int \mathrm{d}u\,F(u) = 1$, we recover the resolution of the identity.

3.2.3. Covariant Quantization and Properties

The F-dependent quantization map $f \mapsto A^{(F)}_f$ is defined as

$$f \mapsto A^{(F)}_f = \int_{\mathbb{R}^2} \frac{\mathrm{d}q\mathrm{d}p}{2\pi} f(q,p)\,\mathcal{P}^{(F)}_{q,p} \tag{36}$$

The usual Wigner-Weyl kernel corresponds to the distribution choice $F(x) = \delta(x)$ and it is, therefore, singular with respect to the functional framework. The case of Born-Jordan corresponds to the choice of the indicator function $F(u) = \mathbf{1}_{[-1/2,1/2]}(u)$. The map $f \mapsto A^{(F)}_f$ is such that whatever F (under the above conditions)

$$A^{(F)}_q = Q \quad \text{and} \quad A^{(F)}_p = P\,, \tag{37}$$

and more generally,

$$A^{(F)}_{f(q)} = f(Q) \quad \text{and} \quad A^{(F)}_{f(p)} = f(P)\,. \tag{38}$$

Therefore, by linearity any classical Hamiltonian $h(q,p) = \frac{1}{2m}p^2 + V(q)$ is mapped into the quantum Hamiltonian $H = \frac{1}{2m}P^2 + V(Q)$ that has the same form. Moreover, with the same conditions on F, we have

$$A^{(F)}_{qp} = \frac{1}{2}(QP + PQ) + c, \quad \text{with} \quad c = -\mathrm{i}\int_{\mathbb{R}} uF(u)\mathrm{d}u\,. \tag{39}$$

The constant c is real due to the condition $\overline{F(u)} = F(-u)$. If $F(u)$ is real then $c = 0$.

Remark 1. *Different quantizations generated by different F cannot be distinguished only using the most common operators involved in non-relativistic quantum mechanics (and corresponding to observables that can be really measured). Therefore there is no reason to privilege a specific one (for example the canonical one).*

3.2.4. Trace Formula

Let us rewrite (28) as:

$$\mathcal{P}^{(F)}_{q,p} = \int_{\mathbb{R}^2} \mathrm{d}x\mathrm{d}y \, \frac{1}{|x-y|} F\left(\frac{x+y-2q}{2(x-y)}\right) e^{\mathrm{i}p(x-y)} |x\rangle\langle y| \, . \tag{40}$$

Using the same kind of transformations as the ones used for the resolution of the identity we have (formally):

$$\mathcal{P}^{(F)}_{q,p} = \int_{\mathbb{R}^2} \mathrm{d}u \, \mathrm{d}z \, F(u) \, e^{\mathrm{i}pz} |q+(u+1/2)z\rangle \, \langle q+(u-1/2)z| \, . \tag{41}$$

Then (still formally)

$$\mathrm{tr}\, \mathcal{P}^{(F)}_{q,p} = \int_{\mathbb{R}^2} \mathrm{d}u \, \mathrm{d}z \, F(u) \, e^{\mathrm{i}pz} \, \delta(z) = 1 \, . \tag{42}$$

For two different functions F and G we obtain the trace formula:

$$\mathrm{tr}\left(\mathcal{P}^{(F)}_{q,p}\mathcal{P}^{(G)}_{q',p'}\right) = \int_{\mathbb{R}} \frac{\mathrm{d}z}{|z|} \, e^{-\mathrm{i}(p-p')z} (F*G)\left(\frac{q-q'}{z}\right) . \tag{43}$$

where $F*G$ is the convolution product of F and G.

3.3. *Invertible W-H Covariant Integral Quantization: Generalization of the Wigner-Weyl Transform*

3.3.1. General Settings

Let us examine the case for which (43) gives the equation $F*G=\delta$. Please note that such an equation has no solution with a pair of summable functions. In this case, we have

$$\mathrm{tr}\left(\mathcal{P}^{(F)}_{q,p}\mathcal{P}^{(G)}_{q',p'}\right) = 2\pi\,\delta(q-q')\,\delta(p-p') \, . \tag{44}$$

Therefore if F possesses a convolution inverse G, the quantization map is invertible. Indeed if G is the inverse of convolution of F then

$$\mathrm{tr}\left(\mathcal{P}^{(G)}_{q,p} A^{(F)}_f\right) = f(q,p) \, . \tag{45}$$

In this regard, the Wigner-Weyl case is *trivial* in the sense that $F=\delta$ is its own inverse and therefore the Wigner-Weyl quantization map is inverted with the same operator. Furthermore since δ is a distribution, the Wigner-Weyl choice is in fact *singular* within this functional framework. Therefore using a true function F can be viewed as a *regularization*. However, this regularization in the quantization map has a cost: the inverse map (if it exists) is more singular than a pure δ.

In the case of Born-Jordan the Fourier transform of the indicator function $F(u)$ is $\hat{F}(k) = \dfrac{\sin(k/2)}{k/2}$ that possesses simple zeros on the real axis. Whence the convolution inverse of F only exists in a distribution sense as a series of principal values.

3.3.2. Generalized Wigner Functions

Given a function F, we now define the *generalized Wigner function* of an operator A as

$$\mathcal{W}^{(F)}_A(q,p) = \mathrm{tr}\left(\mathcal{P}^{(F)}_{q,p} A\right) \, . \tag{46}$$

If A is the pure state $|\psi\rangle\langle\psi|$, this function reads

$$\mathcal{W}_{\psi}^{(F)}(q,p) \equiv \mathcal{W}_{|\psi\rangle\langle\psi|}^{(F)}(q,p) = \langle\psi|\mathcal{P}_{q,p}^{(F)}|\psi\rangle \tag{47}$$

$$= \int_{\mathbb{R}^2} \mathrm{d}u\, \mathrm{d}z\, F(u)\, e^{\mathrm{i}pz}\, \overline{\psi(q+(u+1/2)z)}\, \psi(q+(u-1/2)z)\,. \tag{48}$$

The standard Wigner function corresponds to $\mathcal{W}_{\psi}^{(\delta)}(q,p)$. All functions $\mathcal{W}_{\psi}^{(F)}(q,p)$ share the same marginal properties. Namely the functions $q \mapsto (2\pi)^{-1}\int \mathrm{d}p\mathcal{W}_{\psi}^{(F)}(q,p)$ and $p \mapsto (2\pi)^{-1}\int \mathrm{d}q\mathcal{W}_{\psi}^{(F)}(q,p)$ are the exact quantum probability distributions for position and momentum. This is a direct consequence of (38). Furthermore, because of the invertible character of the corresponding Wigner-Weyl transform, i.e.,

$$\mathcal{W}_{A_f}^{(\delta)}(q,p) := \mathrm{tr}\left(\mathcal{P}_{q,p}^{(\delta)} A_f\right) = f(q,p)\,, \tag{49}$$

we have

$$|\psi\rangle\langle\psi| = \int_{\mathbb{R}^2} \frac{\mathrm{d}q'\,\mathrm{d}p'}{2\pi} \mathcal{W}_{\psi}^{(\delta)}(q',p')\mathcal{P}_{q',p'}^{(\delta)}\,. \tag{50}$$

Therefore

$$\mathcal{W}_{\psi}^{(F)}(q,p) = \int_{\mathbb{R}^2} \frac{\mathrm{d}q'\,\mathrm{d}p'}{2\pi} \mathcal{W}_{\psi}^{(\delta)}(q',p')\,\mathrm{tr}\left(\mathcal{P}_{q,p}^{(F)}\mathcal{P}_{q',p'}^{(\delta)}\right)\,. \tag{51}$$

Using (43) we obtain

$$\mathcal{W}_{\psi}^{(F)} = \mathcal{W}_{\psi}^{(\delta)} * \Lambda(F)\,. \tag{52}$$

where $*$ holds for the $2d$-convolution product with the measure $\dfrac{\mathrm{d}q\,\mathrm{d}p}{2\pi}$ and

$$\Lambda(F)(q,p) = \tilde{F}(qp), \quad \text{with} \quad \tilde{F}(\omega) = \int_{\mathbb{R}} \frac{\mathrm{d}\alpha}{|\alpha|} e^{-\mathrm{i}\omega/\alpha} F(\alpha) \tag{53}$$

Remark 2.

- *The function $\Lambda(F)$ only depends on the variable qp. Therefore it cannot belong to some L^r space on the plane. Hence, the convolution product involved in (52) should be understood in general in the distribution sense.*
- *The function $\tilde{F}$ is defined as an integral only if F belongs to $L^1(\mathbb{R}, |\alpha|^{-1}\mathrm{d}\alpha)$. In other cases an extension in the distribution framework is needed.*
- *An interesting question concerns the positiveness of $\mathcal{W}_{\psi}^{(F)}$. In the genuine Wigner-Weyl case ($F = \delta$), Hudson theorem [36] asserts that only gaussian states ψ lead to positive Wigner functions $\mathcal{W}_{\psi}^{(\delta)}(q,p)$, and so the latter can be interpreted as probability densities on phase space. Beyond the pure Gaussian case, see for instance [37]. The problem now is to formulate a generalized version of the Hudson theorem (involving maybe a different family of states) for the generalized Wigner function $W_{\psi}^{(F)}$). In other words, for a given state ψ, is it possible to "build" a function F such that the corresponding Wigner function $\mathcal{W}_{\psi}^{(F)}$ is positive?*

3.3.3. Examples of Invertible Map

In the following lines, we give an explicit example of invertible map, dependent on two strictly positive parameters α and β and that includes the Wigner-Weyl solution as a special case (this example was found through the use of Fourier transform). Let us define $F_{\alpha,\beta}$ as

$$F_{\alpha,\beta}(x) = \alpha^4\delta(x) + \frac{1}{2}\alpha\beta(1-\alpha^4)e^{-\alpha\beta|x|}\,. \tag{54}$$

Obviously we have $\overline{F(x)} = F(-x)$ (in the distribution sense), and formally $\int F(x)\mathrm{d}x = 1$. Taking into account the elementary result for $a, b > 0$:

$$e^{-a|x|} * e^{-b|x|} = \frac{2}{b^2 - a^2}\left(be^{-a|x|} - ae^{-b|x|}\right), \tag{55}$$

we find that a convolution inverse of $F_{\alpha,\beta}$ is $F_{\alpha',\beta'}$ with $\alpha' = 1/\alpha$ et $\beta' = \beta\alpha^{-2}$. The Wigner-Weyl case corresponds to the degenerate case $F_{1,\beta}(x) = \delta(x)$.

4. Quantization of the Half-Plane With the Affine Group: Wigner-Weyl-Like Scheme

4.1. The Group Background

The half-plane is defined as $\Pi_+ = \{(q,p) \,|\, q > 0, p \in \mathbb{R}\}$. Equipped with the law

$$(q,p)(q',p') = \left(qq', p + \frac{p'}{q}\right), \tag{56}$$

Π_+ is viewed as the affine group $\mathrm{Aff}_+(\mathbb{R})$ of the real line. The left invariant measure is $\mathrm{d}\mu(q,p) = \mathrm{d}q\mathrm{d}p$. Besides a trivial one, the affine group possesses two nonequivalent square integrable UIR's. Equivalent realizations of one of them, say, U, are carried by Hilbert spaces $L^2(\mathbb{R}_+, \mathrm{d}x/x^\mu)$. Nonetheless these multiple possibilities do not introduce noticeable differences. Therefore we choose in the sequel $\mu = 0$, and denote $\mathcal{H} = L^2(\mathbb{R}_+, \mathrm{d}x)$. The UIR of $\mathrm{Aff}_+(\mathbb{R})$, when expressed in terms of the (dimensionless) phase-space variables (q,p), acts on $\mathcal{H}$ as

$$U_{q,p}\psi(x) = \frac{1}{\sqrt{q}}e^{\mathrm{i}px}\psi(x/q). \tag{57}$$

We define the (essentially) self-adjoint operator Q on $\mathcal{H}$ as the multiplication operator $(Q\phi)(x) = x\phi(x)$ and the symmetric operator P as $(P\phi)(x) = -\mathrm{i}\phi'(x)$. Let us note that P has no self-adjoint extension in $\mathcal{H}$ [27].

4.2. Wigner-Weyl-Like Covariant Affine Quantization

General Settings

In the continuation of the procedure exposed in the previous sections, we now investigate special cases of affine covariant integral quantization that leads to remarkable properties. They are analogous to the Wigner-Weyl transform on the plane. As for the plane, the interest of these cases on the physical level is that if we restore physical dimensions for q or x (length) and p (momentum) they only include the Planck constant as a dimensional parameter. The freedom of the quantization map lies again in the choice of a pure mathematical function F. This section generalizes Wigner-like and Weyl-like aspects of affine covariant quantization presented in [11] by introducing families of invertible mappings that look like the Wigner-Weyl case in the plane (see the discussion below).

In this affine context, we define the operators $\mathcal{P}^{(F)}_{q,p}$, $(q,p) \in \Pi_+$, dependent on a possibly complex function $F : \mathbb{R}^+ \ni u \mapsto F(u) \in \mathbb{C}$, by their kernel $\langle x|\mathcal{P}^{(F)}_{q,p}|y\rangle$ in the generalized basis $|x\rangle$, $x \geq 0$, such that $Q\,|x\rangle = x\,|x\rangle$:

$$\langle x|\mathcal{P}^{(F)}_{q,p}\,|y\rangle = \delta(\sqrt{xy} - q)F\left(\sqrt{x/y}\right)e^{\mathrm{i}p(x-y)}, \tag{58}$$

Note the alternative expression, $\delta(\sqrt{xy} - q) = (2q/x)\delta(y - q^2/x)$.

It is easy to verify that the covariance with respect to the affine group holds true. If needed, we remind that the presence of the Planck constant is restored by replacing $e^{\mathrm{i}p(x-y)}$ with $\exp\left(\frac{\mathrm{i}}{\hbar}p(x-y)\right)$.

We prove in Appendix B that the operator $\mathcal{P}^{(F)}_{q,p}$ is bounded if the function $u \mapsto u^2 F(u)$ is bounded. In addition, to impose the self-adjointness of $\mathcal{P}^{(F)}_{q,p}$ we assume that F fulfills the symmetry: $\overline{F(x)} = F(1/x)$.

Remark 3. *We already noticed that the Wigner-Weyl transform on the plane induced by the operators* $\mathcal{P}^{(\delta)}_{q,p}$ *introduced in the previous section involves the arithmetic mean* $(x+y)/2$ *through* $\delta(2^{-1}(x+y)-q)$. *In the present case of the half-plane, its affine symmetry leads us to replace the arithmetic mean by the geometric mean* $\sqrt{xy}$ *appearing in* $\delta(\sqrt{xy}-q)$.

4.3. Resolution of the Identity

The operators $\mathcal{P}^{(F)}_{q,p}$ defined by their kernels (58) solve the identity. Indeed, we check (formally) that

$$\int_{\mathbb{R}} \frac{dp}{2\pi} \langle x|\mathcal{P}^{(F)}_{q,p}|y\rangle = \delta(x-y)\delta(x-q)F(1)\,, \tag{59}$$

and therefore

$$\int_{\mathbb{R}^+\times\mathbb{R}} \frac{dqdp}{2\pi} \langle x|\mathcal{P}^{(F)}_{q,p}|y\rangle = F(1)\delta(x-y) \tag{60}$$

Therefore if we impose $F(1) = 1$ we obtain the resolution of the identity.

In the sequel we assume the function F fulfill both the conditions $F(1) = 1$ and $\overline{F(x)} = F(1/x)$.

4.4. Affine Covariant Quantization and Properties

The F-dependent quantization map $f \mapsto A^{(F)}_f$ is defined as

$$f \mapsto A^{(F)}_f = \int_{\Pi_+} \frac{\mathrm{d}q\,\mathrm{d}p}{2\pi} f(q,p)\,\mathcal{P}^{(F)}_{q,p}\,. \tag{61}$$

This map is such that whatever F (under the above conditions) we have:

$$A^{(F)}_q = Q, \quad A^{(F)}_p = P + \frac{\mathrm{i}}{2Q}F'(1)\,. \tag{62}$$

A_p is symmetric because $\overline{F'(1)} = -F'(1)$. If we impose F to be real, then we have $F(u) = F(1/u)$ and then $F'(1) = 0$, therefore $A^{(F)}_p = P$.

More generally, whatever F we have the following relation which is similar to the Wigner-Weyl quantization map:

$$A^{(F)}_{f(q)} = f(Q)\,. \tag{63}$$

Whatever F we have for the kinetic term p^2,

$$A^{(F)}_{p^2} = P^2 + \frac{\mathrm{i}F'(1)}{2}\left(\frac{1}{Q}P + P\frac{1}{Q}\right) - \frac{F''(1)+F'(1)}{4Q^2}\,. \tag{64}$$

From $\overline{F'(1)} = -F'(1)$, and $\overline{F''(1)} = 2F'(1) + F''(1)$ one deduces that A_{p^2} is symmetric.

If $F(u)$ is real, then $F(u) = F(1/u)$, and $F'(1) = 0$ (but the sign of $F''(1)$ is unspecified). It follows that

$$A^{(F)}_{p^2} = P^2 - \frac{F''(1)}{4Q^2}\,. \tag{65}$$

If $F''(1) < -3$ then $A^{(F)}_{p^2}$ has a unique self-adjoint extension on $\mathcal{H}$ [27,38].

We notice that at the opposite of the Wigner-Weyl case we have not in general $A_{f(p)} = f(P)$. The arbitrary choice of function F allows some regularization at the operator level. For example, in the case of $A^{(F)}_{p^2}$, an adequate choice of F leads to a natural unique self-adjoint extension that uniquely specifies the quantization of p^2.

Trace Formula

The trace of $\mathcal{P}^{(F)}_{q,p}$ reads (formally)

$$\mathrm{tr}\,\mathcal{P}^{(F)}_{q,p} = \int_{\mathbb{R}^+} \mathrm{d}x \langle x|\mathcal{P}^{(F)}_{q,p}|x\rangle = \int_{\mathbb{R}^+} \mathrm{d}x\delta(x-q)F(1) = F(1) = 1\,. \tag{66}$$

Concerning the trace of the product of two different operators $\mathcal{P}^{(F)}_{q,p}$ and $\mathcal{P}^{(G)}_{q',p'}$ we successively have

$$\begin{aligned}
\mathrm{tr}\left(\mathcal{P}^{(F)}_{q,p}\mathcal{P}^{(G)}_{q',p'}\right) &= \int_{\mathbb{R}^+\times\mathbb{R}^+} \mathrm{d}x\,\mathrm{d}y\,\langle x|\mathcal{P}^{(F)}_{q,p}|y\rangle\langle y|\mathcal{P}_{q',p'}(G)|x\rangle \\
&= 2\sqrt{qq'}\delta(q-q')\int_{\mathbb{R}^+}\frac{\mathrm{d}x}{x}\exp\left(\mathrm{i}(p-p')\left(x-\frac{qq'}{x}\right)\right)F\left(\frac{x}{\sqrt{qq'}}\right)G\left(\frac{\sqrt{qq'}}{x}\right) \\
&= 2\sqrt{qq'}\delta(q-q')\int_{\mathbb{R}^+}\frac{\mathrm{d}u}{u}\exp\left(\mathrm{i}(p-p')\sqrt{qq'}(u-1/u)\right)F(u)G(1/u)\,.
\end{aligned} \tag{67}$$

Applying our symmetry assumption $\overline{G(x)} = G(1/x)$ we get

$$\mathrm{tr}\left(\mathcal{P}^{(F)}_{q,p}\mathcal{P}^{(G)}_{q',p'}\right) = 2\sqrt{qq'}\delta(q-q')\int_{\mathbb{R}^+}\frac{\mathrm{d}u}{u}\exp\left(\mathrm{i}(p-p')\sqrt{qq'}(u-1/u)\right)F(u)\overline{G(u)} \tag{68}$$

We now define the function $\phi : \mathbb{R}^+ \ni u \mapsto \xi = u - 1/u \in \mathbb{R}$. We have $\phi'(u) = 1 + u^{-2}$ and $u = \phi^{-1}(\xi) = (\xi/2) + \sqrt{(\xi/2)^2+1}$. Therefore

$$\begin{aligned}
\mathrm{tr}\left(\mathcal{P}^{(F)}_{q,p}\mathcal{P}^{(G)}_{q',p'}\right) &= \sqrt{qq'}\delta(q-q')\int_{\mathbb{R}}\frac{\mathrm{d}\xi}{\xi/2+\sqrt{(\xi/2)^2+1}}\left(1+\frac{\xi}{\sqrt{\xi^2+4}}\right)\times \\
&\quad\times e^{\mathrm{i}(p-p')\sqrt{qq'}\xi}F[\phi^{-1}(\xi)]\overline{G[\phi^{-1}(\xi)]} \\
&= 2\sqrt{qq'}\delta(q-q')\int_{\mathbb{R}}\frac{d\eta}{\sqrt{\eta^2+1}}e^{2\mathrm{i}(p-p')\sqrt{qq'}\eta}F[\phi^{-1}(2\eta)]\overline{G[\phi^{-1}(2\eta)]}\,.
\end{aligned} \tag{69}$$

Defining $\tilde{F}(\eta)$ (and $\tilde{G}(\eta)$) as

$$\tilde{F}(\eta) = \frac{1}{(\eta^2+1)^{1/4}}F[\phi^{-1}(2\eta)]\,, \tag{70}$$

we finally get

$$\mathrm{tr}\left(\mathcal{P}^{(F)}_{q,p}\mathcal{P}^{(G)}_{q',p'}\right) = 2\sqrt{qq'}\delta(q-q')\int_{\mathbb{R}}\mathrm{d}\eta\, e^{2\mathrm{i}(p-p')\sqrt{qq'}\eta}\tilde{F}[\eta]\overline{\tilde{G}[\eta]}\,. \tag{71}$$

4.5. Invertible W-H-like Affine Covariant Quantization

Trivially, if we impose in (71) the relation $\tilde{G}(\eta) = \overline{\tilde{F}(\eta)}^{-1}$, then

$$\mathrm{tr}\left(\mathcal{P}^{(F)}_{q,p}\mathcal{P}^{(G)}_{q',p'}\right) = 2\pi\,\delta(q-q')\,\delta(p-p')\,. \tag{72}$$

This means that the quantization map is invertible. The simplest case is obtained for $\tilde{F}(\eta) = \tilde{G}(\eta) = 1$ which corresponds to

$$F(u) = \frac{1}{\sqrt{2}}\sqrt{u+\frac{1}{u}}\,. \tag{73}$$

We notice that the constraint $F(1) = 1$ is verified. This solution gives an affine counterpart of the Wigner-Weyl transform since we need an unique function to build the quantization map and its inverse. However, we notice that the function F of (73) does not fulfill the boundedness condition $|u^2F(u)| \leq C$ which was requested at the beginning of this section. Therefore the operators $\mathcal{P}^{(F)}_{q,p}$ involved in this case might be unbounded. In fact, this solution is a special case of a larger family of functions: $F_\nu(u)$ with

$$F_\nu(u) = \left(\frac{1}{2}(u + u^{-1})\right)^{\nu+1/2} . \tag{74}$$

The "conjugate function" allowing to build the inverse map due to $F_\nu(u)$ is just $F_{-\nu}(u)$.

The boundedness condition $|u^2F_\nu(u)| \leq C$ is fulfilled only for $\nu \leq -5/2$. Therefore F_ν and $F_{-\nu}$ cannot fulfill this condition at once. However, if we assume $\nu \leq -5/2$ for the quantization mapping, then $F''(1) = \frac{3}{2}(\nu + 1/2) < -3$. Therefore in that case the operator $A^{(F_\nu)}_{p^2}$ has a unique self-adjoint extension. We notice also that for $\nu = 0$ (our analogue of Wigner-Weyl) we obtain an attractive potential in $A^{(F)}_{p^2}$.

4.6. Discussion

Some Wigner-like and Weyl-like aspects of affine covariant quantization are presented in [11]. The calculations developed in Section 7 of [11] correspond to the simplest case $F(u) = 1$ which corresponds to $\nu = -1/2$ in our family F_ν. This choice allows to reproduce in the affine framework the Wigner-Weyl properties $A_{f(q)} = f(Q)$ and $A_{f(p)} = f(P)$. However, in that case the inverse of the quantization mapping cannot be built using the same function (as noticed in Proposition 7.5 of [11]) and there exists different possible self-adjoint extensions of the quantized kinetic operator $A_{p^2} = P^2$ (as noticed below Equation (7.7) of [11]). Therefore this choice is not a complete analogue of the Wigner-Weyl map. In fact, a complete analogue of the Wigner-Weyl map does not exist in the affine framework. In general for $\nu \neq -1/2$ we fail to impose $A_{f(p)} = f(P)$, but for $\nu = 0$ we preserve the use of a unique function (operator) for the inverse map, while for $\nu < -5/2$ we are able to uniquely specify the self-adjoint kinetic operator A_{p^2}.

5. Conclusions

Through the above specifications of covariant integral quantization, in their Wigner-Weyl-like restrictions, to two basic cases, the euclidean plane with its translational symmetry on one hand, the open half-plane with its affine symmetry on the other hand, we have provided an illustration of the crucial role of the Fourier transform, which is needed at each step of the calculations. With these generalizations of the Wigner-Weyl transform we have shown that the Weyl integral quantization, often thought of as the "best" option, has many interesting features shared by a wide panel of other integral quantizations.

We also think that similar features hold far beyond the two elementary symmetries which have been examined here. There exist many versions of the Wigner function or equivalent quasi-distribution for other groups, see for instance [39,40] for $\mathrm{SU}(n)$ and references therein. In the case of non-compact groups, particularly those which are semi-direct products of groups, the existence of square-integrability of the UIR requested by the resolution of the identity lying at the heart of the construction is in general not guaranteed. However, we think that it is possible to get round this issue if square-integrability of the UIR holds with respect to a subgroup. Related concepts and material on the restricted level of coherent states are found for instance in [41] and the chapters 7 and 8 of [10] with references therein.

As a final comment, the methods of quantization which have been exposed here are just a tiny part of a huge variety of ways of building quantum models from a unique classical one. We should always keep in our mind that mathematical models for physical systems are mainly effective, and the freedom one has in picking one specific model should be considered as an attractive feature rather than a drawback [42].

Author Contributions: The original results presented in Sections 3 and 4 have been established by Hervé Bergeron. Jean-Pierre Gazeau's contribution is mainly about the content of other sections.
Funding This research received no external funding.

Conflicts of Interest: The authors declare no conflict of interest.

Appendix A. Quantization of The Plane: Boundedness Of $\mathcal{P}_0^{(F)}$

We prove the bounded character of the operator $\mathcal{P}_0^{(F)}$ when F belongs to $L^1(\mathbb{R}, \mathrm{d}u) \bigcap L^1\left(\mathbb{R}, |u^2-1/4|^{-1/2}\mathrm{d}u\right)$. From the Riesz lemma it is sufficient to prove that $B(\phi,\psi) = \langle \phi|\mathcal{P}_0^{(F)}|\psi\rangle$ is a bounded bilinear form. Using (33) we have

$$|B(\phi,\psi)| \leq \int_{\mathbb{R}} |F(u)|\mathrm{d}u \int_{\mathbb{R}} \mathrm{d}z\, |\overline{\phi((u+1/2)z)}|\, |\psi((u-1/2)z)|\,, \tag{A1}$$

Using Cauchy-Schwarz inequality and a change of variable we obtain

$$\int_{\mathbb{R}} \mathrm{d}z\, |\overline{\phi((u+1/2)z)}|\, |\psi((u-1/2)z)| \leq \frac{1}{\sqrt{|u^2-1/4|}} \|\phi\|\, \|\psi\|\,. \tag{A2}$$

Therefore if F belongs to $L^1(\mathbb{R}, \mathrm{d}u) \bigcap L^1\left(\mathbb{R}, |u^2-1/4|^{-1/2}\mathrm{d}u\right)$ we have $|B(\phi,\psi)| \leq C||\phi||\,||\psi||$ with $C = \int_{\mathbb{R}} |F(u)|\, |u^2-1/4|^{-1/2}\,\mathrm{d}u$ and $B(\phi,\psi)$ is a bounded bilinear functional.

We notice that the same reasoning holds if we replace $F(u)\mathrm{d}u$ by a positive measure $\mathrm{d}\mu(u)$ such that $u \mapsto |u^2-1/4|^{-1/2}$ belongs to $L^1(\mathbb{R}, \mathrm{d}\mu(u))$. This is in particular the case when we choose $F(u) = \delta(u)$ (Wigner-Weyl transform).

Appendix B. Quantization of The Half-Plane: Boundedness of $\mathcal{P}_{q,p}^{(F)}$

We prove the boundedness of the operator $\mathcal{P}_{q,p}^{(F)}$ when $u \mapsto u^2F(u)$ is a bounded function. From the Riesz lemma it is sufficient to prove that $B(\phi,\psi) = \langle \phi|\mathcal{P}_{q,p}^{(F)}|\psi\rangle$ is a bounded bilinear form. From (58) $B(\phi,\psi)$ reads

$$B(\phi,\psi) = \int_{\mathbb{R}^+} \mathrm{d}x\, \frac{2x}{q} F(x/q)\, \overline{\phi(x)}\, \psi(q^2/x)\, e^{\mathrm{i}p(x-q^2/x)}\,. \tag{A3}$$

Therefore we obtain:

$$|B(\phi,\psi)| \leq 2\int_{\mathbb{R}^+} \mathrm{d}x\, \frac{x^2}{q^2} F(x/q)|\phi(x)|\, \frac{q}{x}|\psi(q^2/x)|\,. \tag{A4}$$

Thus if $u \mapsto u^2F(u)$ is a bounded function with $|u^2F(u)| \leq C$ we have

$$|B(\phi,\psi)| \leq 2C \int_{\mathbb{R}^+} \mathrm{d}x\, |\phi(x)|\, \frac{q}{x}|\psi(q^2/x)|\,. \tag{A5}$$

Then using the Cauchy-Schwarz inequality and a change of variable in the integral involving $(q/x)\psi(q^2/x)$ we obtain:

$$|B(\phi,\psi)| \leq 2C\, \|\phi\|\, \|\psi\|\,. \tag{A6}$$

We conclude that the operator $\mathcal{P}_{q,p}^{(F)}$ is bounded.

References

1. Fourier, J. *Théorie Analytique De La Chaleur*; Firmin Didot Père et Fils.: Paris, France, 1822. (In French)
2. Fourier, J.B.J. *The Analytical Theory of Heat*; Cambridge University Press: Cambridge, UK, 1980.
3. Weyl, H. Quantenmechanik und Gruppentheorie. *Z. Phys.* **1927**, 46, 1–46. [CrossRef]
4. Wigner, E. On the Quantum Correction For Thermodynamic Equilibrium. *Phys. Rev.* **1932**, 40, 749. [CrossRef]
5. von Neumann, J. Die eindeutigkeit der Schröderschen Operatoren. *Math. Ann.* **1931**, 104, 570–578. [CrossRef]

6. Grossmann, A. Parity Operator and Quantization of δ-Functions. *Commun. Math. Phys.* **1976**, *48*, 191–194. [CrossRef]
7. Groenewold, H.J. On the Principles of Elementary Quantum Mechanics. *Physica* **1946**, *12*, 405–446. [CrossRef]
8. Moyal, J.E.; Bartlett, M.S. Quantum mechanics as a statistical theory. In *Mathematical Proceedings of the Cambridge Philosophical Society*; Cambridge University Press: Cambridge, UK, 1949; Volume 45, p. 99.
9. Bergeron, H.; Gazeau, J.P. Integral quantizations with two basic examples. *Ann. Phys.* **2014**, *344*, 43–68. [CrossRef]
10. Ali, S.T.; Antoine, J.-P.; Gazeau, J.-P. *Coherent States, Wavelets and their Generalizations*, 2nd ed.; Theoretical and Mathematical Physics; Springer: New York, NY, USA, 2014.
11. Gazeau, J.P.; Murenzi, R. Covariant affine integral quantization(s). *J. Math. Phys.* **2016**, *57*, 052102. [CrossRef]
12. Gazeau, J.P.; Heller, B. POVM Quantization. *Axioms* **2015**, *4*, 1–29. [CrossRef]
13. Bergeron, H.; Curado, E.; Gazeau, J.P.; Ligia, M.C.S. Rodrigues, Weyl-Heisenberg integral quantization(s): A compendium. *arXiv* **2017**, arXiv:1703.08443.
14. Berezin, F.A. Quantization, Mathematics of the USSR-Izvestiya. *Trans. Am. Math. Soc.* **1974**, *8*, 1109–1165.
15. Berezin, F.A. General concept of quantization. *Commun. Math. Phys.* **1975**, *40*, 153–174. [CrossRef]
16. Daubechies, I. On the distributions corresponding to bounded operators in the Weyl quantization. *Commun. Math. Phys.* **1980**, *75*, 229–238. [CrossRef]
17. Daubechies, I.; Grossmann, A. An integral transform related to quantization. I. *J. Math. Phys.* **1980**, *21*, 2080–2090. [CrossRef]
18. Daubechies, I.; Grossmann, A.; Reignier, J. An integral transform related to quantization. II. *J. Math. Phys.* **1983**, *24*, 239–2542. [CrossRef]
19. Bergeron, H.; Dapor, A.; Gazeau, J.P.; Małkiewicz, P. Smooth big bounce from affine quantization. *Phys. Rev. D* **2014**, *89*, 083522. [CrossRef]
20. Bergeron, H.; Dapor, A.; Gazeau, J.P.; Małkiewicz, P. Smooth Bounce in Affine Quantization of Bianchi I. *Phys. Rev. D* **2015**, *91*, 124002. [CrossRef]
21. Bergeron, H.; Czuchry, E.; Gazeau, J.P.; Małkiewicz, P.; Piechocki, W. Smooth quantum dynamics of the mixmaster universe. *Phys. Rev. D* **2015**, *92*, 061302. [CrossRef]
22. Bergeron, H.; Czuchry, E.; Gazeau, J.P.; Małkiewicz, P.; Piechocki, W. Singularity avoidance in a quantum model of the Mixmaster universe. *Phys. Rev. D* **2015**, *92*, 124018. [CrossRef]
23. Bergeron, H.; Czuchry, E.; Gazeau, J.P.; Małkiewicz, P. Vibronic Framework for Quantum Mixmaster Universe. *Phys. Rev. D* **2016**, *93*, 064080. [CrossRef]
24. Klauder, J.R.; Aslaksen, E.W. Elementary Model for Quantum Gravity. *Phys. Rev. D* **1970**, *2*, 272–276. [CrossRef]
25. Klauder, J.R. An affinity for affine quantum gravity. *Proc. Steklov Inst. Math.* **2011**, *272*, 169–176. [CrossRef]
26. Fanuel, M.; Zonetti, S. Affine quantization and the initial cosmological singularity. *Europhys. Lett.* **2013**, *101*, 10001. [CrossRef]
27. Reed, M.; Simon, B. *Methods of Modern Mathematical Physics, II. Fourier Analysis, Self-Adjointness*; Academic Press: New York, NY, USA, 1975; Volume 2.
28. Bergeron, H.; Siegl, P.; Youssef, A. New SUSYQM coherent states for Pöschl-Teller potentials: A detailed mathematical analysis. *J. Phys. A Math. Theor.* **2010**, *45*, 244028. [CrossRef]
29. Stenzel, M.B. Holomorphic Sobolev spaces and the generalized Segal-Bargmann transform. *J. Funct. Anal.* **1994**, *165*, 44–58. [CrossRef]
30. Born, M.; Jordan, P. Zur Quantenmechanik. *Z. Physik* **1925**, *34*, 858–888. [CrossRef]
31. Born, M.; Heisenberg, W.; Jordan, P. Zur Quantenmechanik II. *Z. Physik* **1925**, *35*, 557–615. [CrossRef]
32. De Gosson, M. *Born-Jordan Quantization: Theory and Applications*; Springer: Berlin/Heidelberg, Germany, 2016.
33. Cohen, L. Generalized phase-space distribution functions. *J. Math. Phys.* **1966**, *7*, 781–786. [CrossRef]
34. Cohen, L. *The Weyl Operator and Its Generalization*; Springer Science & Business Media: Berlin/Heidelberg, Germany, 2012.
35. Agarwal, B.S.; Wolf, E. Calculus for Functions of Noncommuting Operators and General Phase-Space Methods in Quantum Mechanics. *Phys. Rev. D* **1970**, *2*, 2161. [CrossRef]
36. Hudson, R.L. When is the Wigner quasi-probability density non-negative? *Rep. Math. Phys.* **1974**, *6*, 249. [CrossRef]
37. Cartwright, N.D. A non-negative Wigner-type distribution. *Phys. Stat. Mech. Appl.* **1976**, *83*, 210. [CrossRef]

38. Gesztesy, F.; Kirsch, W.; Reine, J. Schrödinger Operators in $L^2(\mathbb{R})$ with Pointwise Localized Potential. *Angew. Math.* **1985**, *362*, 28–50.
39. Klimov, A.B.; De Guise, H. General approach to SU(n) quasi-distribution functions. *J. Phys. A Math. Theory* **2010**, *43*, 402001. [CrossRef]
40. Klimov, A.B.; Romero, J.L.; de Guise, H. Generalized SU(2) covariant Wigner functions and some of their applications. *J. Phys. A Math. Theory* **2017**, *50*, 323001. [CrossRef]
41. Healy, D.M., Jr.; Schroeck, F.E., Jr. On informational completeness of covariant localization observables and Wigner coefficients. *J. Math. Phys.* **1995**, *36*, 453–507. [CrossRef]
42. Gazeau, J.P. From Classical to Quantum Models: The Regularising Rôle of Integrals, Symmetry and Probabilities. *Found. Phys.* **2018**. [CrossRef]

entropy

MDPI

Article

Short-Time Propagators and the Born–Jordan Quantization Rule

Maurice A. de Gosson

Faculty of Mathematics (NuHAG), University of Vienna, Oskar-Morgenstern-Platz 1, 1090 Vienna, Austria; maurice.de.gosson@univie.ac.at

Received: 14 October 2018; Accepted: 8 November 2018; Published: 10 November 2018

Abstract: We have shown in previous work that the equivalence of the Heisenberg and Schrödinger pictures of quantum mechanics requires the use of the Born and Jordan quantization rules. In the present work we give further evidence that the Born–Jordan rule is the correct quantization scheme for quantum mechanics. For this purpose we use correct short-time approximations to the action functional, initially due to Makri and Miller, and show that these lead to the desired quantization of the classical Hamiltonian.

Keywords: Born–Jordan quantization; short-time propagators; time-slicing; Van Vleck determinant

---

1. Motivation and Background

1.1. *Weyl versus Born and Jordan*

There have been several attempts in the literature to find the "right" quantization rule for observables using either algebraic or analytical techniques [1–7]. In a recent paper [8] we have analyzed the Heisenberg and Schrödinger pictures of quantum mechanics, and shown that if one postulates that both theories are equivalent, then one must use the Born–Jordan quantization rule

$$\text{(BJ)}\quad x^m p^\ell \longrightarrow \frac{1}{m+1}\sum_{k=0}^{m} \widehat{x}^k \widehat{p}^\ell \widehat{x}^{m-k}, \tag{1}$$

and *not* the Weyl rule (To be accurate, it was McCoy [9] who showed that Weyl's quantization scheme leads to Formula (2)).

$$\text{(Weyl)}\quad x^m p^\ell \longrightarrow \frac{1}{2^m}\sum_{k=0}^{m} \binom{m}{k} \widehat{x}^k \widehat{p}^\ell \widehat{x}^{m-k} \tag{2}$$

for monomial observables. The Born–Jordan and Weyl rules yield the same result only if $m < 2$ or $\ell < 2$; for instance in both cases the quantization of the product xp is $\frac{1}{2}(\widehat{x}\widehat{p} + \widehat{p}\widehat{x})$. One can also show that the product $pf(x)$ is, for any smooth function f of position alone, given in both cases by the symmetric rule

$$pf(x) \longrightarrow \frac{1}{2}(\widehat{p}f(x) + f(x)\widehat{p}).$$

It follows that if H is a Hamiltonian of the type

$$H = \sum_{j=1}^{n} \frac{1}{2m_j}(p_j - A_j(x))^2 + V(x)$$

one can use either the Weyl or the Born–Jordan prescriptions to get the the corresponding quantum operator, which yields the familiar expression

$$\widehat{H} = \sum_{j=1}^{n} \frac{1}{2m_j}\left(-i\hbar\frac{\partial}{\partial x_j} - A(x)\right)^2 + V(x).$$

(See Section 3.3). Since this Hamiltonian is without doubt the one which most often occurs in quantum mechanics one could ask why one should bother about which is the "correct" quantization. It turns out that this question is just a little bit more than academic: There are simple physical observables which yield different quantizations in the Weyl and Born–Jordan schemes. One interesting example is that of the squared angular momentum: Writing $\mathbf{r} = (x, y, z)$ and $\mathbf{p} = (p_x, p_y, p_z)$ the square of the classical angular momentum

$$\ell = (yp_z - zp_y)\mathbf{i} + (zp_x - xp_z)\mathbf{j} + (xp_y - yp_x)\mathbf{k} \tag{3}$$

is the function $\ell^2 = \ell_x^2 + \ell_y^2 + \ell_z^2$ where

$$\ell_x^2 = x^2p_y^2 + y^2p_x^2 - 2xp_xyp_y \tag{4}$$

and so on. The Weyl quantization of ℓ_x^2 is

$$(\widehat{\ell_x^2})_{\mathrm{W}} = \widehat{x}^2\widehat{p}_y^2 + \widehat{x}_y^2\widehat{p}_x^2 - \tfrac{1}{2}(\widehat{x}\widehat{p}_x + \widehat{p}_x\widehat{x})(\widehat{y}\widehat{p}_y + \widehat{p}_y\widehat{y}) \tag{5}$$

while its Born–Jordan quantization is

$$(\widehat{\ell_x^2})_{\mathrm{BJ}} = \widehat{x}^2\widehat{p}_y^2 + \widehat{x}_y^2\widehat{p}_x^2 - \tfrac{1}{2}(\widehat{x}\widehat{p}_x + \widehat{p}_x\widehat{x})(\widehat{y}\widehat{p}_y + \widehat{p}_y\widehat{y}) - \tfrac{1}{6}\hbar^2; \tag{6}$$

similar relations are obtained for ℓ_y^2 and ℓ_z^2 so that, in the end,

$$(\widehat{\ell^2})_{\mathrm{W}} - (\widehat{\ell^2})_{\mathrm{BJ}} = \tfrac{1}{2}\hbar^2. \tag{7}$$

This discrepancy has been dubbed the "angular momentum dilemma "[10]; in [11] we have discussed this apparent paradox and shown that it disappears if one systematically uses Born–Jordan quantization.

1.2. The Kerner and Sutcliffe Approach to Quantization

As we have proven in [8,12], Heisenberg's matrix mechanics [13], as rigorously constructed by Born and Jordan in [14] and Born, Jordan, and Heisenberg in [15], explicitly requires the use of the quantization rule (1) to be mathematically consistent, a fact which apparently has escaped the attention of physicists, and philosophers or historians of Science. In the present paper, we will show that the Feynman path integral approach is another genuinely physical motivation for Born–Jordan quantization of arbitrary observables; it corrects previous unsuccessful attempts involving path integral arguments which *do not work* for a reason that will be explained. One of the most convincing of these attempts is the paper [16] by Kerner and Sutcliffe. Elaborating on previous work of Garrod [17] Kerner and Sutcliffe tried to justify the Born–Jordan rule as the unique possible quantization (see Steven Kauffmann's [18,19] brilliant discussion of this work). Assuming that $\widehat{H}$ is the quantization of some general Hamiltonian H, they write as is usual in the theory of the phase space Feynman integral the propagator as

$$\langle x|e^{-\frac{i}{\hbar}\widehat{H}t}|x'\rangle = \lim_{N\to\infty}\int dx_{N-1}\cdots dx_1\textstyle\prod_{k=1}^{N}\langle x_k|e^{-\frac{i}{\hbar}\widehat{H}\Delta t}|x_{k-1}\rangle \tag{8}$$

where $x_N = x$ and $x_0 = x'$ are fixed and $\Delta t = t/N$. They thereafter use the approximation

$$\langle x_k|e^{-\frac{i}{\hbar}\widehat{H}\Delta t}|x_{k-1}\rangle \approx \frac{1}{2\pi\hbar}\int e^{\frac{i}{\hbar}\overline{S}(x,x',p,\Delta t)}dp \tag{9}$$

the function $\overline{S}$ being given by

$$\overline{S}(x, x', p, \Delta t) = p(x - x') - \overline{H}(x, x', p)\Delta t \tag{10}$$

where $\overline{H}$ is the time average of H over p fixed and $x = x(t)$, that is

$$\overline{H}(x, x', p) = \frac{1}{\Delta t}\int_0^{\Delta t} H(x' + s\frac{x - x'}{\Delta t}, p)ds. \tag{11}$$

Notice that introducing the dimensionless parameter $\tau = s/\Delta t$, Formula (11) can be written in the more convenient form

$$\overline{H}(x, x', p) = \int_0^1 H(\tau x + (1 - \tau)x', p)d\tau \tag{12}$$

which is the usual mathematical definition of Born–Jordan quantization: See de Gosson [12,20] and de Gosson and Luef [21].

Taking the limit $\Delta t \to 0$ the operator $\widehat{H}$ can then be explicitly and uniquely determined, and Kerner and Sutcliffe show that in particular this leads to the Born–Jordan ordering (1) when their Hamiltonian H is a monomial $x^m p^\ell$. Unfortunately (as immediately Cohen's rebuttal was published in the same volume of *J. Math. Phys.* in which Kerner and Sutcliffe published their results. Noted by Cohen [22]) there are many a priori equally good constructions of the Feynman integral, leading to other quantization rules. In fact, argues Cohen, there is a great freedom of choice in calculating the action $p(x - x') - \overline{H}$ appearing in the right-hand side of (11). For instance, one can choose

$$S(x, x', p, \Delta t) = p(x - x') - H(\tfrac{1}{2}(x + x'), p)\Delta t \tag{13}$$

which leads for $x^m p^\ell$ to Weyl's rule (2), or one can choose

$$S(x, x', p, \Delta t) = p(x - x') - \tfrac{1}{2}(H(x, p) + H(x', p))\Delta t, \tag{14}$$

which leads to the symmetric rule

$$x^m p^\ell \longrightarrow \frac{1}{2}(\widehat{x}^m \widehat{p}^\ell + \widehat{p}^\ell \widehat{x}^m). \tag{15}$$

This ambiguity shows—in an obvious way—that Feynman path integral theory does not lead to an uniquely defined quantization scheme for observables. However—and this is the main point of the present paper—while Cohen's remark was mathematically justified, Kerner and Sutcliffe's insight was right (albeit for the wrong reason).

1.3. What We Will Do

It turns out that the Formula (10) for the approximate action that Kerner and Sutcliffe "guessed" has been justified independently (in another context) by Makri and Miller [23,24] and the present author [25] by rigorous mathematical methods. This formula is actually the *correct* approximation to action up to order $O(\Delta t^2)$ (as opposed to the "midpoint rules" commonly used in the theory of the Feynman integral which yield much cruder approximations); it follows that Kerner and Sutcliffe's Formula (9) indeed yields a correct approximation of the infinitesimal propagator $\langle x_k | e^{-\frac{i}{\hbar}\widehat{H}\Delta t} | x_{k-1} \rangle$, in fact the *best one* for calculational purposes since it ensures a swift convergence of numerical schemes. This is because for short times Δt the solution of Schrödinger's equation

$$i\hbar\frac{\partial \psi}{\partial t}(x, t) = \left[\sum_{j=1}^{n} \frac{-\hbar^2}{2m_j}\frac{\partial^2}{\partial x_j^2} + V(x)\right]\psi(x, t) \tag{16}$$

with initial condition $\psi(x,0) = \psi_0(x)$ is given by the asymptotic formula

$$\psi(x,\Delta t) = \int \overline{K}(x,x',\Delta t)\psi_0(x')d^n x' + O(\Delta t^2); \tag{17}$$

the approximate propagator $\overline{K}$ being defined, for arbitrary time t, by

$$\overline{K}(x,x',t) = \left(\tfrac{1}{2\pi\hbar}\right)^n \int \exp\left(\tfrac{i}{\hbar}\left[p(x-x') - (H_{\text{free}}(p) + \overline{V}(x,x'))t\right]\right) d^n p, \tag{18}$$

where, by definition, $H_{\text{free}}(p)$ is the free particle Hamiltonian function, and the two-point function

$$\overline{V}(x,x') = \int_0^1 V(\tau x + (1-\tau)x')d\tau$$

is the average value of the potential V on the line segment $[x', x]$.

- In Section 2 we discuss the accuracy of Kerner and Sutcliffe's propagator by comparing it with the more familiar Van Vleck propagator; we show that for small times both are approximations to order $O(t^2)$ to the exact propagator of Schrödinger's equation.
- In Section 3 we show that if one assume's that short-time evolution of the wavefunction (for an arbitrary Hamiltonian H) is given by the Kerner and Sutcliffe propagator, then H must be quantized following the rule (12); we thereafter show that when H is a monomial $x^m p^\ell$ then the corresponding operator is given by the Born–Jordan rule (1), *not* by the Weyl rule 2.

Notation 1. *The generalized position and momentum vectors are* $x = (x_1, ..., x_n)$ *and* $p = (p_1, ..., p_n)$*; we set* $px = p_1x_1 + \cdots + p_nx_n$*. We denote by* $\widehat{x}_j$ *the operator of multiplication by* x_j *and by* $\widehat{p}_j$ *the momentum operator* $-i\hbar(\partial/\partial x_j)$.

2. On Short-Time Propagators

In this section we only consider Hamiltonian functions of the type "kinetic energy plus potential":

$$H(x,p) = H_{\text{free}}(p) + V(x) \ , \ H_{\text{free}}(p) = \sum_{j=1}^{n} \frac{1}{2m_j} p_j^2. \tag{19}$$

These are the simplest physical Hamiltonians, both from a classical and a quantum perspective.

2.1. The Van Vleck Propagator

Consider a Hamiltonian function of the type (19) above; the corresponding Schrödinger equation is

$$i\hbar \frac{\partial \psi}{\partial t}(x,t) = \left[\sum_{j=1}^{n} \frac{-\hbar^2}{2m_j}\frac{\partial^2}{\partial x_j^2} + V(x)\right]\psi(x,t). \tag{20}$$

We will denote by $K(x,x',t) = \langle x|e^{-\frac{i}{\hbar}\widehat{H}t}|x'\rangle$ the corresponding exact propagator:

$$\psi(x,t) = \int K(x,x',t)\psi_0(x')d^n x' \tag{21}$$

where with $\psi_0(x)$ is the value of ψ at time $t = 0$. The function $K(x,x',t)$ must thus satisfy the boundary condition

$$\lim_{t\to 0} K(x,x',t) = \delta(x-x'). \tag{22}$$

It is well-known (see e.g., Gutzwiller [26], Schulman [27], de Gosson [25], Maslov and Fedoriuk [28]) that for short times an approximate propagator is given by Van Vleck's formula

$$\widetilde{K}(x,x',t) = \left(\tfrac{1}{2\pi i\hbar}\right)^{n/2} \sqrt{\rho(x,x',t)}e^{\frac{i}{\hbar}S(x,x',t)} \tag{23}$$

where

$$S(x,x',t) = \int_0^t \left(\textstyle\sum_{j=1}^n \frac{1}{2}m_j\dot{x}_j(s)^2 - V(x(s)\right) ds \tag{24}$$

is the action along the classical trajectory leading from x' at time $t'=0$ to x at time t (there is no sum over different classical trajectories because only one trajectory contributes in the limit $t\to 0$ [23]) and

$$\rho(x,x',t) = \det\left(-\frac{\partial^2 S(x,x',t)}{\partial x_j \partial x'_{jk}}\right)_{1\leq j,k\leq n} \tag{25}$$

is the Van Vleck density of trajectories [25–27]; the argument of the square root is chosen so that the initial condition (22) is satisfied [25,29]. It should be emphasized that although the Van Vleck propagator is frequently used in semiclassical mechanics, it has nothing "semiclassical" *per se*, since it is genuinely an approximation to the exact propagator for small t – not just in the limit $\hbar\to 0$. In fact:

Theorem 1. *Let $\widetilde{\psi}$ be given by*

$$\widetilde{\psi}(x,t) = \int \widetilde{K}(x,x',t)\psi_0(x')d^n x'$$

where ψ_0 is a tempered distribution. Let ψ be the exact solution of Schrödinger's equation with initial datum ψ_0. We have

$$\psi(x,t) - \widetilde{\psi}(x,t) = O(t^2). \tag{26}$$

In particular, the Van Vleck propagator $\widetilde{K}(x,x',t)$ is an $O(t^2)$ approximation to the exact propagator $K(x,x',t)$:

$$K(x,x',t) - \widetilde{K}(x,x',t) = O(t^2) \tag{27}$$

for $t\to 0$ and hence

$$\lim_{t\to 0}\widetilde{K}(x,x',t) = \delta(x-x').$$

Proof. Referring to de Gosson [25] (Lemma 241) for details, we sketch the main lines in the case $n=1$. Assuming that ψ_0 belongs to the Schwartz space $\mathcal{S}(\mathbb{R}^n)$ of rapidly decreasing functions, one expands the solution ψ of Schrödinger's equation to second order:

$$\psi(x,t) = \psi_0(x) + \frac{\partial\psi}{\partial t}(x,0)t + O(t^2).$$

Taking into account the fact that ψ is a solution of Schrödinger's equation this can be rewritten

$$\psi(x,t) = \left[1 + \frac{t}{i\hbar}\left(-\frac{\hbar^2}{2m}\frac{\partial^2}{\partial x^2} + V(x)\right)\right]\psi_0(x) + O(t^2). \tag{28}$$

Expanding the exponential $e^{iS/\hbar}$ in Van Vleck's Formula (23) at $t=0$ one shows, using the estimate (32) in Theorem 2, that we also have

$$\widetilde{\psi}(x,t) = \left[1 + \frac{t}{i\hbar}\left(-\frac{\hbar^2}{2m}\frac{\partial^2}{\partial x^2} + V(x)\right)\right]\psi_0(x) + O(t^2); \tag{29}$$

comparison with (28) implies that $\psi(x,t) - \widetilde{\psi}(x,t) = O(t^2)$. By density of the Schwartz space in the class of tempered distributions $\mathcal{S}'(\mathbb{R}^n)$ the estimate (26) is valid if one chooses $\psi_0(x) = \delta(x - x_0)$, which yields Formula (27) since we have

$$\int \widetilde{K}(x,x',t)\delta(x-x_0)d^n x' = \widetilde{K}(x,x_0,t)$$

and

$$\int K(x,x',t)\delta(x-x_0)d^n x' = K(x,x_0,t).$$

□

Let us briefly return to the path integral. Replacing the terms $\langle x_k | e^{-\frac{i}{\hbar}\widehat{H}\Delta t} | x_{k-1}\rangle$ in the product Formula (8) with $\widetilde{K}(x_{k-1}, x_{k-1}, \Delta t)$ one shows, using the Lie–Trotter Formula [25,27], that the exact propagator $K(x,x',t) = \langle x | e^{-\frac{i}{\hbar}\widehat{H}t} | x'\rangle$ is given by

$$\langle x | e^{-\frac{i}{\hbar}\widehat{H}t} | x'\rangle = \lim_{N\to\infty} \int dx_{N-1} \cdots dx_1 \textstyle\prod_{k=1}^{N} \widetilde{K}(x_{k-1}, x_{k-1}, \Delta t). \tag{30}$$

This formula is often taken as the starting point of path integral arguments: observing that the expression (23) is in most cases (The free particle and the harmonic oscillator are remarkable particular cases where the action integral can be explicitly calculated and thus yields an explicit formula for the propagator, but mathematically speaking this fact is rather a consequence of the theory of the metaplectic group [25,29]) difficult to calculate (it implies the computation of an action integral, which can be quite cumbersome) people working in the theory of the Feynman integral replace the exact action $S(x,x',t)$ in (23) with approximate expressions, for instance the "midpoint rules" that will be discussed below. Now, one should be aware that this *legerdemain* works, because when taking the limit $N \to \infty$ one indeed obtains the correct propagator, but it does *not* imply that these midpoint rules are accurate approximations to $S(x,x',t)$.

2.2. The Kerner–Sutcliffe Propagator

We showed above that the Van Vleck propagator is an approximation to order $O(t^2)$ to the exact propagator. We now show that the propagator proposed by Kerner and Sutcliffe in [16] approximates the Van Vleck propagator also at order $O(t^2)$. Hence

$$\textit{Van Vleck} = \textit{Kerner–Sutcliffe} + O(t^2).$$

We begin by giving a correct short-time approximation to the action.

Theorem 2. *The function $\overline{S}$ defined by*

$$\overline{S}(x,x',t) = \sum_{j=1}^{n} m_j \frac{(x_j - x'_j)^2}{2t} - \overline{V}(x,x')t \tag{31}$$

where $\overline{V}(x,x')$ is the average of the potential V along the line segment $[x',x]$:

$$\overline{V}(x,x') = \int_0^1 V(\tau x + (1-\tau)x')d\tau.$$

satisfies for $t \to 0$ the estimate

$$S(x,x',t) - \overline{S}(x,x',t) = O(t^2). \tag{32}$$

For detailed proofs we refer to the aforementioned papers [23,24] by Makri and Miller, and to our book [25]; also see de Gosson and Hiley [30,31]. The underlying idea is quite simple (and already appears in germ in Park's book [32], p. 438): one remarks that the function $S = S(x, x', t)$ satisfies the Hamilton–Jacobi equation

$$\frac{\partial S}{\partial t} + \sum_{j=1}^{n} \frac{1}{2m_j} \left(\frac{\partial S}{\partial x_j} \right)^2 + V(x) = 0 \tag{33}$$

and one thereafter looks for an asymptotic solution

$$S(x, x', t) = \frac{1}{t} S_0(x, x') + S_1(x, x')t + S_2(x, x')t^2 + \cdots.$$

Insertion in (33) then leads to

$$S_0(x, x') = \sum_{j=1}^{n} m_j \frac{(x_j - x_j')^2}{2}$$

and $S_1(x, x') = -\overline{V}(x, x')$ hence (31). Notice that this procedure actually allows one to find approximations to S to an arbitrary order of accuracy by solving successively the equations satisfied by S_2 ,S_3, ... (see [23,24] for explicit formulas).

Let us now set

$$\overline{H}(x, x', t) = H_{\text{free}}(p) + \overline{V}(x, x')$$

where

$$\overline{V}(x, x') = \int_0^1 V(\tau x + (1 - \tau)x')d\tau$$

is the averaged potential.

Let us now show that the propagator postulated by Garrod [17] and Kerner and Sutcliffe [16] is as good an approximation to the exact propagator as Van Vleck's is. We recall the textbook Fourier formula

$$\left(\tfrac{1}{2\pi\hbar}\right)^n \int e^{\frac{i}{\hbar}p(x-x')} p_j^\ell d^n p = \left(-i\hbar \tfrac{\partial}{\partial x_j}\right)^\ell \delta(x - x'). \tag{34}$$

Theorem 3. *Let $\overline{K} = \overline{K}(x, x', t)$ be defined (in the distributional sense) by*

$$\overline{K}(x, x', t) = \left(\tfrac{1}{2\pi\hbar}\right)^n \int e^{\frac{i}{\hbar}(p(x-x') - \overline{H}(x,x',p)t)} d^n p. \tag{35}$$

and set

$$\overline{\psi}(x, t) = \int \overline{K}(x, x', t)\psi_0(x')d^n x'. \tag{36}$$

Let ψ be the solution of Schrödinger's equation with initial condition ψ_0. We have

$$\overline{\psi}(x, t) - \psi(x, t) = O(t^2). \tag{37}$$

The function $\overline{K}$ is an $O(t^2)$ approximation to the exact propagator K:

$$K(x, x', t) - \overline{K}(x, x', t) = O(t^2). \tag{38}$$

Proof. It is sufficient to prove (37); Formula (38) follows by the same argument as in the proof of Theorem 1. To simplify notation we assume again $n = 1$; the general case is a straightforward extension. Expanding for small t the exponential in the integrand of (35) we have

$$\begin{aligned} \overline{K}(x, x', t) &= \left(\tfrac{1}{2\pi\hbar}\right)^n \int e^{\frac{i}{\hbar}p(x-x')} (1 - \frac{i}{\hbar}\overline{H}(x, x', p)t)dp + O(t^2) \\ &= \delta(x - x') - \frac{it}{\hbar} \int e^{\frac{i}{\hbar}p(x-x')} \overline{H}(x, x', p)dp + O(t^2) \end{aligned}$$

and hence

$$\overline{\psi}(x,t) = \psi_0(x) - \frac{it}{\hbar} \int e^{\frac{i}{\hbar}p(x-x')} \overline{H}(x,x',p)\psi_0(x')dpdx' + O(t^2).$$

We have

$$\int e^{\frac{i}{\hbar}p(x-x')} \overline{H}(x,x',p)d^n p = \int e^{\frac{i}{\hbar}p(x-x')} \left(\frac{p^2}{2m} + \overline{V}(x,x')\right) dp;$$

using the Fourier Formula (34) we get

$$\left(\tfrac{1}{2\pi\hbar}\right)^n \int e^{\frac{i}{\hbar}p(x-x')} \frac{p^2}{2m} dp = -\frac{\hbar^2}{2m}\frac{\partial^2}{\partial x^2}\delta(x-x')$$

and, noting that $\overline{V}(x,x) = V(x)$,

$$\begin{aligned}\left(\tfrac{1}{2\pi\hbar}\right)^n \int e^{\frac{i}{\hbar}p(x-x')} \overline{V}(x,x')dp &= \overline{V}(x,x')\delta(x-x') \\ &= V(x)\delta(x-x').\end{aligned}$$

Summarizing,

$$\overline{K}(x,x',t) = \delta(x-x') + \frac{it}{\hbar}\left(-\frac{\hbar^2}{2m}\frac{\partial^2}{\partial x^2} + V(x)\right)\delta(x-x') + O(t^2) \tag{39}$$

and hence

$$\overline{\psi}(x,t) = \psi_0(x) - \frac{it}{\hbar}\left(-\frac{\hbar^2}{2m}\frac{\partial^2}{\partial x^2} + V(x)\right)\psi_0(x) + O(t^2).$$

Comparing this expression with (28) yields (38). □

2.3. Comparison of Short-Time Propagators

We have seen above that both the Van Vleck and the Kerner–Sutcliffe propagators are accurate to order $O(t^2)$:

$$K(x,x',t) - \widetilde{K}(x,x',t) = O(t^2). \tag{40}$$

$$K(x,x',t) - \overline{K}(x,x',t) = O(t^2) \tag{41}$$

and hence, of course,

$$\widetilde{K}(x,x',t) - \overline{K}(x,x',t) = O(t^2). \tag{42}$$

Let us now study the case of the most commonly approximations to the action used in the theory of the Feynman integral, namely the mid-point rules

$$S_1(x,x',t,t') = \sum_{j=1}^{n} m_j \frac{(x_j - x_j')^2}{2t} - \frac{1}{2}(V(x) + V(x'))t \tag{43}$$

and

$$S_2(x,x',t) = \sum_{j=1}^{n} m_j \frac{(x_j - x_j')^2}{2t} - V(\tfrac{1}{2}(x+x'))\Delta t. \tag{44}$$

We begin with a simple example, that of the harmonic oscillator

$$H(x,p) = \frac{p^2}{2m} + \frac{1}{2}m^2\omega^2 x^2$$

(we are assuming $n = 1$). The exact value of the action is given by the generating function

$$S(x, x', t) = \frac{m}{2\sin\omega t}((x^2 + x'^2)\cos\omega t - 2xx'); \tag{45}$$

expanding the terms $\sin\omega t$ and $\cos\omega t$ in Taylor series for $t \to 0$ yields the approximation

$$S(x, x', t) = m\frac{(x - x')^2}{2t} - \frac{m\omega^2}{6}(x^2 + xx' + x'^2)t + O(t^2). \tag{46}$$

It is easy to verify, averaging $\frac{1}{2}m^2\omega^2x^2$ over $[x', x]$ that

$$\overline{S}(x, x', t) = m\frac{(x - x')^2}{2t} - \frac{m\omega^2}{6}(x^2 + xx' + x'^2)t$$

is precisely the approximate action provided by (31). If we now instead apply the midpoint rule (43) we get

$$S_1(x, x', t) = m\frac{(x - x')^2}{2t} - \frac{m^2\omega^2}{4}(x^2 + x'^2)t$$

which differs from the correct value (46) by a term $O(\Delta t)$. Similarly, the rule (44) yields

$$S_2(x, x', t) = m\frac{(x - x')^2}{2t} - \frac{m^2\omega^2}{8}(x + x')^2t$$

which again differs from the correct value (45) by a term $O(t)$. It is easy to understand why it is so by examining the case of a general potential function, and to compare $\overline{V}(x, x')$, $\frac{1}{2}(V(x) + V(x'))$, and $V(\frac{1}{2}(x + x')$. Consider for instance $\overline{V}(x, x') - V(\frac{1}{2}(x + x')$. Expanding $V(x)$ in a Taylor series at $\overline{x} = \frac{1}{2}(x + x')$ we get after some easy calculations

$$\begin{aligned}\overline{V}(x, x') &= V(\overline{x}) + V'(\overline{x})(x - x') + \frac{1}{2}V''(\overline{x})(x - x')^2 + O((x - x')^3) \\ &= V(\tfrac{1}{2}(x + x') - \tfrac{1}{12}V''(\tfrac{1}{2}(x + x'))(x - x')^3 + O((x - x')^3)\end{aligned}$$

hence $\overline{V}(x, x') - V(\frac{1}{2}(x + x')$ is different from zero unless $x = x'$ (or if $V(x)$ is linear) and hence the difference between $\overline{S}(x, x', t)$ and $S_2(x, x', t)$ will always generate a term containing t so that $\overline{S}(x, x', t) - S_2(x, x', t) = O(t)$ (and not $O(t^2)$). A similar calculation shows that we will also always have $\overline{S}(x, x', t) - S_1(x, x', t) = O(t)$. Denoting by $K_1(x, x', t)$ and $K_2(x, x', t)$ the approximate propagators obtained from the midpoint rules (43) and (44), respectively, one checks without difficulty that we will have

$$\overline{K}(x, x', t) - K_1(x, x', t) = O(t)$$
$$\overline{K}(x, x', t) - K_2(x, x', t) = O(t)$$

where $\overline{K}(x, x', t)$ is the Kerner–Sutcliffe propagator (35) (in these relations we can of course replace $\overline{K}(x, x', t)$ with the van Vleck propagator $\widetilde{K}(x, x', t)$ since both differ by a quantity $O(t^2)$ in view of Theorem 3.

3. The Case of Arbitrary Hamiltonians

3.1. The Main Result

We now consider the following very general situation: We assume that we are in the presence of a quantum system represented by a state $|\psi\rangle$ whose evolution is governed by a strongly continuous one-parameter group (U_t) of unitary operators acting on $L^2(\mathbb{R}^n)$; the operator U_t takes an initial wavefunction ψ_0 to $\psi = U_t\psi_0$. It follows from Schwartz's kernel theorem [33] that there exists a function

$K = K(x, x'; t)$ such that (This equality is sometimes postulated; it is in fact a mathematical fact which is true in quite general situations.)

$$\psi(x,t) = \int K(x,x';t)\psi_0(x')d^n x' \tag{47}$$

and from Stone's [34] theorem one strongly continuous one-parameter groups of unitary operators that there exists a self-adjoint (generally unbounded) operator $\widehat{H}$ on $L^2(\mathbb{R}^n)$ such that

$$\psi(x,t) = e^{-\frac{i}{\hbar}\widehat{H}t}\psi_0(x); \tag{48}$$

equivalently $\psi(x,t)$ satisfies the abstract Schrödinger equation (Jauch [35])

$$i\hbar\frac{\partial\psi}{\partial t}(x,t) = \widehat{H}\psi(x,t). \tag{49}$$

We now make the following crucial assumption, which extrapolates to the general case what we have done for Hamiltonians of the type classical type "kinetic energy plus potential": the quantum dynamics is again given by the Kerner–Sutcliffe propagator (35) for small times t, i.e.,

$$K(x,x',t) = \overline{K}(x,x',t) + O(t^2) \tag{50}$$

the approximate propagator being given by

$$\overline{K}(x,x',t) = \left(\tfrac{1}{2\pi\hbar}\right)^n \int e^{\frac{i}{\hbar}(p(x-x') - \overline{H}(x,x')t)} d^n p \tag{51}$$

where $\overline{H}$ is this time the averaged Hamiltonian function

$$\overline{H}(x,x',p) = \int_0^1 H(\tau x + (1-\tau)x', p)d\tau. \tag{52}$$

Obviously, when $H = H_{\text{free}} + V$ the function $\overline{H}$ reduces to the function $H_{\text{free}} + \overline{V}$ considered in Section 2.

This assumption can be motivated as follows (see de Gosson [12], Proposition 15, §4.4). Let

$$S(x,x',t) = \int_\gamma pdx - Hdt$$

be Hamilton's two-point function calculated along the phase space path leading from an initial point $(x', p', 0)$ to a final point (x, p, t) (the existence of such a function for small t is guaranteed by Hamilton–Jacobi theory; see e.g., Arnol'd [36] or Goldstein [37]). That function satisfies the Hamilton–Jacobi equation

$$\frac{\partial S}{\partial t} + H(x, \nabla_x S) = 0.$$

One then shows that the function

$$\overline{S}(x,x',t) = p(x-x') - \overline{H}(x,x',p)t$$

where p is the momentum at time t is an approximation to $S(x,x',t)$, in fact

$$\overline{S}(x,x',t) - S(x,x',t) = O(t^2).$$

Here is an example: Choose $H = \frac{1}{2}p^2x^2$ (we are assuming here $n = 1$); then

$$S(x, x', t) = \frac{(\ln(x/x'))^2}{2t}.$$

Using the formula

$$\overline{H}(x, x', p) = \frac{1}{6}p^2(x^2 + xx' + x'^2)$$

one shows after some calculations involving the Hamiltonian equations for H that

$$\overline{S}(x, x', t) = \frac{(\ln(x/x'))^2}{2t} + O(t^2)$$

(see [12], Chapter 4, Examples 10 and 16 for detailed calculations).

We are now going to show that the operator $\widehat{H}$ can be explicitly and uniquely determined from the knowledge of $\overline{K}(x, x', t)$.

Theorem 4. *If we assume that the short-time propagator is given by formula (51) then the operator $\widehat{H}$ appearing in the abstract Schrödinger Equation (49) is given by*

$$\widehat{H}\psi(x) = \left(\tfrac{1}{2\pi\hbar}\right)^n \int e^{\frac{i}{\hbar}p(x-x')}\overline{H}(x, x', p)\psi(x')d^n p d^n x'. \tag{53}$$

Proof. Differentiating both sides of the equality (47) with respect to time we get

$$i\hbar\frac{\partial\psi}{\partial t}(x, t) = i\hbar \int \frac{\partial K}{\partial t}(x, x', t)\psi_0(x')d^n x';$$

since K itself satisfies the Schrödinger Equation (49) we thus have

$$\widehat{H}\psi(x, t) = i\hbar \int \frac{\partial K}{\partial t}(x, x', t)\psi_0(x')d^n x'.$$

It follows, using the assumptions (50) and (51), that

$$\widehat{H}\psi(x, t) = i\hbar \int \frac{\partial \overline{K}}{\partial t}(x, x', t)\psi_0(x')d^n x' + O(t)$$

and hence, letting $t \to 0$,

$$\widehat{H}\psi_0(x) = i\hbar \int \frac{\partial \overline{K}}{\partial t}(x, x', 0)\psi_0(x')d^n x'. \tag{54}$$

Introducing the notation

$$\overline{S}(x, x', t) = p(x - x') - \overline{H}(x, x', p)t$$

we have

$$\begin{aligned}\frac{\partial \overline{K}}{\partial t}(x, x', t) &= \left(\tfrac{1}{2\pi\hbar}\right)^n \tfrac{i}{\hbar} \int e^{\frac{i}{\hbar}\overline{S}(x,x',t)}\frac{\partial \overline{S}}{\partial t}(x, x', t)d^n p' \\ &= \left(\tfrac{1}{2\pi\hbar}\right)^n \tfrac{1}{i\hbar} \int e^{\frac{i}{\hbar}\overline{S}(x,x',t)}\overline{H}(x, x', p')d^n p'.\end{aligned}$$

Taking the limit $t \to 0$ and multiplying both sides of this equality by $i\hbar$ we finally get

$$\widehat{H}\psi_0(x) = \left(\tfrac{1}{2\pi\hbar}\right)^n \int e^{\frac{i}{\hbar}p(x-x')}\overline{H}(x, x', p', t')\psi_0(x')d^n p' d^n x'$$

which proves (53). □

We will call the operator $\widehat{H}$ defined by (53) the Born–Jordan quantization of the Hamiltonian function H. That this terminology is justified is motivated below.

3.2. *The Case of Monomials*

Let us show that (53) reduces to the usual Born–Jordan quantization rule (1) when $H = x^m p^\ell$ (we are thus assuming dimension $n = 1$). We have here

$$H(\tau x + (1-\tau)x', p) = (\tau x + (1-\tau)x')^m p^\ell$$

hence, using the binomial formula,

$$H(\tau x + (1-\tau)x', p) = \sum_{k=0}^{m} \binom{m}{k} \tau^k (1-\tau)^{m-k} x^k p^\ell x'^{m-k}. \tag{55}$$

Integrating from 0 to 1 in τ and noting that

$$\int_0^1 \tau^k (1-\tau)^{m-k} d\tau = \frac{k!(m-k)!}{(m+1)!}$$

we get

$$\overline{H}(x, x', p) = \frac{1}{m+1} \sum_{k=0}^{m} x^k p^\ell x'^{m-k}$$

and hence, using the definition (53) of $\widehat{H}$,

$$\begin{aligned} \widehat{H}\psi(x) &= \frac{1}{2\pi\hbar(m+1)} \sum_{k=0}^{m} \int_{-\infty}^{\infty} e^{\frac{i}{\hbar}p(x-x')} x^k p^\ell x'^{m-k} \psi(x') dp dx' \\ &= \frac{x^k}{2\pi\hbar(m+1)} \sum_{k=0}^{m} \int_{-\infty}^{\infty} \left(\int_{-\infty}^{\infty} e^{\frac{i}{\hbar}p(x-x')} p^\ell dp \right) x'^{m-k} \psi(x') dx'. \end{aligned}$$

In view of the Fourier inversion Formula (34) we have

$$\frac{1}{2\pi\hbar} \int_{-\infty}^{\infty} e^{\frac{i}{\hbar}p(x-x')} p^\ell dp = (-i\hbar)^\ell \delta^{(\ell)}(x - x') \tag{56}$$

so that we finally get

$$\widehat{H}\psi(x) = \frac{1}{m+1} \sum_{k=0}^{m} x^k (-i\hbar)^\ell \frac{\partial^\ell}{\partial x^\ell} (x^{m-k}\psi),$$

which is equivalent to (1) since $\widehat{p}^\ell = (-i\hbar)^\ell \partial^\ell / \partial x^\ell$.

3.3. *Physical Hamiltonians*

Let us now show that the Born–Jordan quantization of a physical Hamiltonian of the type

$$H = \sum_{j=1}^{n} \frac{1}{2m_j} (p_j - A_j(x))^2 + V(x) \tag{57}$$

coincide with the usual operator

$$\widehat{H} = \sum_{j=1}^{n} \frac{1}{2m_j} \left(-i\hbar \frac{\partial}{\partial x_j} - A_j(x) \right)^2 + V(x) \tag{58}$$

obtained by Weyl quantization (the functions A_j and V are assumed to be C^1). Since the quantizations of p_j^2, $A_j(x)$ and $V(x)$ are the same in all quantization schemes (they are respectively $-\hbar^2\partial^2/\partial x_j^2$ and multiplication by $A_j(x)$ and $V(x)$), we only need to bother about the cross-products $p_jA(x)$. We claim that

$$\widehat{p_jA}\psi = -\frac{i\hbar}{2}\left[\frac{\partial}{\partial x_j}(A\leftarrow) + A\frac{\partial\psi}{\partial x_j}\right], \tag{59}$$

from which (58) immediately follows. Let us prove (59); it is sufficient to do this in the case $n = 1$. Denoting by $\overline{pA}$ the Born–Jordan quantization of the function pA we have

$$\overline{pA}(x, x', p) = p\int_0^1 A(\tau x + (1-\tau)x')d\tau = p\overline{A}(x, x')$$

and hence

$$\begin{aligned}\widehat{pA}\psi(x) &= \frac{1}{2\pi\hbar}\int e^{\frac{i}{\hbar}p(x-x')}p\overline{A}(x, x')\psi(x')dx'dp \\ &= \int_{-\infty}^{\infty}\left(\frac{1}{2\pi\hbar}\int_{-\infty}^{\infty} e^{\frac{i}{\hbar}p(x-x')}pdp\right)\overline{A}(x, x')\psi(x')dx'.\end{aligned}$$

In view of (34) the expression between the square brackets is $-i\hbar\delta'(x-x')$ so that

$$\begin{aligned}\widehat{pA}\psi(x) &= -i\hbar\int_{-\infty}^{\infty}\delta'(x-x')\overline{A}(x, x')\psi(x')dx' \\ &= -i\hbar\int_{-\infty}^{\infty}\delta(x-x')\frac{\partial}{\partial x'}(\overline{A}(x, x')\psi(x'))dx' \\ &= -i\hbar\left(\frac{\partial\overline{A}}{\partial x'}(x, x)\psi(x)) + \overline{A}(x, x)\frac{\partial\psi}{\partial x'}(x))\right)\end{aligned}$$

Now, by definition of $\overline{A}(x, x')$ we have $\overline{A}(x, x) = A(x)$ and

$$\frac{\partial\overline{A}}{\partial x'}(x, x) = \int_0^1 (1-\tau)\frac{\partial A}{\partial x}(x)d\tau = \frac{1}{2}\frac{\partial A}{\partial x}(x)$$

and hence

$$\widehat{pA}\psi = -\frac{i\hbar}{2}\frac{\partial A}{\partial x}\psi - i\hbar A\frac{\partial\psi}{\partial x}$$

which is the same thing as (59).

4. Discussion

Both Kerner and Sutcliffe, and Cohen relied on path integral arguments which were doomed to fail because of the multiple possible choices of histories in path integration. However, it follows from our rigorous constructions that Kerner and Sutcliffe's insight was right, even though their construction was not rigorously mathematically justified. While there is, as pointed out by Cohen [22], a great latitude in choosing the short-time propagator, thus leading to different quantizations, our argument did not make use of any path-integral argument; what we did was to propose a short-time propagator which is *exact* up to order $O(t^2)$ (as opposed to those obtained by using midpoint rules), and to show that if one use this propagator, then one must quantize Hamiltonian functions (and in particular monomials) following the prescription proposed by Born and Jordan in the case of monomials.

Funding: This work has been financed by the Grant P 27773–N25 of Austrian Science Fund (FWF).

Conflicts of Interest: The author declares no conflict of interest.

References

1. Dewey, T.G. Numerical mathematics of Feynman path integrals and the operator ordering problem. *Phys. Rev. A* **1990**, *42*, 32–37. [CrossRef] [PubMed]
2. Hall, M. Weyl's rule and Wigner equivalents for phase space monomials. *J. Phys. A Math. Gen.* **1985**, *18*, 29–36. [CrossRef]
3. Khandekar, D.C.; Lawande, S.V. Feynman Path Integrals: Some Exact Results and Applications. *Phys. Reps.* **1986**, *137*, 115–229. [CrossRef]
4. Kumano-go, N.; Fujiwara, D. Phase space Feynman path integrals via piecewise bicharacteristic paths and their semiclassical approximations, *Bull. Sci. Math.* **2008**, *132*, 313–357.
5. Mayes, I.W.; Dowker, J.S. Canonical functional integrals in general coordinates. *Proc. R. Soc. Lond. A* **1972**, *327*, 131–135. [CrossRef]
6. Mayes, I.W.; Dowker, J.S. Hamiltonian orderings and functional integrals. *J. Math. Phys.* **1973**, *14*, 434–439. [CrossRef]
7. Shewell, J.R. On the Formation of Quantum-Mechanical Operators. *Am. J. Phys.* **1959**, *27*, 16–21. [CrossRef]
8. de Gosson, M. Born–Jordan Quantization and the Equivalence of the Schrödinger and Heisenberg Pictures. *Found. Phys.* **2014**, *44*, 1096–1106. [CrossRef] [PubMed]
9. McCoy, N.H. On the function in quantum mechanics which corresponds to a given function in classical mechanics. *Proc. Natl. Acad. Sci. USA* **1932**, *18*, 674–676. [CrossRef] [PubMed]
10. Dahl, J.P.; Springborg, M. Wigner's phase space function and atomic structure: I. The hydrogen atom ground state. *Mol. Phys.* **1982**, *47*, 1001–1019. [CrossRef]
11. de Gosson, M. The Angular Momentum Dilemma and Born–Jordan Quantization. *Found. Phys.* **2017**, *47*, 61–70. [CrossRef]
12. de Gosson, M. *Born–Jordan Quantization: Theory and Applications;* Springer: New York, NY, USA, 2016.
13. Heisenberg, W. Über quantentheoretische Umdeutung kinematischer und mechanischer Beziehungen. *Z. Physik* **1925**, *33*, 879–893. [CrossRef]
14. Born, M.; Jordan, P. Zur Quantenmechanik. *Z. Physik* **1925**, *34*, 858–888. [CrossRef]
15. Born, M.; Heisenberg, W.; Jordan, P. Zur Quantenmechanik II. *Z. Physik* **1925**, *35*, 557–615. [CrossRef]
16. Kerner, E.H.; Sutcliffe, W.G. Unique Hamiltonian Operators via Feynman Path Integrals. *J. Math. Phys.* **1970**, *11*, 391–393. [CrossRef]
17. Garrod, C. Hamiltonian Path-Integral Methods. *Rev. Mod. Phys.* **1966**, *38*, 483–494. [CrossRef]
18. Kauffmann, S.K. Unique Closed-Form Quantization Via Generalized Path Integrals or by Natural Extension of the Standard Canonical Recipe. *arXiv* **1995**, arXiv:hep-th/9505189v2.
19. Kauffmann, S.K. Unambiguous Quantization from the Maximum Classical Correspondence that Is Self-consistent: The Slightly Stronger Canonical Commutation Rule Dirac Missed. *Found. Phys.* **2011**, *41*, 805–819. [CrossRef]
20. de Gosson, M. Symplectic Covariance properties for Shubin and Born-Jordan pseudo-differential operators. *Trans. Amer. Math. Soc.* **2013**, *365*, 3287–3307. [CrossRef]
21. de Gosson, M.; Luef, F. Preferred Quantization Rules: Born–Jordan vs. Weyl; Applications to Phase Space Quantization. *J. Pseudo-Differ. Oper. Appl.* **2011**, *2*, 115–139. [CrossRef]
22. Cohen, L. Hamiltonian Operators via Feynman Path Integrals. *J. Math. Phys.* **1970**, *11*, 3296–3297. [CrossRef]
23. Makri, N.; Miller, W.H. Correct short time propagator for Feynman path integration by power series expansion in Δt. *Chem. Phys. Lett.* **1988**, *15*, 1–8. [CrossRef]
24. Makri, N.; Miller, W.H. Exponential power series expansion for the quantum time evolution operator. *J. Chem. Phys.* **1989**, *90*, 904–911. [CrossRef]
25. de Gosson, M. *The Principles of Newtonian and Quantum Mechanics, the Need for Planck's Constant $\hbar$;* Imperial College Press: London, UK, 2001.
26. Gutzwiller, M.C. *Chaos in Classical and Quantum Mechanics;* Springer Science & Business Media: Berlin/Heidelberg, Germany, 2013.
27. Schulman, L.S. *Techniques and Applications of Path Integration;* Wiley: New York, NY, USA, 1981.
28. Maslov, V.P.; Fedoriuk, M.V. *Semi-Classical Approximation in Quantum Mechanics;* Springer Science & Business Media: Berlin/Heidelberg, Germany, 2001; Volume 7.

29. de Gosson, M. *Symplectic Methods in Harmonic Analysis and in Mathematical Physics;* Springer Science & Business Media: Berlin/Heidelberg, Germany, 2011.
30. de Gosson, M.; Hiley, B.J. Short-time quantum propagator and Bohmian trajectories. *Phys. Lett. A* **2013**, *377*, 3005–3008. [CrossRef] [PubMed]
31. de Gosson, M.; Hiley, B.J. Hamiltonian flows, short-time quantum propagators and the quantum Zeno effect. *J. Phys. Conf. Ser.* **2013**, *504*, 012027. [CrossRef]
32. Park, D. *Introduction to the Quantum Theory;* McGraw-Hill Inc.: New York, NY, USA, 1992.
33. Hörmander, L. *The Analysis of Linear Partial Differential Operators I;* Springer: New York, NY, USA, 1985; Volume 256.
34. Stone, M.H. Linear Transformations in Hilbert Space: III. Operational Methods and Group Theory. *Proc. Natl. Acad. Sci. USA* **1930**, *16*, 172–175. [CrossRef] [PubMed]
35. Jauch, J.M.; Morrow, R.A. Foundations of quantum mechanics. *Am. J. Phys.* **1968**, *36*, 771. [CrossRef]
36. Arnold, V.I. *Mathematical Methods of Classical Mechanics, Graduate Texts in Mathematics*, 2nd ed.; Springer: New York, NY, USA, 1989.
37. Goldstein, H. *Classical Mechanics*, 2nd ed.; Addison–Wesley: Boston, MA, USA, 1980.

entropy

MDPI

Article

Fourier Transform on the Homogeneous Space of 3D Positions and Orientations for Exact Solutions to Linear PDEs

Remco Duits *, Erik J. Bekkers and Alexey Mashtakov

Department of Mathematics and Computer Science (CASA), Eindhoven University of Technology, 5600 MB Eindhoven, The Netherlands; E.J.Bekkers@tue.nl (E.J.B.); A.Mashtakov@tue.nl (A.M.)
* Correspondence: R.Duits@tue.nl; Tel.: +31-40-247-2859

Received: 31 October 2018; Accepted: 18 December 2018; Published: 8 January 2019

Abstract: Fokker–Planck PDEs (including diffusions) for stable Lévy processes (including Wiener processes) on the joint space of positions and orientations play a major role in mechanics, robotics, image analysis, directional statistics and probability theory. Exact analytic designs and solutions are known in the 2D case, where they have been obtained using Fourier transform on $SE(2)$. Here, we extend these approaches to 3D using Fourier transform on the Lie group $SE(3)$ of rigid body motions. More precisely, we define the homogeneous space of 3D positions and orientations $\mathbb{R}^3 \rtimes S^2 := SE(3)/(\{\mathbf{0}\} \times SO(2))$ as the quotient in $SE(3)$. In our construction, two group elements are equivalent if they are equal up to a rotation around the reference axis. On this quotient, we design a specific Fourier transform. We apply this Fourier transform to derive new exact solutions to Fokker–Planck PDEs of α-stable Lévy processes on $\mathbb{R}^3 \rtimes S^2$. This reduces classical analysis computations and provides an explicit algebraic spectral decomposition of the solutions. We compare the exact probability kernel for $\alpha = 1$ (the diffusion kernel) to the kernel for $\alpha = \frac{1}{2}$ (the Poisson kernel). We set up stochastic differential equations (SDEs) for the Lévy processes on the quotient and derive corresponding Monte-Carlo methods. We verified that the exact probability kernels arise as the limit of the Monte-Carlo approximations.

Keywords: fourier transform; rigid body motions; partial differential equations; Lévy processes; Lie Groups; homogeneous spaces; stochastic differential equations

---

1. Introduction

The Fourier transform has had a tremendous impact on various fields of mathematics including analysis, algebra and probability theory. It has a broad range of applied fields such as signal and image processing, quantum mechanics, classical mechanics, robotics and system theory. Thanks to Jean-Baptiste Joseph Fourier (1768–1830), who published his pioneering work "Théory analytique de la chaleur" in 1822, the effective technique of using a Fourier transform to solve linear PDE-systems (with appropriate boundary conditions) for heat transfer evolutions on compact subsets Ω of $\mathbb{R}^d$ was born. The Fourier series representations of the solutions helped to understand the physics of heat transfer. Due to the linearity of the evolution operator that maps the possibly discontinuous square integrable initial condition to the square integrable solution at a fixed time $t > 0$, one can apply a spectral decomposition which shows how each eigenfunction is dampened over time. Thanks to contributions of Johann Peter Gustav Lejeune Dirichlet (1805–1859), completeness of the Fourier basis could then be formalized for several boundary conditions. Indeed, separation of variables (also known as "the Fourier method") directly provides a Sturm–Liouville problem [1] and an orthonormal basis of eigenfunctions for $\mathbb{L}_2(\Omega)$, which is complete due to compactness of the associated self-adjoint kernel operator. When dilating the subset Ω to the full space $\mathbb{R}^d$, the discrete set of eigenvalues start to fill

$\mathbb{R}$ and the discrete spectrum approximates a continuous spectrum (see, e.g., [2]). Then, a diffusion system on $\mathbb{R}^d$ can be solved via a unitary Fourier transform on $\mathbb{L}_2(\mathbb{R}^d)$ (cf. [3]).

Nowadays, in fields such as mechanics/robotics [4–7], mathematical physics/harmonic analysis [8], machine learning [9–13] and image analysis [14–19], it is worthwhile to extend the spatial domain of functions on $M = \mathbb{R}^d$ (or $M = \mathbb{Z}^d$) to groups $G = M \rtimes T$ that are the semi-direct product of an Abelian group M and another matrix group T. This requires a generalization of the Fourier transforms on the Lie group $(\mathbb{R}^d, +)$ towards the groups $G = \mathbb{R}^d \rtimes T$. Then, the Fourier transform gives rise to an invertible decomposition of a square integrable function into irreducible representations. This is a powerful mechanism in view of the Schur's lemma [20,21] and spectral decompositions [22,23]. However, it typically involves regularity constraints ([22], ch:6.6, [24], ch:3.6) on the structure of the dual orbits in order that Mackey's imprimitivity theory [25] can be applied to characterize *all* unitary irreducible representations (UIRs) of G. This sets the Fourier transform on the Lie group G [22,24,26]. Here, we omit technicalities on regularity constraints on the dual orbits and the fact that G may not be of type I (i.e., the quasi-dual group of G may not be equal to the dual group of G (cf. [22], thm.7.6, 7.7, [24], ch:3, [27]), as this does not play a role in our case of interest.

We are concerned with the case $M = \mathbb{R}^3$ and $T = SO(3)$ where $G = SE(3) = M \rtimes SO(3)$ is the Lie group of 3D rigid body motions. It is a (type I) Lie group with an explicit Fourier transform $\mathcal{F}_G$ where the irreducible representations are determined by regular dual orbits (which are spheres in the Fourier domain indexed by their radius $p > 0$) and an integer index $s \in \mathbb{Z}$ (cf. [4,26]).

In this article, we follow the idea of Joseph Fourier: we apply the Fourier transform $\mathcal{F}_G$ on the rigid body motion group $G = SE(3)$ to solve both non-degenerate and degenerate (hypo-elliptic) heat flow evolutions, respectively, on the Lie group G. More precisely, we design a Fourier transform $\mathcal{F}_{G/H}$ on the homogeneous space of positions and orientations G/H with $H \equiv \{\mathbf{0}\} \times SO(2)$ to solve degenerate and non-degenerate heat flow evolutions on the homogeneous space G/H. We also simultaneously solve related PDEs (beyond the diffusion case), as we explain below. For general Fourier theory and harmonic analysis on homogeneous spaces, see the works by Ghaani Farashahi [28–31], of which the work in [31] applies to our setting $G/H = \mathbb{R}^3 \rtimes S^2$. In contrast to ([31], ch:5.2), we consider the subgroup $H \equiv \{\mathbf{0}\} \times SO(2)$ instead of $\{\mathbf{0}\} \times SO(3)$, and we include an extra projection in our design of $\mathcal{F}_{G/H}$.

The idea of applying Fourier transforms to solve linear (degenerate) PDEs on non-commutative groups of the type $\mathbb{R}^d \rtimes T$ is common and has been studied by many researchers. For example, tangible probability kernels for heat transfer (and fundamental solutions) on the Heisenberg group were derived by Gaveau [32]. They can be derived by application ([23], ch:4.1.1) of the Fourier transform on the Heisenberg group ([22], ch:1). This also applies to probability kernels for degenerate, hypo-elliptic diffusions on $SE(2) = \mathbb{R}^2 \rtimes SO(2)$, where three different types (a Fourier series, a rapidly decaying series, and a single analytic formula that equals the rapidly decaying series) of explicit solutions to probability kernels for (convection-)diffusions were derived in previous works by Duits et al. [33–36]. For a concise review, see ([37], ch:5.1). Here, the two fundamental models for contour perception by, respectively, Mumford [38], Petitot [39] and Citti and Sarti [15] formed great sources of inspiration to study the degenerate diffusion problem on $SE(2)$.

The degenerate (hypo-elliptic) diffusion kernel formula in terms of a Fourier series representation was generalized to the much more wide setting of unimodular Lie groups by Agrachev, Boscain, Gauthier and Rossi [23]. This approach was then pursued by Portegies and Duits to achieve explicit exact solutions to (non-)degenerate (convection-)diffusions on the particular $SE(3)$ case (see [40]).

The idea of using Fourier transform on $SE(3)$ to represent solutions to the linear heat equations on $SE(3)$ has been considered by other authors in a wide variety of applications in the last decade. For a concise theoretical survey, see the recent work of Chirikjian [41]; for related articles with convincing applications, see [42,43]. In the recent work by Portegies and Duits [40], exact solutions are expressed in terms of an explicit, converging, eigenfunction decomposition in spheroidal wave-functions via technical, classical analysis techniques. This provides exact, analytic and converging series expressions that hold (and allow for analysis) prior to any numerical approximation. They can be used to compare different numerical techniques, as was done by Zhang and Duits et al. in the $SE(2)$ case [37]. In numerical implementations, the exact series must be truncated, and, as the spectrum is derived analytically, it is easy to control and reduce approximation errors to a neglectable level [44] (as in the $SE(2)$-case ([37], ch:5.1.4, thm 5.2 and 5.3, ch:6) with comparisons to an alternative single formula by Duits ([36], thm 5.2)).

Here, we aim to simplify and generalize the explicit spectral decompositions [40] of degenerate diffusions on $\mathbb{R}^3 \rtimes S^2 = SE(3)/H$, and to put this in the algebraic context of Fourier transform on $G = SE(3)$ [4,26,41], or more efficiently on the algebraic context of a Fourier transform on G/H. To this end, we first propose a specific Fourier transform on G/H in Theorem 1. Then, we use it to derive explicit spectral decompositions of the evolution operator in Theorem 2, from which we deduce explicit new kernel expressions in Theorem 3. Finally, we generalize the exact solutions to other PDE systems beyond the diffusion case: We simultaneously solve the Forward-Kolmogorov PDEs for α-stable Lévy processes on the homogeneous space of positions and orientations. Next, we address their relevance in the fields of image analysis, robotics and probability theory.

In image analysis, left-invariant diffusion PDEs on $SE(3)$ have been widely used for crossing-preserving diffusion and enhancement of fibers in diffusion-weighted MRI images of brain white matter [45–50], or for crossing-preserving enhancements of 3D vasculature in medical images [18]. They extend classical works on multi-scale image representations [51–54] to Lie groups [55].

In robotics, they play a role via the central limit theorem [56] in work-space generation of robot arms ([4], ch.12) and they appear indirectly in Kalman-filtering on $SE(3)$ for tracking [57], motion planning of robotic devices [42], and camera motion estimation [58].

In probability theory, diffusion systems on Lie groups describe Brownian motions [59,60] and they appear as limits in central limit theorem on Lie groups [56].

Both in probability theory [61] and in image analysis [62–65], the spectral decomposition of the evolution operator also allows simultaneously dealing with important variants of the diffusion evolution. These variants of the heat-evolution are obtained by taking fractional powers $-(-\Delta)^{\alpha}$ (cf. [66]) of the minus Laplacian operator $\Delta = \text{div} \circ \text{grad}$ that generates the heat flow (due to Fick's law and the Gauss divergence theorem), where $\alpha \in (0,1]$.

This generalization allows for heavy tailed distributions of α-stable Lévy processes, which arise in a fundamental generalization [61] of the central limit theorem *where one drops the finite variance condition*. Here, we note that recently an extension of the central limit on linear groups (such as $SE(3)$) has been achieved for finite second-order moments [56]. In engineering applications, where (iterative group-)convolutions are applied ([4], ch.12 and 13, [9,12,13,67–71]), the "kernel width" represents the spread of information or the scale of observing the signal. In the case the applications allow for an underlying probabilistic model with finite variances, variance is indeed a good measure for "kernel width". However, often this is not the case. Probability kernels for stochastic Lévy processes (used in directional statistics [72], stock market modeling [73], natural image statistics [65]), and modeling of point-spread functions in acquired images (e.g., in spectroscopy [74])) do require distributions with heavier tails than diffusion kernels. Therefore, "full width at half maximum" is a more generally applicable measure for kernel width than variance, as it applies to all α-stable Lévy processes. The probability distributions for $\alpha < 1$ encode a longer range of interaction via their heavy tails and still allow for unlimitedly sharp kernels.

Finally, regarding entropy, we show that for $\alpha \in \{\frac{1}{2}, 1\}$ we have monotonic increase of entropy $E_{\alpha}(t)$ over evolution time $t > 0$ of our α-stable Lévy processes. For $\alpha = 1$, one arrives at a diffusion system, and a previous result by Chirikjian on Lie groups [75], also applies to the Lie group quotient $G/H = \mathbb{R}^3 \rtimes S^2$. Thereby, $E_1'(t) = \text{trace}\{\mathbf{D} \cdot \mathbf{F}_1(t)\} > 0$, where $\mathbf{F}_1(t)$ is the Fisher information matrix and $\mathbf{D}$ is the diffusion matrix. We show that for $\alpha = \frac{1}{2}$ one arrives at a Poisson system where entropy also increases monotonically over time, again relative to a corresponding Fisher matrix. It is also intriguing, from the perspective of geometric theory of information and heat [76], to study optimal entropy on $\mathbb{R}^3 \rtimes S^2$ and (Fourier) Cramér Transforms building on results [77,78] on $\mathbb{R}^n$. However, such investigations first require a good grip on the spectral decompositions of the PDE-evolution operators for α-stable Lévy processes via a Fourier transform on $\mathbb{R}^3 \rtimes S^2$, which is our primary focus here.

1.1. Structure of the Article

The structure of the article is as follows. In the first part of the Introduction, we briefly discuss the history of the Fourier transform, and its generalization to other groups that are the semi-direct product of the translation group and another matrix group, where we provide an overview of related works. Then, we specify our domain of interest—the Fourier transform on the homogeneous space G/H of positions and orientations, which is a Lie group quotient of the Lie group $G = SE(3)$ with a subgroup H isomorphic to $\{\mathbf{0}\} \times SO(2)$. Then, we address its application of solving PDE systems on G/H, motivated from applications in image analysis, robotics and probability theory.

There are four remaining subsections of the Introduction. In Section 1.2, we provide basic facts on the homogeneous space G/H of positions and orientations and we provide preliminaries for introducing a Fourier transform on G/H. In Section 1.3, we formulate the PDEs of interest on G/H that we solve. In Section 1.4, we formulate the corresponding PDEs on the group G. In Section 1.5, we relate the PDE for $\alpha = \frac{1}{2}$ to a Poisson system and quantify monotonic increase of entropy for $\alpha \in \{\frac{1}{2}, 1\}$. In Section 1.6, we provide a roadmap on the spectral decomposition of the PDE evolutions.

In Section 2, based on previous works, we collect the necessary prior information about the PDEs of interest and the corresponding kernels. We also describe how to extend the case $\alpha = 1$ (the diffusion case) to the general case $\alpha \in (0, 1]$.

In Section 3, we describe the Fourier transform on the Lie group $SE(3)$, where we rely on UIRs of $SE(3)$. In particular, by relating the UIRs to the dual orbits of $SO(3)$ and by using a decomposition with respect to an orthonormal basis of modified spherical harmonics, we recall an explicit formula for the inverse Fourier transform.

In Section 4, we present a Fourier transform $\mathcal{F}_{G/H}$ on the quotient $G/H = \mathbb{R}^3 \rtimes S^2$. Our construction requires an additional constraint—an input function must be bi-invariant with respect to subgroup H, as explained in Remark 3. This extra symmetry constraint is satisfied by the PDE kernels of interest. We prove a theorem, where we present: (1) a matrix representation for the Fourier transform on the quotient; (2) an explicit inversion formula; and (3) a Plancherel formula.

In Section 5, we apply our Fourier transform on the quotient to solve the PDEs of interest. The solution is given by convolution of the initial condition with the specific kernels (which are the probability kernels of α-stable Lévy process). We find the exact formulas for the kernels in the frequency domain relying on a spectral decomposition of the evolution operator (involved in the PDEs). We show that this result can be obtained either via conjugation of the evolution operator with our Fourier transform on $\mathbb{R}^3 \rtimes S^2$ or (less efficiently) via conjugation of the evolution operator with the Fourier transform acting only on the spatial part $\mathbb{R}^3$. Then, we present a numerical scheme to approximate the kernels via Monte-Carlo simulation and we provide a comparison of the exact solutions and their approximations. Finally, in Section 6, we summarize our results and discuss their applications. In the appendices, we address the probability theory and stochastic differential equations (SDEs) regarding Lévy processes on $\mathbb{R}^3 \rtimes S^2$.

The main contributions of this article are:

- We construct $\mathcal{F}_{\mathbb{R}^3 \rtimes S^2}$—the Fourier transform on the quotient $\mathbb{R}^3 \rtimes S^2$, in Equation (43).
- The matrix representations for $\mathcal{F}_{\mathbb{R}^3 \rtimes S^2}$, explicit inversion and Plancherel formulas are shown in Theorem 1.
- The explicit spectral decompositions of PDE evolutions for α-stable Lévy process on $\mathbb{R}^3 \rtimes S^2$, in the Fourier domains of both $\mathbb{R}^3 \rtimes S^2$ and $\mathbb{R}^3$, are shown in Theorem 2; here, the new spectral decomposition in the Fourier domain of $\mathbb{R}^3 \rtimes S^2$ is simpler and involves ordinary spherical harmonics.

- The quantification of monotonic increase of entropy of PDE solutions for α-stable Lévy processes on $\mathbb{R}^3 \rtimes S^2$ for $\alpha \in \{\frac{1}{2}, 1\}$ in terms of Fisher information matrices is shown in Proposition 1.
- the exact formulas for the probability kernels of α-stable Lévy processes on $\mathbb{R}^3 \rtimes S^2$, in Theorem 3. This also includes new formulas for the heat kernels (the case $\alpha = 1$), that are more efficient than the heat kernels presented in previous work [40].
- Simple formulation and verifications (Monte-Carlo simulations) of discrete random walks for α-stable Lévy processes on $\mathbb{R}^3 \rtimes S^2$ in Proposition 3. The corresponding SDEs are in Appendix A.

1.2. Introduction to the Fourier Transform on the Homogeneous Space of Positions and Orientations

Let $G = SE(3)$ denote the Lie group of rigid body motions, equipped with group product:

$$g_1 g_2 = (\mathbf{x}_1, \mathbf{R}_1)(\mathbf{x}_2, \mathbf{R}_2) = (\mathbf{R}_1\mathbf{x}_2 + \mathbf{x}_1, \mathbf{R}_1\mathbf{R}_2), \quad \text{with } g_k = (\mathbf{x}_k, \mathbf{R}_k) \in G,\ k = 1,2. \tag{1}$$

Here, $\mathbf{x}_k \in \mathbb{R}^3$ and $\mathbf{R}_k \in SO(3)$. Note that $SE(3) = \mathbb{R}^3 \rtimes SO(3)$ is a semi-direct product of $\mathbb{R}^3$ and $SO(3)$.

Definition 1. *Let $B(\mathcal{H})$ denote the vector space of bounded linear operators on some Hilbert space $\mathcal{H}$. Within the space $B(\mathcal{H})$, we denote the subspace of bounded linear trace-class operators by*

$$B_2(\mathcal{H}) = \left\{ A : \mathcal{H} \to \mathcal{H} \mid A \text{ linear and } |||A|||^2 := \text{trace}(A^*A) < \infty \right\}.$$

Definition 2. *Consider a mapping $\sigma : G \to B(\mathcal{H}_\sigma)$, where $\mathcal{H}_\sigma$ denotes the Hilbert space on which each σ_g acts. Then, σ is a Unitary Irreducible Representation (UIR) of G if*

1. *$\sigma : G \to B(\mathcal{H}_\sigma)$ is a homomorphism;*
2. *$\sigma_g^{-1} = \sigma_g^*$ for all $g \in G$; and*
3. *there does not exist a closed subspace V of $\mathcal{H}_\sigma$ other than $\{0, \mathcal{H}_\sigma\}$ such that $\sigma_g V \subset V$.*

We denote by $\hat{G}$ the dual group of G. Its elements are equivalence classes of UIRs, where one identifies elements via $\sigma_1 \sim \sigma_2 \Leftrightarrow$ there exists a unitary linear operator v, s.t. $\sigma_1 = v \circ \sigma_2 \circ v^{-1}$. Note that $G = SE(3)$ is a unimodular Lie group of type I, which means that the left and right-invariant Haar measure coincide, and that its dual group and its quasi dual group coincide. Thereby it admits a Plancherel theorem [22,24].

Definition 3. *The Fourier transform $\mathcal{F}_G(f) = ((\mathcal{F}_G f)(\sigma))_{\sigma \in \hat{G}}$ of a square-integrable, measurable and bounded function f on G is a measurable field of bounded operators indexed by unitary irreducible representations (UIR's) σ. Now, $\hat{G}$ can be equipped with a canonical Plancherel measure ν and the Fourier transform $\mathcal{F}_G$ admits an extension unitary operator from $\mathbb{L}_2(G)$ to the direct-integral space $\int_{\hat{G}}^{\oplus} B_2(\mathcal{H}_\sigma) d\nu(\sigma)$. This unitary extension ([22], 4.25) (also known as "Plancherel transform" ([24], thm.3.3.1)) is given by*

$$\begin{aligned} &\mathcal{F}_G(f) = \int_{\hat{G}}^{\oplus} \hat{f}(\sigma)\, d\nu(\sigma), \text{ with} \\ &\hat{f}(\sigma) = (\mathcal{F}_G f)(\sigma) = \int_G f(g)\, \sigma_{g^{-1}} dg \ \in B_2(\mathcal{H}_\sigma), \ \text{for all } \sigma \in \hat{G}, \end{aligned} \tag{2}$$

for all $f \in \mathbb{L}_1(G) \cap \mathbb{L}_2(G)$.

The Plancherel theorem states that $\|\mathcal{F}_G(f)\|^2_{\mathbb{L}_2(\hat{G})} = \int_{\hat{G}} |||\mathcal{F}_G(f)(\sigma)|||^2 d\nu(\sigma) = \int_G |f(g)|^2 dg = \|f\|^2_{\mathbb{L}_2(G)}$ for all $f \in \mathbb{L}_2(G)$, and we have the inversion formula $f = \mathcal{F}_G^{-1}\mathcal{F}_G f = \mathcal{F}_G^* \mathcal{F}_G f$. For details, see [22,24], and, for detailed explicit computations, see [4].

In this article, we constrain and modify the Fourier transform $\mathcal{F}_G$ on $G = SE(3)$ such that we obtain a suitable Fourier transform $\mathcal{F}_{G/H}$ defined on a homogeneous space

$$\mathbb{R}^3 \rtimes S^2 := G/H \text{ with subgroup } H = \{\mathbf{0}\} \times \mathrm{Stab}_{SO(3)}(\mathbf{a}) \tag{3}$$

of left cosets, where $\mathrm{Stab}_{SO(3)}(\mathbf{a}) = \{\mathbf{R} \in SO(3) \mid \mathbf{Ra} = \mathbf{a}\}$ denotes the subgroup of $SO(3)$ that stabilizes an a priori reference axis $\mathbf{a} \in S^2$, say $\mathbf{a} = \mathbf{e}_z = (0,0,1)^T$. In the remainder of this article, we set this choice $\mathbf{a} = \mathbf{e}_z$.

Remark 1. *Although the semi-direct product notation $\mathbb{R}^3 \rtimes S^2$ is formally not correct as S^2 is not a Lie group, it is convenient: it reminds that G/H denotes the homogeneous space of positions and orientations.*

Remark 2. *(notation and terminology)*

Elements in Equation (3) denote equivalence classes of rigid body motions $g = (\mathbf{x}, \mathbf{R_n}) \in SE(3)$ that map $(\mathbf{0}, \mathbf{a})$ to $(\mathbf{x}, \mathbf{n})$:

$$[g] =: (\mathbf{x}, \mathbf{n}) \in \mathbb{R}^3 \rtimes S^2 \quad \Leftrightarrow \quad g \odot (\mathbf{0}, \mathbf{a}) = (\mathbf{x}, \mathbf{n}),$$

under the (transitive) action

$$g \odot (\mathbf{x}', \mathbf{n}') = (\mathbf{Rx}' + \mathbf{x}, \mathbf{Rn}'), \qquad \textit{for all } g = (\mathbf{x}, \mathbf{R}) \in SE(3),\ (\mathbf{x}', \mathbf{n}') \in \mathbb{R}^3 \rtimes S^2. \tag{4}$$

Therefore, we simply denote the equivalence classes $[g]$ by $(\mathbf{x}, \mathbf{n})$. This is similar to the conventional writing $\mathbf{n} \in S^2 = SO(3)/SO(2)$. Throughout this manuscript, we refer to G/H as "the homogeneous space of positions and orientations" and henceforth $\mathbf{R_n}$ denotes any rotation that maps the reference axis $\mathbf{a}$ into $\mathbf{n}$.

The precise definition of the Fourier transform $\mathcal{F}_{G/H}$ on the homogeneous space G/H is presented in Section 4. It relies on the decomposition into unitary irreducible representations in Equation (2), but we must take both a domain and a range restriction into account. This is explained in Section 4. Next, we address an a priori domain constraint that is rather convenient than necessary.

Remark 3. *We constrain the Fourier transform $\mathcal{F}_{G/H}$ to*

$$\mathbb{L}_2^{sym}(G/H) := \left\{ f \in \mathbb{L}_2(G/H) \mid \forall_{\mathbf{R} \in \mathrm{Stab}_{SO(3)}(\mathbf{a})} : f(\mathbf{x}, \mathbf{n}) = f(\mathbf{Rx}, \mathbf{Rn}) \right\}. \tag{5}$$

This constraint is convenient in view of the PDEs of interest (and the symmetries of their kernels) that we formulate in the next subsection, and that solve via Fourier's method in Section 5.

1.3. Introduction to the PDEs of Interest on the Quotient $\mathbb{R}^3 \rtimes S^2$

Our main objective is to use the Fourier transform $\mathcal{F}_{G/H}$ to solve the following PDEs on $\mathbb{R}^3 \rtimes S^2$:

$$\boxed{\begin{cases} \dfrac{\partial}{\partial t} W_\alpha(\mathbf{x}, \mathbf{n}, t) = Q_\alpha W_\alpha(\mathbf{x}, \mathbf{n}, t), \\ W_\alpha(\mathbf{x}, \mathbf{n}, 0) = U(\mathbf{x}, \mathbf{n}), \end{cases}} \tag{6}$$

where $(\mathbf{x}, \mathbf{n}) \in \mathbb{R}^3 \rtimes S^2$, $t \geq 0$, $\alpha \in (0,1]$ and the generator

$$Q_\alpha := -(-Q)^\alpha \tag{7}$$

is expressed via

$$Q = D_{11} \|\mathbf{n} \times \nabla_{\mathbb{R}^3}\|^2 + D_{33} (\mathbf{n} \cdot \nabla_{\mathbb{R}^3})^2 + D_{44} \Delta_{\mathbf{n}}^{S^2},$$

with $D_{33} > D_{11} \geq 0$, $D_{44} > 0$, and with $\Delta_{\mathbf{n}}^{S^2}$ the Laplace–Beltrami operator on $S^2 = \{\mathbf{n} \in \mathbb{R}^3 \mid \|\mathbf{n}\| = 1\}$.

Note that the generator Q is a self-adjoint unbounded operator with domain

$$\mathcal{D}(Q) := \mathbb{H}_2(\mathbb{R}^3) \otimes \mathbb{H}_2(S^2),$$

where $\mathbb{H}_2$ denotes the Sobolev space $\mathbb{W}_2^2$.

The semigroup for $\alpha = 1$ is a strongly continuous semigroup on $\mathbb{L}_2(\mathbb{R}^3 \rtimes S^2)$ with a closed generator, and by taking the fractional power of the generator one obtains another strongly continuous semigroup, as defined and explained in a more general setting in the work by Yosida ([66], ch:11). The fractional power is formally defined by

$$Q_\alpha W = -(-Q)^\alpha W := \frac{\sin \alpha\pi}{\pi} \int_0^\infty \lambda^{\alpha-1}(Q - \lambda I)^{-1}(-QW)\mathrm{d}\lambda \text{ for all } W \in \mathcal{D}(Q). \tag{8}$$

In Section 1.6, we show that the common technical representation Equation (8) is not really needed for our setting. In fact, it is very easy to account for $\alpha \in (0,1]$ in the solutions; by a spectral decomposition, we only need to take fractional powers of certain eigenvalues in the Fourier domain. For the moment, the reader may focus on the case $\alpha = 1$, where the system in Equation (6) becomes an ordinary elliptic diffusion system which is hypo-elliptic (in the sense of Hörmander [79]) even in the degenerate case where $D_{11} = 0$.

The PDEs in Equation (6) have our interest as they are Forward-Kolmogorov equations for α-stable Lévy processes on G/H. See Appendix A for a precise formulation of discrete and continuous stochastic processes. This generalizes previous works on such basic processes [61,64] with applications in financial mathematics [80] and computer vision [65,78,81,82], from Lie group $\mathbb{R}^3$ to the Lie group quotient $\mathbb{R}^3 \rtimes S^2$.

See Figure 1 for a visualization of sample paths from the discrete stochastic processes explained in Appendix A. They represent "drunk man's flights" rather than "drunk man's walks".

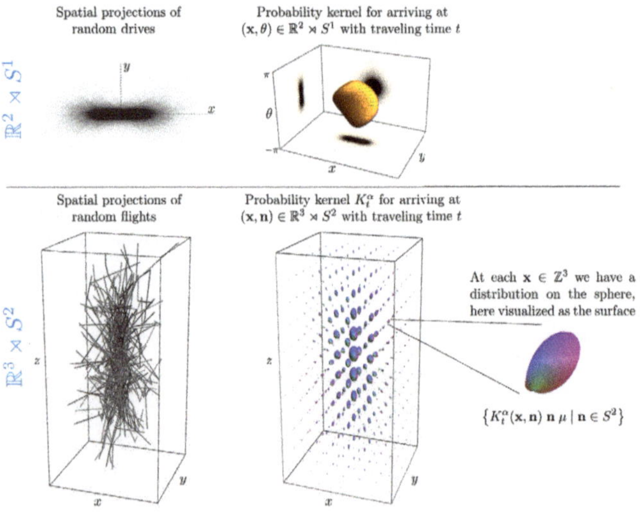

Figure 1. Various visualization of the diffusion process ($\alpha = 1$) on $\mathbb{R}^d \rtimes S^{d-1}$, for $d = 2$ and $d = 3$. (**Top**) random walks (or rather "drunk man's drives") and an iso-contour of the limiting diffusion kernel, for the case $d = 2$ studied in previous works (see, e.g., [15,37,83]); and (**Bottom**) random walks (or rather "drunk man's flights") and a visualization of the limiting distribution for the case $d = 3$. This limiting distribution is a degenerate diffusion kernel $(\mathbf{x},\mathbf{n}) \mapsto K_t^{\alpha=1}(\mathbf{x},\mathbf{n})$ that we study in this article. We visualize kernel $K_t^{\alpha=1}$ by a spatial grid of surfaces, where all surfaces are scaled by the same $\mu > 0$.

1.4. Reformulation of the PDE on the Lie Group SE(3)

Now, we reformulate and extend our PDEs in Equation (6) to the Lie group $G = SE(3)$ of rigid body motions, equipped with group product in Equation (1). This helps us to better recognize symmetries, as we show in Section 2.1. To this end, the PDEs are best expressed in a basis of left-invariant vector fields $\{g \mapsto \mathcal{A}_i|_g\}_{i=1}^6$ on G. Such left-invariant vector fields are obtained by push forward from the left-multiplication $L_{g_1}g_2 := g_1 g_2$ as

$$\mathcal{A}_i|_g = (L_g)_* A_i \in T_g(G),$$

where $A_i := \mathcal{A}_i|_e$ form an orthonormal basis for the Lie algebra $T_e(G)$. We choose such a basis typically such that the first three are spatial generators $A_1 = \partial_x, A_2 = \partial_y, A_3 = \partial_z = \mathbf{a} \cdot \nabla_{\mathbb{R}^3}$ and the remaining three are rotation generators, in such a way that A_6 is the generator of a counter-clockwise rotation around the reference axis $\mathbf{a}$. For all $\tilde{U} \in C^1(G)$ and $g \in G$, one has

$$\mathcal{A}_i \tilde{U}(g) = \lim_{t \downarrow 0} \frac{\tilde{U}(g\, e^{tA_i}) - \tilde{U}(g)}{t}, \tag{9}$$

where $A \mapsto e^A$ denotes the exponent that maps Lie algebra element $A \in T_e(G)$ to the corresponding Lie group element. The explicit formulas for the left-invariant vector fields in Euler-angles (requiring two charts) can be found in Appendix B, or in [4,84].

Now we can re-express the PDEs in Equation (6) on the group $G = SE(3)$ as follows:

$$\boxed{\begin{cases} \frac{\partial}{\partial t}\tilde{W}_\alpha(g,t) = \tilde{Q}_\alpha \tilde{W}_\alpha(g,t), & g \in G, t \geq 0 \\ \tilde{W}_\alpha(g,0) = \tilde{U}(g), & g \in G, \end{cases}} \tag{10}$$

where the generator

$$\tilde{Q}_\alpha := -(-\tilde{Q})^\alpha \tag{11}$$

is again a fractional power ($\alpha \in (0,1]$) of the diffusion generator $\tilde{Q}$ given by

$$\tilde{Q} = D_{11}(\mathcal{A}_1^2 + \mathcal{A}_2^2) + D_{33}\,\mathcal{A}_3^2 + D_{44}(\mathcal{A}_4^2 + \mathcal{A}_5^2), \tag{12}$$

where $\mathcal{A}_i^2 = \mathcal{A}_i \circ \mathcal{A}_i$ for all $i \in \{1, \ldots, 5\}$. The initial condition in Equation (10) is given by

$$\tilde{U}(g) = \tilde{U}(\mathbf{x}, \mathbf{R}) = U(\mathbf{x}, \mathbf{Ra}).$$

Similar to the previous works [40,85], one has

$$\tilde{W}_\alpha(\mathbf{x}, \mathbf{R}, t) = W_\alpha(\mathbf{x}, \mathbf{Ra}, t), \tag{13}$$

that holds for all $t \geq 0$, $(\mathbf{x}, \mathbf{R}) \in SE(3)$.

Remark 4. *Equation (13) relates the earlier PDE formulation in Equation (6) on the quotient G/H to the PDE formulation in Equation (10) on the group G. It holds since we have the relations*

$$\begin{aligned}
&\mathcal{A}_6 \tilde{W}_\alpha(\mathbf{x}, \mathbf{R}, t) = 0, \\
&(\mathcal{A}_5^2 + \mathcal{A}_4^2)\tilde{W}_\alpha(\mathbf{x}, \mathbf{R}, t) = \Delta^{S^2} W_\alpha(\mathbf{x}, \mathbf{Ra}, t), \\
&\mathcal{A}_3 \tilde{W}_\alpha(\mathbf{x}, \mathbf{R_n}, t) = \mathbf{n} \cdot \nabla_{\mathbb{R}^3} W_\alpha(\mathbf{x}, \mathbf{n}, t), \\
&(\mathcal{A}_1^2 + \mathcal{A}_2^2)\, \tilde{W}_\alpha(\mathbf{x}, \mathbf{R}, t) = \left(\Delta^{\mathbb{R}^3} - \mathcal{A}_3^2\right) \tilde{W}_\alpha(\mathbf{x}, \mathbf{R}, t) = \|\mathbf{n} \times \nabla_{\mathbb{R}^3}\|^2\, W_\alpha(\mathbf{x}, \mathbf{Ra}, t)
\end{aligned}$$

so that the generator of the PDE in Equation (10) on G and the generator of the PDE in Equation (6) on G/H indeed stay related via

$$\tilde{Q}_\alpha \tilde{W}_\alpha(\mathbf{x}, \mathbf{R}, t) = Q_\alpha W_\alpha(\mathbf{x}, \mathbf{Ra}, t) \text{ for all } t \geq 0. \tag{14}$$

1.5. Increase of Entropy for the Diffusion System ($\alpha = 1$) and the Poisson System ($\alpha = \frac{1}{2}$) on G/H

The PDE-system in Equation (6) on G/H relates to the PDE-system in Equation (10) on G via Equation (14). Next, we show that for $\alpha = \frac{1}{2}$ the PDE-system boils down to a Poisson system. For $\alpha = 1$ the PDE-system in Equation (10) is a diffusion system on Lie group G, for which one has monotonic increase of entropy [75]. The next theorem quantifies the monotonic increase of entropy for $\alpha \in \{\frac{1}{2}, 1\}$ in terms of Fisher matrices.

Definition 4. *Let $\alpha \in (0,1]$. Let $\tilde{W}_\alpha$ be the solution to Equation (10) with positive initial condition $\tilde{U} > 0$ with $\tilde{U} \in \mathbb{L}_2(G)$ and $\int_G \tilde{U}(g)\mathrm{d}g = 1$. Then, we define the entropy $E_\alpha(t)$ at evolution time $t \geq 0$ as*

$$E_\alpha(t) := -\int_G \tilde{W}_\alpha(g,t) \, \log \tilde{W}_\alpha(g,t) \, \mathrm{d}g. \tag{15}$$

Proposition 1. *For $\alpha = \frac{1}{2}$, the PDE system in (10) yields the same solutions as the following Poisson system:*

$$\begin{cases} \left(\frac{\partial^2}{\partial t^2} + \tilde{Q}\right) \tilde{W}_{\frac{1}{2}}(g,t) = 0 & g \in G, t \geq 0, \text{ with } \forall_{t \geq 0} : \tilde{W}_{\frac{1}{2}}(\cdot, t) \in \mathbb{L}_2(G) \\ \tilde{W}_{\frac{1}{2}}(g,0) = \tilde{U}(g) > 0, & g \in G. \end{cases} \tag{16}$$

The entropy in Equation (15) equals $E_\alpha(t) = -2\pi \int_{G/H} W_\alpha(\mathbf{x},\mathbf{n},t) \, \log W_\alpha(\mathbf{x},\mathbf{n},t) \, \mathrm{d}\mathbf{x}\mathrm{d}\mu_{S^2}(\mathbf{n})$.

For all $t > 0$, one has

$$\begin{aligned} &E_1'(t) = \mathrm{trace}\{\mathbf{D} \cdot \mathbf{F}_1(t)\} > 0, \\ &E_{\frac{1}{2}}''(t) < -\mathrm{trace}\{\mathbf{D} \cdot \mathbf{F}_{\frac{1}{2}}(t)\} < 0 \text{ and } E_{\frac{1}{2}}'(t) = \int_t^\infty \mathrm{trace}\{\mathbf{D} \cdot \mathbf{F}_{\frac{1}{2}}(\tau)\} + F(\tau) \, \mathrm{d}\tau > 0, \end{aligned} \tag{17}$$

for the diffusion matrix $\mathbf{D} = \mathrm{diag}\{D_{ii}\}_{i=1}^6 > 0$, where $D_{11} = D_{22}$, D_{33} and $D_{44} = D_{55}$ are the coefficients in $\tilde{Q}$, and with Fisher matrix $\mathbf{F}_\alpha(t) = \mathrm{diag}\{\int_G \frac{|\mathcal{A}_i \tilde{W}_\alpha(g,t)|^2}{\tilde{W}_\alpha(g,t)} \mathrm{d}g\}_{i=1}^6$, and $F(t) = \int_G \frac{|\partial_\tau \tilde{W}_{1/2}(g,t)|^2}{\tilde{W}_{1/2}(g,t)} \mathrm{d}g \geq 0$.

Proof. For $\alpha = \frac{1}{2}$, one has by the square integrability constraint in Equation (16) and application of the unitary Fourier transform on G that $\left(\frac{\partial^2}{\partial t^2} + \tilde{Q}\right) \tilde{W}_{\frac{1}{2}} = \left(\frac{\partial}{\partial t} - \sqrt{-\tilde{Q}}\right)\left(\frac{\partial}{\partial t} + \sqrt{-\tilde{Q}}\right) \tilde{W}_{\frac{1}{2}} = 0 \Rightarrow \left(\frac{\partial}{\partial t} + \sqrt{-\tilde{Q}}\right) \tilde{W}_{\frac{1}{2}} = 0$ and thereby the PDE system in Equation (10) can be replaced by the Poisson system in Equation (16) on $G \times \mathbb{R}^+$. The formula for the entropy follows from a product decomposition of the (bi-invariant) haar measure on G into measure on the quotient G/H and a measure on the subgroup $H \equiv \{\mathbf{0}\} \times SO(2)$ and the fact that $\tilde{W}_\alpha(gh,t) = \tilde{W}_\alpha(g,t)$ for all $h \in H$, $\alpha \in (0,1]$, due to Equation (14). For $\alpha = \frac{1}{2}$, we have that $\tilde{W}_\alpha$ satisfies Equation (16) and

$$\begin{aligned} E_{\frac{1}{2}}''(t) &= -\int_G \frac{(\partial_t \tilde{W}_{\frac{1}{2}}(g,t))^2}{\tilde{W}_{\frac{1}{2}}(g,t)} - (\log(\tilde{W}_{\frac{1}{2}}(g,t)+1)) \, \partial_t^2 \tilde{W}_{\frac{1}{2}}(g,t) \, \mathrm{d}g \\ &< \int_G (\log \tilde{W}_{\frac{1}{2}}(g,t) + 1) \, \tilde{Q}\tilde{W}_{\frac{1}{2}}(g,t) \, \mathrm{d}g = \int_G (\log \tilde{W}_{\frac{1}{2}}(g,t)) \, \tilde{Q}\tilde{W}_{\frac{1}{2}}(g,t) \, \mathrm{d}g \\ &= -\int_G \sum_{i=1}^6 \frac{D_{ii}|\mathcal{A}_i \tilde{W}_{\frac{1}{2}}(g,t)|^2}{\tilde{W}_{\frac{1}{2}}(g,t)} \, \mathrm{d}g = -\mathrm{trace}(\mathbf{D} \cdot \mathbf{F}_{\frac{1}{2}}(t)), \end{aligned}$$

where we use integration by parts and short notation with $\partial_t := \frac{\partial}{\partial t}$. Now, $E''_{\frac{1}{2}} < 0$ and $E'_{\frac{1}{2}}$ is continuous (due to the Lebesgue dominated convergence principle and continuity of each mapping $t \mapsto \partial_t \tilde{W}(g,t)$ indexed by $g \in G$) and $E_{\frac{1}{2}}(t) \to 0$ when $t \to \infty$, from which we deduce that $E'_{\frac{1}{2}}(t) = -\int_t^\infty E''_{\frac{1}{2}}(\tau)\mathrm{d}\tau > 0$.

For $\alpha = 1$, we follow ([75], Thm.2) and compute (again using the PDE and integration by parts)

$$E'_1(t) = -\int_G (\partial_t \tilde{W}_1(g,t)) \log \tilde{W}_1(g,t) + \tilde{W}_1(g,t)\,\mathrm{d}g = \int_G \sum_{i=1}^{6} D_{ii} \frac{|\mathcal{A}_i \tilde{W}_1(g,t)|^2}{\tilde{W}_1(g,t)}\,\mathrm{d}g = \mathrm{trace}(\mathbf{D} \cdot \mathbf{F}_1(t)) > 0.$$

Regarding the strict positivity in Equation (17), we note that $\tilde{U} > 0 \Rightarrow \tilde{W}_\alpha > 0$ and if $E'_\alpha(t) = 0$ then this would imply that $\tilde{W}_\alpha(\cdot,t)$ is constant, which violates $\tilde{W}_\alpha(\cdot,t) \in \mathbb{L}_2(G)$ as G is not compact. □

1.6. A Preview on the Spectral Decomposition of the PDE Evolution Operator and the Inclusion of α

Let U be in the domain of the generator Q_α given by Equation (7), of our evolution Equation (6). For a formal definition of this domain, we refer to ([86], Equation 9). Let its spatial Fourier transform be given by

$$\overline{U}(\boldsymbol{\omega},\mathbf{n}) = [\mathcal{F}_{\mathbb{R}^3} U(\cdot,\mathbf{n})](\boldsymbol{\omega}) := \frac{1}{(2\pi)^{\frac{3}{2}}} \int_{\mathbb{R}^3} U(\mathbf{x},\mathbf{n})\, e^{-i\boldsymbol{\omega}\cdot\mathbf{x}}\,\mathrm{d}\mathbf{x}. \tag{18}$$

To the operator Q_α, we associate the corresponding operator $-(-\mathcal{B})^\alpha$ in the spatial Fourier domain by

$$-(-\mathcal{B})^\alpha = \left(\mathcal{F}_{\mathbb{R}^3} \otimes 1_{\mathbb{L}_2(S^2)}\right) \circ Q_\alpha \circ \left(\mathcal{F}^{-1}_{\mathbb{R}^3} \otimes 1_{\mathbb{H}_{2\alpha}(S^2)}\right). \tag{19}$$

Then, direct computations show us that

$$-(-\mathcal{B})^\alpha \overline{U}(\boldsymbol{\omega},\mathbf{n}) = \left[-(-\mathcal{B}_{\boldsymbol{\omega}})^\alpha \overline{U}(\boldsymbol{\omega},\cdot)\right](\mathbf{n}), \text{ for all } \mathbf{n} \in S^2, \tag{20}$$

where, for each fixed $\boldsymbol{\omega} \in \mathbb{R}^3$, the operator $-(-\mathcal{B}_{\boldsymbol{\omega}})^\alpha : \mathbb{H}_{2\alpha}(S^2) \to \mathbb{L}_2(S^2)$ is given by

$$-(-\mathcal{B}_{\boldsymbol{\omega}})^\alpha = -\left(-D_{44}\Delta^{S^2}_{\mathbf{n}} + D_{11}\|\boldsymbol{\omega} \times \mathbf{n}\|^2 + D_{33}(\boldsymbol{\omega}\cdot\mathbf{n})^2\right)^\alpha. \tag{21}$$

In this article, we employ Fourier transform techniques to derive a complete orthonormal basis (ONB) of eigenfunctions

$$\left\{\Phi^{l,m}_{\boldsymbol{\omega}} \mid l \in \mathbb{N}_0, m \in \mathbb{Z} \text{ with } |m| \leq l\right\}, \tag{22}$$

in $\mathbb{L}_2(S^2)$ for the operator $-(-\mathcal{B}_{\boldsymbol{\omega}}) := -(-\mathcal{B}_{\boldsymbol{\omega}})^{\alpha=1}$. Then, clearly, this basis is also an ONB of eigenfunctions for $-(-\mathcal{B}_{\boldsymbol{\omega}})^\alpha$, as we only need to take the fractional power of the eigenvalues. Indeed, once the eigenfunctions in Equation (22) and the eigenvalues

$$\mathcal{B}_{\boldsymbol{\omega}} \Phi^{l,m}_{\boldsymbol{\omega}} = \lambda^{l,m}_r\, \Phi^{l,m}_{\boldsymbol{\omega}}, \text{ with } r = \|\boldsymbol{\omega}\|, \tag{23}$$

are known, the exact solution of Equation (6) is given by (shift-twist) convolution with a probability kernel on $\mathbb{R}^3 \rtimes S^2$. More precisely, the solutions of Equation (6) can be expressed as follows:

$$\begin{aligned} W_\alpha(\mathbf{x},\mathbf{n},t) &= (K^\alpha_t * U)(\mathbf{x},\mathbf{n}) := \int_{S^2}\int_{\mathbb{R}^3} K^\alpha_t(\mathbf{R}^T_{\mathbf{n}'}(\mathbf{x}-\mathbf{x}'), \mathbf{R}^T_{\mathbf{n}'}\mathbf{n})\, U(\mathbf{x}',\mathbf{n}')\,\mathrm{d}\mathbf{x}'\mathrm{d}\mu_{S^2}(\mathbf{n}') \\ &= \int_{\mathbb{R}^3} \sum_{l=0}^{\infty}\sum_{m=-l}^{l} \left\langle \overline{U}(\boldsymbol{\omega},\cdot), \Phi^{l,m}_{\boldsymbol{\omega}}(\cdot) \right\rangle_{\mathbb{L}_2(S^2)} \Phi^{l,m}_{\boldsymbol{\omega}}(\mathbf{n})\, e^{-(-\lambda^{l,m}_r)^\alpha t} e^{i\mathbf{x}\cdot\boldsymbol{\omega}}\mathrm{d}\boldsymbol{\omega}, \end{aligned}$$

with the probability kernel given by (24)

$$\begin{aligned} &K^\alpha_t(\mathbf{x},\mathbf{n}) = \left[\mathcal{F}^{-1}_{\mathbb{R}^3}\left(\overline{K^\alpha_t}(\cdot,\mathbf{n})\right)\right](\mathbf{x}), \\ &\quad \text{with } \overline{K^\alpha_t}(\boldsymbol{\omega},\mathbf{n}) = \sum_{l=0}^{\infty}\sum_{m=-l}^{l} \overline{\Phi^{l,m}_{\boldsymbol{\omega}}(\mathbf{a})}\, \Phi^{l,m}_{\boldsymbol{\omega}}(\mathbf{n})\, e^{-(-\lambda^{l,m}_r)^\alpha t}. \end{aligned}$$

Here, the inner product in $\mathbb{L}_2(S^2)$ is given by

$$\langle y_1(\cdot), y_2(\cdot)\rangle_{\mathbb{L}_2(S^2)} := \int_{S^2} y_1(\mathbf{n})\,\overline{y_2(\mathbf{n})}\,\mathrm{d}\mu_{S^2}(\mathbf{n}). \tag{25}$$

where μ_{S^2} is the usual Lebesgue measure on the sphere S^2.

Remark 5. *The eigenvalues $\lambda_r^{l,m}$ only depend on $r = \|\omega\|$ due to the symmetry $\Phi_{\mathbf{R}\omega}^{l,m}(\mathbf{R}\mathbf{n}) = \Phi_{\omega}^{l,m}(\mathbf{n})$ that one directly recognizes from Equations (21) and (23).*

Remark 6. *The kernels K_t^{α} are the probability density kernels of stable Lévy processes on $\mathbb{R}^3 \rtimes S^2$, see Appendix A.1. Therefore, akin to the $\mathbb{R}^n$-case [61,65], we refer to them as the α-stable Lévy kernels on $\mathbb{R}^3 \rtimes S^2$.*

2. Symmetries of the PDEs of Interest

Next, we employ the PDE formulation in Equation (10) on the group $G = SE(3)$ to summarize the symmetries for the probability kernels $K_t^{\alpha} : \mathbb{R}^3 \rtimes S^2 \to \mathbb{R}^+$. For details, see [40,87].

2.1. PDE Symmetries

Consider the PDE system in Equation (10) on the group $G = SE(3)$. Due to left-invariance (or rather left-covariance) of the PDE, linearity of the map $\tilde{U}(\cdot) \mapsto \tilde{W}_{\alpha}(\cdot, t)$, and the Dunford–Pettis theorem [88], the solutions are obtained by group convolution with a kernel $\tilde{K}_t^{\alpha} \in \mathbb{L}_1(G)$:

$$\tilde{W}_{\alpha}(g,t) = \left(\tilde{K}_t^{\alpha} * \tilde{U}\right)(g) := \int_G \tilde{K}_t^{\alpha}(h^{-1}g)\,\tilde{U}(h)\,\mathrm{d}h, \tag{26}$$

where we take the convention that the probability kernel acts from the left. In the special case, $U = \delta_e$ with unity element $e = (\mathbf{0}, \mathbf{I})$ we get $\tilde{W}_{\alpha}(g,t) = \tilde{K}_t^{\alpha}(g)$.

Thanks to the fundamental relation in Equation (13) that holds in general, we have in particular that

$$\forall_{t\geq 0}\,\forall_{(\mathbf{x},\mathbf{R})\in G} \;:\; \tilde{K}_t^{\alpha}(\mathbf{x},\mathbf{R}) = K_t^{\alpha}(\mathbf{x},\mathbf{R}\mathbf{a}). \tag{27}$$

Furthermore, the PDE system given by Equation (10) is invariant under $\mathcal{A}_i \mapsto -\mathcal{A}_i$, and, since inversion on the Lie algebra corresponds to inversion on the group, the kernels must satisfy

$$\forall_{t\geq 0}\,\forall_{g\in G} \;:\; \tilde{K}_t^{\alpha}(g) = \tilde{K}_t^{\alpha}(g^{-1}), \tag{28}$$

and for the corresponding kernel on the quotient this means

$$\forall_{t\geq 0}\,\forall_{(\mathbf{x},\mathbf{n})\in G/H} \;:\; K_t^{\alpha}(\mathbf{x},\mathbf{n}) = K_t^{\alpha}(-\mathbf{R}_{\mathbf{n}}^T\mathbf{x}, \mathbf{R}_{\mathbf{n}}^T\mathbf{a}). \tag{29}$$

Finally, we see invariance of the PDE with respect to right actions of the subgroup H. This is due to the isotropy of the generator $\tilde{Q}_{\alpha}$ in the tangent subbundles $\text{span}\{\mathcal{A}_1, \mathcal{A}_2\}$ and $\text{span}\{\mathcal{A}_4, \mathcal{A}_5\}$. This due to Equation (A11) in Appendix B. Note that invariance of the kernel with respect to right action of the subgroup H and invariance of the kernel with respect to inversion in Equation (28) also implies invariance of the kernel with respect to left-actions of the subgroup H, since $(g^{-1}(h')^{-1})^{-1} = h'g$ for all $h' \in H$ and $g \in G$. Therefore, we have

$$\begin{aligned} \forall_{t\geq 0}\,\forall_{g\in G}\forall_{h,h'\in H} :&\quad \tilde{K}_t^{\alpha}(gh) &= \tilde{K}_t^{\alpha}(g) = \tilde{K}_t^{\alpha}(h'g), \\ \forall_{t\geq 0}\,\forall_{(\mathbf{x},\mathbf{n})\in G/H}\forall_{\overline{\alpha}\in[0,2\pi)} :&\quad K_t^{\alpha}(\mathbf{x},\mathbf{n}) &= K_t^{\alpha}(\mathbf{R}_{\mathbf{a},\overline{\alpha}}\mathbf{x}, \mathbf{R}_{\mathbf{a},\overline{\alpha}}\mathbf{n}). \end{aligned} \tag{30}$$

Remark 7. *(notations, see also the list of abbreviations at the end of the article)*

To avoid confusion between the Euler angle $\overline{\alpha}$ and the α indexing the α-stable Lévy distribution, we put an overline for this specific angle. Henceforth, $\mathbf{R}_{\mathbf{v},\psi}$ denotes a counter-clockwise rotation over axis **v** *with angle ψ. This applies in particular to the case where the axis is the reference axis $\mathbf{v} = \mathbf{a} = (0,0,1)^T$ and $\psi = \overline{\alpha}$. Recall that $\mathbf{R}_{\mathbf{n}}$ (without an angle in the subscript) denotes any 3D rotation that maps reference axis* **a** *onto* **n**.

We write the symbol $\hat{\cdot}$ above a function to indicate its Fourier transform on G and G/H; we use the symbol $\overline{\cdot}$ for strictly spatial Fourier transform; the symbol $\tilde{\cdot}$ above a function/operator to indicate that it is defined on the group G and the function/operator without symbols when it is defined on the quotient G/H.

2.2. Obtaining the Kernels with $D_{11} > 0$ from the Kernels with $D_{11} = 0$

In ([40], cor.2.5), it was deduced that for $\alpha = 1$ the elliptic diffusion kernel ($D_{11} > 0$) directly follows from the degenerate diffusion kernel ($D_{11} = 0$) in the spatial Fourier domain via

$$\overline{K}_t^{1,\text{elliptic}}(\boldsymbol{\omega},\mathbf{n}) = e^{-r^2 D_{11} t}\,\overline{K}_t^{1,\text{degenerate}}\left(\sqrt{\frac{D_{33}-D_{11}}{D_{33}}}\boldsymbol{\omega},\mathbf{n}\right), \quad \text{with } r = \|\boldsymbol{\omega}\|,\ 0 \leq D_{11} < D_{33}.$$

For the general case $\alpha \in (0,1]$, the transformation from the case $D_{11} = 0$ to the case $D_{11} > 0$ is achieved by replacing $-(-\lambda_r^{l,m})^\alpha \mapsto -(-\lambda_r^{l,m} + r^2 D_{11})^\alpha$ and $r \mapsto r\sqrt{\frac{D_{33}-D_{11}}{D_{33}}}$ in Equation (24) for the kernel. Henceforth, we set $D_{11} = 0$.

3. The Fourier Transform on $SE(3)$

The group $G = SE(3)$ is a unimodular Lie group (of type I) with (left- and right-invariant) Haar measure $\mathrm{d}g = \mathrm{d}\mathbf{x}\mathrm{d}\mu_{SO(3)}(\mathbf{R})$ being the product of the Lebesgue measure on $\mathbb{R}^3$ and the Haar measure $\mu_{SO(3)}$ on $SO(3)$. Then, for all $f \in \mathbb{L}_1(G) \cap \mathbb{L}_2(G)$, the Fourier transform $\mathcal{F}_G f$ is given by Equation (2). For more detailsm see [22,24,26]. One has the inversion formula:

$$f(g) = (\mathcal{F}_G^{-1}\mathcal{F}_G f)(g) = \int_{\hat{G}} \text{trace}\left\{(\mathcal{F}_G f)(\sigma)\,\sigma_g\right\} \mathrm{d}\nu(\sigma) = \int_{\hat{G}} \text{trace}\left\{\hat{f}(\sigma)\,\sigma_g\right\} \mathrm{d}\nu(\sigma). \tag{31}$$

In our Lie group case of $SE(3)$, we identify all unitary irreducible representations $\sigma^{p,s}$ having non-zero dual measure with the pair $(p,s) \in \mathbb{R}^+ \times \mathbb{Z}$. This identification is commonly applied (see, e.g., [4]). Using the method ([26], Thm. 2.1, [25]) of induced representations, all unitary irreducible representations (UIRs) of G, up to equivalence, with non-zero Plancherel measure are given by:

$$\boxed{\begin{aligned} &\sigma = \sigma^{p,s} : SE(3) \to B(\mathbb{L}_2(pS^2)), \qquad p > 0,\ s \in \mathbb{Z}, \\ &\left(\sigma^{p,s}_{(\mathbf{x},\mathbf{R})}\phi\right)(\mathbf{u}) = e^{-i\mathbf{u}\cdot\mathbf{x}}\,\phi\left(\mathbf{R}^{-1}\mathbf{u}\right)\,\Delta_s\left(\mathbf{R}_{\frac{\mathbf{u}}{p}}^{-1}\mathbf{R}\mathbf{R}_{\frac{\mathbf{R}^{-1}\mathbf{u}}{p}}\right), \quad \mathbf{u} \in pS^2,\ \phi \in \mathbb{L}_2(pS^2), \end{aligned}} \tag{32}$$

where pS^2 denotes a 2D sphere of radius $p = \|\mathbf{u}\|$; Δ_s is a unitary irreducible representation of $SO(2)$ (or rather of the stabilizing subgroup $\text{Stab}_{SO(3)}(\mathbf{a}) \subset SO(3)$ isomorphic to $SO(2)$) producing a scalar.

In Equation (32), $\mathbf{R}_{\frac{\mathbf{u}}{p}}$ denotes a rotation that maps **a** onto $\frac{\mathbf{u}}{p}$. Thus, direct computation

$$\mathbf{R}_{\frac{\mathbf{u}}{p}}^{-1}\mathbf{R}\mathbf{R}_{\frac{\mathbf{R}^{-1}\mathbf{u}}{p}}\mathbf{a} = \mathbf{R}_{\frac{\mathbf{u}}{p}}^{-1}\mathbf{R}\mathbf{R}^{-1}\left(\frac{\mathbf{u}}{p}\right) = \mathbf{a}$$

shows us that it is a rotation around the z-axis (recall $\mathbf{a} = \mathbf{e}_z$), e.g. about angle $\overline{\alpha}$. This yields character $\Delta_s\left(\mathbf{R}_{\frac{\mathbf{u}}{p}}^{-1}\mathbf{R}\mathbf{R}_{\frac{\mathbf{R}^{-1}\mathbf{u}}{p}}\right) = e^{-is\overline{\alpha}}$, for details, see ([4], ch.10.6). Thus, we can rewrite Equation (32) as

$$\left(\sigma^{p,s}_{(\mathbf{x},\mathbf{R})}\phi\right)(\mathbf{u}) = e^{-i\,(\mathbf{u}\cdot\mathbf{x}+s\overline{\alpha})}\,\phi(\mathbf{R}^{-1}\mathbf{u}), \qquad \text{where } (\mathbf{x},\mathbf{R}) \in G,\ \mathbf{u} \in pS^2,\ \phi \in \mathbb{L}_2(pS^2).$$

Mackey's theory [25] relates the UIR $\sigma^{p,s}$ to the dual orbits pS^2 of $SO(3)$. Thereby, the dual measure ν can be identified with a measure on the family of dual orbits of $SO(3)$ given by $\{pS^2 \mid p > 0\}$, and

$$\left(\mathcal{F}_G^{-1}\hat{f}\right)(g) = \int_{\hat{G}} \operatorname{trace}\left\{\hat{f}(\sigma^{p,s})\,\sigma_g^{p,s}\right\}\,\mathrm{d}\nu(\sigma^{p,s}) = \int_{\mathbb{R}^+} \operatorname{trace}\left\{\hat{f}(\sigma^{p,s})\,\sigma_g^{p,s}\right\} p^2 \mathrm{d}p,$$

for all $p > 0$, $s \in \mathbb{Z}$. For details, see ([24], ch. 3.6.).

The matrix elements of $\hat{f} = \mathcal{F}_G f$ with respect to an orthonormal basis of modified spherical harmonics $\{Y_s^{l,m}(p^{-1}\cdot)\}$, with $|m|, |s| \leq l$ (see ([4], ch.9.8)) for $\mathbb{L}_2(pS^2)$ are given by

$$\hat{f}^{p,s}_{l,m,l',m'} := \int_G f(g)\,\left\langle \sigma^{p,s}_{g^{-1}} Y_s^{l',m'}(p^{-1}\cdot)\,,\, Y_s^{l,m}(p^{-1}\cdot)\right\rangle_{\mathbb{L}_2(pS^2)}\,\mathrm{d}g, \tag{33}$$

where the $\mathbb{L}_2$ inner product is given by $\langle y_1(\cdot)\,,\,y_2(\cdot)\rangle_{\mathbb{L}_2(pS^2)} := \langle y_1(p\cdot)\,,\,y_2(p\cdot)\rangle_{\mathbb{L}_2(S^2)}$ (recall Equation (25)).

For an explicit formula for the modified spherical harmonics $Y_s^{l,m}$ see [4], where they are denoted by $h^l_{m,s}$. The precise technical analytic expansion of the modified spherical harmonics is not important for this article. The only properties of $Y_s^{l,m}$ that we need are gathered in the next proposition.

Proposition 2. *The modified spherical harmonics $Y_s^{l,m}$ have the following properties:*

(1) For $s = 0$ or $m = 0$, they coincide with standard spherical harmonics $Y^{l,m}$, cf. ([89], eq.4.32):
$Y^{l,m}_{s=0} = Y^{l,m}$ and $Y_s^{l,0} = (-1)^s\, Y^{l,s}$, where $Y^{l,m}(\mathbf{n}(\beta,\gamma)) = \frac{\epsilon_m}{\sqrt{2\pi}}\, P_l^m(\cos\beta)\, e^{im\gamma}$,
with $\mathbf{n}(\beta,\gamma) = (\cos\gamma\sin\beta, \sin\gamma\sin\beta, \cos\beta)^T$, with spherical angles $\beta \in [0,\pi], \gamma \in [0,2\pi)$,
with P_l^m the normalized associated Legendre polynomial and $\epsilon_m = (-1)^{\frac{1}{2}(m+|m|)}$.

(2) They have a specific rotation transformation property in view of Equation (32):
$\sigma^{p,s}_{(\mathbf{0},\mathbf{R})}\, Y_s^{l,m} = \sum\limits_{m'=-l}^{l} \mathcal{D}^l_{m'm}(\mathbf{R})\; Y_s^{l,m'}$, where $\mathcal{D}^l_{m'm}(\cdot)$ denotes the Wigner D-function [90].

(3) For each $s \in \mathbb{Z}$ fixed, they form a complete orthonormal basis for $\mathbb{L}_2(S^2)$:
$\left\langle Y_s^{l,m}(\cdot), Y_s^{l',m'}(\cdot)\right\rangle_{\mathbb{L}_2(S^2)} = \delta^{l,l'}\,\delta^{m,m'}$ for all $m, m' \in \mathbb{Z}$, $l, l' \in \mathbb{N}_0$, with $|m| \leq l, |m'| \leq l'$, $l, l' \geq |s|$.

For details and relation between different Euler angle conventions, see ([4], ch:9.4.1). In our convention of ZYZ-Euler angles (see Appendix B), one has

$$\mathcal{D}^l_{m'm}(\mathbf{R}_{\mathbf{e}_z,\tilde{\alpha}}\mathbf{R}_{\mathbf{e}_y,\beta}\mathbf{R}_{\mathbf{e}_z,\gamma}) = e^{-im'\tilde{\alpha}}\,P^l_{m'm}(\cos\beta)e^{-im\gamma}, \tag{34}$$

with $P^l_{m'm}$ a generalized associated Legendre polynomial given in ([4], eq.9.21).

Moreover, we have inversion formula ([4], Equation 10.46):

$$f(g) = \frac{1}{2\pi^2}\sum_{s\in\mathbb{Z}}\sum_{l'=|s|}^{\infty}\sum_{l=|s|}^{\infty}\sum_{m'=-l'}^{l'}\sum_{m=-l}^{l}\int_0^\infty \hat{f}^{p,s}_{l,m,l',m'}\,\left(\sigma_g^{p,s}\right)_{l',m',l,m}\,p^2\mathrm{d}p, \tag{35}$$

with matrix coefficients (independent of f) given by

$$\left(\sigma_g^{p,s}\right)_{l',m',l,m} = \left\langle \sigma_g^{p,s} Y_s^{l,m}(p^{-1}\cdot)\,,\, Y_s^{l',m'}(p^{-1}\cdot)\right\rangle_{\mathbb{L}_2(pS^2)}. \tag{36}$$

Note that $\sigma^{p,s}$ is a UIR so we have

$$\left(\sigma^{p,s}_{g^{-1}}\right)_{l',m',l,m} = \overline{\left(\sigma_g^{p,s}\right)_{l,m,l',m'}}. \tag{37}$$

4. A Specific Fourier Transform on the Homogeneous Space $\mathbb{R}^3 \rtimes S^2$ of Positions and Orientations

Now that we have introduced the notation of Fourier transform on the Lie group $G = SE(3)$, we define the Fourier transform $\mathcal{F}_{G/H}$ on the homogeneous space $G/H = \mathbb{R}^3 \rtimes S^2$. Afterwards, in the subsequent section, we solve the Forward-Kolmogorov/Fokker–Planck PDEs in Equation (6) via application of this transform, or, more precisely, via conjugation with Fourier transform $\mathcal{F}_{G/H}$.

4.1. The Homogeneous Space $\mathbb{R}^3 \rtimes S^2$

Throughout this manuscript, we rely on a Fourier transform on the homogeneous space of positions and orientations that is defined by the partition of left-cosets: $\mathbb{R}^3 \rtimes S^2 := G/H$, given by Equation (3).

Note that subgroup H can be parameterized as follows:

$$H = \{h_{\overline{\alpha}} := (\mathbf{0}, \mathbf{R}_{\mathbf{a},\overline{\alpha}}) \mid \overline{\alpha} \in [0, 2\pi)\}, \tag{38}$$

where we recall that $\mathbf{R}_{\mathbf{a},\overline{\alpha}}$ denotes a (counter-clockwise) rotation around the reference axis $\mathbf{a} = \mathbf{e}_z$. The reason behind this construction is that the group $SE(3)$ acts transitively on $\mathbb{R}^3 \rtimes S^2$ by $(\mathbf{x}', \mathbf{n}') \mapsto g \odot (\mathbf{x}', \mathbf{n}')$ given by Equation (4). Recall that by the definition of the left-cosets one has

$$H = \{\mathbf{0}\} \times SO(2), \text{ and } g_1 \sim g_2 \Leftrightarrow g_1^{-1} g_2 \in H.$$

The latter equivalence simply means that for $g_1 = (\mathbf{x}_1, \mathbf{R}_1)$ and $g_2 = (\mathbf{x}_2, \mathbf{R}_2)$ one has

$$g_1 \sim g_2 \Leftrightarrow \mathbf{x}_1 = \mathbf{x}_2 \text{ and } \exists_{\overline{\alpha} \in [0,2\pi)} \ : \ \mathbf{R}_1 = \mathbf{R}_2 \mathbf{R}_{\mathbf{a},\overline{\alpha}}.$$

The equivalence classes $[g] = \{g' \in SE(3) \mid g' \sim g\}$ are often just denoted by $(\mathbf{x}, \mathbf{n})$ as they consist of all rigid body motions $g = (\mathbf{x}, \mathbf{R}_{\mathbf{n}})$ that map reference point $(\mathbf{0}, \mathbf{a})$ onto $(\mathbf{x}, \mathbf{n}) \in \mathbb{R}^3 \rtimes S^2$:

$$g \odot (\mathbf{0}, \mathbf{a}) = (\mathbf{x}, \mathbf{n}), \tag{39}$$

where we recall $\mathbf{R}_{\mathbf{n}}$ is *any* rotation that maps $\mathbf{a} \in S^2$ onto $\mathbf{n} \in S^2$.

4.2. Fourier Transform on $\mathbb{R}^3 \rtimes S^2$

Now we can define the Fourier transform $\mathcal{F}_{G/H}$ on the homogeneous space G/H. Prior to this, we specify a class of functions where this transform acts.

Definition 5. *Let $p > 0$ be fixed and $s \in \mathbb{Z}$. We denote*

$$\mathbb{L}_2^{sym}(pS^2) = \left\{ f \in \mathbb{L}_2(pS^2) \;\middle|\; \forall_{\overline{\alpha} \in [0,2\pi)} \ \sigma^{p,s}_{h_{\overline{\alpha}}} f = f \right\}$$

the subspace of spherical functions that have the prescribed axial symmetry, with respect to the subgroup H (recall Equation (38)).

Definition 6. *We denote the orthogonal projection from $\mathbb{L}_2(pS^2)$ onto the closed subspace $\mathbb{L}_2^{sym}(pS^2)$ by $\mathbb{P}_p^{sym}$.*

Definition 7. *To the group representation $\sigma^{p,s} : SE(3) \to B(\mathbb{L}_2(pS^2))$ given by Equation (32), we relate a "representation" $\overline{\sigma}^{p,s} : \mathbb{R}^3 \rtimes S^2 \to B(\mathbb{L}_2(pS^2))$ on $\mathbb{R}^3 \rtimes S^2$, defined by*

$$\overline{\sigma}^{p,s}_{[g]} := \frac{1}{(2\pi)^2} \int_0^{2\pi} \int_0^{2\pi} \sigma^{p,s}_{h_{\tilde{\alpha}} g h_{\overline{\alpha}}} \, \mathrm{d}\overline{\alpha} \mathrm{d}\tilde{\alpha} = \mathbb{P}_p^{sym} \circ \sigma_g^{p,s} \circ \mathbb{P}_p^{sym}. \tag{40}$$

Definition 8. *A function* $\tilde{U} : G \to \mathbb{C}$ *is called axially symmetric if*

$$\tilde{U}(\mathbf{x}, \mathbf{R}) = \tilde{U}(\mathbf{x}, \mathbf{R}\mathbf{R}_{\mathbf{a},\overline{\alpha}}) \quad \text{for all } \overline{\alpha} \in [0, 2\pi) \text{ and all } (\mathbf{x}, \mathbf{R}) \in G. \tag{41}$$

To each function $U : G/H \to \mathbb{C}$, we relate an axially symmetric function $\tilde{U} : G \to \mathbb{C}$ by

$$\tilde{U}(\mathbf{x}, \mathbf{R}) := U(\mathbf{x}, \mathbf{R}\mathbf{a}). \tag{42}$$

Definition 9. *We define the Fourier transform of function U on* $G/H = \mathbb{R}^3 \rtimes S^2$ *by*

$$\boxed{\hat{U}(\overline{\sigma}^{p,s}) = (\mathcal{F}_{G/H}U)(\overline{\sigma}^{p,s}) := \mathbb{P}_p^{sym} \circ \mathcal{F}_G\tilde{U}(\sigma^{p,s}) \circ \mathbb{P}_p^{sym}.} \tag{43}$$

Standard properties of the Fourier transform $\mathcal{F}_G$ on $SE(3)$ such as the Plancherel theorem and the inversion formula [4,26] naturally carry over to $\mathcal{F}_{G/H}$ with "simpler formulas". This is done by a domain and range restriction via the projection operators $\mathbb{P}_p^{sym}$ in Equation (43). The reason for the specific construction Equation (43) becomes clear from the next lemmas, and the "simpler formulas" for the Plancherel and inversion formulas are then summarized in a subsequent theorem, where we constrain ourselves to the case $m = m' = 0$ in the formulas. The operator $\mathbb{P}_p^{sym}$ that is most right in Equation (43) constrains the basis $Y_s^{l,m}$ to $m = 0$, whereas the operator $\mathbb{P}_p^{sym}$ that is most left in Equation (43) constrains the basis $Y_s^{l',m'}$ to $m' = 0$.

Lemma 1. *(axial symmetry) Let* $\tilde{U} : G \to \mathbb{C}$ *be axially symmetric. Then,*

1. *it relates to a unique function* $U : G/H \to \mathbb{C}$ *via* $U(\mathbf{x}, \mathbf{n}) = \tilde{U}(\mathbf{x}, \mathbf{R}_\mathbf{n})$;
2. *the matrix coefficients*

$$\hat{\tilde{U}}^{p,s}_{l,m,l',m'} = \left[\mathcal{F}_G\tilde{U}(\sigma^{p,s})\right]_{l,m,l',m'} \text{ of linear operator } \mathcal{F}_G\tilde{U}(\sigma^{p,s})$$

   relative to the modified spherical harmonic basis $\{Y_s^{l,m}\}$ *vanish if* $m \neq 0$; *and*
3. *the matrix coefficients*

$$\hat{U}^{p,s}_{l,m,l',m'} = [\mathcal{F}_{G/H}U(\overline{\sigma}^{p,s})]_{l,m,l',m'} \text{ of linear operator } \mathcal{F}_{G/H}U(\overline{\sigma}^{p,s})$$

   relative to the modified spherical harmonic basis $\{Y_s^{l,m}\}$ *vanish if* $m \neq 0$ *or* $m' \neq 0$.

Conversely, if $\tilde{U} = \mathcal{F}_G^{-1}(\hat{\tilde{U}})$ *and*

$$\forall_{p>0}\forall_{l\in\mathbb{N}_0}\forall_{s\in\mathbb{Z},\text{ with } |s|\leq l}\ \forall_{m'\in\mathbb{Z},\text{ with } |m'|\leq l}\ \forall_{m\neq 0}\ :\ \hat{\tilde{U}}^{p,s}_{l,m,l',m'} = 0, \tag{44}$$

then $\tilde{U}$ *satisfies the axial symmetry in Equation (41).*

Proof. Item 1: Uniqueness of U follows by the fact that the choice of $\mathbf{R}_\mathbf{n}$ of some rotation that maps $\mathbf{a}$ onto $\mathbf{n}$ does not matter. Indeed, $U(\mathbf{x}, \mathbf{n}) = \tilde{U}(\mathbf{x}, \mathbf{R}_\mathbf{n}\mathbf{R}_{\mathbf{a},\overline{\alpha}}) = \tilde{U}(\mathbf{x}, \mathbf{R}_\mathbf{n})$.

Item 2: Assumption Equation (41) can be rewritten as $\tilde{U}(g) = \tilde{U}(gh_{\overline{\alpha}})$ for all $h_{\overline{\alpha}} \in H$, $g \in G$. This gives:

$$\begin{aligned}
\hat{U}^{p,s}_{l,m,l',m'} &= \left\langle (\mathcal{F}_G\tilde{U})(Y^{l',m'}_s(p^{-1}\cdot)), Y^{l,m}_s(p^{-1}\cdot) \right\rangle_{\mathbb{L}_2(pS^2)} \\
&= \int_G \tilde{U}(g) \left\langle \sigma^{p,s}_{g^{-1}} Y^{l',m'}_s(p^{-1}\cdot), Y^{l,m}_s(p^{-1}\cdot) \right\rangle_{\mathbb{L}_2(pS^2)} \mathrm{d}g \\
&= \int_G \tilde{U}(g) \left\langle Y^{l',m'}_s(p^{-1}\cdot), \sigma^{p,s}_g Y^{l,m}_s(p^{-1}\cdot) \right\rangle_{\mathbb{L}_2(pS^2)} \mathrm{d}g \\
&= \int_G \tilde{U}(gh_{\overline{\alpha}}) \left\langle Y^{l',m'}_s(p^{-1}\cdot), \sigma^{p,s}_{gh_{\overline{\alpha}}} Y^{l,m}_s(p^{-1}\cdot) \right\rangle_{\mathbb{L}_2(pS^2)} \mathrm{d}(gh_{\overline{\alpha}}) \\
&= \int_G \tilde{U}(g) \left\langle Y^{l',m'}_s(p^{-1}\cdot), \sigma^{p,s}_g \circ \sigma^{p,s}_{h_{\overline{\alpha}}} Y^{l,m}_s(p^{-1}\cdot) \right\rangle_{\mathbb{L}_2(pS^2)} \mathrm{d}(gh_{\overline{\alpha}}) \\
&= e^{-im\overline{\alpha}}\, \hat{U}^{p,s}_{l,m,l',m'} \text{ for all } \overline{\alpha} \in [0, 2\pi),
\end{aligned} \tag{45}$$

where we recall that σ is a UIR and that the Haar measure on G is bi-invariant. In the first step, we used the third property, whereas in the final step we used the second property of Proposition 2 together with

$$\mathcal{D}^l_{m'm}(\mathbf{R}_{\mathbf{a},\overline{\alpha}}) = e^{-im\overline{\alpha}}\delta_{m'm} \text{ so that } \sigma^{p,s}_{h_{\overline{\alpha}}} Y^{l,m}_s(p^{-1}\cdot) = e^{-im\overline{\alpha}} Y^{l,m}_s(p^{-1}\cdot). \tag{46}$$

We conclude that $(1 - e^{-im\overline{\alpha}})\hat{U}^{p,s}_{l,m,l',m'} = 0$ for all $\overline{\alpha} \in [0, 2\pi)$ so $m \neq 0 \Rightarrow \hat{U}^{p,s}_{l,m,l',m'} = 0$.

Item 3: Due to the second property in Proposition 2, we have

$$\sigma^{p,s}_{(\mathbf{0},\mathbf{R})} Y^{l,m}_s(p^{-1}\cdot) = \sum_{m'=-l}^{l} \mathcal{D}^l_{m'm}(\mathbf{R})\; Y^{l,m'}_s(p^{-1}\cdot).$$

Thereby, the projection $\mathbb{P}^{sym}_p$ is given by

$$\mathbb{P}^{sym}_p \left(\sum_{l=0}^{\infty} \sum_{m=-l}^{l} \alpha_{l,m} Y^{l,m}_s \right) = \sum_{l=0}^{\infty} \alpha_{l,0} Y^{l,0}_s. \tag{47}$$

Now, the projection $\mathbb{P}^{sym}_p$ that is applied first in Equation (43) filters out $m = 0$ as the only possible nonzero component. The second projection filters out $m' = 0$ as the only possible nonzero component.

Conversely, if Equation (44) holds, one has by inversion Equation (35) that

$$\tilde{U}(g) = \frac{1}{2\pi^2} \sum_{s\in\mathbb{Z}} \sum_{l=|s|}^{\infty} \sum_{l'=|s|}^{\infty} \sum_{m'=-l'}^{l'} \int_0^{\infty} \hat{U}^{p,s}_{l,0,l',m'} \left(\sigma^{p,s}_g\right)_{l',m',l,0} p^2 \mathrm{d}p,$$

so then the final result follows by the identity

$$\left(\sigma^{p,s}_{gh_{\overline{\alpha}}}\right)_{l',m',l,0} = \left(\sigma^{p,s}_g\right)_{l',m',l,0}. \tag{48}$$

Thus, it remains to show why Equation (48) holds. It is due to $\sigma^{p,s}_{(\mathbf{x},\mathbf{R})} = \sigma^{p,s}_{(\mathbf{x},\mathbf{I})} \circ \sigma^{p,s}_{(\mathbf{0},\mathbf{R})}$ and Equation (46), as one has

$$\sigma^{p,s}_{gh_{\overline{\alpha}}} = \sigma^{p,s}_{(\mathbf{x},\mathbf{R})(\mathbf{0},\mathbf{R}_{\mathbf{a},\overline{\alpha}})} = \sigma^{p,s}_{(\mathbf{x},\mathbf{R}\mathbf{R}_{\mathbf{a},\overline{\alpha}})} = \sigma^{p,s}_{(\mathbf{x},\mathbf{R})} \circ \sigma^{p,s}_{(\mathbf{0},\mathbf{R}_{\mathbf{a},\overline{\alpha}})}, \text{ and } Y^{l,0}_s(p^{-1}\mathbf{R}^{-1}_{\mathbf{a},\overline{\alpha}}\cdot) = Y^{l,0}_s(p^{-1}\cdot) \tag{49}$$

and thereby Equation (48) follows by Equation (36). □

Lemma 2. *If $\tilde{K} \in \mathbb{L}_2(G)$ is real-valued and satisfies the axial symmetry in Equation (41), and moreover the following holds*

$$\tilde{K}(g^{-1}) = \tilde{K}(g) \tag{50}$$

then the Fourier coefficients satisfy $\hat{K}^{p,s}_{l,m,l',m'} = \overline{\hat{K}^{p,s}_{l',m',l,m}}$ *and they vanish for* $m \neq 0$ *and for* $m' \neq 0$.

Proof. The proof follows by Equation (37) and inversion invariance of the Haar measure on G (see [86]). $\square$

The next lemma shows that Equation (50) is a sufficient but not a necessary condition for the Fourier coefficients to vanish for both the cases $m' \neq 0$ and $m \neq 0$.

Lemma 3. *Let* $\tilde{K} \in \mathbb{L}_2(G)$ *and* $K \in \mathbb{L}_2(G/H)$ *be related by Equation (42). Then, we have the following equivalences:*

$$\begin{array}{c} K(\mathbf{x},\mathbf{n}) = K(\mathbf{R}_{\mathbf{a},\overline{\alpha}}\mathbf{x}, \mathbf{R}_{\mathbf{a},\overline{\alpha}}\mathbf{n}), \quad \text{for all } \overline{\alpha} \in [0,2\pi), (\mathbf{x},\mathbf{n}) \in G/H \\ \Updownarrow \\ \tilde{K}(gh) = \tilde{K}(g) = \tilde{K}(hg), \quad \text{for all } g \in G,\ h \in H \\ \Updownarrow \\ \text{The Fourier coefficients } \hat{K}^{p,s}_{l,m,l',m'} \text{ vanish for } m \neq 0 \text{ and for } m' \neq 0. \end{array} \tag{51}$$

Proof. We show $a \Rightarrow b \Rightarrow c \Rightarrow a$ to get $a \Leftrightarrow b \Leftrightarrow c$.
$a \Rightarrow b$: Denoting $h = h_{\overline{\alpha}} = (\mathbf{0}, \mathbf{R}_{\mathbf{a},\overline{\alpha}})$, $g = (\mathbf{x}, \mathbf{R})$, we have

$$\begin{aligned} \forall_{\overline{\alpha},\overline{\alpha}' \in [0,2\pi)} \forall_{\mathbf{x} \in \mathbb{R}^3} \forall_{\mathbf{R} \in SO(3)} \ : \ \tilde{K}(gh_{\overline{\alpha}}) &= \tilde{K}(\mathbf{x}, \mathbf{R}\mathbf{R}_{\mathbf{a},\overline{\alpha}}) = K(\mathbf{x}, \mathbf{R}\mathbf{R}_{\mathbf{a},\overline{\alpha}}\mathbf{a}) = K(\mathbf{x}, \mathbf{R}\mathbf{a}) = \tilde{K}(\mathbf{x}, \mathbf{R}) = \tilde{K}(g) \\ &= K(\mathbf{R}_{\mathbf{a},\overline{\alpha}}\mathbf{x}, \mathbf{R}_{\mathbf{a},\overline{\alpha}}\mathbf{R}\mathbf{a}) = \tilde{K}(\mathbf{R}_{\mathbf{a},\overline{\alpha}}\mathbf{x}, \mathbf{R}_{\mathbf{a},\overline{\alpha}}\mathbf{R}) = \tilde{K}(h_{\overline{\alpha}}g). \end{aligned}$$

$b \Rightarrow c$: By Lemma 1, we know that the Fourier coefficients vanish for $m \neq 0$. Next, we show they also vanish for $m' \neq 0$. Similar to Equation (49) we have

$$\sigma^{p,s}_{h_{\overline{\alpha}}g} = \sigma^{p,s}_{(\mathbf{R}_{\mathbf{a},\overline{\alpha}}\mathbf{x}, \mathbf{R}_{\mathbf{a},\overline{\alpha}}\mathbf{R})} = \sigma^{p,s}_{(\mathbf{R}_{\mathbf{a},\overline{\alpha}}\mathbf{x}, \mathbf{I})} \circ \sigma^{p,s}_{(\mathbf{0}, \mathbf{R}_{\mathbf{a},\overline{\alpha}}\mathbf{R})}, \tag{52}$$

which gives the following relation for the matrix-coefficients:

$$\begin{aligned} \left(\sigma^{p,s}_{g=(\mathbf{x},\mathbf{R})}\right)_{l',m',l,m} &= \sum_{j=-l}^{l} \left\langle \sigma^{p,s}_{(\mathbf{x},\mathbf{I})} Y^{l,j}_s(p^{-1}\cdot), Y^{l',m'}_s(p^{-1}\cdot) \right\rangle_{\mathbb{L}_2(pS^2)} \mathcal{D}^l_{jm}(\mathbf{R}) \quad \Rightarrow \\ \left(\sigma^{p,s}_{h_{\overline{\alpha}}g}\right)_{l',m',l,m} &= \sum_{j=-l}^{l} e^{-i(m'-j)\overline{\alpha}} \left\langle \sigma^{p,s}_{(\mathbf{x},\mathbf{I})} Y^{l,j}_s(p^{-1}\cdot), Y^{l',m'}_s(p^{-1}\cdot) \right\rangle_{\mathbb{L}_2(pS^2)} e^{-ij\overline{\alpha}} \mathcal{D}^l_{jm}(\mathbf{R}) \Rightarrow \\ \left(\sigma^{p,s}_{h_{\overline{\alpha}}g}\right)_{l',m',l,m} &= e^{-im'\overline{\alpha}} \left(\sigma^{p,s}_g\right)_{l',m',l,m}. \end{aligned} \tag{53}$$

The implication can be directly verified by Proposition 2, Equations (34) and (52), and

$$\begin{aligned} \left\langle Y^{l',m'}_s(p^{-1}\cdot), \sigma^{p,s}_{(\mathbf{R}_{\mathbf{a},\overline{\alpha}}\mathbf{x},\mathbf{I})} Y^{l,j}_s(p^{-1}\cdot) \right\rangle_{\mathbb{L}_2(pS^2)} &= \int_{pS^2} e^{-ip(\mathbf{x}\cdot\mathbf{R}^T_{\mathbf{a},\overline{\alpha}}\mathbf{u})}\, Y^{l,j}_s(\mathbf{u})\, \overline{Y^{l',m'}_s(\mathbf{u})}\, \mathrm{d}\mu_{pS^2}(\mathbf{u}) \\ &= \int_{pS^2} e^{-ip(\mathbf{x}\cdot\mathbf{v})} Y^{l,j}_s(\mathbf{R}_{\mathbf{a},\overline{\alpha}}\mathbf{v})\, \overline{Y^{l',m'}_s(\mathbf{R}_{\mathbf{a},\overline{\alpha}}\mathbf{v})}\, \mathrm{d}\mu_{pS^2}(\mathbf{v}). \end{aligned}$$

From Equation (53), we deduce that:

$$\begin{aligned} \hat{K}^{p,s}_{l,m,l',m'} &= \int_G \tilde{K}(g) \left\langle \sigma^{p,s}_g Y^{l,m}_s(p^{-1}\cdot), Y^{l',m'}_s(p^{-1}\cdot) \right\rangle_{\mathbb{L}_2(pS^2)} \mathrm{d}g \\ &= \int_G \tilde{K}(h_{\overline{\alpha}}g) \left\langle \sigma^{p,s}_{h_{\overline{\alpha}}g} Y^{l,m}_s(p^{-1}\cdot), Y^{l',m'}_s(p^{-1}\cdot) \right\rangle_{\mathbb{L}_2(pS^2)} \mathrm{d}(h_{\overline{\alpha}}g) \\ &= \int_G \tilde{K}(g) \left\langle \sigma^{p,s}_g Y^{l,m}_s(p^{-1}\cdot), \sigma^{p,s}_{h^{-1}_{\overline{\alpha}}} Y^{l',m'}_s(p^{-1}\cdot) \right\rangle_{\mathbb{L}_2(pS^2)} \mathrm{d}g = e^{+im'\overline{\alpha}}\, \hat{K}^{p,s}_{l,m,l',m'}, \end{aligned}$$

which holds for all $\overline{\alpha} \in [0, 2\pi)$. Thereby, if $m' \neq 0$, then $\hat{K}^{p,s}_{l,m,l',m'} = 0$.

$c \Rightarrow a$: By inversion of Equation (35), where the only contributing terms have $m = 0$ and $m' = 0$, we see that $\tilde{K}(gh) = \tilde{K}(hg) = \tilde{K}(g)$ for all $h = (\mathbf{0}, \mathbf{R}_{\mathbf{a},\overline{\alpha}})$. Thereby, $\tilde{K}$ is axially symmetric and by Lemma 1 it relates to a unique kernel on G/H via $K(\mathbf{x}, \mathbf{n}) = \tilde{K}(\mathbf{x}, \mathbf{R}_{\mathbf{n}})$ and the result follows by Equation (30). □

Now that we have characterized all functions $K \in \mathbb{L}_2(G/H)$ for which the Fourier coefficients $\hat{K}^{p,s}_{l,m,l',m'}$ vanish for $m \neq 0$ and $m' \neq 0$ in the above lemma, we considerably simplify the inversion and Plancherel formula for Fourier transform $\mathcal{F}_G$ on the group $G = SE(3)$ to the Fourier transform $\mathcal{F}_{G/H}$ on the homogeneous space $G/H = \mathbb{R}^3 \rtimes S^2$ in the next theorem. This is important to our objective of deriving the kernels for the linear PDEs in Equation (6) that we address in the next section.

Theorem 1. *(matrix-representation for $\mathcal{F}_{G/H}$, explicit inversion and Plancherel formula)*
Let $K \in \mathbb{L}_2^{sym}(G/H)$ and $\tilde{K} \in \mathbb{L}_2(G)$ be related by Equation (42). Then, the matrix elements of $\mathcal{F}_{G/H}K$ are given by

$$\begin{aligned}
\hat{K}^{p,s}_{l',0,l,0} &= \int_G \tilde{K}(g) \left(\sigma^{p,s}_{g^{-1}}\right)_{l',0,l,0} \mathrm{d}g\,, \\
\text{with } \left(\sigma^{p,s}_{g}\right)_{l',0,l,0} &= \sum_{j=-l}^{l} [l',0 \mid p,s \mid l,j]\,(\mathbf{x})\, \mathcal{D}^{l}_{j0}(\mathbf{R}) \quad \text{for all } g = (\mathbf{x}, \mathbf{R}) \in G.
\end{aligned}$$

The constants $[l',0 \mid p,s \mid l,j]\,(\mathbf{x}) := \left\langle \sigma^{p,s}_{(\mathbf{x},\mathbf{I})} Y^{l,j}_s(p^{-1}\cdot), Y^{l',0}_s(p^{-1}\cdot) \right\rangle_{\mathbb{L}_2(pS^2)}$ *admit an analytic expression in terms of elementary functions ([4], Equation10.34) and the Wigner D-functions in Equation (34).*
Furthermore, we have the following Plancherel and inversion formula:

$$\begin{aligned}
\|K\|^2_{\mathbb{L}_2(G/H)} &= \|\mathcal{F}_{G/H}K\|^2 = \sum_{s\in\mathbb{Z}} \int_{\mathbb{R}^+} |||\, (\mathcal{F}_{G/H}K)\,(\overline{\sigma}^{p,s}) |||^2\, p^2 \mathrm{d}p = \int_{\mathbb{R}^+} \sum_{s=-\infty}^{\infty} \sum_{l'=|s|}^{\infty} \sum_{l=|s|}^{\infty} |\hat{K}^{p,s}_{l,0,l',0}|^2\, p^2 \mathrm{d}p, \\
K(\mathbf{x}, \mathbf{n}) &= \left(\mathcal{F}^{-1}_{G/H}\mathcal{F}_{G/H}K\right)(\mathbf{x}, \mathbf{n}) = \sum_{s\in\mathbb{Z}} \int_{\mathbb{R}^+} \text{trace}\left\{ (\mathcal{F}_{G/H}K)(\overline{\sigma}^{p,s})\, \overline{\sigma}^{p,s}_{(\mathbf{x},\mathbf{n})} \right\} p^2 \mathrm{d}p \\
&= \frac{1}{2\pi^2} \sum_{s\in\mathbb{Z}} \sum_{l'=|s|}^{\infty} \sum_{l=|s|}^{\infty} \int_{\mathbb{R}^+} \hat{K}^{p,s}_{l,0,l',0} \left(\overline{\sigma}^{p,s}_{(\mathbf{x},\mathbf{n})}\right)_{l',0,l,0} p^2 \mathrm{d}p,
\end{aligned}$$

with matrix coefficients given by (for analytic formulas, see ([4], eq.10.35))

$$\begin{aligned}
\left(\overline{\sigma}^{p,s}_{(\mathbf{x},\mathbf{n})}\right)_{l',0,l,0} &= \left(\sigma^{p,s}_{g}\right)_{l',0,l,0} = \left\langle \sigma^{p,s}_{g} Y^{l,0}_s(p^{-1}\cdot), Y^{l',0}_s(p^{-1}\cdot) \right\rangle_{\mathbb{L}_2(pS^2)} \\
&= \left\langle \sigma^{p,s}_{g} Y^{l,s}(p^{-1}\cdot), Y^{l',s}(p^{-1}\cdot) \right\rangle_{\mathbb{L}_2(pS^2)} \quad \text{for } g = (\mathbf{x}, \mathbf{R}_{\mathbf{n}}).
\end{aligned} \tag{54}$$

Proof. The above formulas are a direct consequence of Lemma 3 and the Plancherel and inversion formulas (see [4], ch:10.8, [26]) for Fourier transform on $SE(3)$. Recall that a coordinate-free definition of $\overline{\sigma}^{p,s}$ is given in Equation (40). Its matrix coefficients are given by Equation (54), where we recall the first item of Proposition 2 and where we note that they are independent on the choice of $\mathbf{R}_{\mathbf{n}} \in SO(3)$ mapping $\mathbf{a}$ onto $\mathbf{n}$. □

Corollary 1. *Let $K_1, K_2 \in \mathbb{L}_2^{sym}(G/H)$. Then, for shift-twist convolution on $G/H = \mathbb{R}^3 \rtimes S^2$ given by*

$$(K_1 * K_2)(\mathbf{x}, \mathbf{n}) = \int_{S^2}\int_{\mathbb{R}^3} K_1(\mathbf{R}^T_{\mathbf{n}'}(\mathbf{x} - \mathbf{x}'), \mathbf{R}^T_{\mathbf{n}'}\mathbf{n})\, K_2(\mathbf{x}', \mathbf{n}')\, \mathrm{d}\mathbf{x}' \mathrm{d}\mu_{S^2}(\mathbf{n}')$$

we have $\mathcal{F}_{G/H}(K_1 * K_2) = (\mathcal{F}_{G/H}K_1) \circ (\mathcal{F}_{G/H}K_2)$.

Proof. Set $\tilde{K}_1(g) = K_1(g \odot (\mathbf{0}, \mathbf{a}))$. Standard Fourier theory [5] gives $\mathcal{F}_G(\widetilde{K_1 * K_2}) = \mathcal{F}_G(\tilde{K}_1 * \tilde{K}_2)$, so

$$\begin{aligned}
\mathcal{F}_{G/H}(K_1 * K_2) &\stackrel{\text{def}}{=} \mathbb{P}_p^{sym} \circ \mathcal{F}_G(\widetilde{K_1 * K_2}) \circ \mathbb{P}_p^{sym} \\
&= \mathbb{P}_p^{sym} \circ \mathcal{F}_G(\tilde{K}_1) \circ \mathcal{F}_G(\tilde{K}_2) \circ \mathbb{P}_p^{sym} \\
&= \mathbb{P}_p^{sym} \circ \mathcal{F}_G(\tilde{K}_1) \circ \mathbb{P}_p^{sym} \circ \mathbb{P}_p^{sym} \circ \mathcal{F}_G(\tilde{K}_2) \circ \mathbb{P}_p^{sym} \\
&= (\mathcal{F}_{G/H}K_1) \circ (\mathcal{F}_{G/H}K_2),
\end{aligned}$$

where the first equality is given by Equation (43) and the third equality follows by Lemma 3 and Equation (47). □

5. Application of the Fourier Transform on $\mathbb{R}^3 \rtimes S^2$ for Explicit Solutions of the Fokker–Planck PDEs of α-stable Lévy Processes on $\mathbb{R}^3 \rtimes S^2$

Our objective is to solve the PDE system in Equation (6) on the homogeneous space of positions and orientations G/H. Recall that we extended this PDE system to G in Equation (10). As the cases $D_{11} > 0$ follow from the case $D_{11} = 0$ (recall Section 2.2), we consider the case $D_{11} = 0$ in this section. From the symmetry consideration in Section 2, it follows that the solution of Equation (10) is given by $\tilde{W}_\alpha(g,t) = (\tilde{K}_t^\alpha * \tilde{U})(g)$ with a probability kernel $\tilde{K}_t^\alpha : G \to \mathbb{R}^+$, whereas the solution of Equation (6) is given by

$$W_\alpha(\mathbf{x}, \mathbf{n}, t) = (K_t^\alpha * U)(\mathbf{x}, \mathbf{n}) := \int_{S^2} \int_{\mathbb{R}^3} K_t^\alpha(\mathbf{R}_{\mathbf{n}'}^T(\mathbf{x} - \mathbf{x}'), \mathbf{R}_{\mathbf{n}'}^T \mathbf{n}) \; U(\mathbf{x}', \mathbf{n}') \; \mathrm{d}\mathbf{x}' \mathrm{d}\mu_{S^2}(\mathbf{n}'),$$

where the kernels K_t^α are invariant with respect to left-actions of the subgroup H (recall Equation (30)). This invariance means that the condition for application of the Fourier transform $\mathcal{F}_{G/H}$ on $\mathbb{R}^3 \rtimes S^2$ is satisfied (recall Lemma 3) and we can indeed employ Theorem 1 to keep all our computations, spectral decompositions and Fourier transforms in the 5D homogeneous space $\mathbb{R}^3 \rtimes S^2 = G/H$ rather than a technical and less direct approach [40] in the 6D group $G = SE(3)$.

Remark 8. *For the underlying probability theory, and sample paths of discrete random walks of the α-Stable Lévy stochastic processes, we refer to Appendix A. To get a general impression of how Monte Carlo simulations of such stochastic processes can be used to approximate the exact probability kernels K_t^α, see Figure 1. In essence, such a stochastic approximation is computed by binning the endpoints of the random walks. A brief mathematical explanation follows in Section 5.2.*

For now, let us ignore the probability theory details and let us first focus on deriving exact analytic solutions to Equation (6) and its kernel K_t^α via Fourier transform $\mathcal{F}_{G/H}$ on $G/H = \mathbb{R}^3 \rtimes S^2$.

5.1. Exact Kernel Representations by Spectral Decomposition in the Fourier Domain

Let us consider the evolution in Equation (6) for α-stable Lévy process on the quotient $G/H = \mathbb{R}^3 \rtimes S^2$. Then, the mapping from the initial condition $W(\cdot, 0) = U(\cdot) \in \mathbb{L}_2(G/H)$ to the solution $W(\cdot, t)$ at a fixed time $t \geq 0$ is a bounded linear mapping. It gives rise to a strongly continuous (holomorphic) semigroup [66]. We conveniently denote the bounded linear operator on $\mathbb{L}_2(G/H)$ as follows:

$$W_\alpha(\cdot, t) = (e^{tQ_\alpha}U)(\cdot), \qquad \text{for all } t \geq 0. \tag{55}$$

In the next main theorem, we provide a spectral decomposition of the operator using both a direct sum and a direct integral decomposition. Note that definitions of direct integral decompositions (and the underlying measure theory) can be found in ([24], ch:3.3 and 3.4).

5.1.1. Eigenfunctions and Preliminaries

To formulate the main theorem, we need some preliminaries and formalities. First, let us define $\overline{\mathcal{F}}_{\mathbb{R}^3} : \mathbb{L}_2(\mathbb{R}^3 \rtimes S^2) \to \mathbb{L}_2(\mathbb{R}^3 \rtimes S^2)$ by

$$(\overline{\mathcal{F}}_{\mathbb{R}^3} U)(\boldsymbol{\omega}, \mathbf{n}) := [\mathcal{F}_{\mathbb{R}^3} U(\cdot, \mathbf{n})](\boldsymbol{\omega}). \tag{56}$$

Recalling Equation (19), we re-express the generator in the spatial Fourier domain:

$$\begin{aligned} -(-\mathcal{B})^{\alpha} &= \overline{\mathcal{F}}_{\mathbb{R}^3} \circ Q_{\alpha} \circ \overline{\mathcal{F}}_{\mathbb{R}^3}^{-1} \Rightarrow \\ -(-\mathcal{B}_{\boldsymbol{\omega}})^{\alpha} &= -\left(-D_{33}\,(i\boldsymbol{\omega}\cdot\mathbf{n})^2 - D_{44}\,\Delta_{\mathbf{n}}^{S^2}\right)^{\alpha} \\ &= -\left(D_{33}\,r^2 \left(\mathbf{a}\cdot(\mathbf{R}_{r^{-1}\boldsymbol{\omega}}^T\mathbf{n})\right)^2 - D_{44}\,\Delta_{\mathbf{n}}^{S^2}\right)^{\alpha} \\ &= -\left(D_{33}\,r^2\cos^2(\beta^{\boldsymbol{\omega}}) - D_{44}\,\Delta_{\mathbf{n}}^{S^2}\right)^{\alpha}, \quad \text{with } r = \|\boldsymbol{\omega}\|, \alpha \in (0,1], \end{aligned} \tag{57}$$

where $\beta^{\boldsymbol{\omega}}$ denotes the angle between $\mathbf{n}$ and $r^{-1}\boldsymbol{\omega}$ (see Figure 2). This re-expression is the main reason for the following definitions.

Instead of the modified spherical Harmonics $Y_s^{l,m}$ in Proposition 2, which are commonly used as a standard basis to represent each operator in the Fourier transform on $SE(3)$, we use our generalized spherical harmonics, depending on a spatial frequency vector, as this is in accordance with Equation (57).

Definition 10. *Let $l \in \mathbb{N}_0$. Let $m \in \mathbb{Z}$ such that $|m| \leq l$. Let $\boldsymbol{\omega} \in \mathbb{R}^3$ be a frequency vector. We define*

$$Y_{\boldsymbol{\omega}}^{l,m}(\mathbf{n}) = Y^{l,m}(\mathbf{R}_{r^{-1}\boldsymbol{\omega}}^T \mathbf{n}), \qquad \text{with } r = \|\boldsymbol{\omega}\|, \ \mathbf{n} \in S^2, \tag{58}$$

where we take the rotation which maps $\mathbf{a}$ onto $r^{-1}\boldsymbol{\omega}$ whose matrix representation in the standard basis is:

$$\mathbf{R}_{r^{-1}\boldsymbol{\omega}} = \left(\frac{(\boldsymbol{\omega}\times\mathbf{a})\times\boldsymbol{\omega}}{\|(\boldsymbol{\omega}\times\mathbf{a})\times\boldsymbol{\omega}\|} \;\middle|\; \frac{\boldsymbol{\omega}\times\mathbf{a}}{\|\boldsymbol{\omega}\times\mathbf{a}\|} \;\middle|\; r^{-1}\boldsymbol{\omega} \right) \text{ for } r^{-1}\boldsymbol{\omega} \neq \mathbf{a}, \text{ and } \mathbf{R}_{\mathbf{a}} = \mathbf{I}, \text{ and } \mathbf{R}_{\mathbf{0}} = \mathbf{I}.$$

Recall the standard spherical angle formula $\mathbf{n}(\beta,\gamma) = (\sin\beta\cos\gamma, \sin\beta\sin\gamma, \cos\beta)^T$ from Proposition 2. These are Euler-angles relative to the reference axis $\mathbf{a} = \mathbf{e}_z$. For the Euler-angles relative to the (normalized) frequency $r^{-1}\boldsymbol{\omega}$ one has (see also Figure 2):

$$\mathbf{n}^{\boldsymbol{\omega}}(\beta^{\boldsymbol{\omega}}, \gamma^{\boldsymbol{\omega}}) = \mathbf{R}_{r^{-1}\boldsymbol{\omega}} \mathbf{n}(\beta^{\boldsymbol{\omega}}, \gamma^{\boldsymbol{\omega}}). \tag{59}$$

Definition 11. *Let $l \in \mathbb{N}_0$. Let $m \in \mathbb{Z}$ such that $|m| \leq l$. We define the functions $\Phi_{\boldsymbol{\omega}}^{l,m} \in \mathbb{L}_2(S^2)$ by*

$$\Phi_{\boldsymbol{\omega}}^{l,m}(\mathbf{n}) = \sum_{j=0}^{\infty} \frac{d_j^{l,m}(r)}{\|\mathbf{d}^{l,m}(r)\|} \, Y_{\boldsymbol{\omega}}^{|m|+j,m}(\mathbf{n}), \tag{60}$$

where $r = \|\boldsymbol{\omega}\|$ and $\mathbf{d}^{l,m}(r) := \left(d_j^{l,m}(r)\right)_{j=0}^{\infty}$ are coefficients such that

$$\Phi_{\boldsymbol{\omega}}^{l,m}(\mathbf{n}^{\boldsymbol{\omega}}(\beta^{\boldsymbol{\omega}}, \gamma^{\boldsymbol{\omega}})) = S_{\rho}^{l,m}(\cos\beta^{\boldsymbol{\omega}}) \, \frac{e^{im\gamma^{\boldsymbol{\omega}}}}{\sqrt{2\pi}}, \text{ with } \rho = r\sqrt{\frac{D_{33}}{D_{44}}},$$

where $S_{\rho}^{l,m}(\cdot)$ denotes the $\mathbb{L}_2$-normalized spheroidal wave function.

Remark 9. *The spheroidal wave function arises from application of the method of separation on operator $\mathcal{B}_\omega$ in Equation (57) where basic computations (for details, see [40]) lead to the following singular Sturm-Liouville problem:*

$$(Ly)(x) = \frac{d}{dx}\left[p(x)\frac{dy(x)}{dx}\right] + q(x)y(x) = -\lambda(r)\, y(x), \quad x = \cos\beta^\omega \in [-1,1]. \tag{61}$$

with $p(x) = (1-x^2)$, $q(x) = -\rho^2 x^2 - \frac{m^2}{1-x^2}$, and again $\rho = r\sqrt{D_{33}/D_{44}}$. In this formulation, $p(x)$ vanishes at the boundary of the interval, which makes our problem a singular Sturm–Liouville problem. It is sufficient to require boundedness of the solution and its derivative at the boundary points to have nonnegative, distinct, simple eigenvalues $\lambda_r^{l,m}$ and existence of a countable, complete orthonormal basis of eigenfunctions $\{y_j\}_{j=0}^\infty$ [91] for the spheroidal wave equation.

As a result, standard Sturm–Liouville theory (that applies the spectral decomposition theorem for compact self-adjoint operators to a kernel operator that is the right-inverse of L) provides us (for each $\boldsymbol{\omega}$ fixed) a complete orthonormal basis of eigenfunctions $\{\Phi_\omega^{l,m}\}$ in $\mathbb{L}_2(S^2)$ with eigenvalues of our (unbounded) generators:

$$-(-\mathcal{B}_\omega)^\alpha\, \Phi_\omega^{l,m} = -(-\lambda_r^{l,m})^\alpha\, \Phi_\omega^{l,m}, \quad \textit{for all } |m| \leq l. \tag{62}$$

Remark 10. *Define $\mathcal{Y}_{l,m}(\beta,\gamma) := Y^{l,m}(\mathbf{n}(\beta,\gamma))$. Then, Equations (58) and (59) imply $Y_\omega^{l,m}(\mathbf{n}^\omega(\beta^\omega,\gamma^\omega)) = \mathcal{Y}_{l,m}(\beta^\omega,\gamma^\omega)$.*

Remark 11. *The matrix-representation of $-(-\mathcal{B}_\omega)^\alpha$ with respect to orthonormal basis $\left\{Y_\omega^{|m|+j,m}\right\}_{j\in\mathbb{N}_0, m\in\mathbb{Z}}$ equals*

$$\bigoplus_{m\in\mathbb{Z}} -(D_{33}r^2\mathbf{M}^m + D_{44}\boldsymbol{\Lambda}^m)^\alpha,$$

where $\boldsymbol{\Lambda}^m := \mathrm{diag}\{l(l+1)\}_{l=|m|}^\infty = \mathrm{diag}\{(|m|+j)(|m|+j+1)\}_{j=0}^\infty$, $r = \|\boldsymbol{\omega}\|$ and where $\mathbf{M}^m$ is the tri-diagonal matrix (that can be computed analytically ([40], eq. 106)) given by

$$(\cos\beta)^2 Y^{|m|+j,m}(\mathbf{n}(\beta,\gamma)) = \sum_{j'=0}^{\infty}\left((\mathbf{M}^m)^T\right)_{j,j'} Y^{|m|+j',m}(\mathbf{n}(\beta,\gamma)). \tag{63}$$

As a result, we see from Equations (60) and (62) that the coefficients $\mathbf{d}^{l,m}(r)$ for our eigenfunctions are eigenvectors of a matrix

$$-\left(D_{33}r^2\mathbf{M}^m + D_{44}\boldsymbol{\Lambda}^m\right)\mathbf{d}^{l,m}(r) = \lambda_r^{l,m}\mathbf{d}^{l,m}(r), \qquad \textit{for } l \geq |m|. \tag{64}$$

This matrix (and its diagonalization) play a central role for our main spectral decomposition theorem both in the spatial Fourier domain and in the Fourier domain of the homogeneous space of positions and orientations.

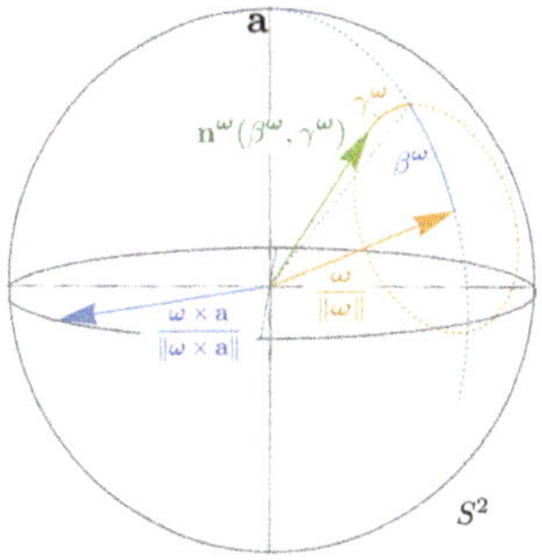

Figure 2. For $\omega \neq \mathbf{a}$, we parameterize every orientation $\mathbf{n}$ (green) by rotations around $r^{-1}\omega$ (orange) and $\frac{\omega \times \mathbf{a}}{\|\omega \times \mathbf{a}\|}$ (blue). In other words, $\mathbf{n}^{\omega}(\beta^{\omega}, \gamma^{\omega}) = \mathbf{R}_{r^{-1}\omega, \gamma^{\omega}} \mathbf{R}_{\frac{\omega \times \mathbf{a}}{\|\omega \times \mathbf{a}\|}, \beta^{\omega}}(r^{-1}\omega)$.

5.1.2. The Explicit Spectral Decomposition of the Evolution Operators

In Theorem 2, we present the explicit spectral decompositions both in the spatial Fourier domain and in the Fourier domain of the homogeneous space of positions and orientations.

Prior to this theorem, we explain the challenges that appear when we apply $\mathcal{F}_{G/H}$ to the PDE of interest in Equation (6) on the quotient G/H. To get a grip on the evolution operator and the corresponding kernel, we set the initial condition equal to a delta distribution at the origin, i.e., we consider

$$U = \delta_{(\mathbf{0},\mathbf{a})} \Rightarrow W_{\alpha}(\cdot, t) = e^{tQ_{\alpha}} U = e^{-t(-Q)^{\alpha}} \delta_{(\mathbf{0},\mathbf{a})} = K_t^{\alpha}.$$

In this case, the necessary condition in Equation (51) in Lemma 3 for application of $\mathcal{F}_{G/H}$ is indeed satisfied, due to the symmetry property of the kernel, given by Equation (30). Now, due to linearity

$$\mathcal{F}_{G/H} \circ e^{tQ_{\alpha}} \circ \mathcal{F}_{G/H}^{-1} = e^{t\left(\mathcal{F}_{G/H} \circ Q_{\alpha} \circ \mathcal{F}_{G/H}^{-1}\right)},$$

we just need to study the generator in the Fourier domain.

For the moment, we set $\alpha = 1$ (the degenerate diffusion case) and return to the general case later on (recall Sections 1.6 and 2.2). Then, it follows that (for details, see ([40], App.D))

$$\begin{aligned} &\left(\mathcal{F}_{G/H} \circ Q \circ \mathcal{F}_{G/H}^{-1} \hat{K}_t^1\right)(\overline{\sigma}^{p,s}) = \left(-D_{33}\,(\mathbf{a} \cdot \mathbf{u})^2 + D_{44}\,\Delta_{\mathbf{u}}^{pS^2}\right) \hat{K}_t^1(\overline{\sigma}^{p,s}), \\ &\text{with the kernel } \hat{K}_t^1 := \mathcal{F}_{G/H} K_t^1(\cdot). \end{aligned} \tag{65}$$

Here, $\Delta_{\mathbf{u}}^{pS^2}$ denotes the Laplace–Beltrami operator on a sphere $pS^2 = \{\mathbf{u} \in \mathbb{R}^3 \,|\, \|\mathbf{u}\| = p\}$ of radius $p > 0$.

We recall that $\mathbf{u} \in pS^2$ is the variable of the functions on which $\overline{\sigma}^{p,s}$ acts. Recalling Equation (32), the first part in the righthand side of Equation (65) denotes a multiplier operator $\mathcal{M}$ given by

$$(\mathcal{M}\phi)(\mathbf{u}) := -\,(\mathbf{a} \cdot \mathbf{u})^2 \phi(\mathbf{u}), \text{ for all } \phi \in \mathbb{L}_2(pS^2), \text{ and almost every } \mathbf{u} \in pS^2.$$

As a result, we obtain the following PDE system for $\hat{K}_t^{\alpha}$ (now for general $\alpha \in (0,1]$):

$$\boxed{\left\{ \begin{array}{ll} \frac{\partial}{\partial t} \hat{K}_t^{\alpha}(\overline{\sigma}^{p,s}) & = -\left(-D_{33}\,\mathcal{M} - D_{44}\,\Delta_{\mathbf{u}}^{pS^2}\right)^{\alpha} \hat{K}_t^{\alpha}(\overline{\sigma}^{p,s}) \\ \hat{K}_0^{\alpha}(\overline{\sigma}^{p,s}) & = 1_{\mathbb{L}_2(pS^2)}. \end{array} \right.}$$

Remark 12. *There is a striking analogy between the operators* $\mathcal{F}_{G/H} \circ Q_\alpha \circ \mathcal{F}_{G/H}^{-1}$ *and* $\overline{\mathcal{F}}_{\mathbb{R}^3} \circ Q_\alpha \circ \overline{\mathcal{F}}_{\mathbb{R}^3}^{-1}$ *given by Equation (57), where the role of* $r\mathbf{R}^T_{\omega/r}\mathbf{n}$ *corresponds to* $\mathbf{u}$. *This correspondence ensures that the multipliers of the multiplier operators in the generator coincide and that the roles of* p *and* r *coincide:*

$$\mathbf{u} = r\mathbf{R}^T_{r^{-1}\omega}\mathbf{n} \Rightarrow (\mathbf{a}\cdot\mathbf{u})^2 = r^2(\mathbf{R}^T_{r^{-1}\omega}\mathbf{a}\cdot\mathbf{n})^2 = (\omega\cdot\mathbf{n})^2 \quad \text{and} \quad \|\mathbf{u}\| = p = r = \|\omega\|.$$

Lemma 4. *Let* $t \geq 0$ *and* $p > 0$ *be fixed. The matrix-representation of operator* $e^{t(D_{33}\mathcal{M}+D_{44}\Delta_{\mathbf{u}}^{pS^2})} : \mathbb{L}_2(pS^2) \to \mathbb{L}_2(pS^2)$ *with respect to the orthonormal basis of spherical harmonics* $\left\{Y^{l=|s|+j\,,\,s}(p^{-1}\cdot)\right\}_{j\in\mathbb{N}_0\,,\,s\in\mathbb{Z}}$ *equals*

$$\bigoplus_{s\in\mathbb{Z}} e^{-t\,(D_{33}p^2\mathbf{M}^s+D_{44}\mathbf{\Lambda}^s)}\,. \tag{66}$$

Proof. Recall Equation (63) that defines matrix $\mathbf{M}^m$ (for analytic formulas of this tri-diagonal matrix, see [40]). This may be re-written as follows:

$$(\mathbf{a}\cdot\mathbf{n})^2 Y^{|m|+j,m}(\mathbf{n}) = \sum_{j'=0}^{\infty}\left((\mathbf{M}^m)^T\right)_{j,j'} Y^{|m|+j',m}(\mathbf{n}).$$

Now, fix $s \in \mathbb{Z}$ and set $m = s$ and $\mathbf{n} = p^{-1}\mathbf{u}$ and we have:

$$\left\langle \left(D_{33}\mathcal{M} + D_{44}\,\Delta_{\mathbf{u}}^{pS^2}\right) Y^{l,s}(p^{-1}\cdot)\,,\, Y^{l',s}(p^{-1}\cdot)\right\rangle_{\mathbb{L}_2(pS^2)} = -p^2\,D_{33}\,(\mathbf{M}^s)_{j',j} - D_{44}l(l+1)\delta_{jj'},$$

where again $l = |s| + j$, $l' = |s| + j'$ and $j, j' \in \mathbb{N}_0$.

Finally, we note that operator $D_{33}\,\mathcal{M} + D_{44}\,\Delta_{\mathbf{u}}^{pS^2}$ is negative definite and maps each subspace $\text{span}\left\{\{Y^{l,s}(p^{-1}\cdot)\}_{l=|s|}^{\infty}\right\}$ for fixed $s \in \mathbb{Z}$ onto itself, which explains direct sum decomposition in Equation (66). □

Next, we formulate the main result, where we apply a standard identification of tensors $\mathbf{a}\otimes\mathbf{b}$ with linear maps:

$$\mathbf{x} \mapsto (\mathbf{a}\otimes\mathbf{b})(\mathbf{x}) = \langle \mathbf{x}\,,\,\mathbf{b}\rangle\,\mathbf{a}. \tag{67}$$

Theorem 2. *We have the following spectral decompositions for the Forward-Kolomogorov evolution operator of* α*-stable Lévy-processes onthe homogeneous space* $G/H = \mathbb{R}^3 \rtimes S^2$:

- *In the Fourier domain of the homogeneous space of positions and orientations, we have:*

$$\boxed{\begin{aligned}&\mathcal{F}_{G/H}\circ e^{-t(-Q)^\alpha}\circ\mathcal{F}_{G/H}^{-1}\\ &= \int_{\mathbb{R}^+}^{\oplus}\bigoplus_{s\in\mathbb{Z}}\sum_{l,l'=|s|}^{\infty}\left[e^{-(D_{33}p^2\mathbf{M}^s+D_{44}\mathbf{\Lambda}^s)^\alpha t}\right]_{l,l'}\left(Y^{l,s}(p^{-1}\cdot)\otimes Y^{l',s}(p^{-1}\cdot)\right)\,p^2\mathrm{d}p\\ &= \int_{\mathbb{R}^+}^{\oplus}\bigoplus_{s\in\mathbb{Z}}\sum_{l=|s|}^{\infty} e^{-(-\lambda_p^{l,s})^\alpha t}\left(\Phi^{l,s}_{p\mathbf{a}}(p^{-1}\cdot)\otimes\Phi^{l,s}_{p\mathbf{a}}(p^{-1}\cdot)\right)\,p^2\mathrm{d}p\end{aligned}} \tag{68}$$

- *In the spatial Fourier domain, we have*

$$\boxed{\begin{aligned}&\left(\overline{\mathcal{F}}_{\mathbb{R}^3}\circ e^{-t(-Q)^\alpha}\circ\overline{\mathcal{F}}_{\mathbb{R}^3}^{-1}\overline{U}\right)(\omega,\cdot) = \overline{W}(\omega,\cdot,t)\\ &= \sum_{m\in\mathbb{Z}}\sum_{l,l'=|m|}^{\infty}\left[e^{-(D_{33}r^2\mathbf{M}^m+D_{44}\mathbf{\Lambda}^m)^\alpha t}\right]_{l,l'}\left(Y_\omega^{l,m}\otimes Y_\omega^{l',m}\right)(\overline{U}(\omega,\cdot))\\ &= \sum_{m\in\mathbb{Z}}\sum_{l=|m|}^{\infty} e^{-(-\lambda_r^{l,m})^\alpha t}\left(\Phi_\omega^{l,m}\otimes\Phi_\omega^{l,m}\right)(\overline{U}(\omega,\cdot))\end{aligned}} \tag{69}$$

where $\overline{W}(\omega,\cdot,t) = \overline{\mathcal{F}}_{\mathbb{R}^3} W(\omega,\cdot,t)$ *and* $\overline{U}(\omega,\cdot) = \overline{\mathcal{F}}_{\mathbb{R}^3} U(\omega,\cdot)$ *(recall Equation (56)).*

In both cases, the normalized eigenfunctions $\Phi_\omega^{l,m}$ *are given by Equation (60) in Definition 11. The eigenvalues* $\lambda_r^{l,m}$ *are the eigenvalues of the spheroidal wave equation, as explained in Remark 9.*

Proof. The first identity in Equation (68) follows by:

$$\begin{aligned}
\mathcal{F}_{G/H} \circ e^{-t(-Q)^\alpha} \circ \mathcal{F}_{G/H}^{-1} &= e^{t\left(\mathcal{F}_{G/H} \circ -(-Q)^\alpha \circ \mathcal{F}_{G/H}^{-1}\right)} \\
&\overset{\text{([40], App.D) and Theorem 1}}{=} \int_{\mathbb{R}^+}^{\oplus} e^{-t\left(-D_{33}\mathcal{M} + D_{44}\Delta_{\mathbf{u}}^{pS^2}\right)^\alpha} p^2 \mathrm{d}p \\
&\overset{\text{Lemma 4 and Theorem 1}}{=} \int_{\mathbb{R}^+}^{\oplus} \bigoplus_{s\in\mathbb{Z}} \sum_{l,l'=|s|}^{\infty} \left[e^{-t\left(D_{33}p^2\mathbf{M}^s + D_{44}\mathbf{\Lambda}^s\right)^\alpha}\right]_{l,l'} \left(Y^{l,s}(p^{-1}\cdot) \otimes Y^{l',s}(p^{-1}\cdot)\right) p^2 \mathrm{d}p \\
&\overset{(60)}{=} \int_{\mathbb{R}^+}^{\oplus} \bigoplus_{s\in\mathbb{Z}} \sum_{l=|s|}^{\infty} e^{-(-\lambda_p^{l,s})^\alpha t} \left(\Phi_{p\mathbf{a}}^{l,s}(p^{-1}\cdot) \otimes \Phi_{p\mathbf{a}}^{l,s}(p^{-1}\cdot)\right) p^2 \mathrm{d}p\,.
\end{aligned}$$

In the last equality, we use the fact that $\Phi_{\mathbf{a}}^{l,m} = Y^{l,m}$. By applying the identification in Equation (67), one observes that Equation (69) is a reformulation of Equation (24), was already been derived for $\alpha = 1$ in previous work by the first author with J.M. Portegies ([40], Thm.2.3 and Equation31). The key idea behind the derivation, the expansion and the completeness of the eigenfunctions $\{\Phi_\omega^{l,m}\}$ is summarized in Remark 9. The general case $\alpha \in (0,1]$ then directly follows by Section 1.6. □

Recently, exact formulas for the (degenerate) heat-kernels on $G = SE(3)$ and on $G/H = \mathbb{R}^3 \rtimes S^2$ (i.e., the case $\alpha = 1$) have been published in [40]. In the next theorem:

- We extend these results to the kernels of PDE in Equation (6), which are Forward-Kolmogorov equations of α-stable Lévy process with $\alpha \in (0,1]$.
- We provide a structured alternative formula via the transform $\mathcal{F}_{G/H}$ characterized in Theorem 1.

Theorem 3. *We have the following formulas for the probability kernels of α-stable Lévy processes on* $\mathbb{R}^3 \rtimes S^2$*:*

- *Via conjugation with* $\mathcal{F}_{\mathbb{R}^3 \rtimes S^2}$*:*

$$\boxed{K_t^\alpha(\mathbf{x},\mathbf{n}) = \frac{1}{(2\pi)^2} \int_0^\infty \sum_{s\in\mathbb{Z}} \sum_{l=|s|}^{\infty} e^{-(-\lambda_p^{l,s})^\alpha t} \left[\overline{\sigma}_{(\mathbf{x},\mathbf{n})}^{p,s}\right]_{l,0,l,0} p^2 \mathrm{d}p,} \tag{70}$$

where $\left[\overline{\sigma}_{(\mathbf{x},\mathbf{n})}^{p,s}\right]_{l,0,l,0} = \left\langle \sigma_{(\mathbf{x},\mathbf{R}_\mathbf{n})}^{p,s} \Phi_{p\mathbf{a}}^{l,s}(p^{-1}\cdot),\, \Phi_{p\mathbf{a}}^{l,s}(p^{-1}\cdot)\right\rangle_{\mathbb{L}_2(pS^2)}$ *can be derived analytically (see ([86], Rem. 16)).*

- *Via conjugation with* $\overline{\mathcal{F}}_{\mathbb{R}^3}$*:*

$$\boxed{K_t^\alpha(\mathbf{x},\mathbf{n}) = \frac{1}{(2\pi)^3} \int_{\mathbb{R}^3} \left(\sum_{l=0}^{\infty} \sum_{m=-l}^{l} e^{-(-\lambda_{\|\omega\|}^{l,m})^\alpha t}\, \overline{\Phi_\omega^{l,m}(\mathbf{a})}\, \Phi_\omega^{l,m}(\mathbf{n})\right) e^{i\mathbf{x}\cdot\omega}\, \mathrm{d}\omega.} \tag{71}$$

Proof. Equation (70) follows by

$$K_t^\alpha(\mathbf{x},\mathbf{n}) = (e^{tQ_\alpha}\delta_{(\mathbf{0},\mathbf{a})})(\mathbf{x},\mathbf{n}) = \left(\mathcal{F}_{G/H}^{-1} \circ e^{t\mathcal{F}_{G/H} \circ Q_\alpha \circ \mathcal{F}_{G/H}^{-1}} \circ \mathcal{F}_{G/H}\delta_{(\mathbf{0},\mathbf{a})}\right)(\mathbf{x},\mathbf{n}).$$

Now, $(\mathcal{F}_{G/H}\delta_{(\mathbf{0},\mathbf{a})})(\sigma^{p,s}) = 1_{\mathbb{L}_2(pS^2)}$ implies $((\mathcal{F}_{G/H}\delta)(\sigma^{p,s})_{(\mathbf{0},\mathbf{a})})(\sigma^{p,s}))_{l,0,l',0} = \delta_{ll'}$ so that the result follows by setting $U = \delta_{(\mathbf{0},\mathbf{a})}$ (or, more precisely, by taking U a sequence that is a bounded approximation of the unity centered around $(\mathbf{0},\mathbf{a})$) in Theorem 2, where we recall the inversion formula from the first part of Theorem 1.

Equation (71) follows similarly by

$$K_t^\alpha(\mathbf{x},\mathbf{n}) = \left(e^{tQ_\alpha}\delta_{(\mathbf{0},\mathbf{a})}\right)(\mathbf{x},\mathbf{n}) = \left(\overline{\mathcal{F}}_{\mathbb{R}^3}^{-1} \circ e^{t\overline{\mathcal{F}}_{\mathbb{R}^3}\circ Q_\alpha \circ \overline{\mathcal{F}}_{\mathbb{R}^3}^{-1}} \circ \overline{\mathcal{F}}_{\mathbb{R}^3}\delta_{(\mathbf{0},\mathbf{a})}\right)(\mathbf{x},\mathbf{n}).$$

Now, $\left(\overline{\mathcal{F}}_{\mathbb{R}^3}\delta_{(\mathbf{0},\mathbf{a})}\right)(\sigma^{p,s}) = \frac{1}{(2\pi)^{\frac{3}{2}}}\delta_{\mathbf{a}}$ and the result follows from the second part of Theorem 1 (again by taking U a sequence that is a bounded approximation of the unity centered around $(\mathbf{0},\mathbf{a})$). □

Remark 13. *There also exist Gaussian estimates for the heat kernel $K_t^{\alpha=1}$ that use a weighted modulus on the logarithm on G, [92]. Such Gaussian estimates can account for the quotient structure G/H [87], and can be reasonably close (cf. [44], Figure 4.4, [93]) to the exact solutions for practical parameter settings in applications [48,94,95].*

5.2. Monte-Carlo Approximations of the Kernels

A stochastic approximation for the kernel K_t^α is computed by binning the endpoints of discrete random walks simulating α-stable processes on the quotient $\mathbb{R}^3 \rtimes S^2$ that we explain next. Let us first consider the case $\alpha = 1$. For $M \in \mathbb{N}$ fixed, we have the discretization

$$\begin{cases} \mathbf{X}_M = \mathbf{X}_0 + \sum\limits_{k=1}^{M} \sqrt{\frac{tD_{33}}{M}}\epsilon_k \mathbf{N}_{k-1}, \\ \mathbf{N}_M = \left(\prod\limits_{k=1}^{M} \mathbf{R}_{\mathbf{a},\gamma_k}\mathbf{R}_{\mathbf{e}_y,\beta_k\sqrt{\frac{tD_{44}}{M}}}\right)\mathbf{N}_0 = \left(\mathbf{R}_{\mathbf{a},\gamma_M}\mathbf{R}_{\mathbf{e}_y,\beta_M\sqrt{\frac{tD_{44}}{M}}} \circ \ldots \circ \mathbf{R}_{\mathbf{a},\gamma_1}\mathbf{R}_{\mathbf{e}_y,\beta_1\sqrt{\frac{tD_{44}}{M}}}\right)\mathbf{N}_0, \end{cases} \tag{72}$$

with $\epsilon_k \sim G_{t=1}^{\mathbb{R}} \sim \mathcal{N}(0, \sigma = \sqrt{2})$ stochastically independent Gaussian distributed on $\mathbb{R}$ with $t = 1$; with uniformly distributed $\gamma_k \sim \mathrm{Unif}(\mathbb{R}/(2\pi\mathbb{Z}) \equiv [-\pi,\pi))$; and $\beta_k \sim g$, where $g : \mathbb{R} \to \mathbb{R}^+$ equals $g(r) = \frac{|r|}{2}e^{-\frac{r^2}{4}}$ in view of the theory of isotropic stochastic processes on Riemannian manifolds by Pinsky [96]. By the central limit theorem for independently distributed variables *with finite variance* it is only the variances of the distributions for the random variables g and $G_{t=1}^{\mathbb{R}}$ that matter. One may also take

$$\epsilon_k \sim \sqrt{3}\,\mathrm{Unif}\left[-\tfrac{1}{2},\tfrac{1}{2}\right] \text{ and } \beta_k \sim \sqrt{6}\,\mathrm{Unif}\left[-\tfrac{1}{2},\tfrac{1}{2}\right] \text{ or } \epsilon_k \sim G_{t=1}^{\mathbb{R}} \text{ and } \beta_k \sim G_{t=2}^{\mathbb{R}}.$$

These processes are implemented recursively; for technical details and background, see Appendix A.

Proposition 3. *The discretization of Equation (72) can be re-expressed, up to order $\frac{1}{M}$ for $M \gg 0$, as follows:*

$$(\mathbf{X}_M, \mathbf{N}_M) \sim \mathbf{G}_M \odot (\mathbf{0},\mathbf{a}), \text{ with } \mathbf{G}_M = \left(\prod_{k=1}^{M} e^{\sum\limits_{i=3}^{5}\sqrt{\frac{tD_{ii}}{M}}\epsilon_k^i A_i}\right)\mathbf{G}_0, \tag{73}$$

with $\epsilon_k^i \sim G_{t=1}^{\mathbb{R}}$ stochastically independent normally distributed variables with $t = \frac{1}{2}\sigma^2 = 1$, and $D_{44} = D_{55}$.

Proof. In our construction, β_k and γ_k can be seen as the polar radius and the polar angle (on a periodic square $[-\pi,\pi] \times [-\pi,\pi]$) of a Gaussian process with $t = 1$ on a plane spanned by rotational generators A_4 and A_5. The key ingredient to obtain Equation (73) from Equation (72) is given by the following relation:

$$e^{u\cos vA_5 - u\sin vA_4} = e^{vA_6}e^{uA_5}e^{-vA_6}, \text{ for all } u,v \in \mathbb{R}, \tag{74}$$

which we use for $u = \beta_k\sqrt{\frac{tD_{44}}{M}}$ and $v = \gamma_k\sqrt{\frac{tD_{44}}{M}}$.

The second ingredient is given by the Campbell–Baker–Hausdorff–Dynkin formula:

$$\text{for all } a_i = O(\frac{1}{\sqrt{M}}) \text{ and for } M \text{ large, we have } e^{a_3 A_3} e^{a_4 A_4} e^{a_5 A_5} = e^{(a_3 A_3 + a_4 A_4 + a_5 A_5)(1+O(\frac{1}{M}))},$$

that allows to decompose the stochastic process in $SE(3)$ into its spatial and angular parts. □

For the binning, we divide $\mathbb{R}^3$ into cubes c_{ijk}, $i, j, k \in \mathbb{Z}$, of size $\Delta s \times \Delta s \times \Delta s$:

$$c_{ijk} := \left[(i-\frac{1}{2})\Delta s, (i+\frac{1}{2})\Delta s\right] \times \left[(j-\frac{1}{2})\Delta s, (j+\frac{1}{2})\Delta s\right] \times \left[(k-\frac{1}{2})\Delta s, (k+\frac{1}{2})\Delta s\right]. \tag{75}$$

We divide S^2 into bins B_l, $l = \{1, \dots, b\}$ for $b \in \mathbb{N}$, with surface area σ_{B_l} and maximal surface area σ_B. The number of random walks in a simulation with traveling time t that have their end point $\mathbf{x}_M \in c_{ijk}$ with their orientation $\mathbf{n}_M \in B_l$ is denoted with $\#_t^{ijkl}$. Furthermore, we define the indicator function

$$1_{c_{ijk},B_l}(\mathbf{x},\mathbf{n}) := \begin{cases} 1 & \mathbf{x} \in c_{ijk}, \mathbf{n} \in B_l, \\ 0 & \text{otherwise.} \end{cases}$$

When the number of paths $N \to \infty$, the number of steps in each path $M \to \infty$ and the bin sizes tend to zero, the obtained distribution converges to the exact kernel:

$$\boxed{\begin{array}{l} \lim\limits_{N\to\infty} \lim\limits_{\Delta s,\sigma_B \to 0} \lim\limits_{M\to\infty} p_t^{\Delta s,\sigma_B,N,M}(\mathbf{x},\mathbf{n}) = K_t^{\alpha=1}(\mathbf{x},\mathbf{n}), \\ \text{with } p_t^{\Delta s,\sigma_B,N,M}(\mathbf{x},\mathbf{n}) = \sum\limits_{l=1}^{b} \sum\limits_{i,j,k\in\mathbb{Z}} 1_{c_{i,j,k},B_l}(\mathbf{x},\mathbf{n}) \frac{\#_t^{ijkl}}{M(\Delta s)^3 \sigma_{B_l}}. \end{array}} \tag{76}$$

The convergence is illustrated in Figure 3.

Monte-Carlo Simulation for $\alpha \in (0,1]$.

Let $q_{t,\alpha} : \mathbb{R}^+ \to \mathbb{R}^+$ be the temporal probability density given by the inverse Laplace transform

$$\begin{array}{l} q_{t,\alpha}(\tau) = \mathcal{L}^{-1}\left(\lambda \to e^{-t\lambda^\alpha}\right)(\tau), \text{ with in particular:} \\ \text{for } \alpha = \frac{1}{2} \text{ it is } q_{t,\frac{1}{2}}(\tau) = \frac{t}{2\tau\sqrt{\pi\tau}} e^{-\frac{t^2}{4\tau}}, \\ \text{for } \alpha \uparrow 1 \text{ we find } q_{t,\alpha}(\cdot) \to \delta_t \text{ in distributional sense}. \end{array} \tag{77}$$

For explicit formulas in the general case $\alpha \in (0,1]$, see [66]. Then, one can deduce from Theorem 3 that

$$K_t^\alpha(\mathbf{x},\mathbf{n}) = \int_0^\infty q_{t,\alpha}(\tau)\, K_\tau^{\alpha=1}(\mathbf{x},\mathbf{n})\, d\tau. \tag{78}$$

This allows us to directly use the Monte-Carlo simulations for the diffusion kernel $\alpha = 1$ for several time instances to compute a Monte-Carlo simulation of the α-stable Lévy kernels for $\alpha \in (0,1]$. To this end, we replace the Monte Carlo approximation in Equation (76) for $\alpha = 1$ in the above Equation (78). See Figure 4, where we compare the diffusion kernel $K_t^{\alpha=1}$ to the Poisson kernel $K_t^{\alpha=\frac{1}{2}}$. See also Appendix A.2.1.

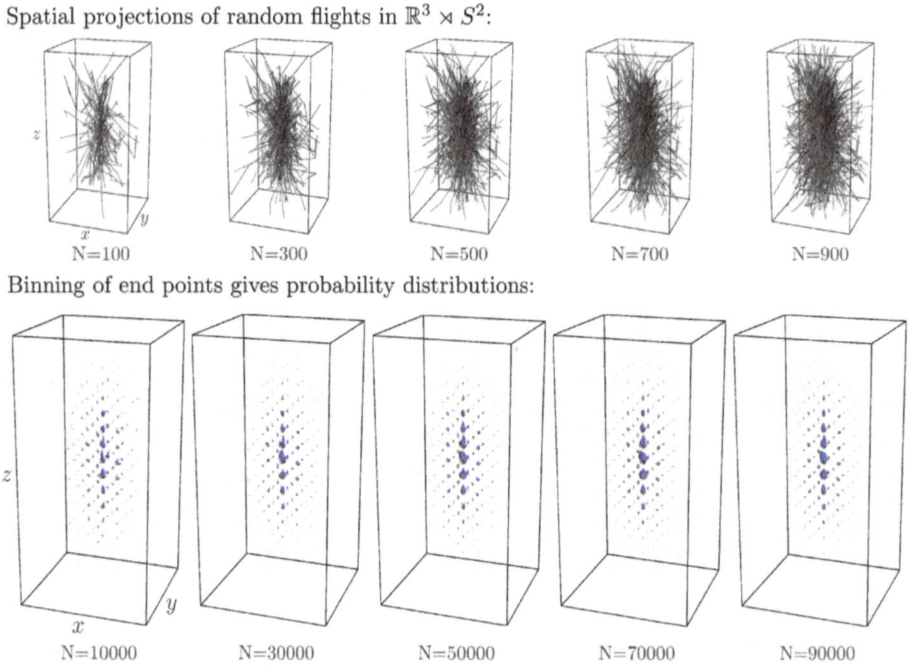

Figure 3. (**Top**) Spatial projections in $\mathbb{R}^3$ of N sample paths of the discrete random walks (or rather "drunk man's flights") in $\mathbb{R}^3 \rtimes S^2$ for $\alpha = 1$, given by Equation (72), for increasing N (with $\sigma = \frac{4\pi}{252}$, $\Delta s = 1$, $M = 40$); and (**Bottom**) convergence of the Monte-Carlo simulation kernel in Equation (76) for $\alpha = 1$ and $N \to \infty$. As N increases, the Monte-Carlo simulation converges towards the exact solution. For a comparison of the exact diffusion kernel in Equation (70) and its Monte-Carlo approximation in Equation (76), see Figure 5.

5.3. Comparison of Monte-Carlo Approximations of the Kernels to the Exact Solutions

In this section, we compute the probability density kernels K_t^α via the analytic approach of Section 5.1.2 (Equation (71), Theorem 3) and via the Monte-Carlo approximation of Section 5.2. The kernels are computed on a regular grid with each (x_i, y_j, z_k) at the center of the cubes c_{ijk} of Equation (75) with $i, j = -3, \ldots, 3$, $k = -5, \ldots, 5$, and $\Delta s = 0.5$. The Monte-Carlo simulations also require spherical sampling which we did by a geodesic polyhedron that sub-divides each mesh triangle of an icosahedron into n^2 new triangles and projects the vertex points to the sphere. We set $n = 4$ to obtain 252 (almost) uniformly sampled points on S^2.

The exact solution is computed using (truncated) spherical harmonics with $l \leq 12$. To obtain the kernel, we first solve the solution in the spatial Fourier domain and then do an inverse spatial Fast Fourier Transform. The resulting kernel K_t^α (where we literally follow Equation (71)) is only spatially sampled and provides for each (x_i, y_j, z_k) an analytic spherical distribution expressed in spherical harmonics.

For the Monte-Carlo approximation, we follow the procedure described in Section 5.2. The kernel K_t^α is obtained by binning the end points of random paths on the quotient $\mathbb{R}^3 \rtimes S^2$ (cf. Equation (72)) and thereby approximate the limit in Equation (76). Each path is discretized with $M = 40$ steps and in total $N = 10^{10}$ random paths were generated. The sphere S^2 is divided into 252 bins with an average surface area of $\sigma_{B_l} \approx \frac{4\pi}{252}$.

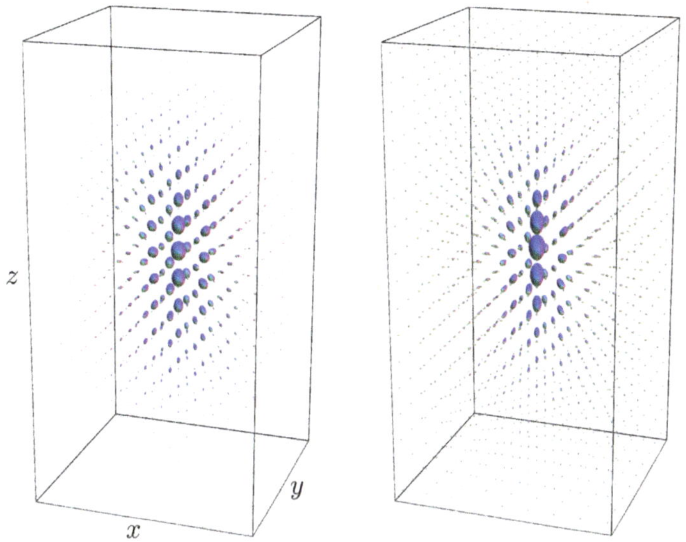

Figure 4. (**Left**) The degenerate diffusion kernel (Equation (70) for $\alpha = 1$ and $t = 2$); and (**Right**) the degenerate Poisson kernel (Equation (70) for $\alpha = \frac{1}{2}$ and $t = 3.5$). Parameters settings: $D_{44} = 0.2, D_{33} = 1, D_{11} = 0$.

In Figures 1 and 3–5, we set $D_{33} = 1$, $D_{44} = 0.2$. In the comparison between the kernels $K_t^{\alpha=1}$ with $K_t^{\alpha=0.5}$, we set $t = 2$ and $t = 3.5$, respectively, to match the full width at half maximum value of the distributions. In Figures 1, 3 and 5, we set $\alpha = 1$ and $t = 2$. In Figures 1, 3 and 4, we sample the grid in Equation (75) with $|i|, |j| \leq 4$, $|k| \leq 8$.

Figure 5 shows that the Monte-Carlo kernel closely approximates the exact solution and since the exact solutions can be computed at arbitrary spherical resolution, it provides a reliable way to validate numerical methods for α-stable Lévy processes on $\mathbb{R}^3 \rtimes S^2$.

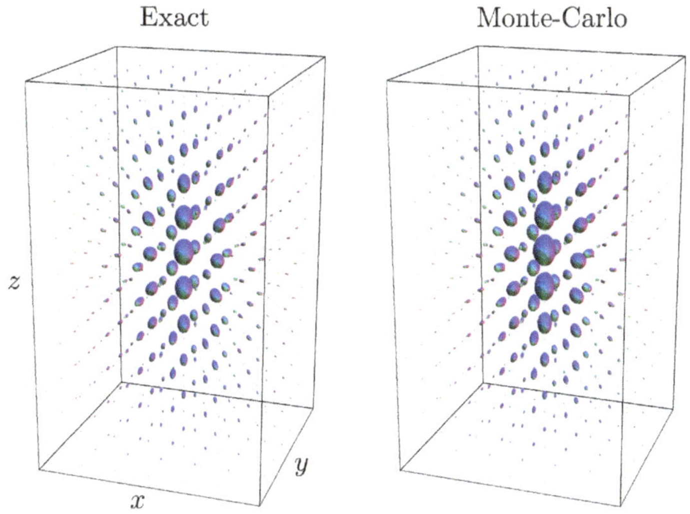

Figure 5. The exact kernel K_t^{α} and its Monte-Carlo approximation for $t = 2$, $\alpha = 1$, $D_{33} = 1$, $D_{44} = 0.2$.

6. Conclusions

We set up a Fourier transform $\mathcal{F}_{G/H}$ on the homogeneous space of positions and orientations. The considered Fourier transform acts on functions that are bi-invariant with respect to the action of subgroup H. We provide explicit formulas (relative to a basis of modified spherical harmonics) for the transform, its inverse, and its Plancherel formula, in Theorem 1.

Then, we use this Fourier transform to derive new exact solutions to the probability kernels of α-stable Lévy processes on G/H, including the diffusion PDE for Wiener processes, which is the special case $\alpha = 1$. They are obtained by spectral decomposition of the evolution operator in Theorem 2.

New formulas for the probability kernels are presented in Theorem 3. There, the general case $0 < \alpha < 1$ follows from the case $\alpha = 1$ by taking the fractional power of the eigenvalues. In comparison to previous formulas in [40] for the special case $\alpha = 1$ obtained via a spatial Fourier transform, we have more concise formulas with a more structured evolution operator in the Fourier domain of G/H, where we rely on ordinary spherical harmonics, and where we reduce the dimension of the manifold over which it is integrated from 3 to 1 (as can be seen in Theorem 3).

We introduce stochastic differential equations (or rather stochastic integral equations) for the α-stable Lévy processes in Appendix A.1, and we provide simple discrete approximations where we rely on matrix exponentials in the Lie group $SE(3)$ in Proposition 3.

We verified the exact solutions and the stochastic process formulations, by Monte-Carlo simulations that confirmed to give the same kernels, as shown in Figure 5. We also observed the expected behavior that the probability kernels for $0 < \alpha < 1$ have heavier tails, as shown in Figure 4.

The PDEs and the probability kernels have a wide variety of applications in image analysis (crossing-preserving, contextual enhancement of diffusion-weighted MRI, cf. [45,46,49,94,97,98] or in crossing-preserving diffusions in 3D scalar images [18]), robotics [4,5,57] and probability theory [56,61]. The generalizations to $\alpha \in (0,1]$ allow for longer range interactions between local orientations (due to the heavy tails). This is also of interest in machine learning, where convolutional neural networks on the homogeneous space of positions and orientations [9,12] can be extended to 3D [67,68], which may benefit from the PDE descriptors and the Fourier transform presented here.

Author Contributions: R.D. led the project, and wrote the main body/general theory of this article. This was done in a close and fruitful collaboration with A.M. (resulting in the final theoretical formulations and the final structure of this article) and with E.J.B. (resulting in the experiments, simulations and discrete stochastic process formulations in the article).

Funding: The research leading to the results of this paper received funding from the European Research Council under the European Community's Seventh Framework Programme (FP7/2007-2013)/ERC grant *Lie Analysis*, agr. nr. 335555. **Acknowledgments:** We gratefully acknowledge former PhD student J.M. Portegies (ASML, The Netherlands) for providing us with the *Mathematica* code for the exact solutions and Monte-Carlo simulations for the diffusion case $\alpha = 1$ that we simplified and generalized to the general case $\alpha \in (0,1]$.

Conflicts of Interest: The authors declare no conflict of interest.

Abbreviations

The following abbreviations and symbols are used in this manuscript:

UIR	Unitary Irreducible Representation	
G	The rigid body motions group $SE(3)$	Equation (1)
$\mathbf{a}$	The reference axis $\mathbf{a} = \mathbf{e}_z = (0,0,1)^T$	Equation (3)
H	The subgroup that stabilizes $(\mathbf{0}, \mathbf{a})$	Equation (3)
G/H	The homogeneous space of positions and orientations $\mathbb{R}^3 \rtimes S^2$	Equation (3)
$\overline{U}$	The spatial Fourier transform of U	Equation (18)
$\hat{U}$	The Fourier transform $\hat{U} = \mathcal{F}_{G/H}U$	Equation (43)
α	Parameter of the α-Stable processes (indexing fractional power of the generator)	Equation (10)

$\overline{\alpha}$	Rotation angle around reference axis $\mathbf{a} = \mathbf{e}_z = (0,0,1)$	Remark 7
$\sigma^{p,s}$	UIR of $G = SE(3)$	Equation (32)
$\overline{\sigma}^{p,s}$	the action on the quotient corresponding to $\sigma^{p,s}$	Definition 7
$\tilde{K}_t^{\alpha}$	The probability kernel on G	Equation (26)
K_t^{α}	The probability kernel on G/H	Equation (27)
$\tilde{W}_{\alpha}$	Solution of the PDE on G	Equation (10)
W_{α}	Solution of the PDE on G/H	Equation (6)
$\tilde{Q}_{\alpha}$	Evolution generator of the PDE on G	Equation (11)
Q_{α}	Evolution generator of the PDE on G/H	Equation (7)
$\mathbf{R}_{\mathbf{n}}$	Any rotation that maps $\mathbf{a}$ onto $\mathbf{n}$	Remark 2
$\mathbf{R}_{\mathbf{v},\phi}$	A counter-clockwise rotation about axis $\mathbf{v}$ with angle ϕ	Remark 2
$\mathbf{P}_t$	Lévy Processes on G/H	Definition A1
$\overline{\mathbf{P}}_t$	Lévy Processes on $\mathbb{R}^3 \times \mathbb{R}^3$	Equation (A4)
$q_{t,\alpha}$	The kernel relating K_t^{α} and K_t^1	Equation (77)
$Y^{l,m}$	The ordinary spherical harmonics	Proposition 2
$Y_s^{l,m}$	The modified spherical harmonics according to [4]	Proposition 2
$Y_{\omega}^{l,m}$	The generalized spherical harmonics according to [40]	Definition 10
$\Phi_{\omega}^{l,m}$	The spheroidal wave basis function for $\mathbb{L}_2(S^2)$	Definition 11
$(\overline{\alpha},\beta,\gamma)$	ZYZ Euler angles.	Equation (A12)

Appendix A. Probability Theory

Appendix A.1. Lévy Processes on $\mathbb{R}^3 \rtimes S^2$

In the next definition, we define Lévy processes on our manifold of interest $G/H = \mathbb{R}^3 \rtimes S^2$. Recall, that the action of $G = SE(3)$ on G/H is given by Equation (4). As a prerequisite, we define the "difference" of two random variables $\mathbf{P}_1 = (\mathbf{X}_1, \mathbf{N}_1)$ and $\mathbf{P}_2 = (\mathbf{X}_2, \mathbf{N}_2)$ in $\mathbb{R}^3 \rtimes S^2$:

$$\mathbf{G}_2^{-1} \odot \mathbf{P}_1 = (\mathbf{X}_2, \mathbf{R}_{\mathbf{N}_2})^{-1} \odot (\mathbf{X}_1, \mathbf{N}_1) = (\mathbf{R}_{\mathbf{N}_2}^T(\mathbf{X}_1 - \mathbf{X}_2), \mathbf{R}_{\mathbf{N}_2}^T \mathbf{N}_1), \tag{A1}$$

where we relate random variables on G/H and in G via $\mathbf{P} = \mathbf{G} \odot (\mathbf{0}, \mathbf{a})$, according to Equation (39).

We assume that $\mathbf{P}_1$ and $\mathbf{P}_2$ are chosen such that the distribution of $\mathbf{G}_2^{-1} \odot \mathbf{P}_1$ is invariant under the choice of rotation variable $\mathbf{R}_{\mathbf{N}_2} \in SO(3)$, which maps reference axis $\mathbf{a}$ onto $\mathbf{N}_2$. This is done in view of the homogeneous space structure in Equation (3) and the fact that Lévy processes on Lie groups such as $G = SE(3)$ require Lie group inversion in their definition (see, e.g., [99]).

Definition A1. *A stochastic process $\{\mathbf{P}_t \,:\, t \geq 0\}$ on G/H is a Lévy process if the following conditions hold:*

1. *For any $n \geq 1$ and $0 \leq t_0 < t_1 < \ldots < t_n$, the variables $\mathbf{P}_{t_0}, \mathbf{G}_{t_0}^{-1} \odot \mathbf{P}_{t_1}, \ldots, \mathbf{G}_{t_{n-1}}^{-1} \odot \mathbf{P}_{t_n}$ are independent.*
2. *The distribution of $\mathbf{G}_s^{-1} \odot \mathbf{P}_{s+t}$ does not depend on $s \geq 0$.*
3. *$\mathbf{P}_0 = (\mathbf{0}, \mathbf{a})$ almost surely.*
4. *It is stochastically continuous, i.e. $\lim_{s \to t} P[d(\mathbf{P}_s, \mathbf{P}_t) > \varepsilon] = 0, \forall \varepsilon > 0$.*
   Here, $d((\mathbf{x}_1, \mathbf{n}_1), (\mathbf{x}_2, \mathbf{n}_2)) = |\mathbf{x}_1 - \mathbf{x}_2|^2 + \arccos^2(\mathbf{n}_1 \cdot \mathbf{n}_2)$.

Let us consider the solutions

$$W_{\alpha}(\mathbf{x}, \mathbf{n}, t) = (K_t^{\alpha} * U)(\mathbf{x}, \mathbf{n})$$

of our linear PDEs of interest in Equation (6) for $\alpha \in (0,1]$ fixed. Let us consider the case where $U \sim \delta_{(\mathbf{0},\mathbf{a})}$, so that the solutions are the probability kernels K_t^{α} themselves. We consider the random variables $\mathbf{P}_t^{\alpha}$ such that their probability densities are given by

$$P(\mathbf{P}_t^{\alpha} = (\mathbf{x}, \mathbf{n})) = K_t^{\alpha}(\mathbf{x}, \mathbf{n}) \text{ for all } t \geq 0, (\mathbf{x}, \mathbf{n}) \in \mathbb{R}^3 \rtimes S^2. \tag{A2}$$

Proposition A1. *The stochastic process $\{\mathbf{P}_t^{\alpha} \,:\, t \geq 0\}$ is a Lévy processes on $\mathbb{R}^3 \rtimes S^2$.*

Proof. We first address Items 1 and 2. On $G = SE(3)$, one has for two stochastically independent variables:

$$P(\mathbf{G}_1 \mathbf{G}_2 = g) = \int_G P(\mathbf{G}_2 = h^{-1} g) P(\mathbf{G}_1 = h) \, \mathrm{d}h.$$

In particular, for $\mathbf{G}_1 = \mathbf{G}_t \sim \tilde{K}_t^{\alpha}$ and $\mathbf{G}_2 = \mathbf{G}_s \sim \tilde{K}_s^{\alpha}$, we have

$$\mathbf{G}_s\mathbf{G}_t \sim \tilde{K}_t^{\alpha} * \tilde{K}_s^{\alpha} = \tilde{K}_{t+s}^{\alpha} \quad \text{and} \quad \mathbf{G}_s^{-1}\mathbf{G}_{t+s} = \mathbf{G}_t \sim \tilde{K}_t^{\alpha},$$

which is due to $e^{t\tilde{Q}_\alpha} \circ e^{s\tilde{Q}_\alpha} = e^{(t+s)\tilde{Q}_\alpha}$ (recall Equation (55)). Similarly, on the quotient G/H, we have

$$\mathbf{G}_s^{-1} \odot \mathbf{P}_{s+t} = \mathbf{P}_t \sim K_t^{\alpha}.$$

Furthermore, the choice of $\mathbf{G}_s$ such that $\mathbf{G}_s \odot (\mathbf{0}, \mathbf{a}) = (\mathbf{0}, \mathbf{a})$ does not matter, since

$$P((\mathbf{0}, \mathbf{R}_{\mathbf{a},\tilde{\alpha}})^{-1}\mathbf{G}_s^{-1} \odot \mathbf{P}_{s+t} = (\mathbf{x}, \mathbf{n})) = K_t^{\alpha}((\mathbf{0}, \mathbf{R}_{\mathbf{a},\tilde{\alpha}}) \odot (\mathbf{x}, \mathbf{n})) = K_t^{\alpha}(\mathbf{x}, \mathbf{n})$$

(recall Equation (30)). Item 3 is obvious since we have $\mathbf{P}_0 = \delta_{(\mathbf{0},\mathbf{a})}$. Item 4 follows by strong continuity of the semigroup operators ([64], Thm. 2), [66]. □

Lemma A1. *The kernels K_t^{α} are infinitely divisible, i.e.*

$$K_t^{\alpha} * K_s^{\alpha} = K_{t+s}^{\alpha} \quad \textit{for all } s,t \geq 0.$$

Proof. The infinite divisibility directly follows from Corollary 1 and $\mathcal{F}_{G/H}(K_t^{\alpha} * K_s^{\alpha}) = \mathcal{F}_{G/H}(K_t^{\alpha}) \circ \mathcal{F}_{G/H}(K_t^{\alpha}) = \mathcal{F}_{G/H}(K_{t+s}^{\alpha})$, which is clear due to Equation (70). □

Remark A1. *Recall that on $\mathbb{R}^n$ a Lévy process $\mathbf{X}_t$ is called α-stable if*

$$a^{-\frac{1}{2\alpha}}\mathbf{X}_{at} \sim \mathbf{X}_t \quad \textit{for all } a > 0. \tag{A3}$$

This convention and property applies to all $n \in \mathbb{N}$, cf. [61]. Next, we come to a generalization of α-stability but then for the processes $\mathbf{P}_t$. Here, an embedding of $\mathbb{R}^3 \rtimes S^2$ into $\mathbb{R}^6 = \mathbb{R}^3 \times \mathbb{R}^3$ is required to give a meaning to α-stability and a scaling relation on $\mathbf{P}_t = (\mathbf{X}_t, \mathbf{N}_t)$ that is similar to Equation (A3).

Appendix A.2. SDE Formulation of α-Stable Lévy Processes on $\mathbb{R}^3 \rtimes S^2$

Consider the Lévy processes $\{\mathbf{P}_t \,:\, t \geq 0\}$ on $\mathbb{R}^3 \rtimes S^2$ given by Equation (A2). They give rise to the Forward Kolmogorov PDEs in Equation (6) in terms of their stochastic differential equation (SDE) according to the book of Hsu on Stochastic Analysis on Manifolds [60].

We apply ([60], Prop.1.2.4) on the embedding map $\Phi : \mathbb{R}^3 \times \mathbb{R}^3 \to \mathbb{R}^3 \rtimes S^2$ given by

$$\Phi : (\mathbf{x}, \overline{\mathbf{n}}) \mapsto \Phi(\mathbf{x}, \overline{\mathbf{n}}) = \left(\mathbf{x}, \frac{\overline{\mathbf{n}}}{\|\overline{\mathbf{n}}\|}\right) = (\mathbf{x}, \mathbf{n}).$$

Note that $\Phi_* = \mathcal{D}\Phi = \left(I, \frac{1}{\|\overline{\mathbf{n}}\|}\left(I - \frac{\overline{\mathbf{n}}}{\|\overline{\mathbf{n}}\|} \otimes \frac{\overline{\mathbf{n}}}{\|\overline{\mathbf{n}}\|}\right)\right)$. Here, I denotes the identity map on $\mathbb{R}^3$.

Let us first concentrate on $\alpha = 1$. In this case, our PDE in Equation (6) becomes a diffusion PDE that is the forward Kolmogorov equation of a Wiener process $\mathbf{P}_t = (\mathbf{X}_t, \mathbf{N}_t)$ on $\mathbb{R}^3 \rtimes S^2$. Next, we relate this Wiener process to a Wiener process $(\mathbf{W}_t^{(1)}, \overline{\mathbf{W}}_t^{(2)})$ in the embedding space $\mathbb{R}^3 \times \mathbb{R}^3$. We write down the stochastic differential equation (SDE) and show that Equation (72) boils down to discretization of the stochastic integral (in Îto sense) solving the SDE.

Next, we define $\overline{\mathbf{P}}_t = (\mathbf{X}_t, \overline{\mathbf{N}}_t)$ by the SDE in the embedding space:

$$\mathrm{d}\overline{\mathbf{P}}_t = \bar{\mathfrak{s}}|_{\overline{\mathbf{P}}_t} \circ \mathrm{d}\mathbf{W}_t, \tag{A4}$$

where $\mathbf{W}_t = (\mathbf{W}_t^{(1)}, \overline{\mathbf{W}}_t^{(2)})$, with $\mathbf{W}_t^{(1)}$ and $\overline{\mathbf{W}}_t^{(2)}$ being Wiener processes in $\mathbb{R}^3$; and where

$$\bar{\mathfrak{s}}|_{\overline{\mathbf{P}}}(\mathrm{d}\mathbf{x}, \mathrm{d}\overline{\mathbf{n}}) = \begin{pmatrix} \bar{\mathfrak{s}}^{(1)}|_{\overline{\mathbf{P}}}(\mathrm{d}\mathbf{x}, \mathrm{d}\overline{\mathbf{n}}) \\ \bar{\mathfrak{s}}^{(2)}|_{\overline{\mathbf{P}}}(\mathrm{d}\mathbf{x}, \mathrm{d}\overline{\mathbf{n}}) \end{pmatrix} = \begin{pmatrix} \sqrt{D_{33}}\, \frac{\overline{\mathbf{N}}}{\|\overline{\mathbf{N}}\|}\left(\frac{\overline{\mathbf{N}}}{\|\overline{\mathbf{N}}\|} \cdot \mathrm{d}\mathbf{x}\right) \\ \sqrt{D_{44}}\, \mathrm{d}\overline{\mathbf{n}} \end{pmatrix}.$$

Here, Index (1) stands for the spatial part and Index (2) stands for the angular part.

Now, we define a corresponding process on $\mathbb{R}^3 \rtimes S^2$:

$$\mathbf{P}_t = \Phi(\overline{\mathbf{P}}_t).$$

Then, the SDE for $\mathbf{P}_t = (\mathbf{X}_t, \mathbf{N}_t)$ becomes (see ([60], Prop.1.2.4))

$$\mathrm{d}\mathbf{P}_t = \mathrm{d}\left(\Phi \circ \overline{\mathbf{P}}_t\right) \Leftrightarrow \begin{cases} \mathrm{d}\mathbf{X}_t = \bar{\mathfrak{s}}^{(1)}\Big|_{\overline{\mathbf{P}}_t} \circ \mathrm{d}\mathbf{W}_t^{(1)}, \\ \mathrm{d}\mathbf{N}_t = \mathbb{P}_{\langle \mathbf{N}_t\rangle^\perp}\, \bar{\mathfrak{s}}^{(2)}\Big|_{\overline{\mathbf{P}}_t} \circ \mathrm{d}\overline{\mathbf{W}}_t^{(2)}, \end{cases}$$

where $\mathbf{N}_t = \frac{\overline{\mathbf{N}}_t}{\|\overline{\mathbf{N}}_t\|}$; and where $\mathbb{P}_{\langle \mathbf{N}_t\rangle^\perp} = (I - \mathbf{N}_t \otimes \mathbf{N}_t)$ denotes the orthogonal projection to the tangent plane perpendicular to $\mathbf{N}_t$.

Therefore, we have the following SDE on $\mathbb{R}^3 \rtimes S^2$:

$$\begin{cases} \mathrm{d}\mathbf{X}_t = \sqrt{D_{33}}\,\mathbf{N}_t(\mathbf{N}_t \cdot \mathrm{d}\mathbf{W}_t^{(1)}), \\ \mathrm{d}\mathbf{N}_t = \sqrt{D_{44}}\,\mathbb{P}_{\langle \mathbf{N}_t\rangle^\perp} \mathrm{d}\overline{\mathbf{W}}_t^{(2)} \end{cases} \tag{A5}$$

Thus, integrating the SDE, we obtain the following stochastic integral (in Îto form):

$$\begin{cases} \mathbf{X}_t = \mathbf{X}_0 + \sqrt{D_{33}} \int\limits_0^t \mathbf{N}_s(\mathbf{N}_s \cdot \mathrm{d}\mathbf{W}_s^{(1)}) = \mathbf{X}_0 + \sqrt{D_{33}}\, \underset{M\to\infty}{\text{ms-lim}} \sum\limits_{k=1}^{M} \mathbf{N}_{t_{k-1}}\left(\mathbf{N}_{t_{k-1}} \cdot \left(\mathbf{W}_{t_k}^{(1)} - \mathbf{W}_{t_{k-1}}^{(1)}\right)\right), \\ \mathbf{N}_t = \underset{M\to\infty}{\text{ms-lim}} \prod\limits_{k=1}^{M} \exp_{S^2}\left(\sqrt{D_{44}}\,(I - \mathbf{N}_{t_{k-1}} \otimes \mathbf{N}_{t_{k-1}})\left(\overline{\mathbf{W}}_{t_k}^{(2)} - \overline{\mathbf{W}}_{t_{k-1}}^{(2)}\right)\right)\mathbf{N}_0. \end{cases} \tag{A6}$$

Here, $\exp_{S^2}(V)\mathbf{n}_0$ denotes the exponential map on a sphere, i.e., its value is the end point (for $t = 1$) of a geodesic starting from $\mathbf{n}_0 \in S^2$ with the tangent vector $V \in T_{\mathbf{n}_0}S^2$. Note that, in the formula above, the symbol $\prod$ denotes the composition

$$\prod_{k=1}^{M} \exp_{S^2}(V_k)\mathbf{n}_0 = \left(\exp_{S^2}(V_M) \circ \ldots \circ \exp_{S^2}(V_1)\right)\mathbf{n}_0.$$

Note that $\sqrt{D_{33}}\left(\mathbf{W}_{t_k}^{(1)} - \mathbf{W}_{t_{k-1}}^{(1)}\right) = \sqrt{D_{33}}\mathbf{W}_{t_k - t_{k-1}}^{(1)} = \sqrt{\frac{tD_{33}}{M}}\epsilon_k$, where $\epsilon_k \sim \mathbf{W}_1^{(1)}$, i.e., $\epsilon_k \sim G_{t=1}$.

For $M \in \mathbb{N}$ fixed, we propose a discrete approximation for the stochastic integrals in Equation (A6):

$$\begin{cases} \mathbf{X}_M = \mathbf{X}_0 + \sum\limits_{k=1}^{M} \sqrt{\frac{tD_{33}}{M}}\epsilon_k \mathbf{N}_{k-1}, \\ \mathbf{N}_M = \left(\prod\limits_{k=1}^{M} \mathbf{R}_{\mathbf{a},\gamma_k} \mathbf{R}_{\mathbf{e}_y, \beta_k \sqrt{\frac{tD_{44}}{M}}}\right)\mathbf{N}_0, \end{cases} \tag{A7}$$

with $\epsilon_k \sim G_{t=1}^{\mathbb{R}} \sim \mathcal{N}(0, \sigma = \sqrt{2})$ stochastically independent Gaussian distributed on $\mathbb{R}$ with $t = 1$; with uniformly distributed $\gamma_k \sim \mathrm{Unif}\,(\mathbb{R}/(2\pi\mathbb{Z}) \equiv [-\pi, \pi))$; and with $\beta_k \sim g$, where $g : \mathbb{R} \to \mathbb{R}^+$ equals $g(\beta) = \frac{|\beta|}{2}\, e^{-\frac{\beta^2}{4}}$. The choice of g is done by application of the theory of isotropic stochastic processes on Riemannian manifolds by Pinsky [96], where we note that

$$G_t^{\mathbb{R}^2}(\beta\cos\gamma, \beta\sin\gamma) = g(\beta)\;\mathrm{Unif}\,([-\pi,\pi))\,(\gamma), \quad \beta \in \mathbb{R}, \gamma \in [-\pi, \pi).$$

Now, in the numerical simulation, we can replace g by $G_{t=2}^{\mathbb{R}}$ due to the central limit theorem on $\mathbb{R}$ and

$$\mathrm{Var}(\beta) = \int\limits_{-\infty}^{\infty} \beta^2 g(\beta)\mathrm{d}\beta = 2\int\limits_0^{\infty} \beta^2 g(\beta)\mathrm{d}\beta = 2.$$

Appendix A.2.1. From the Diffusion Case $\alpha = 1$ to the General Case $\alpha \in (0,1]$

For the case $\alpha \in (0,1]$, we define the (fractional) random processes by their probability densities

$$\begin{aligned} P(\mathbf{P}_t^{\alpha} = (\mathbf{x},\mathbf{n})) &= \int_0^{\infty} q_{t,\alpha}(\tau)\; P(\mathbf{P}_\tau = (\mathbf{x},\mathbf{n}))\, \mathrm{d}\tau, \\ P(\overline{\mathbf{P}}_t^{\alpha} = (\mathbf{x},\overline{\mathbf{n}})) &= \int_0^{\infty} q_{t,\alpha}(\tau)\; P(\overline{\mathbf{P}}_\tau = (\mathbf{x},\overline{\mathbf{n}}))\, \mathrm{d}\tau. \end{aligned} \tag{A8}$$

Recal that the kernel $q_{t,\alpha}(\tau)$ is given by Equation (77). For Monte-Carlo simulations, one can use Equation (78), or alternatively use $\mathbf{P}_{t_M}^{\alpha} \approx \prod_{i=1}^{M} \mathbf{G}_{T_i} \odot \mathbf{P}_0$, for $M \gg 0$, where $\mathbf{P}_0$ is almost surely $(\mathbf{0},\mathbf{a})$, with T_i a temporal random variable with $P(T_i = \tau) = q_{t_i,\alpha}(\tau)$, with $t_i = \frac{i}{M} t$ and $\mathbf{G}_{t_i}$ given by Equation (73).

Appendix A.2.2. α-Stability of the Lévy Process

Due to the absence of suitable dilations on G/H, we resort to the embedding space where α-stability is defined. The Lévy process $\{\overline{\mathbf{P}}_t^{\alpha} = (\mathbf{X}_t^{\alpha}, \overline{\mathbf{N}}_t^{\alpha}) \mid t \geq 0\}$ associated to the Lévy process $\{\mathbf{P}_t^{\alpha} = (\mathbf{X}_t^{\alpha}, \mathbf{N}_t^{\alpha}) \mid t \geq 0\}$ in $\mathbb{R}^3 \rtimes S^2$ is α-stable, i.e., for all $a, t > 0$ we have (by Equations (A5) and (78))

$$a^{-\frac{1}{2\alpha}} \mathbf{X}_{at}^{\alpha} \sim \mathbf{X}_t^{\alpha} \text{ and } a^{-\frac{1}{2\alpha}} \overline{\mathbf{N}}_{at}^{\alpha} \sim \overline{\mathbf{N}}_t^{\alpha}.$$

Appendix B. Left-Invariant Vector Fields on SE(3) via Two Charts

We need two charts to cover $SO(3)$. When using the following coordinates (ZYZ-Euler angles) for $SE(3) = \mathbb{R}^3 \rtimes SO(3)$ for the first chart:

$$g = (x, y, z, \mathbf{R}_{\mathbf{e}_z,\gamma} \mathbf{R}_{\mathbf{e}_y,\beta} \mathbf{R}_{\mathbf{e}_z,\overline{\alpha}}), \text{ with } \beta \in (0,\pi), \overline{\alpha}, \gamma \in [0, 2\pi), \tag{A9}$$

Equation (9) yields the following formulas for the left-invariant vector fields:

$$\begin{aligned} \mathcal{A}_1|_g &= (\cos\overline{\alpha}\cos\beta\cos\gamma - \sin\overline{\alpha}\sin\gamma)\partial_x + (\sin\overline{\alpha}\cos\gamma + \cos\overline{\alpha}\cos\beta\sin\gamma)\partial_y - \cos\overline{\alpha}\sin\beta\,\partial_z \\ \mathcal{A}_2|_g &= (-\sin\overline{\alpha}\cos\beta\cos\gamma - \cos\overline{\alpha}\sin\gamma)\partial_x + (\cos\overline{\alpha}\cos\gamma - \sin\overline{\alpha}\cos\beta\sin\gamma)\partial_y + \sin\overline{\alpha}\sin\beta\,\partial_z \\ \mathcal{A}_3|_g &= \sin\beta\cos\gamma\,\partial_x + \sin\beta\sin\gamma\,\partial_y + \cos\beta\,\partial_z, \\ \mathcal{A}_4|_g &= \cos\overline{\alpha}\cot\beta\,\partial_{\overline{\alpha}} + \sin\overline{\alpha}\,\partial_\beta - \tfrac{\cos\overline{\alpha}}{\sin\beta}\,\partial_\gamma, \\ \mathcal{A}_5|_g &= -\sin\overline{\alpha}\cot\beta\,\partial_{\overline{\alpha}} + \cos\overline{\alpha}\,\partial_\beta + \tfrac{\sin\overline{\alpha}}{\sin\beta}\,\partial_\gamma, \\ \mathcal{A}_6|_g &= \partial_{\overline{\alpha}}. \end{aligned} \tag{A10}$$

We observe that

$$\underline{\mathcal{A}}_{gh_{\overline{\alpha}}} \equiv (\mathbf{R}_{\mathbf{e}_z,\overline{\alpha}} \oplus \mathbf{R}_{\mathbf{e}_z,\overline{\alpha}})^T \underline{\mathcal{A}}_g, \quad \text{where } \underline{\mathcal{A}}_g = (\mathcal{A}_1|_g, \ldots, \mathcal{A}_6|_g). \tag{A11}$$

The above formulas do not hold for $\beta = \pi$ or $\beta = 0$. Thus, we even lack expressions for our left-invariant vector fields at the unity element $(\mathbf{0}, \mathbf{I}) \in SE(3)$ when using the standard ZYZ-Euler angles. Therefore, one formally needs a second chart, for example the XYZ-coordinates in [84,87,100]:

$$g = (x, y, z, \mathbf{R}_{\mathbf{e}_x,\tilde{\gamma}} \mathbf{R}_{\mathbf{e}_y,\tilde{\beta}} \mathbf{R}_{\mathbf{e}_z,\overline{\alpha}}), \text{ with } \tilde{\beta} \in [-\pi,\pi), \overline{\alpha} \in [0,2\pi),\ \tilde{\gamma} \in (-\pi/2, \pi/2), \tag{A12}$$

Equation (9) yields the following formulas for the left-invariant vector fields (only for $|\tilde{\beta}| \neq \frac{\pi}{2}$):

$$\begin{aligned} \mathcal{A}_1|_g &= \cos\overline{\alpha}\cos\tilde{\beta}\,\partial_x + (\cos\tilde{\gamma}\sin\overline{\alpha} + \cos\overline{\alpha}\sin\tilde{\beta}\sin\tilde{\gamma})\,\partial_y + (\sin\overline{\alpha}\sin\tilde{\gamma} - \cos\overline{\alpha}\sin\tilde{\beta}\cos\tilde{\gamma})\,\partial_z \\ \mathcal{A}_2|_g &= -\sin\overline{\alpha}\cos\tilde{\beta}\,\partial_x + (\cos\overline{\alpha}\cos\tilde{\gamma} - \sin\overline{\alpha}\sin\tilde{\beta}\sin\tilde{\gamma})\partial_y + (\sin\overline{\alpha}\sin\tilde{\beta}\cos\tilde{\gamma} + \cos\overline{\alpha}\sin\tilde{\gamma})\,\partial_z \\ \mathcal{A}_3|_g &= \sin\tilde{\beta}\,\partial_x - \cos\tilde{\beta}\sin\tilde{\gamma}\,\partial_y + \cos\tilde{\beta}\cos\tilde{\gamma}\,\partial_z, \\ \mathcal{A}_4|_g &= -\cos\overline{\alpha}\tan\tilde{\beta}\,\partial_{\overline{\alpha}} + \sin\overline{\alpha}\,\partial_{\tilde{\beta}} + \tfrac{\cos\overline{\alpha}}{\cos\tilde{\beta}}\partial_{\tilde{\gamma}}, \\ \mathcal{A}_5|_g &= \sin\overline{\alpha}\tan\tilde{\beta}\,\partial_{\overline{\alpha}} + \cos\overline{\alpha}\partial_{\tilde{\beta}} - \tfrac{\sin\overline{\alpha}}{\cos\tilde{\beta}}\,\partial_{\tilde{\gamma}}, \\ \mathcal{A}_6|_g &= \partial_{\overline{\alpha}}. \end{aligned} \tag{A13}$$

References

1. Zettl, A. *Sturm-Liouville Theory*; Mathematical Surveys and Monographs; American Mathematical Society: Providence, RI, USA, 2005; Volume 121.
2. Kato, T. Operators in Hilbert spaces. In *Perturbation Theory for Linear Operators*; Classics in Mathematics; Springer: Berlin/Heidelberg, Germany, 1976; pp. 251–308.
3. Rudin, W. *Functional Analysis*, 2nd ed.; McGraw-Hill, Inc.: New York, NY, USA, 1991.
4. Chirikjian, G.S.; Kyatkin, A.B. *Engineering Applications of Noncommutative Harmonic Analysis: With Emphasis on Rotation and Motion Groups*; CRC Press: Boca Raton, FL, USA, 2000.
5. Chirikjian, G.S. *Stochastic Models, Information Theory, and Lie Groups: Analytic Methods and Modern Applications*; Springer Science & Business Media: Berlin, Germany, 2011; Volume 2.
6. Saccon, A.; Aguiar, A.P.; Hausler, A.J.; Hauser, J.; Pascoal, A.M. Constrained motion planning for multiple vehicles on SE(3). In Proceedings of the 2012 IEEE 51st IEEE Conference on Decision and Control (CDC), Maui, HI, USA, 10–13 December 2012; pp. 5637–5642.
7. Henk Nijmeijer, A.V.D.S. *Nonlinear Dynamical Control Systems*; Springer: Berlin/Heidelberg, Germany, 1990; p. 426.
8. Ali, S.; Antoine, J.; Gazeau, J. *Coherent States, Wavelets and Their Generalizations*; Springer: New York, NY, USA; Berlin/Heidelberg, Germany, 1999.
9. Bekkers, E.; Lafarge, M.; Veta, M.; Eppenhof, K.; Pluim, J.; Duits, R. Roto-Translation Covariant Convolutional Networks for Medical Image Analysis. In *Medical Image Computing and Computer Assisted Intervention—MICCAI 2018*; Springer International Publishing: Cham, Switzerland, 2018; pp. 440–448.
10. Bekkers, E.; Loog, M.; ter Haar Romeny, B.; Duits, R. Template matching via densities on the roto-translation group. *IEEE Trans. Pattern Anal. Mach. Intell.* **2017**, *40*, 452–466. [CrossRef] [PubMed]
11. Cohen, T.S.; Geiger, M.; Weiler, M. Intertwiners between Induced Representations (with Applications to the Theory of Equivariant Neural Networks). *arXiv* **2018**, arXiv:1803.10743.
12. Cohen, T.; Welling, M. Group equivariant convolutional networks. In Proceedings of the International Conference on Machine Learning, New York, NY, USA, 19–24 June 2016; pp. 2990–2999.
13. Sifre, L.; Mallat, S. Rotation, scaling and deformation invariant scattering for texture discrimination. In Proceedings of the 2013 IEEE Conference on Computer Vision and Pattern Recognition, Portland, OR, USA, 23–28 June 2013; pp. 1233–1240.
14. Duits, R.; Felsberg, M.; Granlund, G.; ter Haar Romeny, B. Image Analysis and Reconstruction using a Wavelet Transform Constructed from a Reducible Representation of the Euclidean Motion Group. *Int. J. Comput. Vis.* **2006**, *72*, 79–102. [CrossRef]
15. Citti, G.; Sarti, A. A Cortical Based Model of Perceptual Completion in the Roto-Translation Space. *J. Math. Imaging Vis.* **2006**, *24*, 307–326. [CrossRef]
16. Duits, R.; Fuehr, H.; Janssen, B.; Florack, L.; van Assen, H. Evolution equations on Gabor transforms and their applications. *ACHA* **2013**, *35*, 483–526. [CrossRef]
17. Prandi, D.; Gauthier, J.P. *A Semidiscrete Version of the Citti-Petitot-Sarti Model as a Plausible Model for Anthropomorphic Image Reconstruction and Pattern Recognition*; Springer International Publishing: Cham, Switzerland, 2018; p. 113.
18. Janssen, M.H.J.; Janssen, A.J.E.M.; Bekkers, E.J.; Bescós, J.O.; Duits, R. Design and Processing of Invertible Orientation Scores of 3D Images. *J. Math. Imaging Vis.* **2018**, *60*, 1427–1458. [CrossRef]
19. Boscain, U.; Duplaix, J.; Gauthier, J.; Rossi, F. Anthropomorphic Image Reconstruction via Hypoelliptic Diffusion. *SIAM J. Control Optim.* **2012**, *50*, 1309–1336. [CrossRef]
20. Schur, I. *Vorlesungen über Invariantentheorie*; P. Noordhoff: Groningen, The Netherlands, 1968.
21. Dieudonné, J. *Treatise on Analysis*; Academic Press: New York, NY, USA, 1977; Volume V.
22. Folland, G.B. *A Course in Abstract Harmonic Analysis*; CRC Press: Boca Raton, FL, USA, 1994.
23. Agrachev, A.; Boscain, U.; Gauthier, J.P.; Rossi, F. The intrinsic hypoelliptic Laplacian and its heat kernel on unimodular Lie groups. *J. Funct. Anal.* **2009**, *256*, 2621–2655. [CrossRef]
24. Führ, H. *Abstract Harmonic Analysis of Continuous Wavelet Transforms*; Springer Science & Business Media: Berlin, Germany, 2005.
25. Mackey, G.W. Imprimitivity for Representations of Locally Compact Groups I. *Proc. Natl. Acad. Sci. USA* **1949**, *35*, 537–545. [CrossRef]

26. Sugiura, M. *Unitary Representations and Harmonic Analysis: An Introduction*; Elsevier: Amsterdam, The Netherlands, 1990.
27. Dixmier, J. *C*-algebras*; North Holland: Amsterdam, Switzerland, 1981.
28. Ghaani Farashani, A. Operator-valued Fourier transforms over homogeneous spaces of compact groups. *Groups Geom. Dyn.* **2017**, *11*, 1437–1467. [CrossRef]
29. Ghaani Farashani, A. Poisson summation formulas over homogeneous spaces of compact groups. *Anal. Math. Phys.* **2017**, *4*, 493–508. [CrossRef]
30. Ghaani Farashani, A. Plancherel (trace) formulas over homogeneous spaces of compact groups. *Can. Math. Bull.* **2017**, *60*, 111–121. [CrossRef]
31. Ghaani Farashahi, A. Relative Fourier transforms over canonical homogeneous spaces of semi-direct product groups with abelian normal factor. *J. Korean Math. Soc.* **2017**, *54*, 117–139. [CrossRef]
32. Gaveau, B. Principe de moindre action, propagation de la chaleur et estimees sous elliptiques sur certains groupes nilpotents. *Acta Math.* **1977**, *139*, 95–153. [CrossRef]
33. Duits, R.; van Almsick, M. The explicit solutions of linear left-invariant second order stochastic evolution equations on the 2D Euclidean motion group. *Q. Appl. Math.* **2008**, *66*, 27–67. [CrossRef]
34. Duits, R.; Franken, E. *Line Enhancement and Completion via Linear Left Invariant Scale Spaces on SE(2)*; SSVM; Springer: Berlin/Heidelberg, Germany, 2009; pp. 795–807.
35. Duits, R.; van Almsick, M. *The Explicit Solutions of Linear Left-Invariant Second Order Stochastic Evolution Equations on the 2D-Euclidean Motion Group*; Technical Report CASA-Report, nr.43; Department of Mathematics and Computer Science, Eindhoven University of Technology: Eindhoven, The Netherlands, 2005; 37p. Available online: http://www.win.tue.nl/analysis/reports/rana05-43.pdf (accessed on 14 December 2005).
36. Duits, R.; Franken, E. Left-invariant parabolic evolutions on SE(2) and contour enhancement via invertible orientation scores Part II: Nonlinear left-invariant diffusions on invertible orientation scores. *Q. Appl. Math.* **2010**, *68*, 293–331. [CrossRef]
37. Zhang, J.; Duits, R.; Sanguinetti, G.; ter Haar Romeny, B.M. Numerical Approaches for Linear Left-invariant Diffusions on SE(2), their Comparison to Exact Solutions, and their Applications in Retinal Imaging. *Numer. Methods Theory Appl.* **2016**, *9*, 1–50. [CrossRef]
38. Mumford, D. Elastica and Computer Vision. In *Algebraic Geometry and its Applications*; Springer: New York, NY, USA, 1994; pp. 491–506.
39. Petitot, J. The neurogeometry of pinwheels as a sub-Riemannian contact structure. *J. Physiol. Paris* **2003**, *97*, 265–309. [CrossRef]
40. Portegies, J.M.; Duits, R. New exact and numerical solutions of the (convection–)diffusion kernels on SE(3). *Differ. Geom. Appl.* **2017**, *53*, 182–219. [CrossRef]
41. Chirikjian, G. Degenerate Diffusions and Harmonic Analysis on SE(3): A Tutorial. In *Stochastic Geometric Mechanics*; Albeverio, S., Cruzeiro, A., Holm, D., Eds.; Springer International Publishing: Cham, Switzerland, 2017; pp. 77–99.
42. Park, W.; Liu, Y.; Zhou, Y.; Moses, M.; Chirikjian, G. Kinematic State Estimation and Motion Planning for Stochastic Nonholonomic Systems Using the Exponential Map. *Robotica* **2008**, *26*, 419–434. [CrossRef]
43. Chirikjian, G.; Wang, Y. Conformational Statistics of Stiff Macromolecules as Solutions to PDEs on the Rotation and Motion Groups. *Phys. Rev. E* **2000**, *62*, 880–892. [CrossRef]
44. Portegies, J. PDEs on the Lie Group SE(3). Ph.D. Thesis, Department of Mathematics and Computer Science, Eindhoven University of Technology, Eindhoven, The Netherlands, 2018. Available online: www.bmia.bmt.tue.nl/people/RDuits/PHD/Jorg.pdf (accessed on 12 March 2018).
45. Portegies, J.M.; Fick, R.H.J.; Sanguinetti, G.R.; Meesters, S.P.L.; Girard, G.; Duits, R. Improving Fiber Alignment in HARDI by Combining Contextual PDE Flow with Constrained Spherical Deconvolution. *PLoS ONE* **2015**, *10*, e0138122. [CrossRef] [PubMed]
46. Momayyez-Siahkal, P.; Siddiqi, K. 3D Stochastic Completion Fields for Fiber Tractography. In Proceedings of the 2009 IEEE Computer Society Conference on Computer Vision and Pattern Recognition Workshops, Miami, FL, USA, 20–25 June 2009; pp. 178–185.
47. Skibbe, H.; Reisert, M. Spherical Tensor Algebra: A Toolkit for 3D Image Processing. *J. Math. Imaging Vis.* **2017**, *58*, 349–381. [CrossRef]

48. Meesters, S.; Ossenblok, P.; Wagner, L.; Schijns, O.; Boon, P.; Florack, L.; Vilanova, A.; Duits, R. Stability metrics for optic radiation tractography: Towards damage prediction after resective surgery. *J. Neurosci. Methods* **2017**, *288*, 34–44. [CrossRef] [PubMed]
49. Reisert, M.; Kiselev, V.G. Fiber Continuity: An Anisotropic Prior for ODF Estimation. *IEEE Trans. Med. Imaging* **2011**, *30*, 1274–1283. [CrossRef]
50. Prčkovska, V.; Rodrigues, P.; Duits, R.; Haar Romenij, B.T.; Vilanova, A. Extrapolating fiber crossings from DTI data: Can we infer similar fiber crossings as in HARDI? In Proceedings of the Workshop on Computational Diffusion MRI, MICCA, Beijing, China, 16 January 2010.
51. Iijima, T. *Basic Theory of Pattern Observation*; Papers of Technical Group on Automata and Automatic Control, IECE: Chiba, Japan, 1959.
52. Koenderink, J.J. The structure of images. *Biol. Cybern.* **1984**, *50*, 363–370. [CrossRef]
53. ter Haar Romeny, B.M. *Front-End Vision and Multi-Scale Image Analysis: Multi-Scale Computer Vision Theory and Applications, Written in Mathematica*; Kluwer Academic Publishers: Dordrecht, The Netherlands, 2003.
54. Weickert, J. *Anisotropic Diffusion in Image Processing*; ECMI, B.G. Teubner: Stuttgart, Germany, 1998.
55. Duits, R.; Burgeth, B. Scale Spaces on Lie Groups. In *SSVM*; Lecture Notes in Computer Science; Springer: Berlin/Heidelberg, Germany, 2007; Volume 4485, pp. 300–312.
56. Benoist, Y.; Quint, J.F. Central limit theorem for linear groups. *Ann. Probab.* **2016**, *44*, 1306–1340. [CrossRef]
57. Pilte, M.; Bonnabel, S.; Barbaresco, F. Maneuver Detector for Active Tracking Update Rate Adaptation. In Proceedings of the 2018 19th International Radar Symposium (IRS), Bonn, Germany, 20–22 June 2018; pp. 1–10.
58. Berger, J.; Neufeld, A.; Becker, F.; Lenzen, F.; Schnörr, C. Second Order Minimum Energy Filtering on SE(3) with Nonlinear Measurement Equations. In *Scale Space and Variational Methods in Computer Vision*; Aujol, J.F., Nikolova, M., Papadakis, N., Eds.; Springer: Cham, Switzerland, 2015; pp. 397–409.
59. Oksendal, B. *Stochastic Differential Equations*; Springer: Berlin/Heidelberg, Germany, 1998.
60. Hsu, E. *Stochastic Analysis on Manifolds*; Contemporary Mathematics; American Mathematical Society: Providence, RI, USA, 2002.
61. Feller, W. *An Introduction to Probability Theory and Its Applications*; Wiley Series in Probability and Mathematical Statistics; Wiley: Hoboken, NJ, USA, 1966; Volume II.
62. Felsberg, M.; Duits, R.; Florack, L. The Monogenic Scale Space on a Bounded Domain and its Applications. In *Scale Space Methods in Computer Vision. Scale-Space 2003*; Lecture Notes of Computer Science; Springer: Berlin/Heidelberg, Germany, 2003; Volume 2695, pp. 209–224.
63. Duits, R.; Felsberg, M.; Florack, L.M.J. α Scale Spaces on a Bounded Domain. In *Scale Space Methods in Computer Vision. Scale-Space 2003*; Lecture Notes of Computer Science; Springer: Berlin/Heidelberg, Germany, 2003; Volume 2695, pp. 494–510.
64. Duits, R.; Florack, L.; Graaf, J.D.; Romeny, B.T.H. On the Axioms of Scale Space Theory. *J. Math. Imaging Vis.* **2004**, *20*, 267–298. [CrossRef]
65. Pedersen, K.S.; Duits, R.; Nielsen, M. On α Kernels, Lévy Processes, and Natural Image Statistics. In *Scale Space and PDE Methods in Computer Vision*; Kimmel, R., Sochen, N.A., Weickert, J., Eds.; Springer: Berlin/Heidelberg, Germany, 2005; pp. 468–479.
66. Yosida, K. *Functional Analysis*; Springer: Berlin/Heidelberg, Germany, 1980.
67. Winkels, M.; Cohen, T.S. 3D G-CNNs for Pulmonary Nodule Detection. *arXiv* **2018**, arXiv:1804.04656.
68. Worrall, D.; Brostow, G. CubeNet: Equivariance to 3D Rotation and Translation. *arXiv* **2018**, arXiv:1804.04458.
69. Weiler, M.; Geiger, M.; Welling, M.; Boomsma, W.; Cohen, T. 3D Steerable CNNs: Learning Rotationally Equivariant Features in Volumetric Data. *arXiv* **2018**, arXiv:1807.02547.
70. Montobbio, N.; Sarti, A.; Citti, G. A metric model for the functional architecture of the visual cortex. *arXiv* **2018**, arXiv:1807.02479.
71. Oyallon, E.; Mallat, S.; Sifre, L. Generic deep networks with wavelet scattering. *arXiv* **2013**, arXiv:1312.5940.
72. Kanti. V.; Mardia, P.E.J. *Directional Statistics*; John Wiley and Sons Ltd.: Hoboken, NJ, USA, 1999.
73. Wu, L. Chapter 3 Modeling Financial Security Returns Using Lévy Processes. In *Handbooks in Operations Research and Management Science*; Birge, J.R., Linetsky, V., Eds.; Elsevier: Amsterdam, The Netherlands, 2007; Volume 15, pp. 117–162.
74. Belkic, D.D.; Belkic, K. *Signal Processing in Magnetic Resonance Spectroscopy with Biomedical Applications*; CRC Press: Boca Raton, FL, USA, 2010.

75. Chirikjian, G. Information-theoretic inequalities on unimodular Lie groups. *J. Geom. Mech.* **2010**, *2*, 119–158. [CrossRef] [PubMed]
76. Barbaresco, F. Higher Order Geometric Theory of Information and Heat Based on Poly-Symplectic Geometry of Souriau Lie Groups Thermodynamics and Their Contextures: The Bedrock for Lie Group Machine Learning. *Entropy* **2018**, *20*, 840. [CrossRef]
77. Akian, M.; Quadrat, J.; Viot, M. Bellman processes. *Lect. Notes Control Inf. Sci.* **1994**, *199*, 302–311.
78. Schmidt, M.; Weickert, J. Morphological Counterparts of Linear Shift-Invariant Scale-Spaces. *J. Math. Imaging Vis.* **2016**, *56*, 352–366. [CrossRef]
79. Hörmander, L. Hypoelliptic second order differential equations. *Acta Math.* **1967**, *119*, 147–171. [CrossRef]
80. Misiorek, A.; Weron, R. Heavy-Tailed Distributions in VaR Calculations. In *Handbook of Computational Statistics: Concepts and Methods*; Gentle, J.E., Härdle, W.K., Mori, Y., Eds.; Springer: Berlin/Heidelberg, Germany, 2012; pp. 1025–1059.
81. Felsberg, M.; Sommer, G. The Monogenic Scale-Space: A Unifying Approach to Phase-Based Image Processing in Scale-Space. *J. Math. Imaging Vis.* **2004**, *21*, 5–26. [CrossRef]
82. Kanters, F.; Florack, L.; Duits, R.; Platel, B.; ter Haar Romeny, B. ScaleSpaceViz: α-Scale spaces in practice. *Pattern Recognit. Image Anal.* **2007**, *17*, 106–116. [CrossRef]
83. Duits, R.; Franken, E. Left-invariant parabolic evolutions on SE(2) and contour enhancement via invertible orientation scores Part I: Linear left-invariant diffusion equations on SE(2). *Q. Appl. Math.* **2010**, *68*, 255–292. [CrossRef]
84. Duits, R.; Franken, E. Left-Invariant Diffusions on the Space of Positions and Orientations and their Application to Crossing-Preserving Smoothing of HARDI images. *Int. J. Comput. Vis.* **2011**, *92*, 231–264. [CrossRef]
85. Duits, R.; Dela Haije, T.; Creusen, E.; Ghosh, A. Morphological and Linear Scale Spaces for Fiber Enhancement in DW-MRI. *J. Math. Imaging Vis.* **2012**, *46*, 326–368. [CrossRef]
86. Duits, R.; Bekkers, E.; Mashtakov, A. Fourier Transform on the Homogeneous Space of 3D Positions and Orientations for Exact Solutions to PDEs. *arXiv* **2018**, arXiv:1811.00363.
87. Portegies, J.; Sanguinetti, G.; Meesters, S.; Duits, R. New Approximation of a Scale Space Kernel on SE(3) and Applications in Neuroimaging. In *SSVM*; Aujol, J.F., Nikolova, M., Papadakis, N., Eds.; Number 9087 in LNCS; Springer International Publishing: Cham, Switzerland, 2015; pp. 40–52.
88. Arendt, W.; Bukhvalov, A.V. Integral representation of resolvent and semigroups. *Forum Math.* **1994**, *6*, 111–137. [CrossRef]
89. Griffiths, D. *Introduction to Quantum Mechanics*; Prentice-Hall: Upper Saddle River, NJ, USA, 1994.
90. Wigner, E. Gruppentheorie und ihre Anwendungen auf die Quantenmechanik der Atomspektren. In *Braunschweig: Vieweg Verlag*; 1931 Translated into English by Griffin, J.J.; Group Theory and Its Application to the Quantum Mechanics of Atomic Spectra; Academic Press: New York, NY, USA, 1959.
91. Margenau, H.; Murphy, G.M. *The Mathematics of Physics and Chemistry*; David Van Nostrand: New York, NY, USA, 1956.
92. ter Elst, A.F.M.; Robinson, D.W. Weighted Subcoercive Operators on Lie Groups. *J. Funct. Anal.* **1998**, *157*, 88–163. [CrossRef]
93. Dong, H.; Chirikjian, G. A Comparison of Gaussian and Fourier Methods for Degenerate Diffusions on $SE(2)$. In Proceedings of the 2015 IEEE Conference on Decision and Control, Osaka, Japan, 15–18 December 2015; pp. 15–18.
94. Meesters, S.P.L.; Sanguinetti, G.R.; Garyfallidis, E.; Portegies, J.M.; Duits, R. Fast Implementations of Contextual PDE'S for HARDI Data Processing in DIPY; Abstract; Presented at 24th ISMRM Annual Meeting and Exhibition, Singapore, 7–13 May 2016; pp.1–4.
95. Chirikjian, G. Group Theory and Biomolecular Conformation, I.: Mathematical and computational models. *J. Phys. Condens. Matter* **2010**, *22*, 323103. [CrossRef] [PubMed]
96. Pinsky, M.A. Isotropic transport process on a Riemannian manifold. *Trans. Am. Math. Soc.* **1976**, *218*, 353–360. [CrossRef]
97. Prčkovska, V.; Andorra, M.; Villoslada, P.; Martinez-Heras, E.; Duits, R.; Fortin, D.; Rodrigues, P.; Descoteaux, M. Contextual Diffusion Image Post-processing Aids Clinical Applications. In *Visualization and Processing of Higher Order Descriptors for Multi-Valued Data*; Hotz, I., Schultz, T., Eds.; Mathematics and Visualization; Springer International Publishing: Cham, Switzerland, 2015; pp. 353–377. doi:10.1007/978-3-319-15090-1_18.

98. Meesters, S.P.L.; Sanguinetti, G.R.; Garyfallidis, E.; Portegies, J.M.; Ossenblok, P.; Duits, R. Cleaning Output of Tractography via Fiber to Bundle Coherence, a New Open Source Implementation; Abstract; Presented at Organization for Human Brain Mapping Annual Meeting, Geneve, Switzerland, 26–30 June 2016.
99. Liao, M. *Lévy Processes in Lie Groups*; Cambridge Tracts in Mathematics, Cambridge University Press: Cambridge, UK, 2004.
100. Duits, R.; Ghosh, A.; Dela Haije, T.C.J.; Mashtakov, A. On Sub-Riemannian Geodesics in SE(3) Whose Spatial Projections do not Have Cusps. *J. Dyn. Control Syst.* **2016**, *22*, 771–805. [CrossRef]

entropy

MDPI

Article

Discrete Transforms and Orthogonal Polynomials of (Anti)symmetric Multivariate Sine Functions

Adam Brus, Jiří Hrivnák * and Lenka Motlochová

Department of Physics, Faculty of Nuclear Sciences and Physical Engineering, Czech Technical University in Prague, Břehová 7, 115 19 Prague 1, Czech Republic; adam.brus@fjfi.cvut.cz (A.B.); lenka.motlochova@fjfi.cvut.cz (L.M.)

* Correspondence: jiri.hrivnak@fjfi.cvut.cz; Tel.: +420-2-2435-8264

Received: 29 October 2018; Accepted: 3 December 2018; Published: 6 December 2018

Abstract: Sixteen types of the discrete multivariate transforms, induced by the multivariate antisymmetric and symmetric sine functions, are explicitly developed. Provided by the discrete transforms, inherent interpolation methods are formulated. The four generated classes of the corresponding orthogonal polynomials generalize the formation of the Chebyshev polynomials of the second and fourth kinds. Continuous orthogonality relations of the polynomials together with the inherent weight functions are deduced. Sixteen cubature rules, including the four Gaussian, are produced by the related discrete transforms. For the three-dimensional case, interpolation tests, unitary transform matrices and recursive algorithms for calculation of the polynomials are presented.

Keywords: discrete multivariate sine transforms, orthogonal polynomials, cubature formulas

1. Introduction

The goal of this article is to develop discrete transforms of the multivariate symmetric and antisymmetric sine functions [1] together with the related Fourier interpolation and Chebyshev polynomial methods. The eight symmetric and eight antisymmetric discrete sine transforms form multivariate generalizations of the standard univariate discrete sine transforms [2] and correspond to their multivariate cosine counterparts from [3]. The cubature formulas of multivariate generalizations of the classical Chebyshev polynomials of the second and fourth kind are induced by the discrete transforms.

The multivariate (anti)symmetric trigonometric functions are constructed in [1] as trigonometric analogues of the Weyl orbit functions which are symmetrized or antisymetrized sums of exponential functions. The (anti)symmetrization is performed with respect to the Weyl group, which is a finite group generated by reflections and uniquely connected to a root system of a simple Lie algebra [4,5]. The continuous and discrete transforms together with interpolation tests of the bivariate cosine and sine cases are detailed in [6,7]. The specific connection between the Weyl orbit functions [4,5,8] and the (anti)symmetric trigonometric functions is deduced in [9]. For instance, since the Weyl group of simple Lie algebra C_n is isomorphic to $(\mathbb{Z}/2\mathbb{Z})^n \rtimes S_n$, the antisymetric sine functions coincide, up to a constant, with the antisymmetric Weyl orbit functions. The discrete transforms of the Weyl orbit functions on lattice fragments are induced by group theoretic arguments in [10–13]. On the other hand, the discrete transforms of the multivariate (anti)symmetric cosine transforms [3] are generated by their eight underlying standard univariate cosine transforms [2]. The eight univariate discrete sine transforms represent solutions of the discretized harmonic oscillator equation with distinct boundary conditions applied on the grid or mid-grid points [2]. These boundary conditions are inherited by the multivariate generalizations and combined with the behavior on the (anti)symmetrization boundaries determine overall properties of each discrete transform. The (anti)symmetric discrete sine transform developed from the discrete cosine transform of type I is derived in [1] and the antisymmetric discrete sine transforms are special cases of the transforms in [14]. However, the symmetric discrete transforms of types II–VIII in n variables have not been studied elsewhere. The multivariate (anti)symmetric trigonometric functions lead to the generalizations of the Chebyshev polynomials and the discrete transforms produce effective interpolation methods along with the related cubature integration formulas.

The four kinds of the classical Chebyshev polynomials serve as widely utilized orthogonal polynomials intertwined with powerful methods of numerical integration and approximation [15,16]. The (anti)symmetric cosine functions grant a multidimensional generalization of the Chebyshev polynomials of the first and third kinds [3], the present (anti)symmetric sine functions provide generalization of the second and fourth kinds firstly introduced in the present paper. The bivariate polynomials form special cases of analogues of the Jacobi polynomials [17]. The multivariate (anti)symmetric sine polynomials acquire essential characteristics from the (anti)symmetric sine functions and this connection grants required apparatus for generalization of efficient cubature formulas of the univariate Chebyshev polynomials. The fundamental goal of cubature formulas, replacing integration by optimal finite summation, is achieved via the suitable point sets of the generalized Chebyshev nodes [18]. Similarly to the classical univariate Chebyshev case, the resulting finite sum over the nodes equals exactly the approximated integral for polynomials up to a specific degree. Gaussian cubature formulas involve the lowest bound of the number of nodes and attain the maximal degree of precision [19–21]. Among the presented sixteen types of the symmetric sine cubature formulas four types are Gaussian which turn out to be special cases of formulas studied from different point of view in [22]. The other cubature formulas completing the set of integration formulas connected with the discrete multivariate sine transforms of types I–VIII are novel.

The successful interpolation tests for the 2D and 3D (anti)symmetric trigonometric functions are accomplished in [3,6,7]. The (anti)symmetric sine functions form solutions of the Laplace equation that satisfy a specific combination of the Dirichlet and von Neumann conditions on the boundaries of the fundamental domain [1]. The 2D and 3D (anti)symmetric sine functions as eigenfunctions of the discretized Laplace operator potentially represent solutions to lattice vibration models in solid state physics as well as foundation for description of the corresponding models in quantum field theory [23]. Boundary conditions of these models are determined by the boundary behavior of the multivariate discrete transforms and the spectral analysis provided by the developed transforms contributes to explicit solutions of the time evolution of the mechanical models. The approximation capability of the cubature formulas in the Weyl group setting that includes 2D cases of the (anti)symmetric trigonometric functions is successfully tested in [24]. The applications of Weyl orbit functions in image processing are developed in [25]. The potential physical applications of the studied cubature formulas encompass calculations in laser optics [26], stochastic dynamics [27], quantum dynamics [28], fluid flows [29], magnetostatic modeling [30], electromagnetic wave propagation [31], micromagnetic simulations [32], liquid crystal colloids [33] and porous materials [34,35].

The paper is organized as follows. In Section 2, the multivariate (anti)symmetric sine functions, their symmetry properties and continuous orthogonality are recalled. In Section 3, the sixteen types of the discrete (anti)symmetric sine transforms are listed and the interpolation method along, with the form of the unitary transform matrices, is presented. In Section 4, the multivariate generalization of the Chebyshev polynomials of the second and fourth kinds are introduced and the corresponding continuous orthogonality and cubature rules are deduced.

2. Multivariate (Anti)symmetric Sine Functions

2.1. *Definitions and Symmetry Properties*

The multivariate symmetric and antisymmetric generalizations of trigonometric functions are introduced in [1], symmetry properties of the (anti)symmetric discrete cosine transforms are detailed in [3]. The antisymmetric sine functions $\sin^-_\lambda(x)$ and the symmetric sine functions $\sin^+_\lambda(x)$, labeled by the parameter $\lambda = (\lambda_1, \lambda_2, \ldots, \lambda_n) \in \mathbb{R}^n$ and of the variable $x = (x_1, x_2, \ldots, x_n) \in \mathbb{R}^n$, are defined via the determinants and permanents of matrices with entries $\sin(\pi\lambda_i x_j)$. Denoting the group of permutations by S_n and the sign homomorphism on S_n by sgn, the (anti)symmetric sine functions of several variables [1] are given explicitly by

$$\begin{aligned}\sin^-_\lambda(x) &= \sum_{\sigma\in S_n} \operatorname{sgn}(\sigma) \sin(\pi\lambda_{\sigma(1)}x_1)\sin(\pi\lambda_{\sigma(2)}x_2)\ldots\sin(\pi\lambda_{\sigma(n)}x_n),\\ \sin^+_\lambda(x) &= \sum_{\sigma\in S_n} \sin(\pi\lambda_{\sigma(1)}x_1)\sin(\pi\lambda_{\sigma(2)}x_2)\ldots\sin(\pi\lambda_{\sigma(n)}x_n).\end{aligned} \tag{1}$$

The (anti)symmetric cosine functions [1] $\cos^\pm_\lambda(x)$ are for parameter $\lambda = (\lambda_1, \lambda_2, \ldots, \lambda_n) \in \mathbb{R}^n$ and variable $x = (x_1, x_2, \ldots, x_n) \in \mathbb{R}^n$ given similarly by the following formulas,

$$\begin{aligned}\cos^-_\lambda(x) &= \sum_{\sigma\in S_n} \operatorname{sgn}(\sigma) \cos(\pi\lambda_{\sigma(1)}x_1)\cos(\pi\lambda_{\sigma(2)}x_2)\ldots\cos(\pi\lambda_{\sigma(n)}x_n),\\ \cos^+_\lambda(x) &= \sum_{\sigma\in S_n} \cos(\pi\lambda_{\sigma(1)}x_1)\cos(\pi\lambda_{\sigma(2)}x_2)\ldots\cos(\pi\lambda_{\sigma(n)}x_n).\end{aligned} \tag{2}$$

The (anti)symmetric sine functions posses several crucial symmetry properties [1]. Directly from definition (1), the multivariate sine functions $\sin^\pm_\lambda(x)$ are (anti)symmetric with respect to the action of a permutation $\sigma \in S_n$,

$$\begin{aligned}\sin^-_\lambda(\sigma(x)) &= \operatorname{sgn}(\sigma)\sin^-_\lambda(x), &\quad \sin^-_{\sigma(\lambda)}(x) &= \operatorname{sgn}(\sigma)\sin^-_\lambda(x),\\ \sin^+_\lambda(\sigma(x)) &= \sin^+_\lambda(x), &\quad \sin^+_{\sigma(\lambda)}(x) &= \sin^+_\lambda(x),\end{aligned} \tag{3}$$

where $\sigma(x) = (x_{\sigma(1)}, x_{\sigma(2)}, \ldots, x_{\sigma(n)})$ and $\sigma(\lambda) = (\lambda_{\sigma(1)}, \lambda_{\sigma(2)}, \ldots, \lambda_{\sigma(n)})$. Furthermore, the sine functions are anti–invariant with respect to sign alternations of both variables and parameters. For a change of sign τ_i of any i-th coordinate of the variable $x \in \mathbb{R}^n$ or the parameter $\lambda \in \mathbb{R}^n$,

$$\begin{aligned}\tau_i(x_1, \ldots, x_i, \ldots, x_n) &= (x_1, \ldots, -x_i, \ldots, x_n),\\ \tau_i(\lambda_1, \ldots, \lambda_i, \ldots, \lambda_n) &= (\lambda_1, \ldots, -\lambda_i, \ldots, \lambda_n),\end{aligned} \tag{4}$$

it holds that

$$\begin{aligned}\sin^\pm_\lambda(\tau_i(x)) &= -\sin^\pm_\lambda(x),\\ \sin^\pm_{\tau_i(\lambda)}(x) &= -\sin^\pm_\lambda(x).\end{aligned} \tag{5}$$

Therefore, the functions $\sin^\pm_\lambda$ vanish if any coordinate of the variable x_i or the parameter λ_i are equal to zero.

Setting the $\varrho-$vector as

$$\varrho = \left(\frac{1}{2}, \frac{1}{2}, \ldots, \frac{1}{2}\right), \tag{6}$$

the functions $\sin^\pm_k$ and $\sin^\pm_{k-\varrho}$, $k \in \mathbb{Z}^n$, admit additional symmetries with respect to multivariate integer shifts $t = (t_1, t_2, \ldots, t_n) \in \mathbb{Z}^n$ that stem from the periodicity of the univariate sine function,

$$\begin{aligned}\sin^\pm_k(x + 2t) &= \sin^\pm_k(x),\\ \sin^\pm_{k-\varrho}(x + 2t) &= (-1)^{t_1+\cdots+t_n}\sin^\pm_{k-\varrho}(x).\end{aligned} \tag{7}$$

The sets of integer parameters $P^\pm_1$ are introduced as

$$\begin{aligned}P^+_1 &= \{(k_1, k_2, \ldots, k_n) \in \mathbb{Z}^n \mid k_1 \geq k_2 \geq \cdots \geq k_n \geq 1\},\\ P^-_1 &= \{(k_1, k_2, \ldots, k_n) \in \mathbb{Z}^n \mid k_1 > k_2 > \cdots > k_n \geq 1\}.\end{aligned} \tag{8}$$

The relations (3) and (5) imply that it suffices to parametrize the functions $\sin_k^{\pm}$ and $\sin_{k-\varrho}^{\pm}$ only by the following values,

$$\begin{aligned} &\sin_k^+, \sin_{k-\varrho}^+ : \quad k \in P_1^+, \\ &\sin_k^-, \sin_{k-\varrho}^- : \quad k \in P_1^-. \end{aligned} \tag{9}$$

Due to relations (3), (5) and (7), the functions $\sin_k^{\pm}$ and $\sin_{k-\varrho}^{\pm}$ are restricted to the closure of the fundamental domain $F(\widetilde{S}_n^{\text{aff}})$,

$$F(\widetilde{S}_n^{\text{aff}}) = \{(x_1, x_2, \ldots, x_n) \in \mathbb{R}^n \,|\, 1 \geq x_1 \geq x_2 \geq \ldots \geq x_n \geq 0\}. \tag{10}$$

Furthermore, it follows from the symmetry relations (3), (5) and the identity $\sin(\pi k_i) = 0, k_i \in \mathbb{Z}$, that the functions $\sin_k^{\pm}$ and $\sin_{k-\varrho}^{\pm}$ are identically equal to zero on certain parts of the boundary of the domain $F(\widetilde{S}_n^{\text{aff}})$. In particular, the following points are omitted from $F(\widetilde{S}_n^{\text{aff}})$,

$$\begin{aligned} &\sin_k^-(x): \quad && x_i = x_{i+1}, i \in \{1, \ldots, n-1\};\; x_1 = 1;\; x_n = 0, \\ &\sin_{k-\varrho}^-(x): && x_i = x_{i+1}, i \in \{1, \ldots, n-1\};\; x_n = 0, \\ &\sin_k^+(x): && x_1 = 1;\; x_n = 0, \\ &\sin_{k-\varrho}^+(x): && x_n = 0. \end{aligned} \tag{11}$$

In order to analyse polynomials of several variables in Section 4, four special cases of multivariate sine functions $\sin_\lambda^{\pm}$, labeled by the generalized $\varrho-$vectors

$$\varrho_2^- = (n, n-1, \ldots, 1), \quad \varrho_2^+ = (1, 1, \ldots, 1), \quad \varrho_4^- = \left(n - \frac{1}{2}, n - \frac{3}{2}, \ldots, \frac{1}{2}\right), \quad \varrho_4^+ = \varrho, \tag{12}$$

are expressed in their product forms.

Proposition 1. *Let $k \in \mathbb{N}$ be given by*

$$k = \begin{cases} \frac{n-1}{2} & \text{for } n \text{ odd}, \\ \frac{n}{2} & \text{for } n \text{ even}. \end{cases} \tag{13}$$

Then it holds that

$$\sin_{\varrho_2^-}^-(x_1, \ldots, x_n) = (-1)^k 2^{n(n-1)} \prod_{i=1}^n \sin(\pi x_i) \prod_{1 \leq i < j \leq n} \sin\left(\frac{\pi}{2}(x_i + x_j)\right) \sin\left(\frac{\pi}{2}(x_i - x_j)\right), \tag{14}$$

$$\sin_{\varrho_2^+}^+(x_1, \ldots, x_n) = n! \prod_{i=1}^n \sin(\pi x_i), \tag{15}$$

$$\sin_{\varrho_4^-}^-(x_1, \ldots, x_n) = (-1)^k 2^{n(n-1)} \prod_{i=1}^n \sin\left(\frac{\pi}{2} x_i\right) \prod_{1 \leq i < j \leq n} \sin\left(\frac{\pi}{2}(x_i + x_j)\right) \sin\left(\frac{\pi}{2}(x_i - x_j)\right), \tag{16}$$

$$\sin_\varrho^+(x_1, \ldots, x_n) = n! \prod_{i=1}^n \sin\left(\frac{\pi}{2} x_i\right). \tag{17}$$

Proof. The Formulas (15) and (17) follow directly from definition. The Equality (14) is derived in [3]. From the definition (1) and the symmetry property (3), the function $\sin^-_{\varrho_4^-}$ is given by

$$\sin^-_{\varrho_4^-}(x_1,\dots,x_n) = (-1)^k \sin^-_{\left(\frac{1}{2},\frac{3}{2},\dots,n-\frac{1}{2}\right)} = (-1)^k \det \begin{pmatrix} \sin\left(\frac{\pi}{2}x_1\right) & \sin\left(\frac{3\pi}{2}x_1\right) & \cdots & \sin\left(\frac{(2n-1)\pi}{2}x_1\right) \\ \sin\left(\frac{\pi}{2}x_2\right) & \sin\left(\frac{3\pi}{2}x_2\right) & \cdots & \sin\left(\frac{(2n-1)\pi}{2}x_2\right) \\ \vdots & \vdots & \ddots & \vdots \\ \sin\left(\frac{\pi}{2}x_n\right) & \sin\left(\frac{3\pi}{2}x_n\right) & \cdots & \sin\left(\frac{(2n-1)\pi}{2}x_n\right) \end{pmatrix}. \quad (18)$$

Basic properties of determinants together with the trigonometric identity for powers of the sine function

$$\sin((2m-1)\theta) = (-1)^{m-1}2^{2(m-1)}\sin^{2m-1}(\theta) - \textstyle\sum_{i=1}^{m-1}(-1)^i\binom{2n-1}{i}\sin((2n-1-2i)\theta), \quad m \in \mathbb{N}, \quad (19)$$

and the power-reduction formula

$$\cos(2\theta) = 1 - 2\sin^2(\theta) \quad (20)$$

imply that the determinant (18) is of the following form,

$$\sin^-_{\varrho_4^-}(x_1,\dots,x_n) = (-1)^k \det \begin{pmatrix} \sin\left(\frac{\pi}{2}x_1\right) & 2\sin\left(\frac{\pi}{2}x_1\right)\cos(\pi x_1) & \cdots & 2^{n-1}\sin\left(\frac{\pi}{2}x_1\right)\cos^{n-1}(\pi x_1) \\ \sin\left(\frac{\pi}{2}x_2\right) & 2\sin\left(\frac{\pi}{2}x_1\right)\cos(\pi x_2) & \cdots & 2^{n-1}\sin\left(\frac{\pi}{2}x_1\right)\cos^{n-1}(\pi x_2) \\ \vdots & \vdots & \ddots & \vdots \\ \sin\left(\frac{\pi}{2}x_n\right) & 2\sin\left(\frac{\pi}{2}x_1\right)\cos(\pi x_n) & \cdots & 2^{n-1}\sin\left(\frac{\pi}{2}x_1\right)\cos^{n-1}(\pi x_n) \end{pmatrix}. \quad (21)$$

The Formula (21) is rewritten as

$$\sin^-_{\varrho_4^-}(x_1,\dots,x_n) = (-1)^k 2^{\frac{n(n-1)}{2}} \prod_{i=1}^{n} \sin\left(\frac{\pi}{2}x_i\right) \det \begin{pmatrix} 1 & \cos(\pi x_1) & \cdots & \cos^{n-1}(\pi x_1) \\ 1 & \cos(\pi x_2) & \cdots & \cos^{n-1}(\pi x_2) \\ \vdots & \vdots & \ddots & \vdots \\ 1 & \cos(\pi x_n) & \cdots & \cos^{n-1}(\pi x_n) \end{pmatrix}. \quad (22)$$

Taking into account that the last determinant is of the Vandermonde type, it holds that

$$\sin^-_{\varrho_4^-}(x_1,\dots,x_n) = (-1)^k 2^{\frac{n(n-1)}{2}} \prod_{i=1}^{n} \sin\left(\frac{\pi}{2}x_i\right) \prod_{1\le i<j\le n} \cos(\pi x_i) - \cos(\pi x_j). \quad (23)$$

The sum-to-product trigonometric identity

$$\cos(\pi x_i) - \cos(\pi x_j) = 2\sin\left(\frac{\pi}{2}(x_i+x_j)\right)\sin\left(\frac{\pi}{2}(x_i-x_j)\right) \quad (24)$$

substituted in relation (23) yields Formula (16). □

Due to identities (14)–(17), the functions $\sin^\pm_{\varrho_2^\pm}$ and $\sin^-_{\varrho_4^\pm}$ vanish only on the parts of the boundary points of $F(\widetilde{S}_n^{\mathrm{aff}})$ specified by (11).

Corollary 1. *The functions $\sin^\pm_{\varrho_2^\pm}$ and $\sin^-_{\varrho_4^\pm}$ are non-zero in the interior $F(\widetilde{S}_n^{\mathrm{aff}})^\circ$ of the fundamental domain $F(\widetilde{S}_n^{\mathrm{aff}})$.*

Example 1. *Contour plots of the graph cuts $z = 1/5$ of the symmetric trivariate sine function $\sin^+_k(x,y,z)$ and $\sin^+_{k-\varrho}(x,y,z)$ are for parameters $k = (2,1,1), (3,1,1), (3,3,1)$ and $k = (3,2,2), (4,2,2), (4,4,2)$ depicted in Figures 1 and 2, respectively. Contour plots of the graph cuts $z = 1/5$ of the antisymmetric trivariate sine function $\sin^-_k(x,y,z)$ and $\sin^-_{k-\varrho}(x,y,z)$ are for parameters $k = (4,2,1), (5,2,1), (5,4,1)$ and $k = (5,3,2), (6,3,2), (6,5,2)$ depicted in Figures 3 and 4, respectively. The specific values of parameters are dispersed to visualize a wide range of different generalized trigonometric functions that possess common* *symmetry properties within each family. The plotting of the figures in this example, as well as plotting, numerical calculations and integrations in subsequent examples is performed by Wolfram Mathematica.*

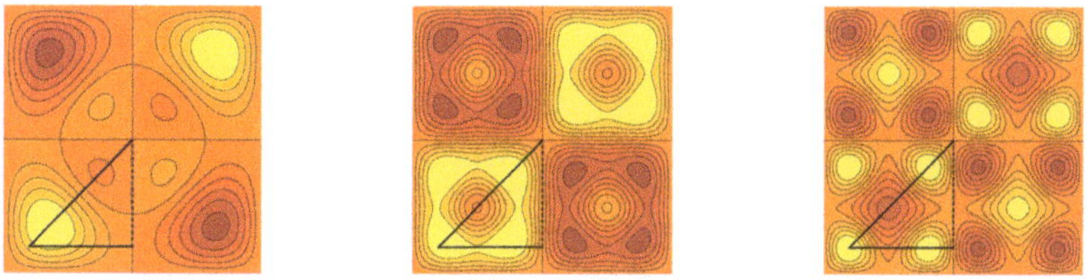

Figure 1. The contour plots of the symmetric trivariate sine function $\sin_k^+(x,y,1/5)$ with $k=(2,1,1)$, $(3,1,1)$ and $(3,3,1)$. The border of the fundamental domain is depicted by the black line.

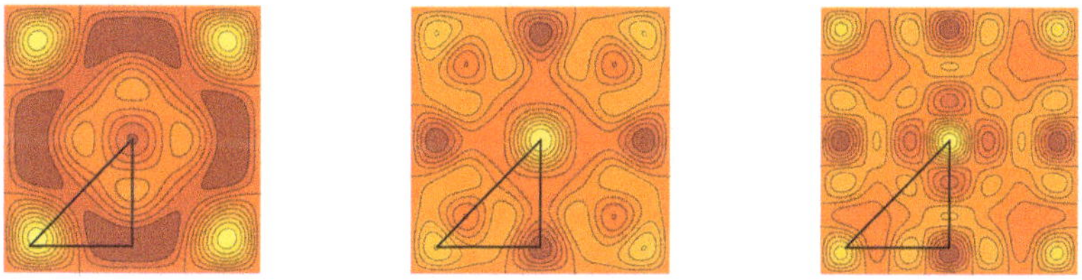

Figure 2. The contour plots of the symmetric trivariate sine function $\sin_{k-\varrho}^+(x,y,1/5)$ with $k=(3,2,2)$, $(4,2,2)$ and $(4,4,2)$. The border of the fundamental domain is depicted by the black line.

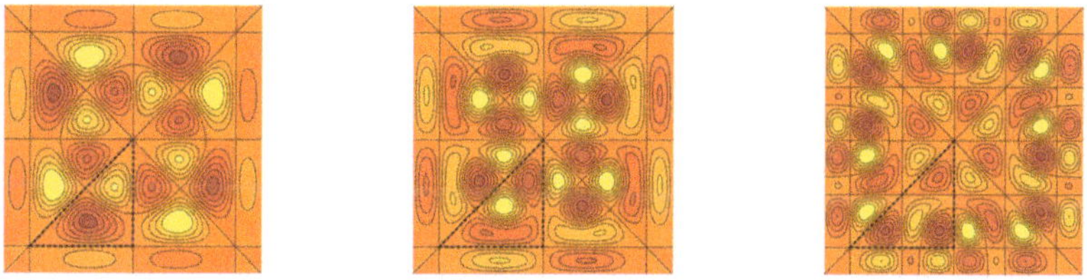

Figure 3. The contour plots of the antisymmetric trivariate sine function $\sin_k^-(x,y,1/5)$ with $k=(4,2,1),(5,2,1),(5,4,1)$. The border of the fundamental domain is depicted by the black line.

Figure 4. The contour plots of the antisymmetric trivariate sine function $\sin_{k-\varrho}^-(x,y,1/5)$ with $k=(5,3,2),(6,3,2),(6,5,2)$. The border of the fundamental domain is depicted by the black line.

2.2. *Continuous Orthogonality*

The antisymmetric and symmetric sine functions (9) are pairwise continuously orthogonal within each family when integrated over $F(\widetilde{S}_n^{\text{aff}})$. Denoting the order of the stabilizer subgroup $\text{Stab}_{S_n}(\lambda)$ of S_n with respect to a point $\lambda \in \mathbb{R}^n$ by H_λ,

$$H_\lambda = |\text{Stab}_{S_n}(\lambda)|, \tag{25}$$

the continuous orthogonality relations of the (anti)symmetric sine functions are given by

$$\int_{F(\widetilde{S}_n^{\text{aff}})} \sin_k^-(x) \sin_{k'}^-(x)\, dx = 2^{-n}\delta_{kk'}, \quad k,k' \in P_1^-, \tag{26}$$

$$\int_{F(\widetilde{S}_n^{\text{aff}})} \sin_{k-\varrho}^-(x) \sin_{k'-\varrho}^-(x)\, dx = 2^{-n}\delta_{kk'}, \quad k,k' \in P_1^-, \tag{27}$$

$$\int_{F(\widetilde{S}_n^{\text{aff}})} \sin_k^+(x) \sin_{k'}^+(x)\, dx = 2^{-n}H_k\delta_{kk'}, \quad k,k' \in P_1^+, \tag{28}$$

$$\int_{F(\widetilde{S}_n^{\text{aff}})} \sin_{k-\varrho}^+(x) \sin_{k'-\varrho}^+(x)\, dx = 2^{-n}H_k\delta_{kk'}, \quad k,k' \in P_1^+, \tag{29}$$

where $\delta_{kk'}$ denotes the Kronecker delta.

The orthogonality relations (26) and (28) are deduced in [1] from the continuous orthogonality of univariate sine functions $\sin(\pi m\theta)$, $m \in \mathbb{N}$ over the interval $[0,1]$. The remaining relations (27) and (29) follow similarly from the continuous orthogonality of the shifted sine functions $\sin(\pi(m-1/2)\theta)$, $m \in \mathbb{N}$,

$$\int_0^1 \sin\left(\pi\left(m-\frac{1}{2}\right)\theta\right)\sin\left(\pi\left(m'-\frac{1}{2}\right)\theta\right)d\theta = \frac{1}{2}\delta_{mm'}, \quad m,m' \in \mathbb{N}. \tag{30}$$

3. Discrete Transforms

The standard discrete sine transforms (DSTs) arise naturally from discretized solution of the harmonic oscillator equation with certain choices of boundary conditions [2]. The Dirichlet boundary condition is required at the beginning of the interval whereas the Neumann and Dirichlet boundary conditions are both allowed at the other end of the interval. Application of the boundary conditions at the grid or mid-grid points produces eight different transforms DST-I, ..., DST-VIII. The antisymmetric and symmetric generalizations of DSTs result in 16 various multivariate discrete transforms denoted by AMDST and SMDST respectively. The (anti-)symmetric multivariate sine transforms of type I are derived in [1] by employing DST-I. A similar method is used to complete the list of AMDST and SMDST.

In order to describe the generalized discrete transforms, two sets of labels $D_{1,N}$ and $D_{1,N,\varrho}$ are introduced for an arbitrary scaling factor $N \in \mathbb{N}$,

$$\begin{aligned} D_{1,N} &= \{(k_1,k_2,\ldots,k_n) \in \mathbb{Z}^n \mid N \geq k_i \geq 1, i=1,\ldots,n\}, \\ D_{1,N,\varrho} &= -\varrho + D_{1,N} = \left\{\left(k_1-\frac{1}{2}, k_2-\frac{1}{2},\ldots,k_n-\frac{1}{2}\right) \mid k_i \in \mathbb{Z}^n,\ N \geq k_i \geq 1, i=1,\ldots,n\right\}. \end{aligned} \tag{31}$$

The normalization function d assigns to each label $k \in D_{1,N}$ the value d_k determined by

$$d_k = c_{k_1}c_{k_2}\cdots c_{k_n}, \qquad c_{k_i} = \begin{cases} \frac{1}{2} & \text{if } k_i = N, \\ 1 & \text{otherwise.} \end{cases} \tag{32}$$

Similarly, to each label $k \in D_{1,N,\varrho}$ is assigned the value $\widetilde{d}_k$ equal to the previous discrete function evaluated at non-shifted point $k+\varrho$ from $D_{1,N}$,

$$\widetilde{d}_k = d_{k+\varrho}. \tag{33}$$

The generalized discrete transforms are developed on specific finite sets of points contained in $F(\widetilde{S}_n^{\mathrm{aff}})$. The point sets $C_{1,N}^m$ and $C_{1,N,\varrho}^m$ are subsets of two types of cubic lattices defined for four cases $m \in \{N+1, N, (2N+1)/2, (2N-1)/2\}$ by

$$C_{1,N}^m = \frac{1}{m} D_{1,N}, \qquad C_{1,N,\varrho}^m = \frac{1}{m} D_{1,N,\varrho}. \tag{34}$$

The discrete weight function ε, defined for each point $s \in C_{1,N}^m$, is specified by the value of the function d on the point $ms \in D_{1,N}$,

$$\varepsilon_s = d_{ms}. \tag{35}$$

The discrete function $\widetilde{\varepsilon}$ is for each point $s \in C_{1,N,\varrho}^m$ given by

$$\widetilde{\varepsilon}_s = \widetilde{d}_{ms}. \tag{36}$$

3.1. Antisymmetric Multivariate Discrete Sine Transforms

For an arbitrary scaling factor $N \in \mathbb{N}$ greater than or equal to n, AMDSTs express a discrete-valued function as a linear combination of antisymmetric sine functions. The functions $\sin_k^-$ are labeled by a finite set $D_{1,N}^-$ of labels in P_1^- with coordinates not exceeding the value N,

$$D_{1,N}^- = D_{1,N} \cap P_1^- = \{(k_1, k_2, \ldots, k_n) \in \mathbb{Z}^n \mid N \geq k_1 > k_2 > \ldots > k_n \geq 1\}, \tag{37}$$

and by the label set $D_{1,N,\varrho}^-$ containing all labels of $D_{1,N}^-$ shifted by $-\varrho$,

$$\begin{aligned} D_{1,N,\varrho}^- &= -\varrho + D_{1,N}^- = D_{1,N,\varrho} \cap \{-\varrho + P_1^-\} \\ &= \left\{ \left(k_1 - \frac{1}{2}, k_2 - \frac{1}{2}, \ldots, k_n - \frac{1}{2}\right) \mid k_i \in \mathbb{Z},\, N \geq k_1 > k_2 > \ldots > k_n \geq 1 \right\}. \end{aligned} \tag{38}$$

In particular, Table 1 determines the finite set of labels $D_N^{\star,-}$ and the corresponding finite set of points $F_N^{\star,-} \subset F(\widetilde{S}_n^{\mathrm{aff}})$, on which an expanded discrete function is evaluated, for each type $\star \in \{\mathrm{I}, \mathrm{II}, \ldots, \mathrm{VIII}\}$ of AMDST. Each antisymmetric $\star-$type transform requires the inherent weights $\varepsilon^\star$ and normalization coefficients $h^\star$ listed also in Table 1. The antisymmetric sine functions labeled by $k, k' \in D_N^{\star,-}$ form an orthogonal basis of real-valued functions defined on the finite point set $F_N^{\star,-}$ of each type,

$$\sum_{s \in F_N^{\star,-}} \varepsilon_s^\star \sin_k^-(s) \sin_{k'}^-(s) = h_k^\star \delta_{kk'}. \tag{39}$$

The discrete orthogonality (39) implies that any function $f : F_N^{\star,-} \to \mathbb{R}$ is expanded in terms of antisymmetric sine functions labeled by $k \in D_N^{\star,-}$ as

$$f(s) = \sum_{k \in D_N^{\star,-}} A_k^\star \sin_k^-(s), \qquad A_k^\star = \frac{1}{h_k^\star} \sum_{s \in F_N^{\star,-}} \varepsilon_s^\star f(s) \sin_k^-(s). \tag{40}$$

The eight types of AMDSTs specialize for $n=1$ to the corresponding standard DSTs.

Table 1. The sets of labels $D_N^{\star,-}, D_N^{\star,+}$ and sets of points $F_N^{\star,-}, F_N^{\star,+}$ together with the weights $\varepsilon^\star$ and normalization coefficients $h^\star$ are specified for each type $\star \in \{\mathrm{I}, \mathrm{II}, \ldots, \mathrm{VIII}\}$ of antisymmetric generalizations of discrete sine transforms (DSTs) (AMDST), and symmetric generalizations of DSTs (SMDST), respectively.

$\star$	$D_N^{\star,-}$	$F_N^{\star,-}$	$D_N^{\star,+}$	$F_N^{\star,+}$	$h_k^\star$	$\varepsilon_s^\star$
I	$D_{1,N}^-$	$\frac{1}{N+1}D_{1,N}^-$	$D_{1,N}^+$	$\frac{1}{N+1}D_{1,N}^+$	$\left(\frac{N+1}{2}\right)^n$	1
II	$D_{1,N}^-$	$\frac{1}{N}D_{1,N,\varrho}^-$	$D_{1,N}^+$	$\frac{1}{N}D_{1,N,\varrho}^+$	$d_k^{-1}\left(\frac{N}{2}\right)^n$	1
III	$D_{1,N,\varrho}^-$	$\frac{1}{N}D_{1,N}^-$	$D_{1,N,\varrho}^+$	$\frac{1}{N}D_{1,N}^+$	$\left(\frac{N}{2}\right)^n$	ε_s
IV	$D_{1,N,\varrho}^-$	$\frac{1}{N}D_{1,N,\varrho}^-$	$D_{1,N,\varrho}^+$	$\frac{1}{N}D_{1,N,\varrho}^+$	$\left(\frac{N}{2}\right)^n$	1
V	$D_{1,N}^-$	$\frac{2}{2N+1}D_{1,N}^-$	$D_{1,N}^+$	$\frac{2}{2N+1}D_{1,N}^+$	$\left(\frac{2N+1}{4}\right)^n$	1
VI	$D_{1,N}^-$	$\frac{2}{2N+1}D_{1,N,\varrho}^-$	$D_{1,N}^+$	$\frac{2}{2N+1}D_{1,N,\varrho}^+$	$\left(\frac{2N+1}{4}\right)^n$	1
VII	$D_{1,N,\varrho}^-$	$\frac{2}{2N+1}D_{1,N}^-$	$D_{1,N,\varrho}^+$	$\frac{2}{2N+1}D_{1,N}^+$	$\left(\frac{2N+1}{4}\right)^n$	1
VIII	$D_{1,N,\varrho}^-$	$\frac{2}{2N-1}D_{1,N,\varrho}^-$	$D_{1,N,\varrho}^+$	$\frac{2}{2N-1}D_{1,N,\varrho}^+$	$\widetilde{d}_k^{-1}\left(\frac{2N-1}{4}\right)^n$	$\widetilde{\varepsilon}_s$

Remark 1. *The antisymmetric sine functions labeled by the parameters k of the form* $(N+1, k_2, \ldots, k_n)$ *and* $(N+1/2, k_2, \ldots, k_n)$ *are identically equal to zero for all points from the sets* $F_N^{\mathrm{I},-}$ *and* $F_N^{\mathrm{VII},-}$*, respectively. Therefore, the discrete orthogonality relations (39) for* $\star = \mathrm{I}$ *and* $\star = \mathrm{VII}$ *remain valid if either k or k′, but not both, are of such form.*

3.2. Symmetric Multivariate Discrete Sine Transforms

For an arbitrary scaling factor $N \in \mathbb{N}$, SMDSTs express a discrete-valued function as a linear combination of symmetric sine functions. The functions $\sin_k^+$ are labeled by a finite set $D_{1,N}^+$ of points in P_1^+ with coordinates not exceeding the value N,

$$D_{1,N}^+ = D_{1,N} \cap P_1^+ = \{(k_1, k_2, \ldots, k_n) \in \mathbb{Z}^n \mid N \geq k_1 \geq k_2 \geq \ldots \geq k_n \geq 1\}, \tag{41}$$

or by the set $D_{1,N,\varrho}^+$ containing all points of $D_{1,N}^+$ shifted by $-\varrho$,

$$\begin{aligned} D_{1,N,\varrho}^+ &= -\varrho + D_{1,N}^+ = D_{1,N,\varrho} \cap \{-\varrho + P_1^+\} \\ &= \left\{ \left(k_1 - \frac{1}{2}, k_2 - \frac{1}{2}, \ldots, k_n - \frac{1}{2}\right) \mid N \geq k_1 \geq k_2 \geq \ldots \geq k_n \geq 1, k_i \in \mathbb{Z} \right\}. \end{aligned} \tag{42}$$

In particular, Table 1 determines the finite set of labels $D_N^{\star,+}$ and the corresponding finite set of points $F_N^{\star,+} \subset F(\widetilde{S}_n^{\mathrm{aff}})$, on which an expanded discrete function is evaluated, for each type $\star \in \{\mathrm{I}, \mathrm{II}, \ldots, \mathrm{VIII}\}$ of SMDST. Besides the weights $\varepsilon^\star$ and normalization coefficients $h^\star$ from Table 1, the stabilizer function H_k, defined by (25), enters each symmetric transform. The symmetric sine functions, labeled by $k, k' \in D_N^{\star,+}$, form an orthogonal basis of real-valued functions defined on the finite point set $F_N^{\star,+}$ of each type,

$$\sum_{s \in F_N^{\star,+}} \varepsilon_s^\star H_s^{-1} \sin_k^+(s) \sin_{k'}^+(s) = h_k^\star H_k \delta_{kk'}. \tag{43}$$

The discrete orthogonality (43) implies that any function $f : F_N^{\star,+} \to \mathbb{R}$ is expanded in terms of symmetric sine functions labeled by $k \in D_N^{\star,+}$ as

$$f(s) = \sum_{k \in D_N^{\star,+}} A_k^\star \sin_k^+(s), \qquad A_k^\star = \frac{1}{h_k^\star H_k} \sum_{s \in F_N^{\star,+}} \varepsilon_s^\star H_s^{-1} f(s) \sin_k^+(s). \tag{44}$$

The eight types of SMDSTs specialize for $n = 1$ to the corresponding standard DSTs.

Remark 2. *The symmetric sine functions, labeled by the parameters k of the form* $(N+1, k_2, \dots, k_n)$ *and* $(N+1/2, k_2, \dots, k_n)$*, are identically equal to zero for all points from the sets* $F_N^{\mathrm{I},+}$ *and* $F_N^{\mathrm{VII},+}$*, respectively. Therefore, the discrete orthogonality relations (43) for* $\star = \mathrm{I}$ *and* $\star = \mathrm{VII}$ *remain valid if k or k′, but not both, are of such form.*

3.3. Interpolation by (Anti)symmetric Sine Functions

The developed formalism of discrete transforms provides solution of an interpolation problem formulated for a real-valued function over the fundamental domain $F(\widetilde{S}_n^{\mathrm{aff}})$. The interpolation problem for $f : F(\widetilde{S}_n^{\mathrm{aff}}) \to \mathbb{R}$ involves formation of an interpolation polynomial in terms of multivariate sine functions, labeled by parameters $k \in D_N^{\star,\pm}$, $\star \in \{\mathrm{I}, \mathrm{II}, \dots, \mathrm{VIII}\}$,

$$\psi_N^{\star,\pm}(x) = \sum_{k \in D_N^{\star,\mp}} A_k^{\star} \sin_k^{\pm}(x). \tag{45}$$

The values of function f are required to coincide with the values of the interpolation polynomial on the corresponding finite set of points $F_N^{\star,\pm}$,

$$\psi_N^{\star,\pm}(s) = f(s), \qquad s \in F_N^{\star,\pm}. \tag{46}$$

Eight different types of antisymmetric interpolation polynomials $\psi_N^{\star,-}$ and eight different types of symmetric interpolation polynomials $\psi_N^{\star,+}$ are formed. Since the functions $\sin_k^{\pm}$ labeled by $k \in D_N^{\star,\pm}$ form an orthogonal basis of all real-valued discrete functions on $F_N^{\star,\pm}$, the coefficients $A_k^{\star}$ are calculated by (40) and (44), respectively.

Example 2. *For* $n = 3$*, the following function f is chosen as a model function,*

$$\begin{aligned} f(x,y,z) = {} & \exp\left(\frac{(x-0{,}7)^2 + (y-0{,}5)^2 + (z-0{,}15)^2}{0{,}005} + 3\right) \\ & + \frac{1}{3}\exp\left(\frac{(x-0{,}87)^2 + (y-0{,}7)^2 + (z-0{,}15)^2}{0{,}005} + 3\right). \end{aligned} \tag{47}$$

The graph of the cut of $f(x,y,z)$ *by the plane* $z = 1/5$ *in the fundamental domain* $F(\widetilde{S}_3^{\mathrm{aff}})$ *is plotted in Figure 5. The model function f is interpolated by the antisymmetric and symmetric polynomials of the sixth type* $\psi_N^{\mathrm{VI},-}$ *and* $\psi_N^{\mathrm{VI},+}$*, with* $N = 7, 12, 17$*. The graph cuts of the interpolating polynomials are depicted in Figures 6 and 7. Table 2 shows the integral error estimates for the polynomial approximations of the model function f by the antisymmetric and symmetric interpolation polynomials of type II, III, VI and VII for* $N = 7, 12, 17, 22, 27$*.*

Figure 5. The graph cut of the model function (47) for fixed $z = 1/5$.

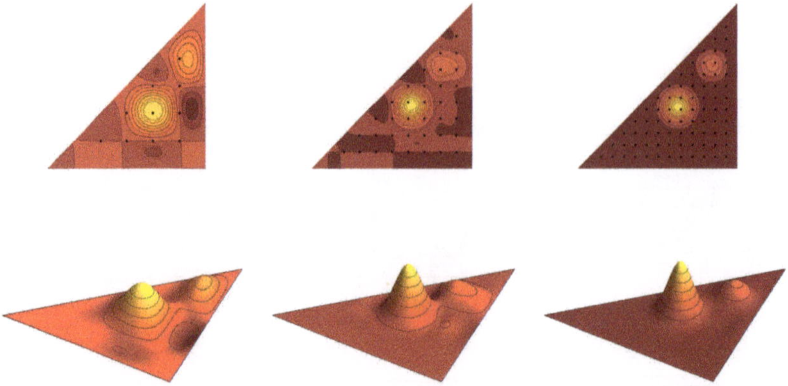

Figure 6. The antisymmetric interpolation polynomial $\psi_N^{\mathrm{VI},-}(x,y,1/5)$ of the model function (47) with $N = 7, 12, 17$. The sets of points $F_N^{\mathrm{VI},-}$ are depicted as the black dots.

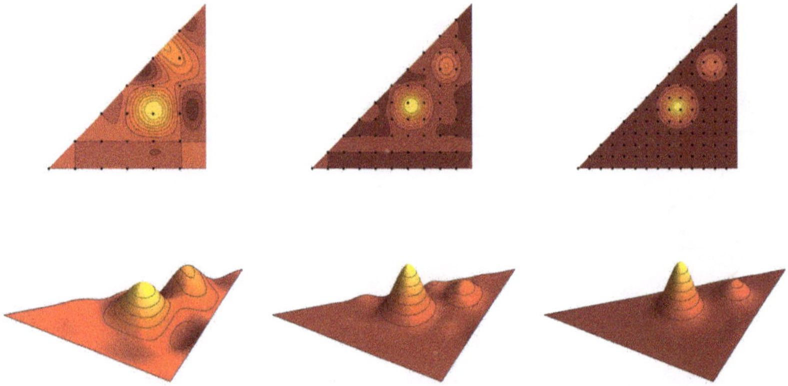

Figure 7. The symmetric interpolation polynomial $\psi_N^{\mathrm{VI},+}(x,y,1/5)$ of the model function (47) with $N = 7, 12, 17$. The sets of points $F_N^{\mathrm{VI},+}$ are depicted as the black dots.

Table 2. Integral error estimates of the polynomial approximations of the model function (47) by $\psi_N^{\mathrm{II},\pm}, \psi_N^{\mathrm{III},\pm}, \psi_N^{\mathrm{VI},\pm}$ and $\psi_N^{\mathrm{VII},\pm}$ for $N = 7, 12, 17, 22, 27$.

N	$\int_{F(\tilde{S}_n^{\mathrm{aff}})} \lvert f - \psi_N^{\mathrm{II},-}\rvert^2$	$\int_{F(\tilde{S}_n^{\mathrm{aff}})} \lvert f - \psi_N^{\mathrm{III},-}\rvert^2$	$\int_{F(\tilde{S}_n^{\mathrm{aff}})} \lvert f - \psi_N^{\mathrm{VI},-}\rvert^2$	$\int_{F(\tilde{S}_n^{\mathrm{aff}})} \lvert f - \psi_N^{\mathrm{VII},-}\rvert^2$
7	$144{,}637 \times 10^{-6}$	$204{,}640 \times 10^{-6}$	$123{,}618 \times 10^{-6}$	$121{,}379 \times 10^{-6}$
12	8850×10^{-6}	$11{,}715 \times 10^{-6}$	7540×10^{-6}	5554×10^{-6}
17	238×10^{-6}	358×10^{-6}	113×10^{-6}	96×10^{-6}
22	21×10^{-6}	21×10^{-6}	20×10^{-6}	11×10^{-6}
27	18×10^{-6}	17×10^{-6}	2×10^{-6}	7×10^{-6}
N	$\int_{F(\tilde{S}_n^{\mathrm{aff}})} \lvert f - \psi_N^{\mathrm{II},+}\rvert^2$	$\int_{F(\tilde{S}_n^{\mathrm{aff}})} \lvert f - \psi_N^{\mathrm{III},+}\rvert^2$	$\int_{F(\tilde{S}_n^{\mathrm{aff}})} \lvert f - \psi_N^{\mathrm{VI},+}\rvert^2$	$\int_{F(\tilde{S}_n^{\mathrm{aff}})} \lvert f - \psi_N^{\mathrm{VII},+}\rvert^2$
7	$146{,}367 \times 10^{-6}$	$146{,}252 \times 10^{-6}$	$190{,}670 \times 10^{-6}$	$158{,}619 \times 10^{-6}$
12	5639×10^{-6}	$11{,}502 \times 10^{-6}$	9971×10^{-6}	8792×10^{-6}
17	192×10^{-6}	297×10^{-6}	238×10^{-6}	191×10^{-6}
22	11×10^{-6}	14×10^{-6}	20×10^{-6}	11×10^{-6}
27	8×10^{-6}	7×10^{-6}	16×10^{-6}	6×10^{-6}

3.4. Matrices of the Normalized Discrete Trigonometric Transforms

The orthogonal matrices $S_N^{\star,\pm}$, that correspond to the eight types of the normalized discrete sine transforms (40) and (44), are defined by relations

$$\left(S_N^{\star,+}\right)_{k,s} = \sqrt{\frac{\varepsilon_s^\star}{h_k^\star H_k H_s}}\,\sin_k^+(s), \quad \left(S_N^{\star,-}\right)_{k,s} = \sqrt{\frac{\varepsilon_s^\star}{h_k^\star}}\,\sin_k^-(s). \tag{48}$$

The ordering inside the point and label sets is chosen as lexicographic.

Example 3. *The orthogonal matrices $S_3^{\mathrm{VI},+}$ and $S_5^{\mathrm{VI},-}$ which realize the trivariate normalized transforms SMDST-VI and AMDST-VI are of the following explicit form,*

$$S_3^{\mathrm{VI},+} = \begin{pmatrix}
0.035 & 0.137 & 0.309 & 0.400 & 0.110 & 0.350 & 0.556 & 0.198 & 0.446 & 0.206 \\
0.110 & 0.321 & 0.480 & 0.309 & 0.150 & 0.202 & -0.115 & -0.079 & -0.527 & -0.446 \\
0.198 & 0.385 & 0.321 & 0.137 & -0.079 & -0.404 & -0.293 & -0.337 & 0.115 & 0.556 \\
0.206 & 0.198 & 0.110 & 0.035 & -0.446 & -0.350 & -0.137 & 0.556 & 0.309 & -0.400 \\
0.137 & 0.293 & 0.115 & -0.556 & 0.321 & 0.404 & -0.337 & 0.385 & 0.079 & 0.198 \\
0.350 & 0.404 & -0.202 & -0.350 & 0.202 & -0.143 & 0.404 & -0.404 & 0.202 & -0.350 \\
0.446 & 0.079 & -0.150 & -0.110 & -0.527 & 0.202 & 0.321 & 0.115 & -0.480 & 0.309 \\
0.309 & 0.115 & -0.527 & 0.446 & 0.480 & -0.202 & -0.079 & 0.321 & -0.150 & 0.110 \\
0.556 & -0.337 & -0.079 & 0.198 & -0.115 & 0.404 & -0.385 & -0.293 & 0.321 & -0.137 \\
0.400 & -0.556 & 0.446 & -0.206 & 0.309 & -0.350 & 0.198 & 0.137 & -0.110 & 0.035
\end{pmatrix},$$

$$S_5^{\mathrm{VI},-} = \begin{pmatrix}
-0.047 & -0.152 & -0.279 & -0.365 & -0.166 & -0.383 & -0.535 & -0.256 & -0.445 & -0.199 \\
-0.166 & -0.365 & -0.383 & -0.279 & -0.256 & -0.199 & 0.047 & 0.152 & 0.535 & 0.445 \\
-0.256 & -0.279 & -0.199 & -0.383 & 0.152 & 0.445 & 0.166 & 0.365 & -0.047 & -0.535 \\
-0.199 & -0.047 & -0.256 & -0.166 & 0.445 & 0.152 & 0.383 & -0.535 & -0.279 & 0.365 \\
-0.279 & -0.445 & -0.047 & 0.535 & -0.383 & -0.166 & 0.365 & -0.199 & -0.152 & -0.256 \\
-0.535 & -0.256 & 0.365 & 0.152 & 0.047 & 0.279 & -0.445 & 0.166 & -0.199 & 0.383 \\
-0.445 & 0.166 & -0.152 & 0.256 & 0.535 & -0.365 & -0.199 & -0.047 & 0.383 & -0.279 \\
-0.365 & 0.199 & 0.535 & -0.445 & -0.279 & -0.047 & 0.152 & -0.383 & 0.256 & -0.166 \\
-0.383 & 0.535 & -0.166 & -0.047 & -0.199 & -0.256 & 0.279 & 0.445 & -0.365 & 0.152 \\
-0.152 & 0.383 & -0.445 & 0.199 & -0.365 & 0.535 & -0.256 & -0.279 & 0.166 & -0.047
\end{pmatrix}.$$

4. Chebyshev-Like Multivariate Orthogonal Polynomials

The Chebyshev polynomials of the first and third kind are generalized to multivariate orthogonal polynomials via the antisymmetric and symmetric cosine Function (2) in [3]. The multivariate generalizations of the Chebyshev polynomials of the second and fourth kind are built on the (anti)symmetric sine functions. Analogously to Chebyshev polynomials, the polynomial variables $X_1, X_2, \ldots, X_n$ are associated with the functions of n variables given by

$$X_1 = \cos^+_{(1,0,\ldots,0)}, \quad X_2 = \cos^+_{(1,1,0\ldots,0)}, \quad X_3 = \cos^+_{(1,1,1,0,\ldots,0)}, \quad \ldots, \quad X_n = \cos^+_{(1,1,\ldots,1)}. \tag{49}$$

The common label set

$$P^+ = \{(k_1, k_2, \ldots, k_n) \in \mathbb{Z}^n \,|\, k_1 \geq k_2 \geq \cdots \geq k_n \geq 0\}, \tag{50}$$

labels all four classes of orthogonal polynomials $\mathcal{P}_k^{\mathrm{II},\pm}, \mathcal{P}_k^{\mathrm{IV},\pm}$, $k \in P^+$, in variables $X_1, X_2, \ldots, X_n$ of degree k_1, induced for $x \in F(\widetilde{S}_n^{\mathrm{aff}})^\circ$ by the following four rational functions,

$$\begin{aligned} \mathcal{P}_k^{\mathrm{II},-}(X_1(x), X_2(x), \ldots, X_n(x)) &= \frac{\sin^-_{k+\varrho_2^-}(x)}{\sin^-_{\varrho_2^-}(x)}, & \mathcal{P}_k^{\mathrm{II},+}(X_1(x), X_2(x), \ldots, X_n(x)) &= \frac{\sin^+_{k+\varrho_2^+}(x)}{\sin^+_{\varrho_2^+}(x)}, \\ \mathcal{P}_k^{\mathrm{IV},-}(X_1(x), X_2(x), \ldots, X_n(x)) &= \frac{\sin^-_{k+\varrho_4^-}(x)}{\sin^-_{\varrho_4^-}(x)}, & \mathcal{P}_k^{\mathrm{IV},+}(X_1(x), X_2(x), \ldots, X_n(x)) &= \frac{\sin^+_{k+\varrho_4^+}(x)}{\sin^+_{\varrho_4^+}(x)}. \end{aligned} \tag{51}$$

Corollary 1 implies that the rational functions in (51) are well-defined on the interior of $F(\widetilde{S}_n^{\mathrm{aff}})$. The polynomials (51) are totally ordered by the lexicographic ordering $>$ on P^+.

4.1. Recurrence Relations

The recursive construction of the polynomials is based on the generalized trigonometric identity that is derived from the classical product-to-sum trigonometric identity and valid for any $\lambda, \mu \in \mathbb{R}^n$,

$$\sin^\pm_\lambda(x)\cos^+_\mu(x) = \frac{1}{2^n}\sum_{\sigma\in S_n}\sum_{\left\{\substack{a_i=-1,1 \\ i=1,\ldots,n}\right\}} \sin^\pm_{(\lambda_1+a_1\mu_{\sigma(1)},\ldots,\lambda_n+a_n\mu_{\sigma(n)})}(x). \tag{52}$$

Proposition 2. *The functions $\mathcal{P}_k^{\mathrm{II},\pm}$, $\mathcal{P}_k^{\mathrm{IV},\pm}$, $k \in P^+$ are polynomials in variables $X_1, X_2, \ldots, X_n$, given by relation (49), of degree k_1.*

Proof. Only the following special choices of parameters λ and μ appearing in (52) are considered,

$$\lambda \in \{\varrho_2^\pm, \varrho_4^\pm\}, \quad \mu = k \in P^+.$$

Referring by minus sign to antisymmetric sign functions and by plus sign to symmetric sine functions, the labels from the right-side of (52) are denoted by $l^\pm$,

$$l^\pm = (\lambda_1 + a_1 k_{\sigma(1)}, \lambda_2 + a_2 k_{\sigma(2)}, \ldots, \lambda_n + a_n k_{\sigma(n)}), \quad \sigma \in S_n, \quad a_i = -1, 1. \tag{53}$$

All labels $l^\pm$ for which the generalized sine functions are identically equal to zero are excluded from consideration. Therefore, any label $l^\pm$ with at least one zero-valued coordinate and any label l^- having at least two coordinates with the same absolute value are omitted.

For the remaining labels satisfy $l_i^\pm \neq 0$, there exists an alternation of signs τ of the negative coordinates such that all coordinates of $\tau(l^\pm)$ are positive. Any sequence of (different) positive numbers can be rearranged by a certain permutation into a (decreasing) non-increasing sequence, i.e. there exists $\sigma' \in S_n$ such that

$$\begin{aligned} &[\tau(l^-)]_{\sigma'(1)} > [\tau(l^-)]_{\sigma'(2)} > \cdots > [\tau(l^-)]_{\sigma'(n)} > 0, \\ &[\tau(l^+)]_{\sigma'(1)} \geq [\tau(l^+)]_{\sigma'(2)} \geq \cdots \geq [\tau(l^+)]_{\sigma'(n)} > 0. \end{aligned} \tag{54}$$

This implies that $\sigma'\tau(l^\pm)$ lies in the set $\lambda + P^+$.

Setting

$$\widetilde{k} = \sigma'\tau(l^\pm) - \lambda \in P^+, \tag{55}$$

either $\widetilde{k} = k$ or it holds that $k_i + \lambda_i > \widetilde{k}_i + \lambda_i = [\sigma'\tau(l^\pm)]_i = |\lambda_j + a_j k_{\sigma(j)}|$ for the first i for which the coordinates of $k + \lambda$ and $\widetilde{k} + \lambda$ differ and some $j \in \{1, \ldots, n\}$. Thus, k is lexicographically higher or equal to $\widetilde{k}$. The equality is fulfilled if and only if $a_i = +1$ for all i for which $k_i > 0$ and, in the case of the antisymmetric sine functions, if the additional condition that σ stabilizes k is satisfied. Considering the possible values of a_i for which $k = \widetilde{k}$, a function T_k is defined by

$$T_k = t_{k_1} t_{k_2} \dots t_{k_n}, \quad t_{k_i} = \begin{cases} 2 & \text{if } k_i = 0, \\ 1 & \text{otherwise.} \end{cases} \tag{56}$$

Reflecting the possible values of σ, functions $H_k^{\pm}$ are introduced by

$$H_k^- = H_k, \quad H_k^+ = n!.$$

By symmetry properties of multivariate sine functions, the Formula (52) is rewritten as

$$\sin_\lambda^{\pm}(x) \cos_k^{+}(x) = \frac{1}{2^n} T_k H_k^{\pm} \sin_{k+\lambda}^{\pm}(x) + \sum_{k > \widetilde{k} \in P^+} A_{\widetilde{k}} \sin_{\widetilde{k}+\lambda}^{\pm}(x), \quad A_{\widetilde{k}} \in \mathbb{R}. \tag{57}$$

It follows from the Formula (57) that each function defined by (51) is built recurrently as certain linear combination of $\cos_{\widetilde{k}}^{+}(x)$ parametrized by $\widetilde{k} \leq k$ and that the coefficient of $\cos_{\widetilde{k}}^{+}(x)$ with $\widetilde{k} = k$ is non-zero. From Proposition 4.1 in [3], $\cos_{\widetilde{k}}^{+}(x)$ is expressible as a polynomial function in variables $X_1, X_2, \dots, X_n$ of degree $\widetilde{k}_1$. □

Remark 3. *Proposition 2 implies that all the points for which the denominator is zero-valued are removable singularities. Therefore, it is possible to extend the domain of definition on the whole* $F(\widetilde{S}_n^{\text{aff}})$.

Special choices of the labels in Formula (52) together with basic properties of multivariate sine functions suffice to deduce recurrence relations for the polynomials (51). In particular, the recurrence relations are deduced from the proof of Proposition 2, or by setting $\lambda = k \in P^+$ and equating successively μ to $(1,0,\dots,0), (1,1,0,\dots,0), \dots, (1,1,\dots,1)$ in (52). Thus, the recurrence algorithm is based on the following formulas,

$$\begin{aligned}
\sin_k^{\pm} &= \frac{2^1}{1!(n-1)!} \sin_{k-l_1}^{\pm} X_1 - \sin_{k-2l_1}^{\pm} - \sum_{i=2}^{n} \sin_{k-l_1-l_i}^{\pm}, \\
\sin_k^{\pm} &= \frac{2^2}{2!(n-2)!} \sin_{k-l_1-l_2}^{\pm} X_2 - \sin_{k-2l_1-2l_2}^{\pm} - \sin_{k-2l_1}^{\pm} - \sin_{k-2l_2}^{\pm} \\
&\quad - \sum_{i=3}^{n} \left(\sin_{k-l_2+l_i}^{\pm} + \sin_{k-2l_1-l_2-l_i}^{\pm} + \sin_{k-l_2-l_i}^{\pm} + \sin_{k-2l_1-l_2+l_i}^{\pm} \right) \\
&\quad - \sum_{\substack{i,j=2 \\ i<j}}^{n} \left(\sin_{k-l_1-l_2+l_i+l_j}^{\pm} + \sin_{k-l_1-l_2-l_i-l_j}^{\pm} + \sin_{k-l_1-l_2+l_i-l_j}^{\pm} + \sin_{k-l_1-l_2-l_i+l_j}^{\pm} \right), \\
&\;\;\vdots \\
\sin_k^{\pm} &= \frac{2^n}{n!} \sin_{k-l_1-l_2-\dots-l_n}^{\pm} X_n - \sum_{i}^{n} \sin_{k-2l_i}^{\pm} - \sum_{\substack{i,j=1 \\ i<j}}^{n} \sin_{k-2l_i-2l_j}^{\pm} - \dots - \sin_{k-2l_1-2l_2-\dots-2l_n}^{\pm},
\end{aligned} \tag{58}$$

where l_i is a vector with i-th coordinate equal to 1 and others to 0.

Example 4. *For* $n = 3$, *the lowest polynomial of type* $\mathcal{P}_k^{\text{II},+}$ *is constant,* $\mathcal{P}_{(0,0,0)}^{\text{II},+} = 1$, *and the first degree polynomials are given by*

$$\mathcal{P}_{(1,0,0)}^{\text{II},+} = \frac{1}{3} X_1, \quad \mathcal{P}_{(1,1,0)}^{\text{II},+} = \frac{2}{3} X_2, \quad \mathcal{P}_{(1,1,1)}^{\text{II},+} = \frac{4}{3} X_3. \tag{59}$$

The consecutive polynomials are then determined by the following recurrence relations.

$$
\begin{aligned}
k_1 \geq 2,\ k_2 = k_3 = 0: \quad & \mathcal{P}^{\mathrm{II},+}_{(k_1,0,0)} = \mathcal{P}^{\mathrm{II},+}_{(k_1-1,0,0)} X_1 - \mathcal{P}^{\mathrm{II},+}_{(k_1-2,0,0)} - 2\mathcal{P}^{\mathrm{II},+}_{(k_1-1,1,0)} \\
k_1 - 1 > k_2 > k_3 = 0: \quad & \mathcal{P}^{\mathrm{II},+}_{(k_1,k_2,0)} = \mathcal{P}^{\mathrm{II},+}_{(k_1-1,k_2,0)} X_1 - \mathcal{P}^{\mathrm{II},+}_{(k_1-2,k_2,0)} - \mathcal{P}^{\mathrm{II},+}_{(k_1-1,k_2+1,0)} \\
& \quad - \mathcal{P}^{\mathrm{II},+}_{(k_1-1,k_2-1,0)} - \mathcal{P}^{\mathrm{II},+}_{(k_1-1,k_2,1)} \\
k_1 - 1 > k_2 = k_3 > 0: \quad & \mathcal{P}^{\mathrm{II},+}_{(k_1,k_2,k_2)} = \mathcal{P}^{\mathrm{II},+}_{(k_1-1,k_2,k_2)} X_1 - \mathcal{P}^{\mathrm{II},+}_{(k_1-2,k_2,k_2)} \\
& \quad - 2\mathcal{P}^{\mathrm{II},+}_{(k_1-1,k_2+1,k_2)} - 2\mathcal{P}^{\mathrm{II},+}_{(k_1-1,k_2,k_2-1)} \\
k_1 - 1 > k_2 > k_3 > 0: \quad & \mathcal{P}^{\mathrm{II},+}_{(k_1,k_2,k_3)} = \mathcal{P}^{\mathrm{II},+}_{(k_1-1,k_2,k_3)} X_1 - \mathcal{P}^{\mathrm{II},+}_{(k_1-2,k_2,k_3)} - \mathcal{P}^{\mathrm{II},+}_{(k_1-1,k_2+1,k_3)} \\
& \quad - \mathcal{P}^{\mathrm{II},+}_{(k_1-1,k_2-1,k_3)} - \mathcal{P}^{\mathrm{II},+}_{(k_1-1,k_2,k_3+1)} - \mathcal{P}^{\mathrm{II},+}_{(k_1-1,k_2,k_3-1)} \\
k_1 - 1 = k_2 > k_3 = 0: \quad & \mathcal{P}^{\mathrm{II},+}_{(k_1,k_1-1,0)} = \frac{1}{2}\mathcal{P}^{\mathrm{II},+}_{(k_1-1,k_1-1,0)} X_1 - \mathcal{P}^{\mathrm{II},+}_{(k_1-1,k_1-2,0)} - \frac{1}{2}\mathcal{P}^{\mathrm{II},+}_{(k_1-1,k_1-1,1)} \\
k_1 - 1 = k_2 > k_3 > 0: \quad & \mathcal{P}^{\mathrm{II},+}_{(k_1,k_1-1,k_3)} = \frac{1}{2}\mathcal{P}^{\mathrm{II},+}_{(k_1-1,k_1-1,k_3)} X_1 - \mathcal{P}^{\mathrm{II},+}_{(k_1-1,k_1-2,k_3)} \\
& \quad - \frac{1}{2}\mathcal{P}^{\mathrm{II},+}_{(k_1-1,k_1-1,k_3+1)} - \frac{1}{2}\mathcal{P}^{\mathrm{II},+}_{(k_1-1,k_1-1,k_3-1)}
\end{aligned}
$$

$$
\begin{aligned}
k_1 - 1 = k_2 = k_3 > 0: \quad & \mathcal{P}^{\mathrm{II},+}_{(k_1,k_1-1,k_1-1)} = \frac{1}{3}\mathcal{P}^{\mathrm{II},+}_{(k_1-1,k_1-1,k_1-1)} X_1 - \mathcal{P}^{\mathrm{II},+}_{(k_1-1,k_1-1,k_1-2)} \\
k_1 = k_2 = 2,\ k_3 = 0: \quad & \mathcal{P}^{\mathrm{II},+}_{(2,2,0)} = 2\mathcal{P}^{\mathrm{II},+}_{(1,1,0)} X_2 - 2\mathcal{P}^{\mathrm{II},+}_{(1,0,0)} X_1 - \frac{2}{3}\mathcal{P}^{\mathrm{II},+}_{(1,1,1)} X_1 + \mathcal{P}^{\mathrm{II},+}_{(0,0,0)} + 4\mathcal{P}^{\mathrm{II},+}_{(1,1,0)} \\
k_1 = k_2 > 2,\ k_3 = 0: \quad & \mathcal{P}^{\mathrm{II},+}_{(k_1,k_1,0)} = 2\mathcal{P}^{\mathrm{II},+}_{(k_1-1,k_1-1,0)} X_2 - 2\mathcal{P}^{\mathrm{II},+}_{(k_1-1,k_1-2,0)} X_1 - \mathcal{P}^{\mathrm{II},+}_{(k_1-1,k_1-1,1)} X_1 \\
& \quad + \mathcal{P}^{\mathrm{II},+}_{(k_1-2,k_1-2,0)} + 3\mathcal{P}^{\mathrm{II},+}_{(k_1-1,k_1-1,0)} + 2\mathcal{P}^{\mathrm{II},+}_{(k_1-1,k_1-2,1)} \\
& \quad + 2\mathcal{P}^{\mathrm{II},+}_{(k_1-1,k_1-3,0)} + \mathcal{P}^{\mathrm{II},+}_{(k_1-1,k_1-1,2)} \\
k_1 = k_2 > k_3 + 2 > 2: \quad & \mathcal{P}^{\mathrm{II},+}_{(k_1,k_1,k3)} = 2\mathcal{P}^{\mathrm{II},+}_{(k_1-1,k_1-1,k3)} X_2 - 2\mathcal{P}^{\mathrm{II},+}_{(k_1-1,k_1-2,k3)} X_1 \\
& \quad - \mathcal{P}^{\mathrm{II},+}_{(k_1-1,k_1-1,k_3+1)} X_1 - \mathcal{P}^{\mathrm{II},+}_{(k_1-1,k_1-1,k_3-1)} X_1 + \mathcal{P}^{\mathrm{II},+}_{(k_1-2,k_1-2,k_3)} \\
& \quad + 2\mathcal{P}^{\mathrm{II},+}_{(k_1-1,k_1-2,k_3+1)} + 2\mathcal{P}^{\mathrm{II},+}_{(k_1-1,k_1-2,k_3-1)} + 4\mathcal{P}^{\mathrm{II},+}_{(k_1-1,k_1-1,k_3)} \\
& \quad + 2\mathcal{P}^{\mathrm{II},+}_{(k_1-1,k_1-3,k_3)} + \mathcal{P}^{\mathrm{II},+}_{(k_1-1,k_1-1,k_3+2)} + \mathcal{P}^{\mathrm{II},+}_{(k_1-1,k_1-1,k_3-2)} \\
k_1 = k_2 = k_3 + 2 = 3: \quad & \mathcal{P}^{\mathrm{II},+}_{(3,3,1)} = 2\mathcal{P}^{\mathrm{II},+}_{(2,2,1)} X_2 - 2\mathcal{P}^{\mathrm{II},+}_{(2,1,1)} X_1 - \frac{2}{3}\mathcal{P}^{\mathrm{II},+}_{(2,2,2)} X_1 \\
& \quad - \mathcal{P}^{\mathrm{II},+}_{(2,2,0)} X_1 + \mathcal{P}^{\mathrm{II},+}_{(1,1,1)} + 5\mathcal{P}^{\mathrm{II},+}_{(2,2,1)} + 4\mathcal{P}^{\mathrm{II},+}_{(2,1,0)}
\end{aligned}
$$

$$\begin{aligned}
k_1 = k_2 = k_3+2 > 3: \quad & \mathcal{P}^{\mathrm{II},+}_{(k_1,k_1,k_1-2)} = 2\mathcal{P}^{\mathrm{II},+}_{(k_1-1,k_1-1,k_1-2)}X_2 - 2\mathcal{P}^{\mathrm{II},+}_{(k_1-1,k_1-2,k_1-2)}X_1 \\
& - \frac{2}{3}\mathcal{P}^{\mathrm{II},+}_{(k_1-1,k_1-1,k_1-1)}X_1 - \mathcal{P}^{\mathrm{II},+}_{(k_1-1,k_1-1,k_1-3)}X_1 + \mathcal{P}^{\mathrm{II},+}_{(k_1-2,k_1-2,k_1-2)} \\
& + 5\mathcal{P}^{\mathrm{II},+}_{(k_1-1,k_1-1,k_1-2)} + 4\mathcal{P}^{\mathrm{II},+}_{(k_1-1,k_1-2,k_1-3)} + \mathcal{P}^{\mathrm{II},+}_{(k_1-1,k_1-1,k_1-4)} \\
k_1 = k_2 = k_3+1 = 2: \quad & \mathcal{P}^{\mathrm{II},+}_{(2,2,1)} = \frac{2}{3}\mathcal{P}^{\mathrm{II},+}_{(1,1,1)}X_2 - \mathcal{P}^{\mathrm{II},+}_{(1,1,0)}X_1 + \mathcal{P}^{\mathrm{II},+}_{(1,0,0)} + \mathcal{P}^{\mathrm{II},+}_{(1,1,1)} \\
k_1 = k_2 = k_3+1 > 2: \quad & \mathcal{P}^{\mathrm{II},+}_{(k_1,k_1,k_1-1)} = \frac{2}{3}\mathcal{P}^{\mathrm{II},+}_{(k_1-1,k_1-1,k_1-1)}X_2 - \mathcal{P}^{\mathrm{II},+}_{(k_1-1,k_1-1,k_1-2)}X_1 \\
& + \mathcal{P}^{\mathrm{II},+}_{(k_1-1,k_1-2,k_1-2)} + \mathcal{P}^{\mathrm{II},+}_{(k_1-1,k_1-1,k_1-1)} + \mathcal{P}^{\mathrm{II},+}_{(k_1-1,k_1-1,k_1-3)} \\
k_1 = k_2 = k_3 = 2: \quad & \mathcal{P}^{\mathrm{II},+}_{(2,2,2)} = \frac{4}{3}\mathcal{P}^{\mathrm{II},+}_{(1,1,1)}X_3 - 6\mathcal{P}^{\mathrm{II},+}_{(1,1,0)}X_2 + 3\mathcal{P}^{\mathrm{II},+}_{(1,0,0)}X_1 \\
& + 2\mathcal{P}^{\mathrm{II},+}_{(1,1,1)}X_1 - \mathcal{P}^{\mathrm{II},+}_{(0,0,0)} - 6\mathcal{P}^{\mathrm{II},+}_{(1,1,0)} \\
k_1 = k_2 = k_3 = 3: \quad & \mathcal{P}^{\mathrm{II},+}_{(3,3,3)} = \frac{4}{3}\mathcal{P}^{\mathrm{II},+}_{(2,2,2)}X_3 - 6\mathcal{P}^{\mathrm{II},+}_{(2,2,1)}X_2 + 3\mathcal{P}^{\mathrm{II},+}_{(2,1,1)}X_1 + 2\mathcal{P}^{\mathrm{II},+}_{(2,2,2)}X_1 \\
& + 3\mathcal{P}^{\mathrm{II},+}_{(2,2,0)}X_1 - \mathcal{P}^{\mathrm{II},+}_{(1,1,1)} - 9\mathcal{P}^{\mathrm{II},+}_{(2,2,1)} - 6\mathcal{P}^{\mathrm{II},+}_{(2,1,0)} \\
k_1 = k_2 = k_3 > 3: \quad & \mathcal{P}^{\mathrm{II},+}_{(k_1,k_1,k_1)} = \frac{4}{3}\mathcal{P}^{\mathrm{II},+}_{(k_1-1,k_1-1,k_1-1)}X_3 - 6\mathcal{P}^{\mathrm{II},+}_{(k_1-1,k_1-1,k_1-2)}X_2 \\
& + 3\mathcal{P}^{\mathrm{II},+}_{(k_1-1,k_1-2,k_1-2)}X_1 + 2\mathcal{P}^{\mathrm{II},+}_{(k_1-1,k_1-1,k_1-1)}X_1 \\
& + 3\mathcal{P}^{\mathrm{II},+}_{(k_1-1,k_1-1,k_1-3)}X_1 - \mathcal{P}^{\mathrm{II},+}_{(k_1-2,k_1-2,k_1-2)} - 9\mathcal{P}^{\mathrm{II},+}_{(k_1-1,k_1-1,k_1-2)} \\
& - 6\mathcal{P}^{\mathrm{II},+}_{(k_1-1,k_1-2,k_1-3)} - 3\mathcal{P}^{\mathrm{II},+}_{(k_1-1,k_1-1,k_1-4)}.
\end{aligned} \tag{60}$$

Similarly, the recurrence relations can be found for polynomials $\mathcal{P}^{\mathrm{II},-}_{(k_1,k_2,k_3)}$ *and* $\mathcal{P}^{\mathrm{IV},\pm}_{(k_1,k_2,k_3)}$*. The polynomials* $\mathcal{P}^{\mathrm{II},\pm}_{(k_1,k_2,k_3)}$ *and* $\mathcal{P}^{\mathrm{IV},\pm}_{(k_1,k_2,k_3)}$ *of degree at most two are listed in Tables 3–6.*

Table 3. The coefficients of the polynomials $\mathcal{P}^{\mathrm{II},\pm}_{(k_1,k_2,k_3)}$ with $k_1 \leq 2$ and $k_1+k_2+k_3$ even.

$\mathcal{P}^{\mathrm{II},-}_{(k_1,k_2,k_3)}$	1	X_2	X_1^2	X_1X_3	X_2^2	X_3^2	$\mathcal{P}^{\mathrm{II},+}_{(k_1,k_2,k_3)}$	1	X_2	X_1^2	X_1X_3	X_2^2	X_3^2
$\mathcal{P}^{\mathrm{II},-}_{(0,0,0)}$	1						$\mathcal{P}^{\mathrm{II},+}_{(0,0,0)}$	1					
$\mathcal{P}^{\mathrm{II},-}_{(1,1,0)}$	2	2					$\mathcal{P}^{\mathrm{II},+}_{(1,1,0)}$	0	$\frac{2}{3}$				
$\mathcal{P}^{\mathrm{II},-}_{(2,0,0)}$	−3	−2	1				$\mathcal{P}^{\mathrm{II},+}_{(2,0,0)}$	−1	$-\frac{4}{3}$	$\frac{1}{3}$			
$\mathcal{P}^{\mathrm{II},-}_{(2,1,1)}$	−2	−2	1	$\frac{4}{3}$			$\mathcal{P}^{\mathrm{II},+}_{(2,1,1)}$	0	$-\frac{2}{3}$	0	$\frac{4}{9}$		
$\mathcal{P}^{\mathrm{II},-}_{(2,2,0)}$	6	10	−2	$-\frac{4}{3}$	4		$\mathcal{P}^{\mathrm{II},+}_{(2,2,0)}$	1	$\frac{8}{3}$	$-\frac{2}{3}$	$-\frac{8}{9}$	$\frac{4}{3}$	
$\mathcal{P}^{\mathrm{II},-}_{(2,2,2)}$	−4	−8	2	$\frac{12}{3}$	−4	$\frac{16}{9}$	$\mathcal{P}^{\mathrm{II},+}_{(2,2,2)}$	−1	−4	1	$\frac{8}{3}$	−4	$\frac{16}{9}$

Table 4. The coefficients of the polynomials $\mathcal{P}^{\mathrm{II},\pm}_{(k_1,k_2,k_3)}$ with $k_1 \leq 2$ and $k_1+k_2+k_3$ odd.

$\mathcal{P}^{\mathrm{II},-}_{(k_1,k_2,k_3)}$	X_1	X_3	X_1X_2	X_2X_3	$\mathcal{P}^{\mathrm{II},+}_{(k_1,k_2,k_3)}$	X_1	X_3	X_1X_2	X_2X_3
$\mathcal{P}^{\mathrm{II},-}_{(1,0,0)}$	1				$\mathcal{P}^{\mathrm{II},+}_{(1,0,0)}$	$\frac{1}{3}$			
$\mathcal{P}^{\mathrm{II},-}_{(1,1,1)}$	1	$\frac{4}{3}$			$\mathcal{P}^{\mathrm{II},+}_{(1,1,1)}$	0	$\frac{4}{3}$		
$\mathcal{P}^{\mathrm{II},-}_{(2,1,0)}$	0	$-\frac{4}{3}$	2		$\mathcal{P}^{\mathrm{II},+}_{(2,1,0)}$	$-\frac{1}{3}$	$-\frac{2}{3}$	$\frac{1}{3}$	
$\mathcal{P}^{\mathrm{II},-}_{(2,2,1)}$	1	4	0	$\frac{8}{3}$	$\mathcal{P}^{\mathrm{II},+}_{(2,2,1)}$	$\frac{1}{3}$	$\frac{4}{3}$	$-\frac{2}{3}$	$\frac{8}{9}$

Table 5. The coefficients of the polynomials $\mathcal{P}^{IV,-}_{(k_1,k_2,k_3)}$ with $k_1 \leq 2$.

$\mathcal{P}^{IV,-}_{(k_1,k_2,k_3)}$	**1**	X_1	X_2	X_3	X_1^2	X_1X_2	X_1X_3	X_2^2	X_2X_3	X_3^2
$\mathcal{P}^{IV,-}_{(0,0,0)}$	1									
$\mathcal{P}^{IV,-}_{(1,0,0)}$	1	1								
$\mathcal{P}^{IV,-}_{(1,1,0)}$	3	1	2							
$\mathcal{P}^{IV,-}_{(1,1,1)}$	3	2	2	$\frac{4}{3}$						
$\mathcal{P}^{IV,-}_{(2,0,0)}$	-3	1	-2	0	1					
$\mathcal{P}^{IV,-}_{(2,1,0)}$	-1	1	0	$-\frac{4}{3}$	1	2				
$\mathcal{P}^{IV,-}_{(2,1,1)}$	-3	2	-2	0	2	2	$\frac{4}{3}$			
$\mathcal{P}^{IV,-}_{(2,2,0)}$	8	0	12	$-\frac{4}{3}$	-2	2	$-\frac{4}{3}$	4		
$\mathcal{P}^{IV,-}_{(2,2,1)}$	6	2	10	4	-1	2	0	4	$\frac{8}{3}$	
$\mathcal{P}^{IV,-}_{(2,2,2)}$	-6	2	-10	$\frac{16}{3}$	3	0	$\frac{16}{3}$	-4	$\frac{8}{3}$	$\frac{16}{9}$

Table 6. The coefficients of the polynomials $\mathcal{P}^{IV,+}_{(k_1,k_2,k_3)}$ with $k_1 \leq 2$.

$\mathcal{P}^{IV,+}_{(k_1,k_2,k_3)}$	**1**	X_1	X_2	X_3	X_1^2	X_1X_2	X_1X_3	X_2^2	X_2X_3	X_3^2
$\mathcal{P}^{IV,+}_{(0,0,0)}$	1									
$\mathcal{P}^{IV,+}_{(1,0,0)}$	1	$\frac{1}{3}$								
$\mathcal{P}^{IV,+}_{(1,1,0)}$	1	$\frac{2}{3}$	$\frac{2}{3}$							
$\mathcal{P}^{IV,+}_{(1,1,1)}$	1	1	2	$\frac{4}{3}$						
$\mathcal{P}^{IV,+}_{(2,0,0)}$	-1	$\frac{1}{3}$	$-\frac{4}{3}$	0	$\frac{1}{3}$					
$\mathcal{P}^{IV,+}_{(2,1,0)}$	-1	0	$-\frac{2}{3}$	$-\frac{2}{3}$	$\frac{1}{3}$	$\frac{1}{3}$				
$\mathcal{P}^{IV,+}_{(2,1,1)}$	-1	$-\frac{1}{3}$	$-\frac{2}{3}$	0	$\frac{1}{3}$	$\frac{2}{3}$	$\frac{4}{9}$			
$\mathcal{P}^{IV,+}_{(2,2,0)}$	1	$-\frac{2}{3}$	$\frac{10}{3}$	$-\frac{4}{3}$	$-\frac{2}{3}$	$\frac{2}{3}$	$-\frac{8}{9}$	$\frac{4}{3}$		
$\mathcal{P}^{IV,+}_{(2,2,1)}$	1	$-\frac{1}{3}$	2	$\frac{4}{3}$	$-\frac{2}{3}$	0	0	$\frac{4}{3}$	$\frac{8}{9}$	
$\mathcal{P}^{IV,+}_{(2,2,2)}$	-1	1	-6	$\frac{16}{3}$	1	-2	4	-4	$\frac{8}{3}$	$\frac{16}{9}$

4.2. Continuous Orthogonality

The key notion that induces continuous orthogonality relations of the polynomials $\mathcal{P}^{II,\pm}_k$ and $\mathcal{P}^{IV,\pm}_k$ is the change of variables (49) in Formulas (26)–(28). The determinant of the Jacobian matrix

$$J(x_1,\ldots,x_n) = \det \frac{\partial(X_1,\ldots,X_n)}{\partial(x_1,\ldots,x_n)} \tag{61}$$

for the change from the polynomial variables $(X_1,\ldots,X_n)$ to $(x_1,\ldots,x_n)$ is calculated in [3]. The absolute value of J is given by

$$|J(x_1,\ldots,x_n)| = c\left|\sin^-_{\varrho_2}(x_1,x_2,\ldots,x_n)\right|, \tag{62}$$

with the constant c determined as

$$c = \pi^n \left(\frac{1}{2}\right)^{\frac{n(n-1)}{2}} \left(\prod_{i=1}^{n}(n-i)!i!\right). \tag{63}$$

The absolute value $|J(x_1,\ldots,x_n)|$ is shown in [3] to be expressible as a function $\mathcal{J}$ in the polynomial variables $(X_1,\ldots,X_n)$,

$$\mathcal{J}(X_1,\ldots,X_n) = |J(x_1,\ldots,x_n)|\,. \tag{64}$$

Positivity of the Jacobian $\mathcal{J}(X_1(x),\ldots,X_n(x)) > 0$ is for all $x \in F(\widetilde{S}_n^{\mathrm{aff}})^\circ$ guaranteed by Corollary 1.

In order to define the underlying weight functions, properties of the three additional products of multivariate sine functions $\sin^+_{\varrho_2^+}(x)\cdot\sin^+_{\varrho_2^+}(x)$, $\sin^-_{\varrho_4^-}(x)\cdot\sin^-_{\varrho_4^-}(x)$ and $\sin^+_{\varrho_4^+}(x)\cdot\sin^+_{\varrho_4^+}(x)$ are determined. Similarly to Formula (52), products of the (anti)symmetric sine functions are by a classical product-to-sum identity decomposed into a sum of the symmetric cosine functions. Denoting the number of positive a_i by $\alpha(a_1,\ldots,a_n)$, it holds that

$$\begin{aligned}
\sin^-_\lambda(x)\sin^-_\mu(x) &= \frac{1}{2^n}\sum_{\sigma\in S_n}\operatorname{sgn}(\sigma)\sum_{\substack{a_i=-1,1\\ i=1,\ldots,n}}(-1)^{\alpha(a_1,\ldots,a_n)}\cos^+_{(\lambda_1+a_1\mu_{\sigma(1)},\ldots,\lambda_n+a_n\mu_{\sigma(n)})}(x),\\
\sin^+_\lambda(x)\sin^+_\mu(x) &= \frac{1}{2^n}\sum_{\sigma\in S_n}\sum_{\substack{a_i=-1,1\\ i=1,\ldots,n}}(-1)^{\alpha(a_1,\ldots,a_n)}\cos^+_{(\lambda_1+a_1\mu_{\sigma(1)},\ldots,\lambda_n+a_n\mu_{\sigma(n)})}(x).
\end{aligned} \tag{65}$$

Therefore, all three products are expressed as polynomials in the variables $X_1,\ldots,X_n$,

$$\begin{aligned}
\mathcal{J}^{\mathrm{II},+}(X_1(x),\ldots,X_n(x)) &= \sin^+_{\varrho_2^+}(x)\sin^+_{\varrho_2^+}(x),\\
\mathcal{J}^{\mathrm{IV},-}(X_1(x),\ldots,X_n(x)) &= \sin^-_{\varrho_4^-}(x)\sin^-_{\varrho_4^-}(x),\\
\mathcal{J}^{\mathrm{IV},+}(X_1(x),\ldots,X_n(x)) &= \sin^+_{\varrho_4^+}(x)\sin^+_{\varrho_4^+}(x).
\end{aligned} \tag{66}$$

In the case of polynomials $\mathcal{P}_k^{\mathrm{II},-}$, the following polynomial function is introduced,

$$\mathcal{J}^{\mathrm{II},-}(X_1,\ldots,X_n) = \sin^-_{\varrho_2^-}(x)\sin^-_{\varrho_2^-}(x) = \left(\frac{\mathcal{J}}{c}\right)^2. \tag{67}$$

Due to Corollary 1, all four Functions (66) and (67) do not vanish in the interior of $F(\widetilde{S}_n^{\mathrm{aff}})$. The Equalities (14)–(17) imply that $\mathcal{J}^2$ is divisible by $\mathcal{J}^{\mathrm{II},\pm}$ and $\mathcal{J}^{\mathrm{IV},\pm}$.

The final weights $w^{\mathrm{II},\pm}$ and $w^{\mathrm{IV},\pm}$ in the continuous orthogonality relations are given for $x \in F(\widetilde{S}_n^{\mathrm{aff}})^\circ$ by

$$w^{\mathrm{II},\pm} = \frac{\mathcal{J}^{\mathrm{II},\pm}}{\mathcal{J}},\qquad w^{\mathrm{IV},\pm} = \frac{\mathcal{J}^{\mathrm{IV},\pm}}{\mathcal{J}}. \tag{68}$$

Example 5. *For $n = 3$, the Jacobian $\mathcal{J}$ in the polynomial variables is of the form*

$$\begin{aligned}
\mathcal{J}(X_1,X_2,X_3) = &\sqrt{\pi^6\left(-8X_2^3 + X_1^2X_2^2 - 12X_3^2 + 12X_1X_2X_3 - \frac{4}{3}X_1^3X_3\right)}\\
&\times\sqrt{(X_3+3X_2+3X_1+6)(-X_3+3X_2-3X_1+6)}.
\end{aligned} \tag{69}$$

The explicit formulas of the polynomials $\mathcal{J}^{\mathrm{II},\pm}$ *and* $\mathcal{J}^{\mathrm{II},\pm}$ *are given by*

$$
\begin{aligned}
\mathcal{J}^{\mathrm{II},-}(X_1,X_2,X_3) =& \frac{1}{9}\left(-8X_2^3+X_1^2X_2^2-12X_3^2+12X_1X_2X_3-\frac{4}{3}X_1^3X_3\right)\\
&\times(X_3+3X_2+3X_1+6)(-X_3+3X_2-3X_1+6),\\
\mathcal{J}^{\mathrm{II},+}(X_1,X_2,X_3) =& (X_3+3X_2+3X_1+6)(-X_3+3X_2-3X_1+6),\\
\mathcal{J}^{\mathrm{IV},-}(X_1,X_2,X_3) =& \frac{1}{12}\left(-8X_2^3+X_1^2X_2^2-12X_3^2+12X_1X_2X_3-\frac{4}{3}X_1^3X_3\right)\\
&\times(-X_3+3X_2-3X_1+6),\\
\mathcal{J}^{\mathrm{IV},+}(X_1,X_2,X_3) =& \frac{3}{4}(-X_3+3X_2-3X_1+6).
\end{aligned}
\tag{70}
$$

Proposition 3. *The polynomials* $\mathcal{P}_k^{\mathrm{II},\pm}$ *and* $\mathcal{P}_k^{\mathrm{IV},\pm}$, $k\in P^+$, *form within each family an orthogonal polynomial sequence on the integration domain* $\mathfrak{F}(\widetilde{S}_n^{\mathrm{aff}})$ *given by*

$$
\mathfrak{F}(\widetilde{S}_n^{\mathrm{aff}}) = \left\{(X_1(x),\dots,X_n(x))\in\mathbb{R}^n \mid x\in F(\widetilde{S}_n^{\mathrm{aff}})\right\}
\tag{71}
$$

and with respect to the weights $w^{\mathrm{II},\pm}$ *and* $w^{\mathrm{IV},\pm}$. *For* $X=(X_1,\dots,X_n)$, $dX=dX_1\dots dX_n$ *and* $k,k'\in P^+$, *the polynomial orthogonality relations are of the explicit form*

$$
\begin{aligned}
\int_{\mathfrak{F}(\widetilde{S}_n^{\mathrm{aff}})} \mathcal{P}_k^{\mathrm{II},-}(X)\mathcal{P}_{k'}^{\mathrm{II},-}(X)w^{\mathrm{II},-}(X)\,dX &= 2^{-n}\delta_{kk'},\\
\int_{\mathfrak{F}(\widetilde{S}_n^{\mathrm{aff}})} \mathcal{P}_k^{\mathrm{II},+}(X)\mathcal{P}_{k'}^{\mathrm{II},+}(X)w^{\mathrm{II},+}(X)\,dX &= 2^{-n}H_k\delta_{kk'},\\
\int_{\mathfrak{F}(\widetilde{S}_n^{\mathrm{aff}})} \mathcal{P}_k^{\mathrm{IV},-}(X)\mathcal{P}_{k'}^{\mathrm{IV},-}(X)w^{\mathrm{IV},-}(X)\,dX &= 2^{-n}\delta_{kk'},\\
\int_{\mathfrak{F}(\widetilde{S}_n^{\mathrm{aff}})} \mathcal{P}_k^{\mathrm{IV},+}(X)\mathcal{P}_{k'}^{\mathrm{IV},+}(X)w^{\mathrm{IV},+}(X)\,dX &= 2^{-n}H_k\delta_{kk'}.
\end{aligned}
\tag{72}
$$

Proof. The correspondence between the domain $F(\widetilde{S}_n^{\mathrm{aff}})$ and $\mathfrak{F}(\widetilde{S}_n^{\mathrm{aff}})$ given by

$$
\varphi:\ x\in F(\widetilde{S}_n^{\mathrm{aff}}) \to (X_1(x),\dots,X_n(x))\in\mathfrak{F}(\widetilde{S}_n^{\mathrm{aff}})
\tag{73}
$$

is proved to be one-to-one in [3]. Therefore, the corresponding change of variables is applicable on the integration Formulas (26)–(29). □

Corollary 2. *The polynomials* $\mathcal{P}_k^{\mathrm{II},\pm}$ *and* $\mathcal{P}_k^{\mathrm{IV},\pm}$, $k\in P^+$ *form an orthogonal basis of all polynomials* $f,g\in\mathbb{R}[X]$ *of n variables with respect to the scalar product defined by*

$$
\langle f,g\rangle_w = \int_{\mathfrak{F}(\widetilde{S}_n^{\mathrm{aff}})} f(X)g(X)w(X)\,dX,\quad w\in\{w^{\mathrm{II},\pm},w^{\mathrm{IV},\pm}\}.
\tag{74}
$$

Proof. Proposition 4.1 in [3] grants that the number of polynomials (51) of degree d is equal to the number of monomials of n variables of degree d. □

4.3. Cubature Formulas

The main objective of cubature formulas is to estimate weighted integrals of multivariate integrable functions $f(Y)$, $Y\in\mathbb{R}^n$, over an integration domain $\Omega\subset\mathbb{R}^n$ with a weight w by linear combinations of function values at a suitable finite set of points $\Omega_n\subset\Omega$,

$$
\int_\Omega f(Y)w(Y)dY \approx \sum_{Y\in\Omega_n} c_Y f(Y).
\tag{75}
$$

The points from Ω_n are generally called nodes. It is required that the cubature formulas hold exactly for polynomials up to a certain degree. The optimal cubature formulas in a sense of minimal number of points on which functions need to be evaluated are called Gaussian. In the following, the integrals over the domain $\mathcal{F}(\widetilde{S}_n^{\mathrm{aff}})$ with weights $w \in \{w^{\mathrm{II},\pm}, w^{\mathrm{IV},\pm}\}$ are replaced by finite summing.

The finite sets $\mathfrak{F}_N^{\star,\pm}$, $\star \in \mathrm{I}, \mathrm{II}, \dots, \mathrm{VIII}$ of generalized Chebyshev nodes, on which cubature formulas are evaluated, are connected to the points in the discrete sets $F_N^{\star,\pm}$ from Table 1 by the φ-transform (73),

$$\mathfrak{F}_N^{\star,\pm} = \left\{ \varphi(s) \mid s \in F_N^{\star,\pm} \right\}. \tag{76}$$

Recall from [3] that the $\varphi-$transform is one-to-one correspondence between $F_N^{\star,\pm}$ and $\mathfrak{F}_N^{\star,\pm}$ and therefore, for each $Y \in \mathfrak{F}_N^{\star,\pm}$ there exists a unique $s \in F_N^{\star,\pm}$ such that $Y = \varphi(s)$. Thus, the following three weight symbols are well-defined,

$$\mathcal{H}_Y = H_s, \quad \mathcal{E}_Y = \varepsilon_s, \quad \widetilde{\mathcal{E}}_Y = \widetilde{\varepsilon}_s. \tag{77}$$

Each family of orthogonal polynomials (51) gives rise to four different cubature formulas. The formulas related to $\mathcal{P}^{\mathrm{II},-}$ are derived from AMDST of type I, II, V and VI whereas the formulas related to $\mathcal{P}^{\mathrm{II},+}$ arise from the symmetric discrete transforms of the same types. Similarly, the formulas related to $\mathcal{P}^{\mathrm{IV},-}$ are deduced from AMDST of type III, IV, VII and VIII and the formulas related to $\mathcal{P}^{\mathrm{IV},+}$ arise from the symmetric discrete transforms of the same types.

Theorem 1.

1. *For $N \in \mathbb{N}$, $N \geq n$ and any polynomial $f(Y)$ of degree at most $2(N-n)+1$, the following formula holds exactly,*

$$\int_{\mathfrak{F}(\widetilde{S}_n^{\mathrm{aff}})} f(Y) w^{\mathrm{II},-}(Y)\, dY = \left(\frac{1}{N+1}\right)^n \sum_{Y \in \mathfrak{F}_N^{\mathrm{I},-}} f(Y) \mathcal{J}^{\mathrm{II},-}(Y). \tag{78}$$

2. *For $N \in \mathbb{N}$, $N \geq n+1$ and any polynomial $f(Y)$ of degree at most $2(N-n)-1$, the following formula holds exactly,*

$$\int_{\mathfrak{F}(\widetilde{S}_n^{\mathrm{aff}})} f(Y) w^{\mathrm{II},-}(Y)\, dY = \left(\frac{1}{N}\right)^n \sum_{Y \in \mathfrak{F}_N^{\mathrm{II},-}} f(Y) \mathcal{J}^{\mathrm{II},-}(Y). \tag{79}$$

3. *For $N \in \mathbb{N}$, $N \geq n$ and any polynomial $f(Y)$ of degree at most $2(N-n)$, the following formulas hold exactly,*

$$\begin{aligned} \int_{\mathfrak{F}(\widetilde{S}_n^{\mathrm{aff}})} f(Y) w^{\mathrm{II},-}(Y)\, dY &= \left(\frac{2}{2N+1}\right)^n \sum_{Y \in \mathfrak{F}_N^{\mathrm{V},-}} f(Y) \mathcal{J}^{\mathrm{II},-}(Y), \\ \int_{\mathfrak{F}(\widetilde{S}_n^{\mathrm{aff}})} f(Y) w^{\mathrm{II},-}(Y)\, dY &= \left(\frac{2}{2N+1}\right)^n \sum_{Y \in \mathfrak{F}_N^{\mathrm{VI},-}} f(Y) \mathcal{J}^{\mathrm{II},-}(Y). \end{aligned} \tag{80}$$

Proof. The linearity of Equation (78) implies that it is sufficient to consider $f(Y)$ in form of a monomial of degree at most $2(N-n)+1$. Any such monomial is expressible as a product of a monomial $p(Y)$ of degree not exceeding $N-n+1$ and a monomial $q(Y)$ of degree at most $N-n$. From Proposition 2 and Corollary 2 it follows that the monomials p and q are expressible as linear combinations of polynomials $\mathcal{P}_k^{\mathrm{II},-}$ with $k_1 \leq N-n+1$ and $\mathcal{P}_{k'}^{\mathrm{II},-}$ with $k_1' \leq N-n$, respectively. From the transform AMDST-I, given by (40), and Remark 1, these polynomials satisfy the following discrete orthogonality relation,

$$\left(\frac{1}{N+1}\right)^n \sum_{Y \in \mathfrak{F}_N^{\mathrm{I},-}} \mathcal{P}_k^{\mathrm{II},-}(Y) \mathcal{P}_{k'}^{\mathrm{II},-}(Y) \mathcal{J}^{\mathrm{II},-}(Y) = 2^{-n} \delta_{kk'}. \tag{81}$$

Comparing (72) and (81), the cubature Formula (78) is derived. Other cubature formulas are deduced similarly from AMDST of type II, V and VI. □

Theorem 2.

1. *For $N \in \mathbb{N}$ and any polynomial $f(Y)$ of degree at most $2N-1$, the following formula holds exactly,*

$$\int_{\mathfrak{F}(\widetilde{S}_n^{\mathrm{aff}})} f(Y) w^{\mathrm{II},+}(Y)\, dY = \left(\frac{1}{N+1}\right)^n \sum_{Y \in \mathfrak{F}_N^{\mathrm{I},+}} \mathcal{H}_Y^{-1} f(Y) \mathcal{J}^{\mathrm{II},+}(Y). \tag{82}$$

2. *For $N \in \mathbb{N}$, $N > 1$ and any polynomial $f(Y)$ of degree at most $2N-3$, the following formula holds exactly,*

$$\int_{\mathfrak{F}(\widetilde{S}_n^{\mathrm{aff}})} f(Y) w^{\mathrm{II},+}(Y)\, dY = \left(\frac{1}{N}\right)^n \sum_{Y \in \mathfrak{F}_N^{\mathrm{II},+}} \mathcal{H}_Y^{-1} f(Y) \mathcal{J}^{\mathrm{II},+}(Y). \tag{83}$$

3. *For $N \in \mathbb{N}$ and any polynomial $f(Y)$ of degree at most $2N-2$, the following formulas hold exactly,*

$$\begin{aligned}\int_{\mathfrak{F}(\widetilde{S}_n^{\mathrm{aff}})} f(Y) w^{\mathrm{II},+}(Y)\, dY &= \left(\frac{2}{2N+1}\right)^n \sum_{Y \in \mathfrak{F}_N^{\mathrm{V},+}} \mathcal{H}_Y^{-1} f(Y) \mathcal{J}^{\mathrm{II},+}(Y),\\ \int_{\mathfrak{F}(\widetilde{S}_n^{\mathrm{aff}})} f(Y) w^{\mathrm{II},+}(Y)\, dY &= \left(\frac{2}{2N+1}\right)^n \sum_{Y \in \mathfrak{F}_N^{\mathrm{VI},+}} \mathcal{H}_Y^{-1} f(Y) \mathcal{J}^{\mathrm{II},+}(Y).\end{aligned} \tag{84}$$

Theorem 3.

1. *For $N \in \mathbb{N}$, $N \geq n$ and any polynomial $f(Y)$ of degree at most $2(N-n)+1$, the following formula holds exactly,*

$$\int_{\mathfrak{F}(\widetilde{S}_n^{\mathrm{aff}})} f(Y) w^{\mathrm{IV},-}(Y)\, dY = \left(\frac{2}{2N+1}\right)^n \sum_{Y \in \mathfrak{F}_N^{\mathrm{VII},-}} f(Y) \mathcal{J}^{\mathrm{IV},-}(Y). \tag{85}$$

2. *For $N \in \mathbb{N}$, $N \geq n$ and any polynomial $f(Y)$ of degree at most $2(N-n)$, the following formulas hold exactly,*

$$\begin{aligned}\int_{\mathfrak{F}(\widetilde{S}_n^{\mathrm{aff}})} f(Y) w^{\mathrm{IV},-}(Y)\, dY &= \left(\frac{1}{N}\right)^n \sum_{Y \in \mathfrak{F}_N^{\mathrm{III},-}} \mathcal{E}_Y f(Y) \mathcal{J}^{\mathrm{IV},-}(Y),\\ \int_{\mathfrak{F}(\widetilde{S}_n^{\mathrm{aff}})} f(Y) w^{\mathrm{IV},-}(Y)\, dY &= \left(\frac{1}{N}\right)^n \sum_{Y \in \mathfrak{F}_N^{\mathrm{IV},-}} f(Y) \mathcal{J}^{\mathrm{IV},-}(Y).\end{aligned} \tag{86}$$

3. *For $N \in \mathbb{N}$, $N \geq n+1$ and any polynomial $f(Y)$ of degree at most $2(N-n)-1$, the following formula holds exactly,*

$$\int_{\mathfrak{F}(\widetilde{S}_n^{\mathrm{aff}})} f(Y) w^{\mathrm{IV},-}(Y)\, dY = \left(\frac{2}{2N-1}\right)^n \sum_{Y \in \mathfrak{F}_N^{\mathrm{VIII},-}} \widetilde{\mathcal{E}}_Y f(Y) \mathcal{J}^{\mathrm{IV},-}(Y). \tag{87}$$

Theorem 4.

1. *For $N \in \mathbb{N}$ and any polynomial $f(Y)$ of degree at most $2N-1$, the following formula holds exactly,*

$$\int_{\mathfrak{F}(\widetilde{S}_n^{\mathrm{aff}})} f(Y) w^{\mathrm{IV},+}(Y)\, dY = \left(\frac{2}{2N+1}\right)^n \sum_{Y \in \mathfrak{F}_N^{\mathrm{VII},+}} \mathcal{H}_Y^{-1} f(Y) \mathcal{J}^{\mathrm{IV},+}(Y). \tag{88}$$

2. *For $N \in \mathbb{N}$ and any polynomial $f(Y)$ of degree at most $2N-2$, the following formulas hold exactly,*

$$\begin{aligned} \int_{\mathfrak{F}(\widetilde{S}_n^{\mathrm{aff}})} f(Y) w^{\mathrm{IV},+}(Y)\, dY &= \left(\frac{1}{N}\right)^n \sum_{Y \in \mathfrak{F}_N^{\mathrm{III},+}} \mathcal{E}_Y \mathcal{H}_Y^{-1} f(Y) \mathcal{J}^{\mathrm{IV},+}(Y), \\ \int_{\mathfrak{F}(\widetilde{S}_n^{\mathrm{aff}})} f(Y) w^{\mathrm{IV},+}(Y)\, dY &= \left(\frac{1}{N}\right)^n \sum_{Y \in \mathfrak{F}_N^{\mathrm{IV},+}} \mathcal{H}_Y^{-1} f(Y) \mathcal{J}^{\mathrm{IV},+}(Y). \end{aligned} \tag{89}$$

3. *For $N \in \mathbb{N}$, $N > 1$ and any polynomial $f(Y)$ of degree at most $2N-3$, the following formulas hold exactly,*

$$\int_{\mathfrak{F}(\widetilde{S}_n^{\mathrm{aff}})} f(Y) w^{\mathrm{IV},+}(Y)\, dY = \left(\frac{2}{2N-1}\right)^n \sum_{Y \in \mathfrak{F}_N^{\mathrm{VIII},+}} \widetilde{\mathcal{E}}_Y \mathcal{H}_Y^{-1} f(Y) \mathcal{J}^{\mathrm{IV},+}(Y). \tag{90}$$

4.4. Gaussian Cubature Formulas

Theorem 5. *The cubature Formulas (78), (82), (85) and (88) are optimal Gaussian cubature formulas. Furthermore, it holds that*

1. *the orthogonal polynomials $\mathcal{P}_k^{\mathrm{II},-}$ of degree $k_1 = N - n + 1$ vanish for all nodes $\mathfrak{F}_N^{\mathrm{I},-}$,*
2. *the orthogonal polynomials $\mathcal{P}_k^{\mathrm{II},+}$ of degree $k_1 = N$ vanish for all nodes $\mathfrak{F}_N^{\mathrm{I},+}$,*
3. *the orthogonal polynomials $\mathcal{P}_k^{\mathrm{IV},-}$ of degree $k_1 = N - n + 1$ vanish for all nodes $\mathfrak{F}_N^{\mathrm{VII},-}$,*
4. *the orthogonal polynomials $\mathcal{P}_k^{\mathrm{IV},+}$ of degree $k_1 = N$ vanish for nodes $\mathfrak{F}_N^{\mathrm{VII},+}$.*

Proof. The number of nodes in the set $\mathfrak{F}_N^{\mathrm{I},-}$ is equal to the number of points in $F_N^{\mathrm{I},-}$ since φ is injective on $F_N^{\mathrm{I},-}$. The definition of $F_N^{\mathrm{I},-}$ in Table 1 and definition (37) imply that

$$\left|F_N^{\mathrm{I},-}\right| = \left|D_{1,N}^{-}\right| \tag{91}$$

and the cardinality of these sets corresponds to the number of polynomials $\mathcal{P}_k^{\mathrm{II},-}$ of degree $N-n$. Therefore, the cubature Formula (78) is Gaussian. Similar counting arguments prove that the remaining listed formulas are also Gaussian. The fact that the nodes are common zeros of the corresponding polynomials of a specific degree follows directly from definition (51) and Remarks 1 and 2. □

The Gaussian cubatures (78), (82), (85) and (88) are special cases of the general cubature Formulas (1) and (2) from [22], where the values of integrals

$$\int f(u)[\Delta(u)]^{\pm\frac{1}{2}}\, dv(u) \tag{92}$$

are studied. The variables u_i are connected to the elementary symmetric functions by relation

$$u_i = u_i(y_1, \ldots, y_n) = \sum_{1 \le k_1 < \cdots < k_i \le n} y_{k_1} \cdots y_{k_i} \tag{93}$$

and coincide, up to a multiplication by a constant, with the current polynomial variables X_i,

$$u_i = \frac{X_i}{(n-i)!i!}. \tag{94}$$

The measures $w(X)\,dX$, $w \in \{w^{\mathrm{II},\pm}, w^{\mathrm{IV},\pm}\}$, given by (68), correspond to the special choices of the positive measure $[\Delta(u)]^{\pm\frac{1}{2}}\,d\nu(u)$ of the form $\mu(u)\,du$ with

$$\mu(u) = \prod_{i=1}^{n}(1-y_i)^{\alpha}(1+y_i)^{\beta}\left(\prod_{1\leq i<j\leq n}(y_i-y_j)^2\right)^{\gamma}, \tag{95}$$

and

$$\alpha,\beta,\gamma \in \left\{\pm\frac{1}{2}\right\}, \quad -1 < y_1 < \cdots < y_n < 1. \tag{96}$$

In particular, the parameters α, β and γ have the following values depending on the chosen family of polynomials,

$$\begin{aligned}
\mathcal{P}_k^{\mathrm{II},-}: &\quad \alpha=\beta=\gamma=\frac{1}{2},\\
\mathcal{P}_k^{\mathrm{II},+}: &\quad \alpha=\beta=\frac{1}{2},\ \gamma=-\frac{1}{2},\\
\mathcal{P}_k^{\mathrm{IV},-}: &\quad \alpha=\gamma=\frac{1}{2},\ \beta=-\frac{1}{2},\\
\mathcal{P}_k^{\mathrm{IV},+}: &\quad \alpha=\frac{1}{2},\ \beta=\gamma=-\frac{1}{2}.
\end{aligned} \tag{97}$$

The cubature Formulas (78), (82), (85) and (88) together with the Gaussian cubatures from [3] form the set of cubatures with all possible values of parameters in (95).

5. Conclusions

- The present fully explicit expression of the cubature rules allows straightforward implementation of the numerical integration and approximation methods. Compared to the abstract variables of the symmetric polynomials (93) from [22], the additional relation (49) established via the fundamental symmetric cosine function connects directly, like in the classical Chebyshev polynomials, the underlying lattice with the generalized Chebyshev nodes. The antisymmetric discrete sine transforms from Table 1 are special cases of the discrete transforms derived in [14] from generalized Schur polynomials associated with Bernstein-Szegö polynomials and parametrized by $(a_-,b_-) \in \{(0,0),(0,-1)\}$. On the other hand, the symmetric discrete sine transforms from Table 1 extend the set of discrete transforms connected to the Chebyshev polynomials of the second and fourth kind.
- The symmetry group of the (anti)symmetric sine functions $(\mathbb{Z}/2\mathbb{Z})^n \rtimes S_n$ is isomorphic to the Weyl groups of the classical series of the simple Lie algebras B_n and C_n. The correspondence between the (anti)symmetric sine and cosine functions and the four types of the Weyl orbit functions is explicitly developed in [9]. The present point sets of the discrete (anti)symmetric sine transforms and the generalized Chebyshev nodes differ from the weight and dual weight lattice point sets on which the discrete transforms and cubature rules of the Weyl orbit functions are formulated. The topology of the current point sets is, however, similar for some cases to the root lattices of the series B_n and C_n and the explicit formulation of the comparison poses an open problem.
- The Lebesgue constant estimates of the polynomial cubatures and integral error estimates for the interpolation formulas together with criteria for the convergence of the polynomial series deserve further study. The developed cubature formulas as well as the rules from [3,8] reveal that the shifted lattice transforms carry high capacity to produce cubature formulas of Gaussian type. Versions of the Clenshaw–Curtis methods of numerical integration [36], developed for the C_2 and A_2 root systems in [37,38], also need to be further investigated. The formation of the hyperinterpolation methods [39,40], which straightforwardly employ the standard polynomial cubature rules, poses an open problem for the presented cubature rules.

- The existence and explicit forms of generating functions for the related Weyl group polynomials, developed in [41,42], further increase the relevance of the presented Chebyshev polynomial methods. The generating functions form a powerful tool for investigating symmetries and parity relations of the generated orthogonal polynomials and represent practical tool for efficient computer implementation and handling of the generated polynomials. The recurrence relations algorithms for the calculation of the trivariate polynomials are potentially superseded by explicit evaluation formulas derived from the generating functions. The form of the generating functions and the explicit evaluation formulas for the current polynomials pose open problems.

Author Contributions: All authors participated in the development of the mathematical concepts and composition of the manuscript. Moreover, A.B. performed numerical verifications and calculations. The authors, identified in alphabetical order, approve the final version of the manuscript.

Funding: This work was supported by the Grant Agency of the Czech Technical University in Prague, grant number SGS16/239/OHK4/3T/14. L.M. and J.H. gratefully acknowledge the support of this work by RVO14000.

Conflicts of Interest: The authors declare no conflict of interest.

References

1. Klimyk, A.; Patera, J. (Anti)symmetric multivariate trigonometric functions and corresponding Fourier transforms. *J. Math. Phys.* **2007**, *48*, 093504. [CrossRef]
2. Britanak, V.; Rao, K.; Yip, P. *Discrete Cosine and Sine Transforms: General Properties, Fast Algorithms And Integer Approximation*; Elsevier/Academic Press: Amsterdam, The Netherlands, 2007.
3. Hrivnák, J.; Motlochová, L. Discrete transforms and orthogonal polynomials of (anti)symmetric multivariate cosine functions. *SIAM J. Numer. Anal.* **2010**, *51*, 073509. [CrossRef]
4. Klimyk, A.; Patera, J. Antisymmetric Orbit Functions. *SIGMA* **2007**, *3*, 023. [CrossRef]
5. Klimyk, A.; Patera, J. Orbit Functions. *SIGMA* **2006**, *2*, 006.
6. Hrivnák, J.; Motlochová, L.; Patera, J. Two dimensional symmetric and antisymmetric generalization of sine functions. *J. Math. Phys.* **2010**, *51*, 073509. [CrossRef]
7. Hrivnák, J.; Patera, J. Two dimensional symmetric and antisymmetric generalization of exponential and cosine functions. *J. Math. Phys.* **2010**, *51*, 023515. [CrossRef]
8. Moody, R.V.; Motlochová, L.; Patera, J. Gaussian cubature arising from hybrid characters of simple lie groups. *J. Fourier Anal. Appl.* **2014**, *20*, 1257–1290. [CrossRef]
9. Hrivnák, J.; Motlochová, L. On connecting Weyl-orbit functions to Jacobi polynomials and multivariate (anti)symmetric trigonometric functions. *Acta Polytech.* **2016**, *56*, 283–290. [CrossRef]
10. Hrivnák, J.; Patera, J. On discretization of tori compact simple Lie groups. *J. Phys. A Math. Theor.* **2009**, *42*, 385208. [CrossRef]
11. Hrivnák, J.; Patera, J. On E-discretization of tori compact simple Lie groups. *J. Phys. A Math. Theor.* **2010**, *43*, 165206. [CrossRef]
12. Hrivnák, J.; Walton, M.A. Weight-lattice discretization of Weyl-orbit functions. *J. Math. Phys.* **2016**, *57*, 083512. [CrossRef]
13. Moody, R.V.; Patera, J. Orthogonality within Families of C-, S- and E- Functions of Any Compact Semisimple Lie Group. *SIGMA* **2006**, *2*, 076. [CrossRef]
14. van Diejen, J.F.; Emsiz, E. Discrete Fourier transform associated with generalized Schur polynomials. *Proc. Am. Math. Soc.* **2018**, *146*, 3459–3472. [CrossRef]
15. Handscomb, D.C.; Mason, J.C. *Chebyshev Polynomials*; Champman & Hall/CRC: New York, NY, USA, 2002.
16. Rivlin, T.J. *The Chebyshev Polynomials*; John Wiley & Sons: New York, NY, USA, 1990.
17. Koornwinder, T.H. Two-variable analogues of the classical orthogonal polynomials. *Theory Appl. Spec. Funct.* **1975**, 435–495. [CrossRef]
18. Cools, R. An encyklopedia of cubature formulas. *J. Complex.* **2003**, *19*, 445–453. [CrossRef]
19. Dunkl, C.F.; Xu, Y. *Orthogonal Polynomials of Several Variables*; Cambridge University Press: Cambridge, UK, 2001.

20. Li, H.; Sun, J.; Xu, Y. Discrete Fourirer analysis and Chebyshev polynomials with G_2 group. *SIGMA* **2012**, *8*, 67.
21. Li, H.; Sun, J.; Xu, Y. Discrete Fourirer analysis cubature and interpolation on hexagon and triangle. *SIAM J. Numer. Anal.* **2008**, *46*, 1653–1681. [CrossRef]
22. Berens, H.; Schmid, H.J.; Xu, Y. Multivariate Gaussian cubature formulae. *Arch. Math.* **1995**, *64*, 26–32. [CrossRef]
23. Drissi, L.B.; Saidi, E.H.; Bousmina, M. Graphene, Lattice Field Theory and Symmetries. *J. Math. Phys.* **2011**, *52*, 022306. [CrossRef]
24. Hrivnák, J.; Motlochová, L.; Patera, J. Cubature formulas of multivariate polynomials arising from symmetric orbit functions. *Symmetry* **2016**, *8*, 63. [CrossRef]
25. Chadzitaskos, G.; Háková, L.; Kájínek, O. Weyl Group Orbit Functions in Image Processing. *Appl. Math.* **2014**, *5*, 501–511. [CrossRef]
26. Çapoğlu, I.R.; Taflove, A.; Backman, V. Computation of tightly-focused laser beams in the FDTD method. *Opt. Express* **2013**, *21*, 87–101. [CrossRef] [PubMed]
27. Xu, J.; Chen, J.; Li, J. Probability density evolution analysis of engineering structures via cubature points. *Comput. Mech.* **2012**, *50*, 135–156. [CrossRef]
28. Lauvergnata, D.; Nauts, A. Quantum dynamics with sparse grids: A combination of Smolyak scheme and cubature. Application to methanol in full dimensionality. *Spectrochim. Acta Part A* **2014**, *119*, 18–25. [CrossRef] [PubMed]
29. Crivellini, A.; D'Alessandro, V.; Bassi, F. High-order discontinuous Galerkin solutions of three-dimensional incompressible RANS equations. *Comput. Fluids* **2013**, *81*, 122–133. [CrossRef]
30. Young, J.C.; Gedney, S.D.; Adams, R.J. Quasi-Mixed-Order Prism Basis Functions for Nyström-Based Volume Integral Equations. *IEEE Trans. Magn.* **2012**, *48*, 2560–2566. [CrossRef]
31. Sfevanovic, I.; Merli, F.; Crespo-Valero, P.; Simon, W. S. Holzwarth, M. Mattes, J. R. Mosig, Integral Equation Modeling of Waveguide-Fed Planar Antennas. *IEEE Antennas Propag. Mag.* **2009**, *51*, 82–92. [CrossRef]
32. Chernyshenko, D.; Fangohr, H. Computing the demagnetizing tensor for finite difference micromagnetic simulations via numerical integration. *J. Magn. Magn. Mater.* **2015**, *381*, 440–445. [CrossRef]
33. Tasinkevych, M.; Silvestre, N.M.; de Gama, M.M.T. Liquid crystal boojum-colloids. *New J. Phys.* **2012**, *14*, 073030. [CrossRef]
34. Marin, M. Weak Solutions in Elasticity of Dipolar Porous Materials. *Math. Probl. Eng.* **2008**, *2008*, 158908. [CrossRef]
35. Marin, M.; Florea, O. On temporal behaviour of solutions in thermoelasticity of porous micropolar bodies. *An. St. Univ. Ovidius Constanta, Ser. Mat.* **2014**, *22*, 169–188. [CrossRef]
36. Clenshaw, C.W.; Curtis, A.R. A method for numerical integration on an automatic computer. *Numer. Math.* **1960**, *2*, 197–205. [CrossRef]
37. Háková, L.; Hrivnák, J.; Motlochová, L. On cubature rules associated to Weyl group orbit functions. *Acta Polytech.* **2016**, *56*, 202–213. [CrossRef]
38. Munthe-Kaas, H.; Ryland, B.N. On multivariate Chebyshev polynomials and spectral approximations on triangles. *Spectral and High Order Methods for Partial Differential Equations*; Lecture Notes in Computational Science and Engineering; Springer: Berlin/Heidelberg, Germany, 2010; Volume 76, pp. 19–41.
39. Sommariva, A.; Vianello, M.; Zanovello, R. Nontensorial Clenshaw-Curtis cubature. *Numer. Algorithms* **2008**, *49*, 409–427. [CrossRef]
40. Caliari, M.; de Marchi, S.; Vianello, M. Hyperinterpolation in the cube. *Comput. Math. Appl.* **2008**, *55*, 2490–2497. [CrossRef]
41. Czyżycki, T.; Hrivnák, J.; Patera, J. Generating functions for orthogonal polynomials of A_2, C_2 and G_2. *Symmetry* **2018**, *10*, 354. [CrossRef]
42. Damaskinsky, E.V.; Kulish, P.P.; Sokolov, M.A. On calculation of generating functions of Chebyshev polynomials in several variables. *J. Math. Phys.* **2015**, *56*, 063507. [CrossRef]

entropy

MDPI

Article

Hermite Functions, Lie Groups and Fourier Analysis

Enrico Celeghini [1,2], Manuel Gadella [2,3] and Mariano A. del Olmo [2,3,*]

1 Dipartimento di Fisica, Università di Firenze and INFN-Sezione di Firenze, 50019 Sesto Fiorentino, Firenze, Italy; celeghini@fi.infn.it

2 Departamento de Física Teórica, Atómica y Optica, Universidad de Valladolid, 47011 Valladolid, Spain; manuelgadella1@gmail.com

3 Instituto de Matemáticas, Universidad de Valladolid (IMUVA), 47011 Valladolid, Spain

* Correspondence: marianoantonio.olmo@uva.es

Received: 30 August 2018; Accepted: 19 October 2018; Published: 23 October 2018

Abstract: In this paper, we present recent results in harmonic analysis in the real line $\mathbb{R}$ and in the half-line $\mathbb{R}^+$, which show a closed relation between Hermite and Laguerre functions, respectively, their symmetry groups and Fourier analysis. This can be done in terms of a unified framework based on the use of rigged Hilbert spaces. We find a relation between the universal enveloping algebra of the symmetry groups with the fractional Fourier transform. The results obtained are relevant in quantum mechanics as well as in signal processing as Fourier analysis has a close relation with signal filters. In addition, we introduce some new results concerning a discretized Fourier transform on the circle. We introduce new functions on the circle constructed with the use of Hermite functions with interesting properties under Fourier transformations.

Keywords: Fourier analysis; special functions; rigged Hilbert spaces; quantum mechanics; signal processing

1. Introduction

The seminal work by Fourier of 1807, published in 1822 [1], about the solution of the heat equation had a deep impact in physics and mathematics as is well known. Roughly speaking, the Fourier method decomposes functions into a superposition of "special functions" [2,3]. In particular, trigonometric functions were used by Fourier himself for this purpose . In addition, the Fourier method makes use of other types of special functions; each of these types is often related with a group. Then, these special functions have symmetry properties, which are inherited from the corresponding group. For instance, this is the way in which harmonic analysis appears in group representation theory [4]. An interesting aspect of Fourier analysis is the decomposition of Hilbert space vectors, quite often represented by square integrable functions on some domain, into an orthogonal basis. This generalizes both the standard Fourier analysis of trigonometric series and the decomposition of a vector in terms of an algebraic basis of linearly independent vectors. Another generalization is the decomposition of a self-adjoint or normal operator on a Hilbert space in terms of spectral measures, say through the spectral representation theorem. We are mainly interested in these generalizations concerning Hilbert space vectors and operators.

In recent works [5,6], we started an attempt to reformulate the harmonic analysis on the real line to obtain a global description of the Hermite functions, the Weyl–Heisenberg Lie algebra and the Fourier analysis in the framework of rigged Hilbert spaces (RHS) that we present here in a more formal way. As is well known, the Fourier transform relates two continuous bases which are used in the description of one-dimensional quantum systems on the whole real line. These are the coordinate and momentum representations, naturally connected with the position and the momentum operator [7,8], respectively. They span the Weyl–Heisenberg algebra together with the identity operator. Moreover, these two continuous bases can be related with a discrete orthonormal basis labeled by the natural numbers via the Hermite functions. In consequence, we have continuous and discrete bases within the same framework. However, only discrete bases as complete orthonormal sets have a precise meaning in Hilbert space. If we have a structure allowing to work with these types of bases and to find relations among them, one needs to extend the Hilbert space to a more general structure called the rigged Hilbert space.

The fundamental message of the present paper is to show how a class of different and apparently unrelated mathematical objects, such as classical orthogonal polynomials, Lie algebras, Fourier analysis, continuous and discrete bases and RHS, can be fully wived as a branch of harmonic analysis, with applications in quantum mechanics and signal processing, among other possible applications.

We have mentioned that the mathematical concept of RHS is very important in our work. It has been introduced by Gelfand and collaborators [9] proving (although Maurin [10]) the nuclear spectral theorem as was heuristically introduced by Dirac [11]. It is also generally accepted that the eigenfunction expansions and the Dirac formalism are generalizations of the Fourier analysis for which we need RHS [10,12]. It is also known that the spectral theory of infinitesimal operators of an arbitrary unitary representation of a Lie group also need RHS [10,12]. In the physics literature, the similarities between the Dirac formalism, classical Fourier analysis and generalized Fourier transforms have been discussed within the RHS framework [13,14]. Another application of RHS, which has a particular importance in our presentation, is signal processing. In particular, in the electrical engineering literature, these aspects have been discussed in [15–18].

Since the average physicists may not be acquainted with the concept of RHS, let us give a definition and some remarks on this concept. A rigged Hilbert space or Gelfand triple is a set of three vector spaces

$$\Phi \subset \mathcal{H} \subset \Phi^\times ,$$

where $\mathcal{H}$ is an infinite dimensional separable Hilbert space, Φ is a topological vector space endowed with a topology finer than the Hilbert space topology and dense on $\mathcal{H}$ with the Hilbert space topology, and $\Phi^\times$ is the dual of Φ (i.e., the space of linear (or antilinear) continuous mappings from Φ into the complex numbers $\mathbb{C}$) and it is endowed with a topology compatible with the pair $(\Phi, \Phi^\times)$.

The formulation of quantum mechanics in terms of RHS was introduced by Bohm and Roberts in the sixties of the last century and further developed later [19–26]. Continuous bases are not well defined in Φ and $\mathcal{H}$ but only in $\Phi^\times$. The action of a functional $F \in \Phi^\times$ on a vector $\varphi \in \Phi$ is written as $\langle \varphi | F \rangle$ for keeping up with the Dirac notation. Since we will consider the scalar product on Hilbert space antilinear to the left, we shall assume the antilinearity of the elements in $\Phi^\times$.

The first part of the present paper is devoted to a review of a previous work by the authors [6] concerning to the above-mentioned extension of Fourier analysis on the real line with the use of special functions such us Hermite functions, which will be here for our main example. This is studied in Section 2. The use of the Fractional Fourier transform (FFT) in this analysis is discussed in Section 3.

In addition, we give a second example in which the real line has been replaced by the semi-axis $\mathbb{R}^+ \equiv [0, \infty)$ and Hermite functions by Laguerre functions. In this latter case, we construct two different Fourier-like transforms $\mathcal{T}^\pm$ and their eigenvectors are functions on the positive half-line. This is given in Sections 4 and 5. Extensions to $\mathbb{R}^n$ using or not spherical coordinates are also possible, although we shall not consider this option in the present manuscript [27]. In Reference [28], we revisited the harmonic analysis on the group $SO(2)$ using RHS. Furthermore, in Reference [29], we introduce a new realization of the group $SU(2)$ in the plane in terms of the associated Laguerre polynomials.

In Section 4, we introduce some new results concerning harmonic analysis on the circle. We construct new functions on the circle using Hermite functions and taking advantage of their properties. Again, these new functions give a unitary view of different mathematical objects that are often considered as unrelated: Fourier transform, discrete Fourier transform, Hermite functions and RHS.

To understand the importance of the present research, let us remark that Hermite and Laguerre functions are bases of spaces of square integrable functions, no matter whether real or complex, defined on $\mathbb{R}$ and $\mathbb{R}^+$, respectively. Square integrable real and complex (wave) functions play a similar role in signal processing and quantum mechanics, respectively. In addition, the interest of signal processing comes after the definition of two new types of filters. One is based in restrictions to subspaces of $L^2(\mathbb{R})$ or $L^2(\mathbb{R}^+)$. We have systematically constructed these filters by the use of the FFT. The other requires choosing low values of the index n in the span of a given function by either Hermite or Laguerre functions (we may also use a combination thereof). These filters remove noise or other spurious effects from the signal or the wave function.

In addition, since the basic operators related with these functions span some Lie algebras, such as the $io(2)$ [30] for the Hermite functions and the $su(1,1)$ for the Laguerre functions, we can introduce a richer space of operators on $L^2(\mathbb{R})$ or $L^2(\mathbb{R}^+)$, related to the universal enveloping algebra (UEA) of $io(2)$ or $su(1,1)$, respectively [31].

These operator spaces connect functions describing the time evolution of the states under filters or some kind of interaction.

Finally, we would like to add that this discussion may be related with some integral transforms of the type Fourier-like, Laplace-like or Sumudu-like transforms [32–36].

2. Harmonic Analysis on $\mathbb{R}$

The first example of Fourier analysis and its relation with group theory is provided by the translation group in one spatial dimension $T_1 \simeq \mathbb{R}$ (for the group $SO(2)$ see [28]). The action of its unitary irreducible representations, $\mathcal{R}$, on the continuous basis $\{|p\rangle\}_{p\in\mathbb{R}}$, given by the eigenvectors of the infinitesimal generator P of the group, is given by

$$\mathcal{R}(x)|p\rangle = e^{-iPx}\,|p\rangle = e^{-ipx}\,|p\rangle\,; \qquad P\,|p\rangle = p\,|p\rangle\,, \quad \forall p \in \mathbb{R}, \forall x \in T_1\,. \tag{1}$$

The vectors of the basis $\{|p\rangle\}$ verify

$$\langle p|p'\rangle = \sqrt{2\pi}\,\delta(p-p')\,, \qquad \frac{1}{\sqrt{2\pi}}\int_{-\infty}^{\infty} |p\rangle\langle p|\,dp = \mathbb{I}\,. \tag{2}$$

Considering the position operator X and a continuous basis $\{|x\rangle\}_{x\in\mathbb{R}}$ of its eigenvectors, i.e.,

$$X\,|x\rangle = x\,|x\rangle\,, \qquad \forall x \in \mathbb{R} \simeq T_1\,. \tag{3}$$

Via the Fourier transform, we can relate both (conjugate) bases $\{|p\rangle\}$ and $\{|x\rangle\}$

$$|x\rangle = \frac{1}{\sqrt{2\pi}}\int_{-\infty}^{\infty} e^{-ipx}\,|p\rangle\,dp, \qquad |p\rangle = \frac{1}{\sqrt{2\pi}}\int_{-\infty}^{\infty} e^{ipx}\,|x\rangle\,dx, \tag{4}$$

such that we find for the basis $\{|x\rangle\}$ that

$$\langle x|x'\rangle = \sqrt{2\pi}\,\delta(x-x')\,, \qquad \frac{1}{\sqrt{2\pi}}\int_{-\infty}^{\infty} |x\rangle\langle x|\,dx = \mathbb{I}\,. \tag{5}$$

Moreover, X, P together with $\mathbb{I}$ determine the Weyl–Heisenberg algebra

$$[X,P] = i\,\mathbb{I}\,, \qquad [\cdot,\mathbb{I}] = 0\,. \tag{6}$$

For more details, see Reference [37].

2.1. Hermite Functions and the Group $IO(2)$

Now, we consider the inhomogeneous orthogonal group $IO(2)$ which is isomorphic to the Euclidean group in the plane, $E(2)$. In the study of the ray representations [38–40], we have to deal with the central extended group [30]. Here, we use a non-standard technique related to the projective representations of $IO(2)$ by considering the algebra of the harmonic oscillator that it is isomorphic to the central extension mentioned above. To proceed, let us consider the operators

$$a := \frac{1}{\sqrt{2}}\,(X+iP)\,, \qquad a^+ := \frac{1}{\sqrt{2}}\,(X-iP)\,, \qquad N := a\,a^+\,, \qquad \mathbb{I}\,,$$

which determine the Lie commutators

$$[a,a^+] = \mathbb{I}\,, \qquad [N,a] = -a\,, \qquad [N,a^+] = a^+\,, \qquad [\mathbb{I},\cdot] = 0\,,$$

and the quadratic Casimir

$$\mathcal{C} = \{a,a^+\} - 2(N+1/2)\,\mathbb{I}\,.$$

In the representation with $\mathcal{C} = 0$, we obtain the differential equation

$$\mathcal{C}\,K_n(x) \equiv \left(-D_x^2 + X^2 - (2N+1)\right)\,K_n(x) = 0\,, \tag{7}$$

where $P = -i\,D_X = -i\,d/dx$ and N is a kind of number operator such that for each index $n \in \mathbb{N}$, $N\,K_n(x) = n\,K_n(x)$, where $K_n(x)$ are solutions of the differential Equation (7). These solutions are the Hermite functions

$$K_n(x) := \frac{e^{-x^2/2}}{\sqrt{2^n n!\sqrt{\pi}}}\,H_n(x)\,, \tag{8}$$

with $H_n(x)$ the Hermite polynomials. Thus, $\{K_n(x)\}_{n\in\mathbb{N}}$ is an orthonormal basis in $L^2(\mathbb{R})$. As is well known,

$$\int_{-\infty}^{\infty} K_n(x)\,K_m(x)\,dx = \delta_{nm}\,, \qquad \sum_{n=0}^{\infty} K_n(x)\,K_n(x') = \delta(x-x')\,. \tag{9}$$

Note that we denote by $\mathbb{N}$ the set of positive integers or natural numbers together with 0 and by $\mathbb{N}^* = \mathbb{N} - \{0\}$.

We see that the spectrum of the operator N is countably infinite, so that we may construct a countable orthonormal basis of eigenvectors of N, $\{|n\rangle\}_{n\in\mathbb{N}}$, in terms of the continuous basis related to X and the Hermite functions. This is given by the following relation:

$$|n\rangle := (2\pi)^{-1/4} \int_{-\infty}^{\infty} K_n(x)\,|x\rangle\,dx\,, \qquad n = 0,1,2,\ldots\,. \tag{10}$$

From the properties of the continuous basis as well as of the Hermite functions, we obtain that

$$\langle n|m\rangle = \delta_{nm}\,, \qquad \sum_{n=0}^{\infty} |n\rangle\langle n| = \mathbb{I}\,. \tag{11}$$

It is worth noticing that the Hermite functions are eigenfunctions of the Fourier transform:

$$[\mathcal{F}\,K_n](p) \equiv \widetilde{K}_n(p) = \frac{1}{\sqrt{2\pi}} \int_{-\infty}^{\infty} e^{ipx}\,K_n(x)\,dx = i^n\,K_n(p)\,. \tag{12}$$

This expression allows us to write relations between the three bases: one discrete and two continuous, which have been defined in this section. These relations are

$$|n\rangle = i^n (2\pi)^{-1/4} \int_{-\infty}^{\infty} K_n(p)\, |p\rangle\, dp\,, \tag{13}$$

$$|x\rangle = (2\pi)^{1/4} \sum_{n=0}^{\infty} K_n(x)\, |n\rangle\,, \tag{14}$$

$$|p\rangle = (2\pi)^{1/4} \sum_{n=0}^{\infty} i^n\, K_n(p)\, |n\rangle\,. \tag{15}$$

We see that the Hermite functions are the elements of the "transition matrices" between the continuous ad the discrete bases. We can express any ket $|f\rangle$ in any of the three bases in terms of the following equations:

$$|f\rangle = \frac{1}{\sqrt{2\pi}} \int_{-\infty}^{\infty} dx\, f(x)\, |x\rangle\,, \qquad |f\rangle = \frac{1}{\sqrt{2\pi}} \int_{-\infty}^{\infty} dp\, \tilde{f}(p)^*\, |p\rangle\,, \qquad |f\rangle = (2\pi)^{-1/4} \sum_{n=0}^{\infty} a_n\, |n\rangle\,, \tag{16}$$

with

$$f(x) := \langle x|f\rangle = \sum_{n=0}^{\infty} c_n\, K_n(x)\,, \qquad \tilde{f}(p)^* := \langle p|f\rangle = \sum_{n=0}^{\infty} (-i)^n\, c_n\, K_n(p)\,, \tag{17}$$

and

$$c_n = (2\pi)^{1/4} \langle n|f\rangle = \int_{-\infty}^{\infty} dx\, K_n(x)\, f(x) = i^n\, (2\pi)^{-1/4} \int_{-\infty}^{\infty} dp\, K_n(p)\, \tilde{f}(p)^*\,. \tag{18}$$

Therefore, we have obtained three different manners of expressing a quantum state $|f\rangle$ in terms of three different bases: two of them are continuous and non-countable , $\{|x\rangle\}_{x\in\mathbb{R}}$ and $\{|p\rangle\}_{p\in\mathbb{R}}$, and the other one, $\{|n\rangle\}_{n\in\mathbb{N}}$, is countably infinite. The framework to deal together with all three of these bases is the RHS [21].

In particular, the set $\{|n\rangle \equiv K_n(x)\}_{n\in\mathbb{N}}$ is a discrete basis of $\Phi \equiv \mathcal{S}$ (the Schwartz space) and $\mathcal{H} \equiv L^2(\mathbb{R})$ and the continuous bases belong to $\Phi^\times \equiv \mathcal{S}^\times$ (the space of tempered distributions). More precisely, we have two equivalent RHS: one is abstract $\Phi \subset \mathcal{H} \subset \Phi^\times$ and the other admits a realization in terms of functions, $\mathcal{S} \subset L^2(\mathbb{R}) \subset \mathcal{S}^\times$. They are related through the unitary map $U : \mathcal{H} \longmapsto L^2(\mathbb{R})$ defined by $U|n\rangle = K_n(x)$. There is another interesting fact related with the use of RHS: the space $\mathcal{S}$ belongs to the domain of the operators in $UEA[io(2)]$. All of these operators can be extended by duality to continuous (under any topology on $\mathcal{S}^\times$ compatible with the dual pair) operators on $\mathcal{S}^\times$. For a detailed exposition of the actual case, see [6] and references therein.

2.2. $UEA[io(2)]$ and Fractional Fourier Transform

Let us consider the kets $|n\rangle$ that form a complete orthonormal system in the abstract Hilbert space $\mathcal{H}$. For any $n \in \mathbb{N}$ and $0 < k \leq n \in \mathbb{N}$, we consider the natural numbers, q and r such that $n = k\,q + r$, where $r = 0, 1, 2, \ldots, k-1$. For k fixed, the set $\{|k\,q + r\rangle\}$ is a complete orthonormal system in $\mathcal{H}$. Let us define the operators Q and R as

$$Q\,|k\,q + r\rangle := q\,|k\,q + r\rangle\,, \qquad R\,|k\,q + r\rangle := r\,|k\,q + r\rangle\,. \tag{19}$$

These operators also act on $\Phi \subset \mathcal{H}$ and can be extended by duality to $\Phi^\times$.

Infinitely many copies of the Lie algebra $io(2)$ are contained in $UEA[io(2)]$. Thus, for any positive integer k, each of the pairs (k, r) with $0 \leq r \leq k-1$ labels a copy of $io(2)$, here denoted as $io_{k,r}(2)$. Furthermore,

$$\bigoplus_{r=0}^{k-1} io_{k,r}(2) \subset UEA[io(2)]\,.$$

We define the family of operators $A^\dagger_{k,r}$ and $A_{k,r}$ in $UEA[io(2)]$ by

$$A^\dagger_{k,r} := (a^\dagger)^k \frac{\sqrt{N+k-r}}{\sqrt{k\prod_{j=1}^{k}(N+j)}}\,, \qquad A_{k,r} := \frac{\sqrt{N+k-r}}{\sqrt{k\prod_{j=1}^{k}(N+j)}}(a)^k\,,$$

where $A^\dagger_{k,r}$ is the formal adjoint of $A_{k,r}$ and viceversa. They are continuous on Φ and can be continuously extended by duality to $\Phi^\times$. Their action on the vectors $|k\,q+r\rangle$ is

$$A^\dagger_{k,r}\,|k\,q+r\rangle = \sqrt{q+1}\,|k\,(q+1)+r\rangle\,, \qquad A_{k,r}\,|k\,q+r\rangle = \sqrt{q}\,|k\,(q-1)+r\rangle\,.$$

For each pair of integers k and r with $0 \le r < k$, the operators Q, $A^\dagger_{k,r}$, $A_{k,r}$ and $\mathbb{I}$ close a $io(2)$ Lie algebra, here denoted as $io_{k,r}(2)$. The commutation relations are

$$\begin{aligned} [Q,\,A^\dagger_{k,r}] &= +A^\dagger_{k,r}\,, & [Q,A_{k,r}] &= -A_{k,r}\,, \\ [A_{k,r},\,A^\dagger_{k,r}] &= \mathbb{I}\,, & [\mathbb{I},\cdot] &= 0\,. \end{aligned}$$

Note that, for any pair (k,r), the kets $|k\,q+r\rangle$ span subspaces $\mathcal{H}_{k,r}$ of $\mathcal{H}$ and $L^2_{k,r}(\mathbb{R})$ of $L^2(\mathbb{R})$. Hence, we have that

$$\mathcal{H} = \bigoplus_{r=0}^{k-1}\mathcal{H}_{k,r}\,, \qquad L^2(\mathbb{R}) = \bigoplus_{r=0}^{k-1} L^2_{k,r}(\mathbb{R})\,. \tag{20}$$

We can easily obtain the spaces $\Phi_{k,r}$ and $\mathcal{S}_{k,r}$. A vector $|\phi\rangle$ belongs to $\Phi_{k,r}$ if and only if

$$|\phi\rangle = \sum_{q=0}^{\infty} c_q\,|k\,q+r\rangle\,, \tag{21}$$

such that

$$\sum_{q=0}^{\infty}(q+1)^{2p}\,|c_q|^2 < \infty\,, \qquad \forall p \in \mathbb{N}\,.$$

A similar result can be obtained for any $\mathcal{S}_{k,r}$, just replacing $|k\,q+r\rangle$ by $K_{k\,q+r}(x)$ in Label (21). Moreover, the corresponding RHS can be obtained:

$$\begin{aligned} &\Phi_{k,r} \subset \mathcal{H}_{k,r} \subset \Phi^\times_{k,r}\,, \\ &\mathcal{S}_{k,r} \subset L^2_{k,r}(\mathbb{R}) \subset \mathcal{S}^\times_{k,r}\,. \end{aligned} \tag{22}$$

One can also prove that an operator $\mathcal{O}$ belongs to $UEA[io_{k,r}(2)]$ if and only if $\mathcal{O}$ is an operator on $\mathcal{H}_{k,r}$.

The split of $L^2(\mathbb{R})$ as a direct sum of subspaces $L^2_{k,r}(\mathbb{R})$ is connected with the FFT, which is is a generalization of the Fourier transform [36]. It is very interesting that we can also relate the FFT with the Hermite functions $K_n(x)$ (8) in a simple manner. Let us first define the *fractional Fourier transform* of $f \in L^2(\mathbb{R})$ associated to $a \in \mathbb{R}$, $\mathcal{F}^a f$, as

$$[\mathcal{F}^a f](p) := \sum_{n=0}^{\infty} c_n\, e^{i\,n\,a\,\pi/2}\, K_n(p)\,, \tag{23}$$

where

$$f(x) = \sum_{n=0}^{\infty} c_n\, K_n(x)\,, \qquad c_n = \int_{-\infty}^{\infty} f^*(x)\, K_n(x)\, dx\,. \tag{24}$$

The convergence of the series in (23) is in the $L^2(\mathbb{R})$ norm as well as in the more generalized sense given in (21) if $f(x) \in \mathcal{S}$, so that $\mathcal{F}^a f \in \mathcal{S}$ if $f \in \mathcal{S}$.

When $a = 4/k$, with $k \in \mathbb{N}^*$, we have

$$\tilde{f}^k(p) := [\mathcal{F}^{4/k} f](p) = \sum_{n=0}^{\infty} c_n \, e^{2\pi i n/k} \, K_n(p) \,. \tag{25}$$

In this case, we recover the standard Fourier transform for $k = 4$, which means that $a = 1$. Since for every $k \in \mathbb{N}^*$, any $n \in \mathbb{N}$ can be decomposed as $n = k\,q + r$ with $q \in \mathbb{N}$ and $0 \leq r \leq k-1$, we have the following decomposition of $\tilde{f}^k$ given by

$$\begin{aligned} \tilde{f}^k(p) &= \sum_{q=0}^{\infty} c_{kq} \, e^{2\pi(kq)i/k} \, K_{kq}(p) + \sum_{q=0}^{\infty} c_{kq+1} \, e^{2\pi(kq+1)i/k} \, K_{kq+1}(p) \\ &\quad + \cdots + \sum_{q=0}^{\infty} c_{kq+k-1} \, e^{2\pi(kq+k-1)i/k} \, K_{kq+k-1}(p) \\ &= \tilde{f}_0^k(p) + \tilde{f}_1^k(p) + \cdots + \tilde{f}_{k-1}^k(p) \,, \end{aligned} \tag{26}$$

where

$$f_r^k(x) := \sum_{q=0}^{\infty} c_{kq+r} \, K_{kq+r}(x) \,, \qquad \tilde{f}_r^k(p) := e^{2\pi r i/k} \, f_r^k(p) \,.$$

Relation (26) gives a split of $L^2(\mathbb{R})$ into an orthonormal direct sum of subspaces because the vectors $\tilde{f}_r^k$, $(r = 0, 1, \ldots, k-1)$ are mutually orthogonal. Moreover, each term in the direct sum is an eigen-subspace of $\mathcal{F}^{4/k}$ with eigenvalue $e^{i2\pi r/k}$ since $\tilde{f}_r^k(p) \equiv [\mathcal{F}^{4/k} f_r^k](p)$. The decomposition is given by

$$L^2(\mathbb{R}) = L^2_{k,0}(\mathbb{R}) \oplus L^2_{k,1}(\mathbb{R}) \oplus \cdots \oplus L^2_{k,k-1}(\mathbb{R}) \,,$$

so that we have recovered the decomposition (20).

3. Harmonic Analysis on $\mathbb{R}^+$

Let $L^2(\mathbb{R}^+)$ be the space of square integrable functions on $\mathbb{R}^+ \equiv [0, +\infty)$. As is well known a basis in $L^2(\mathbb{R}^+)$ is determined by the Laguerre functions

$$M_n^\alpha(y) = \sqrt{\frac{\Gamma(n+1)}{\Gamma(n+\alpha+1)}} \, y^{\alpha/2} \, e^{-y/2} \, L_n^\alpha(y) \,, \tag{27}$$

with $-1 < \alpha < +\infty$, $n = 0, 1, 2, \ldots$, and $L_n^\alpha(y)$ the associated Laguerre polynomials [41,42]. Indeed, $M_n^\alpha(y)$ verify the following orthonormality and completeness relations

$$\int_0^\infty M_n^\alpha(y) \, M_m^\alpha(y) \, dy = \delta_{nm} \,, \qquad \sum_{n=0}^{\infty} M_n^\alpha(y) \, M_n^\alpha(y') = \delta(y - y') \,. \tag{28}$$

3.1. Harmonic Analysis on $su(1,1)$

Let us define the following operators on $L^2(\mathbb{R}^+)$

$$\begin{aligned} N \, M_n^\alpha(y) &:= n \, M_n^\alpha(y) \,, & \mathbb{I} \, M_n^\alpha(y) &:= M_n^\alpha(y) \,, \\ Y \, M_n^\alpha(y) &:= y \, M_n^\alpha(y) \,, & D_y \, M_n^\alpha(y) &:= M_n^\alpha(y)' = \frac{d}{dy} M_n^\alpha(y) \,. \end{aligned} \tag{29}$$

Using the operators defined in (29), we may define some others:

$$J_+ := \left(Y \, D_y + N + 1 + \frac{\alpha - Y}{2} \right) , \qquad J_- := \left(-Y \, D_y + N + \frac{\alpha - Y}{2} \right) . \tag{30}$$

These operators act on $M_n^\alpha(y)$ as

$$J_+ M_n^\alpha(y) = \sqrt{(n+1)(n+\alpha+1)}\, M_{n+1}^\alpha(y)\,, \\ J_- M_n^\alpha(y) = \sqrt{n(n+\alpha)}\, M_{n-1}^\alpha(y). \tag{31}$$

These two operators, together with

$$J_3 := N + \frac{\alpha+1}{2}\,\mathbb{I}\,, \qquad J_3\, M_n^\alpha(y) = \left(n + \frac{\alpha+1}{2}\right)\, M_n^\alpha(y)\,,$$

define the Lie algebra $su(1,1)$ because their commutation relations are

$$[J_3, J_\pm] = \pm J_\pm\,, \qquad [J_+, J_-] = -2\, J_3\,. \tag{32}$$

The Casimir operator $\mathcal{C}$ of $su(1,1)$ is

$$\mathcal{C} = J_3^2 - \frac{1}{2}\{J_+, J_-\} = \frac{\alpha^2 - 1}{4}\,\mathbb{I}\,. \tag{33}$$

From (30), we obtain that $Y = -(J_+ + J_-) + 2N + (\alpha+1)\,\mathbb{I}$, and from the Casimir we may obtain the differential equation defining the associated Laguerre polynomials.

Omitting the technical details which can be found in Reference [6], let us say that there exists a set of generalized eigenvectors of Y, $\{|y\rangle\}_{y\in\mathbb{R}^+}$, (or more strictly of $U^{-1}YU$, where $U : \mathcal{H} \longmapsto L^2(\mathbb{R}^+)$ is a unitary operator and $\mathcal{H}$ is a separable Hilbert space) such that

$$Y|y\rangle = y|y\rangle\,, \qquad \langle y|y'\rangle = \delta(y-y')\,, \qquad \int_{-\infty}^{+\infty} |y\rangle\langle y|\, dy = \mathbb{I}\,. \tag{34}$$

Actually, we have two families, depending on α, of equivalent RHS $\Phi_\alpha \subset \mathcal{H} \subset \Phi_\alpha^\times$ and $\mathcal{D}_\alpha \subset L^2(\mathbb{R}^+) \subset \mathcal{D}_\alpha^\times$. All the elements and their extensions of the $UEA(su(1,1))$ are continuous on both RHS.

In analogy with the case of the whole real line, a decomposition like (20) for any $k \neq 0 \in N$, we also obtain here that

$$\mathcal{H} = \bigoplus_{r=0}^{k-1} \mathcal{H}_{k,r}\,, \quad L^2(\mathbb{R}^+) = \bigoplus_{r=0}^{k-1} L^2_{k,r}(\mathbb{R}^+)\,. \tag{35}$$

We define the vectors $|n,\alpha\rangle \in \Phi_\alpha$ as

$$|n,\alpha\rangle := \int_0^\infty dy\, M_n^\alpha(y)|y\rangle\,, \qquad \forall n \in \mathbb{N}\,,\;\; \alpha \in (-1,+\infty)\,, \tag{36}$$

which after (28) and (34), they have the properties

$$\langle n,\alpha|m,\alpha\rangle = \delta_{nm}\,, \qquad \sum_{n=0}^{\infty} |n,\alpha\rangle\langle n,\alpha| = \mathbb{I}\,. \tag{37}$$

Hence, $|n,\alpha\rangle$ with $n \in \mathbb{N}$ (and α fix) is an orthonormal basis in $\mathcal{H}$. Taking into account the unitarity of the operator U, we have that $U|n,\alpha\rangle = M_n^\alpha$. For $y \geq 0$, we easily obtain

$$\langle y|n,\alpha\rangle = \int_0^\infty dy'\, M_n^\alpha(y')\,\langle y|y'\rangle = M_n^\alpha(y)\,. \tag{38}$$

In analogy with the previous case in which we have considered functions on the whole real line, we also have two different bases spanning the vectors in $\Phi_\alpha \subset \mathcal{H}$: a continuous one $\{|y\rangle\}_{y\in\mathbb{R}^+}$ and another discrete $\{|n,\alpha\rangle\}_{n\in\mathbb{N}}$ whose elements are eigenvectors of the operator $U^{-1}NU$ where N was defined in (29).

3.2. Fourier-Like Transformations on $\mathbb{R}^+$

In Section 2.2, we have introduced the FFT related to the Hermite functions. Now, after the results displayed in the previous section that show a close analogy between the formalisms on $\mathbb{R}$ and on R^+, we may consider an extension of the FFT valid for the generalized Laguerre functions. However, this is not possible since the Laguerre functions, $M_n^\alpha(y)$, are not eigenfunctions of the Fourier transform unlike the Hermite functions. Fortunately, there exists a partial way out due to the relations among Hermite and Laguerre polynomials

$$H_{2n}(x) = (-1)^n\, 2^{2n}\, n!\, L_n^{-1/2}(x^2)\,, \qquad H_{2n+1}(x) = (-1)^n\, 2^{2n+1}\, n!\, x\, L_n^{+1/2}(x^2)$$

that allow us to relate the above-mentioned functions as

$$K_{2n}(x) = (-1)^n\, (x^2)^{1/4}\, M_n^{-1/2}(x^2)\,, \qquad K_{2n+1}(x) = (-1)^n\, x\, (x^2)^{-1/4}\, M_n^{1/2}(x^2)\,.$$

Thus, we can define the transforms $\mathcal{T}_\pm$ on functions $f(y) \in L^2(\mathbb{R}^+)$ by

$$[\mathcal{T}_\pm f](s) := \frac{1}{\sqrt{2\pi}} \int_0^\infty dy\, \frac{S_\pm(\sqrt{s\,y})}{(s\,y)^{1/4}}\, f(y)\,, \qquad S_+(\cdot) = \cos(\cdot),\;\; S_-(\cdot) = \sin(\cdot)\,, \tag{39}$$

such that they verify the relation

$$[\mathcal{T}_\pm M_n^{\pm 1/2}](s) = (-1)^n\, M_n^{\pm 1/2}(s)\,, \tag{40}$$

which means that $M_n^{\pm 1/2}(s)$ are eigenfunctions with eigenvalues $(-1)^n$ of $\mathcal{T}_\pm$. In consequence, we have two relevant values of the label α: $\pm 1/2$. Then, since, for any $f(y) \in L^2(\mathbb{R}^+)$,

$$f(y) = \sum_{n=0}^\infty c_n^\pm\, M_n^{\pm 1/2}(y)\,, \qquad c_n^\pm = \int_0^\infty f^*(y) M_n^{\pm 1/2}(y)\, dy\,. \tag{41}$$

We may introduce two new families of FFT $\mathcal{T}_\pm^a$ ($a \in \mathbb{R}$) by

$$[\mathcal{T}_\pm^a f](s) := \sum_{n=0}^\infty c_n^\pm\, e^{i\,n\,a\,\pi/2}\, M_n^{\pm 1/2}(s)\,.$$

Thus, if we choose $a = 4/k$ with $k \in \mathbb{N}^*$, we have

$$\begin{aligned} f_\pm^k(s) \;&:=\; [\mathcal{T}_\pm^{4/k} f](s) = \sum_{q=0}^\infty c_{kq}^\pm\, e^{-2\pi(kq)i/k}\, M_{kq}^{\pm 1/2}(s) + \sum_{q=0}^\infty c_{kq+1}^\pm\, e^{-2\pi(kq+1)i/k}\, M_{kq+1}^{\pm 1/2}(s) + \ldots \\ &\qquad \cdots + \sum_{q=0}^\infty c_{kq+k-1}^\pm\, e^{-2\pi(kq+k-1)i/k}\, M_{kq+k-1}^{\pm 1/2}(s) \qquad (42) \\ &= f_{0,\pm}^k(s) + e^{-2\pi i/k}\, f_{1,\pm}^k(s) + \cdots + e^{-2\pi(k-1)i/k}\, f_{k-1,\pm}^k(s)\,, \end{aligned}$$

with

$$f_{r,\pm}^k(s) := \sum_{q=0}^\infty c_{kq+r}^\pm\, M_{kq+r}^{\pm 1/2}(s)\,.$$

We have recovered the splitting (35) of $L^2(\mathbb{R}^+)$ for the particular cases of $\alpha = \pm 1/2$

$$L^2(\mathbb{R}^+) = L_{k,0}^2(\mathbb{R}^+)^\pm \oplus L_{k,1}^2(\mathbb{R}^+)^\pm \oplus \cdots \oplus L_{k,k-1}^2(\mathbb{R}^+)^\pm\,,$$

where each of the closed subspaces $L_{k,r}^2(\mathbb{R}^+)^\pm$ is an eigen-subspace of $\mathcal{T}_\pm$ with eigenvalue $e^{-2\pi r i/k}$.

4. A New Harmonic Analysis on the Circle

The set Hermite functions $K_n(x)$ is a good tool so as to construct a countable set of periodic functions, which is a system of generators of the space of square integrable functions on the unit circle $L^2(C)$, i.e., the functions $f(\phi) : C \longmapsto \mathbb{C}$ with norm $||f(\phi)||$ defined by

$$||f(\phi)||^2 := \frac{1}{2\pi} \int_{-\pi}^{\pi} |f(\phi)|^2 \, d\theta < \infty . \tag{43}$$

Let us define the periodic functions (with period 2π)

$$\mathcal{K}_n(\phi) := \sum_{k=-\infty}^{\infty} K_n(\phi + 2k\pi) , \qquad -\pi \leq \phi < \pi , \; n = 0, 1, 2, \ldots . \tag{44}$$

It can be proven that the series defining the $\mathcal{K}_n(\phi)$ are absolutely convergent and also that every $\mathcal{K}_n(\phi)$ is bounded and square integrable on the interval $-\pi \leq \phi < \pi$. Using this property and the Lebesgue theorem, we may also prove that

$$\int_{-\pi}^{\pi} e^{im\phi} \, d\phi \sum_{k=-\infty}^{\infty} K_n(\phi + 2k\pi) = \sum_{k=-\infty}^{\infty} \int_{-\pi}^{\pi} e^{im\phi} \, K_n(\phi + 2k\pi) \, d\phi . \tag{45}$$

A Discretized Fourier Transform on the Circle

Let us compare the space $L^2(C)$, which we may also denote as $L^2[-\pi, \pi]$, to the space $l_2(\mathbb{Z})$ of 2-power summable sequences. As is well known, an orthonormal basis on $L^2(C)$ is $\{(2\pi)^{-1} e^{in\phi}\}$ with $n \in \mathbb{Z}$. Hence, any $f(\phi) \in L^2(C)$ admits the following span into exponential Fourier series given by

$$f(\phi) = \frac{1}{2\pi} \sum_{n \in \mathbb{Z}} f_n \, e^{in\phi} , \qquad f_n \in \mathbb{C} , \tag{46}$$

with

$$f_n = \frac{1}{2\pi} \int_{-\pi}^{\pi} f(\phi) \, e^{-in\phi} \, d\phi . \tag{47}$$

The sum (46) converges in the sense of the norm (43). Moreover, for any continuous function $f(\phi)$, the series also converge pointwise. The properties of orthonormal basis in Hilbert spaces show that

$$\frac{1}{2\pi} \sum_{n \in \mathbb{Z}} |f_n|^2 = ||f(\phi)||^2 . \tag{48}$$

We may call to the sequence of complex numbers $\{f_n\}_{n \in \mathbb{Z}}$, the components of f.

The Hilbert space $l_2(\mathbb{Z})$ is a space of sequences of complex numbers $A \equiv \{a_n\}_{n \in \mathbb{Z}}$ such that

$$||A||^2 := \frac{1}{2\pi} \sum_{n \in \mathbb{Z}} |a_n|^2 < \infty , \tag{49}$$

with scalar product given by

$$\langle A|B \rangle := \frac{1}{2\pi} \sum_{n \in \mathbb{Z}} a_n^* \, b_n . \tag{50}$$

An orthonormal basis for $l_2(\mathbb{Z})$ is given by the sequences $\mathcal{E}_k = \{\delta_{k,n}\}_{n \in \mathbb{Z}}$ with $k \in \mathbb{Z}$. Any $f \in l_2(\mathbb{Z})$ with components $\{f_n\}_{n \in \mathbb{Z}}$ may be written as

$$f = \frac{1}{2\pi} \sum_{n \in \mathbb{Z}} f_n \, \mathcal{E}_n , \qquad \frac{1}{2\pi} \sum_{n \in \mathbb{Z}} |f_n|^2 = ||f||^2 < +\infty . \tag{51}$$

Therefore, there exists a unitary correspondence between $L^2(\mathcal{C})$ and $l_2(\mathbb{Z})$ which maps any $f(\phi) \in L^2(\mathcal{C})$ as in (46) into f as in (51), provided that in both cases the sequence $\{f_n\}_{n\in\mathbb{Z}}$ is the same.

Expression (46) gives the expansion into Fourier series of the functions in $L^2(\mathcal{C})$. From this point of view, we may say that the Fourier series is a unitary mapping, $\mathcal{F}$, from $L^2(\mathcal{C})$ onto $l_2(\mathbb{Z})$. It admits an inverse, $\mathcal{F}^{-1}$, from $l_2(\mathbb{Z})$ onto $L^2(\mathcal{C})$, which is also unitary and is sometimes called the discrete Fourier transform, i.e.,

$$\mathcal{F}[f(\phi)] = \frac{1}{2\pi} \sum_{n\in\mathbb{Z}} f_n \, e^{in\phi} \equiv \{f_n\}_{n\in\mathbb{N}}\,, \qquad \mathcal{F}^{-1}[\{a_n\}_{n\in\mathbb{N}}] = \frac{1}{2\pi} \sum_{n\in\mathbb{Z}} a_n \, e^{in\phi} \equiv a(\phi)\,, \tag{52}$$

with $f_n \in \mathbb{C}$ and given by (47).

As we mention in the introduction, we will give a unitary version of concepts that are often introduced separately, like Fourier transform, Fourier series and discrete Fourier transform in one side and the Hermite functions on the other.

We start by constructing a set of functions in $l_2(\mathbb{Z})$ using the Hermite functions $K_n(x)$. We introduce the sequences χ_n associated to $K_n(x)$ as follows:

$$\chi_n := \{K_n(m)\}_{m\in\mathbb{Z}}\,, \qquad n \in \mathbb{N}\,. \tag{53}$$

These sequences χ_n are in $l_2(\mathbb{Z})$. Moreover, they are linearly independent and span $l_2(\mathbb{Z})$. The proof can be find in [43] and they are based on the fact that

$$\begin{vmatrix} H_0(-N) & \dots & H_0(0) & \dots & H_0(N) \\ H_1(-N) & \dots & H_1(0) & \dots & H_1(N) \\ \dots & \dots & \dots & \dots & \dots \\ H_{2N}(-N) & \dots & H_{2N}(0) & \dots & H_{2N}(N) \end{vmatrix} \neq 0\,, \tag{54}$$

and

$$\begin{vmatrix} H_0(0) & H_0(1) & \dots & H_0(N) \\ H_1(0) & H_1(1) & \dots & H_1(N) \\ \dots & \dots & \dots & \dots \\ H_N(0) & H_N(1) & \dots & H_N(N) \end{vmatrix} \neq 0\,, \tag{55}$$

for any $N \in \mathbb{N}$, where $H_n(k)$ is the Hermite polynomial $H_n(x)$ evaluated at the integer k (remember that $K_n(x) = e^{-x^2/2}\, H_n(x)/\sqrt{2^n n! \sqrt{\pi}}$).

Since the functions $\mathcal{K}_n(\phi)$ are in $L^2[-\pi,\pi]$, they admit a span in terms of the orthonormal basis $\{(2\pi)^{-1}\, e^{im\phi}\}_{m\in\mathbb{Z}}$ in $L^2[-\pi,\pi]$. Thus, we can write

$$\mathcal{K}_n(\phi) = \frac{1}{\sqrt{2\pi}} \sum_{m=-\infty}^{\infty} k_n^m \, e^{-im\phi}\,, \tag{56}$$

with

$$k_n^m = \frac{1}{\sqrt{2\pi}} \int_{-\pi}^{\pi} e^{im\phi}\, \mathcal{K}_n(\phi)\, d\phi\,. \tag{57}$$

The continuity of the functions $\mathcal{K}_n(\phi)$ on $[-\pi,\pi]$ guarantees the pointwise convergence of (56) and since the $\mathcal{K}_n(\phi)$ are periodic with period 2π, hence (56) is valid for all $\phi \in \mathbb{R}$.

We recall that the Hermite functions $K_n(x)$ are eigenfunctions of the Fourier transform with eigenvalue $(-i)^n$ (12), i.e., $[\mathcal{F}\, K_n](p) = (-i)^n\, K_n(p)$. Thus, $K_n(x)$ are eigenfunctions of the inverse Fourier transform with eigenvalue i^n, i.e., $[\mathcal{F}^{-1}\, K_n](x) = i^n\, K_n(x)$. From this fact, we can find an explicit expression of the coefficients k_n^m (57) in terms of the values of the Hermite functions at the integers

$$
\begin{aligned}
k_n^m &= \frac{1}{\sqrt{2\pi}} \int_{-\pi}^{\pi} e^{im\phi}\, d\phi \left[\sum_{k=-\infty}^{\infty} K_n(\phi + 2k\pi) \right] = \frac{1}{\sqrt{2\pi}} \sum_{k=-\infty}^{\infty} \int_{-\pi}^{\pi} e^{im\phi}\, K_n(\phi+2k\pi)\, d\phi \\
&= \frac{1}{\sqrt{2\pi}} \sum_{k=-\infty}^{\infty} \int_{-\pi+2k\pi}^{\pi+2k\pi} e^{ims}\, K_n(s)\, ds = \frac{1}{\sqrt{2\pi}} \int_{-\infty}^{\infty} e^{ims}\, K_n(s)\, ds = i^n\, K_n(m)\,,
\end{aligned}
$$

where $s := \phi + 2k\pi$ and $e^{im\phi} = e^{im(\phi+2k\pi)} = e^{ims}$. Hence, (56) and (57) can be written, respectively, as

$$
\mathcal{K}_n(\phi) = \frac{i^n}{\sqrt{2\pi}} \sum_{m=-\infty}^{\infty} K_n(m)\, e^{-im\phi}\,, \qquad K_n(m) = \frac{(-i)^n}{\sqrt{2\pi}} \int_{-\pi}^{\pi} \mathcal{K}_n(\phi)\, e^{im\phi}\, d\phi\,, \tag{58}
$$

where $k_n^m = i^n\, K_n(m)$.

The systems of generators in $L^2[-\pi,\pi] \equiv L^2(\mathcal{C})$, $\{\mathcal{K}_n(\phi)\}_{n\in\mathbb{Z}}$, and in $l_2(\mathbb{Z})$, given by the set of sequences $\{\chi_n\}_{n\in\mathbb{Z}}$, are not orthonormal basis. The scalar product on $L^2[-\pi,\pi]$ is related with the scalar product in $l_2(\mathbb{Z})$

$$
\begin{aligned}
\langle \mathcal{K}_n | \mathcal{K}_m \rangle &= \int_{-\pi}^{\pi} \mathcal{K}_n^*(\phi)\, \mathcal{K}_m(\phi)\, d\phi = \frac{1}{2\pi} \int_{-\pi}^{\pi} d\phi \sum_{j=-\infty}^{\infty} \sum_{k=-\infty}^{\infty} (-i)^n\, i^m\, K_n^*(j)\, K_m(k)\, e^{-i(k-j)\phi} \\
&= \sum_{j=-\infty}^{\infty} \sum_{k=-\infty}^{\infty} \delta_{j,k}\, i^{m-n}\, K_n^*(j)\, K_m(k) = \sum_{j=-\infty}^{\infty} i^{m-n}\, K_n^*(j)\, K_m(j) \\
&= i^{m-n}\, (\chi_n, \chi_m)\,.
\end{aligned}
$$

The Gramm–Schmidt procedure allows us to obtain orthogonal bases in both spaces.

5. Conclusions

In this paper, we have presented a unified framework where Hermite functions, or alternatively Laguerre functions, their symmetry groups, Fourier analysis and RHS fit in a perfect manner. Hermite functions are basic in the study of quantum mechanics and signal processing on the real line $\mathbb{R}$, while Laguerre functions play the same role on the half-line $\mathbb{R}^+$. We have also studied the particular relation between both situations. In both cases, these functions are eigenvectors of the Fourier transform and this is an essential property.

It is precisely the use of RHS that allows the use of bases of different cardinality on a simple and interchangeable manner. This makes RHS the correct mathematical formulation that encompasses both quantum mechanics and signal processing. Here, Hermite functions act as transition elements of transition matrices between continuous and discrete bases. This is not strictly new as was already discussed in [21], although we introduce a general point of view which could be relevant for computational and epistemological purposes in quantum theory.

We have shown how Fourier analysis allows for the decomposition of RHS into direct sums of RHS. This may permit the filtering of noise or any other undesirable signal. The same applies to operators as we may restrict their evolution to a sub-algebra, which has been chosen among infinite other possibilities in the universal enveloping algebra of the corresponding symmetry group. The decomposition of RHS is consistent with the FFT. This is the cornerstone of the filtering procedure. We have extended the formalism to functions over the semi-axis $\mathbb{R}^+$ by the construction of a pair of "Fourier-like" transformations which play the role before reserved to the Fourier transform on $\mathbb{R}$. FFTs may be defined after these Fourier-like transforms and also serve for filtering. Moreover, the algebraic approach associated to the Lie symmetry algebra and its universal enveloping algebra extends the discussion from the vector spaces to the space of operators acting on them.

All of these results can be also translated, in some sense, to the circle. We have constructed some special functions on the circle out of Hermite functions and have taken advantage of the properties of Hermite functions in order to use Fourier analysis on the circle as well. This work is still in process.

As a final remark, let us insist that we have given a unitary point of view of mathematical objects that are often considered as unrelated such as Fourier transform, discrete Fourier transform, Hermite and Laguerre functions and RHS.

Author Contributions: This paper is a fully collaborative work and all authors have equally contributed to all technical details as well as to the final presentation of the manuscript.

Funding: This research is supported in part by the Ministerio de Economía y Competitividad of Spain under grant MTM2014-57129-C2-1-P and the Junta de Castilla y León (grant VA137G18).

Conflicts of Interest: The authors declare no conflicts of interest.

References

1. Fourier, J.B.J. *Théorie Analytique de la Chaleur*; Firmin Didot: Paris, France, 1822. (In French)
2. Berry, M. Physics in a New Era. *Phys. Today* **2001**, *54*, 11. [CrossRef]
3. Andrews, G.E.; Askey, R.; Roy, R. *Special Functions*; Cambridge University Press: Cambridge, UK, 1999.
4. Folland, G.B. *A Course in Abstract Harmonic Analysis*; CRC Press: Boca Raton, FL, USA, 2015.
5. Celeghini, E.; del Olmo, M.A. Quantum physics and signal processing in rigged Hilbert spaces by means of special functions, Lie algebras and Fourier and Fourier-like transforms. *J. Phys. Conf. Ser.* **2015**, *597*, 012022. [CrossRef]
6. Celeghini, E.; Gadella, M.; del Olmo, M.A. Applications of rigged Hilbert spaces in quantum mechanics and signal processing. *J. Math. Phys.* **2016**, *57*, 072105. [CrossRef]
7. Coutinho, S.C. *A Premier on Algebraic D-Modulus*; Cambridge University Press: Cambridge, UK, 1995.
8. Folland, G.B. *Fourier Analysis and Its Applications*; Wadsworth: Pacific Grove, CA, USA, 1992.
9. Gelfand, I.M.; Vilenkin, N.Y. *Generalized Functions: Applications to Harmonic Analysis*; Academic Press: New York, NY, USA, 1964.
10. Maurin, K. *General Eigenfunction Expansions and Unitary Representations of Topological Groups*; Polish Scientific Publishers: Warszawa, Poland, 1968.
11. Dirac, P.A.M. *The Principles of Quantum Mechanics*; Clarendon Press: Oxford, UK, 1958.
12. Maurin, K. *Methods of Hilbert Spaces*; Polish Scientific Publishers: Warszawa, Poland, 1967.
13. De la Madrid, R.; Bohm, A.; Gadella, M. Rigged Hilbert Space Treatment of Continuous Spectrum. *Fort. Phys.* **2002**, *50*, 185–216. [CrossRef]
14. De la Madrid, R. The role of the rigged Hilbert space in Quantum Mechanics. *Eur. J. Phys.* **2005**, *26*, 287–312. [CrossRef]
15. Heredia-Juesas, J.; Gago-Ribas, E. A new view of spectral analysis of linear systems. In *PIERS Proceedings*; Progress in Electromagnetics Research Symposium; Electromagnetics Academy: Cambridge, MA, USA, 2012.
16. Heredia-Juesas, J.; Gago-Ribas, E.; Ganoza-Quintana, J.L. A new version of a generalized signals & systems scheme to parameterize and analyze physical problems. In Proceedings of the International Cost Estimating and Analysis Association (ICEAA) 2014, Palm Beach, Aruba, 3–9 August 2014.
17. Heredia-Juesas, J.; Gago-Ribas, E.; Vidal-García, P. Application of the rigged Hilbert spaces into the generalized signals & systems theory. In Proceedings of the ICEAA 2015, Turin, Italy, 7–11 September 2015.
18. Heredia-Juesas, J.; Gago-Ribas, E.; Vidal-García, P. Application of the Rigged Hilbert Spaces into the Generalized Signals and Systems Theory: Practical Example. In Proceedings of the PhotonIcs and Electromagnetics Research Symposium (PIERS) 2016, Shanghai, China, 8–11 August 2016.
19. Roberts, J.E. Rigged Hilbert spaces in quantum mechanics. *Comm. Math. Phys.* **1966**, *3*, 98–119. [CrossRef]
20. Antoine, J.P. Dirac Formalism and Symmetry Problems in Quantum Mechanics. I. General Dirac Formalism. *J. Math. Phys.* **1969**, *10*, 53–69. [CrossRef]
21. Bohm, A. *The Rigged Hilbert Space and Quantum Mechanics*; Springer Lecture Notes in Physics; Springer: Berlin, Germany, 1978; Volume 78.
22. Melsheimer, O. Rigged Hilbert space formalism as an extended mathematical formalism for quantum systems. I. General theory. *J. Math. Phys.* **1974**, *15*, 902–916. [CrossRef]

23. Gadella, M; Gómez, F. A Unified Mathematical Formalism for the Dirac Formulation of Quantum Mechanics. *Found. Phys.* **2002**, *32*, 815–869. [CrossRef]
24. Gadella, M; Gómez, F. On the Mathematical Basis of the Dirac Formulation of Quantum Mechanics. *Int. J. Theor. Phys.* **2003**, *42*, 2225–2254. [CrossRef]
25. Gadella, M; Gómez, F. Eigenfunction Expansions and Transformation Theory. *Acta Appl. Math.* **2010**, *109*, 721–742. [CrossRef]
26. Bohm, A.; Gadella, M.; Kielanowski, P. Time asymmetric quantum mechanics. *SIGMA* **2011**, *7*, 86–98. [CrossRef]
27. Celeghini, E.; Gadella, M.; del Olmo, M.A. Spherical harmonics and rigged Hilbert spaces. *J. Math. Phys.* **2018**, *59*, 053502. [CrossRef]
28. Celeghini, E.; Gadella, M.; del Olmo, M.A. Lie algebra representations and rigged Hilbert spaces: The $SO(2)$ case. *Acta Polytech.* **2017**, *57*, 379–384. [CrossRef]
29. Celeghini, E.; Gadella, M.; del Olmo, M.A. $SU(2)$, associated Laguerre polynomials and rigged Hilbert spaces. In *Quantum Theory and Symmetries with Lie Theory and Its Applications in Physics Volume 2*; Dobrev, V., Ed.; Springer: Singapore, 2018; Volume 255.
30. Gadella, M.; Martín, M.A.; Nieto, L.M.; del Olmo, M.A. The Stratonovich-Weyl correspondence for one dimensional kinematical groups. *J. Math. Phys.* **1991**, *32*, 1182–1192. [CrossRef]
31. Celeghini, E.; del Olmo, M.A. Coherent orthogonal polynomials. *Ann. Phys.* **2013**, *335*, 78–85. [CrossRef]
32. Yang, X.-J. A new integral transform operator for solving the heat-diffusion problem. *Appl. Math. Lett.* **2017**, *64*, 193–197. [CrossRef]
33. Yang, X.-J. A new integral transform method for solving steady heat-transfer problem. *Therm. Sci.* **2016**, *20*, S639–S642. [CrossRef]
34. Yang, X.-J. A new integral transform with an application in heat-transfer problem. *Therm. Sci.* **2016**, *20*, S677–S681. [CrossRef]
35. Yang, X.-J. New integral transforms for solving a steady heat transfer problem. *Therm. Sci.* **2017**, *21*, S79–S87. [CrossRef]
36. Ozaktas, H.M.; Zalevsky, Z.; Alper Kutay, M. *The Fractional Fourier Transform*; Wiley: Chichester, UK, 2001.
37. Tung, W.-K. *Group Theory in Physics*; World Scientific: Philadelphia, PA, USA, 1985.
38. Hammermesh, M. *Group Theory*; Addison-Wesley: Reading, MA, USA, 1962.
39. Bargmann, V. On unitary ray representations of continuous groups. *Ann. Math.* **1954**, *59*, 1–46. [CrossRef]
40. Celeghini E.; Tarlini, M. Contractions of group representations.—II. *Nuovo Cimento B* **1981**, *65*, 172–180. [CrossRef]
41. Abramovich, M.; Stegun, I.A. *Handbook of Mathematical Functions with Formulas, Graphs, and Mathematical Tables*; Dover: New York, NY, USA, 1972.
42. Roy, R.; Olver, F.W.J.; Askey, R.A.; Wong, R.S.C.; Temme, N.M.; Paris, R.B.; Maximon, L.C.; Daalhuis, A.B.O.; Dunster, T.M.; Andrews, G.E.; et al. *NIST Handbook of Mathematical Functions*; Olver, F.W.J., Lozier, D.W., Boisiert, R.F., Clark, C.W., Eds.; Cambridge University Press: Cambridge, UK, 2010.
43. Celeghini, E.; Gadella, M.; del Olmo, M.A. *Hermite Functions and Fourier Series in the Circle*; 2018, in preparation.

MDPI
St. Alban-Anlage 66
4052 Basel
Switzerland
Tel. +41 61 683 77 34
Fax +41 61 302 89 18
www.mdpi.com

Entropy Editorial Office
E-mail: entropy@mdpi.com
www.mdpi.com/journal/entropy

www.ingramcontent.com/pod-product-compliance
Lightning Source LLC
Chambersburg PA
CBHW051725210326
41597CB00032B/5614